Environmental Science and Engineering

Environmental Science

More information about this series at http://www.springer.com/series/3234

Sunil Nautiyal · Ruediger Schaldach
K.V. Raju · Harald Kaechele · Bill Pritchard
Kottapalli Sreenivasa Rao
Editors

Climate Change Challenge (3C) and Social-Economic-Ecological Interface-Building

Exploring Potential Adaptation Strategies for Bio-resource Conservation and Livelihood Development

Springer

Editors
Sunil Nautiyal
Centre for Ecological Economics
and Natural Resources (CEENR)
Institute for Social and Economic Change
Nagarabhavi
India

Bill Pritchard
School of Geosciences
University of Sydney
Sydney
Australia

and

Institute for Geography
University of Cologne
Köln
Germany

Ruediger Schaldach
Centre for Environmental Systems Research
University of Kassel
Kassel
Germany

K.V. Raju
International Crops Research Institute for the
Semi-Arid Tropics (ICRISAT)
Patancheru
India

Kottapalli Sreenivasa Rao
Department of Botany
University of Delhi
Delhi
India

Harald Kaechele
Institute of Socio-Economics
ZALF e.V.
Müncheberg
Germany

ISSN 1863-5520 ISSN 1863-5539 (electronic)
Environmental Science and Engineering
ISSN 1431-6250
Environmental Science
ISBN 978-3-319-31013-8 ISBN 978-3-319-31014-5 (eBook)
DOI 10.1007/978-3-319-31014-5

Library of Congress Control Number: 2016936281

Printed on acid-free paper

This Springer imprint is published by Springer Nature
The registered company is Springer International Publishing AG Switzerland

Foreword

The vast majority of scientists now agree that if global warming exceeds a mean temperature of 2 °C it will lead to dangerous, irreversible and practically uncontrollable consequences for both nature and mankind. This was thoroughly discussed during recent COP 21 in Paris. Research shows that there is only a realistic chance of restricting global warming to 2 °C if a limit is set on the total amount of CO_2 emitted globally between now and 2050 (CO_2 global budget). Continuation of the current trends of greenhouse gas emissions would mean that humankind will be confronted with dramatic challenges in future, with the loss of natural resources as well as impaired security of societies. From a scientific perspective, climate change is a global risk multiplier. Unabated climate change would substantially add to existing vulnerabilities of poor and indigenous communities in many parts of the world, who are inadequately prepared for adapting to unforeseen changes in their economic, social and environmental context. Evidence of these vulnerabilities is already visible in India, which has faced extreme weather events over the past ten years and witnessed a decrease in food grain production.

This publication emerged from two International Conferences held at ISEC in Bangalore, India: The International Conference on "Climate Change and Social-Ecological-Economical Interface-Building: Modelling Approach to Exploring Potential Adaptation Strategies for Bio-resource Conservation and Livelihood Development" held during 20–21 May 2015, organized by Centre for Ecological Economics and Natural Resources (CEENR), Institute for Social and Economic Change (ISEC) in association with Centre for Environmental System Research (CESR), University of Kassel, Germany, and the International Conference "Climate Change and Food Security—the Global and Indian Contexts", jointly convened by CEENR, ISEC and the School of Geosciences, University of Sydney, on 18–19 February 2015. The participants in these International Conferences presented and discussed their broad expertise from many different subject domains. Thereby, the human resource capacity present in India, Australia, Germany and many other countries as a collaborative activity were explored. This volume brings in selected presentations from the International Conferences, which addressed the issues of social, economic, policy and technological adaptation strategies that could be drawn from knowledge systems that existed in the scientific manpower of the countries. The presentations in this book mainly address the issues related to climate change challenges and adaptation to climate change as major concern in the scientific knowledge systems.

A new kind of development is imaginable: the great transformation towards sustainability. The necessity for and possible ways towards this transformation, focusing on climate change and global decarbonization, was described by the German Advisory Council on Global Change in 2011. One important contribution to this transformation is the ongoing process of the "German Energiewende", a rapid decarbonization of the national energy system. The German Advisory Council on Global Change depicts the important role for research and science in the great transformation and the need for intensified inter- as well as transdisciplinary cooperation. These cannot be overestimated. In addition, strongly intensified international cooperation in business and politics as well as research, education and science is a prerequisite for the great transformation. Bilateral interactions are a substantial contribution to the required intensification of international scientific exchange, and a splendid effort in bringing out the best in various countries scientific capacity to address the issues of global importance.

March 2016

Klaus Mueller
Leibniz Centre for Agricultural
Landscape Research
Müncheberg, Germany

Contents

Climate Change Challenge (3C) and Social-Economic-Ecological Interface-Building—Exploring Potential Adaptation Strategies for Bio-resource Conservation and Livelihood Development: Prologue

Sunil Nautiyal, Ruediger Schaldah, K.V. Raju, Harald Kaechele, Bill Pritchard and K.S. Rao

1 Human-Environment Systems

Since their beginning, human societies and their ecosystems have been inter-dependent. While the early hunting-gathering way of life relied on the provisioning of wild animals and plants, with the emergence of agriculture and permanent

S. Nautiyal (✉)
Centre for Ecological Economics and Natural Resources (CEENR), IInstitute for Social and Economic Change (ISEC), Dr. VKRV Rao Road, Nagarabhavi, Bangalore 560072, India
e-mail: nautiyal_sunil@yahoo.com

R. Schaldah
Centre for Environmental System Research (CESR), IUniversity of Kassel, Kassel, Germany
e-mail: schaldach@usf.uni-kassel.de

K.V. Raju
ICRISAT, Patancheru, India
e-mail: kvraju2008@gmail.com

H. Kaechele
Institute of Socio-Economics, Leibniz-Centre for Agricultural Landscape Research (ZALF), Muencheberg, Germany
e-mail: Harald.Kaechele@zalf.de

B. Pritchard
School of Geosciences, University of Sydney, Sydney, Australia
e-mail: bill.pritchard@sydney.edu.au

B. Pritchard
Institute for Geography, University of Cologne, Köln, Germany

K.S. Rao
Department of Botany, University of Delhi, Delhi, India
e-mail: srkottapalli@yahoo.com

S. Nautiyal et al. (eds.), *Climate Change Challenge (3C) and Social-Economic-Ecological Interface-Building*, Environmental Science,
DOI 10.1007/978-3-319-31014-5_1

settlements humans began to actively shape the land surface of the Earth in order to produce biomass for food, materials and energy conversion. First evidence of the existence of villages in India can be dated back to the seventh millennium B.C. (Misra 2001). In the following millennia, advancements of technologies, medicine and the access to biotic and mineral resources led to continuous increase of the world population and consequential transformations to natural land-cover. After the Industrial Revolution in the eighteenth century, human impacts on nature have reached an unprecedented level of magnitude. In this light some scientists are speaking of a new geologic era, the Anthropocene where humanity alters actively the functioning of the whole Earth System (e.g. Crutzen 2006).

The interactivity between humans and ecosystems can be understood through the concept of ecosystem services. This presumes that ecosystems provide essential goods and services to society, and that as humanity uses these services they may be depleted and hence their sustainability requires mechanisms for their management. This includes resources such as food, energy and space for living, but also processes such as climate regulation and water retention to avoid flooding (Millennium Ecosystem Assessment 2005). The portfolio of goods and services provided by a particular ecosystem depends on its structure and how it is managed by humans.

A scientific framework to describe and to better understand interactions between human societies and ecosystems is the concept of human-environment systems (e.g. Turner et al. 2007). These are composed of a human sub-system and an environment sub-system which interact with each other (Fig. 1). The human sub-system includes all elements of societal life such as economy, politics and technology while the environment sub-system comprises ecosystems and their inherent processes such as biomass production, food webs and biogeochemical cycles. The interface between the sub-systems has two-way interactivity. On the one hand human actions (agricultural management, urbanization etc.) directly change the structure and functioning of ecosystems, and therefore the portfolio of provided ecosystem services (land-use management). For example, the conversion of a tropical rainforest to cropland replaces a forest ecosystem with its related goods and services (e.g. carbon storage, preservation of genetic diversity) by an agricultural system with a strong focus on food production by intensive crop cultivation and/or cattle ranching. On the other hand the human sub-system has the ability to react to unfavourable changes of

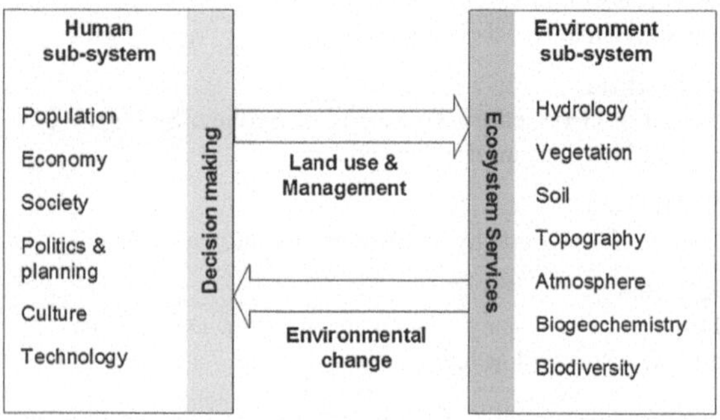

Fig. 1 Human-environment system (based on Turner et al. 2007)

the environment, e.g. pollution or loss of fertile land by erosion by adaptation of land management practices. Under this framework, scientists can design analytical tools to study and to better understand the functioning of human-environment interactions on different spatial and temporal scale levels. This is a prerequisite to define trade-offs that are required to achieve a balance between the provision of biomass resources for food and energy and the stabilization of other ecosystem services in the long run (Foley et al. 2005) and to identify pathways to steer the system towards a more sustainable development trajectory.

2 The Climate Change Challenge

Climate change is among the greatest challenges to humanity (Rockström et al. 2009). Human activity is the primary driver of climate change (IPCC 2014), via the emission of greenhouse gases through the burning of fossil fuels, the release of soil carbon because of agriculture, the change of surface albedo due to land-use change and other processes. At the same time, climate change and its manifestations, particularly through rising temperatures, changing rainfall, sea-level rise and increasing droughts and floods have the potential to adversely impact natural ecosystems (such as forests, grasslands, rivers and oceans) and socio-economic systems (such as food production, fisheries and coastal settlements). This is also adding additional stress to ecosystems services, which form substantial source of income in particular to the rural poor in many world regions (Milder et al. 2010).

Therefore, climate change is most immediately and inextricably linked to well-being, development and economic growth which are in the ambit of the 17 Sustainable Development Goals (SDGs) that address the root causes of poverty and the universal need for sustainable development that works for all people (UN 2015). In particular SDG 2 "Zero hunger", SDG 3 "Good Health and Well-being" as well as the SDGs 14 "Life on land" and SDG 15 "Life in Water" are directly affected by negative consequences of climate change.

At the Conference of the Parties (COP) 21 of the United Nations Framework Convention on Climate Change (UNFCCC), it was noted that the chief bearers and victims of climate change are vulnerable groups in developing countries, particularly those whose livelihood is dependent on land use. Therefore, solving the climate dilemma through is an ethical concern for science and societies. Combating climate change and its impacts as defined by SDG 13 requires the implementation of mitigation and adaptation strategies. While mitigation mainly aims at reducing the emissions of greenhouse gases in order to lower the intensity of climate change, adaptation includes actions that help societies to cope with negative climate impacts. Both approaches are directly related to the different components of the human-environment system. To give an example, the changing climate might have negative effects of the provision of ecosystem services such as food production which in turn triggers actions in the human sub-system, e.g. to adapt agricultural management to the new climate conditions. This requires policy formulation, research, technology transfer, financing and enhancing adaptive capacity of the poor at national, regional as well as local level.

3 How Will India Be Affected by Climate Change?

According to the Intergovernmental Panel on Climate Change (IPCC 2014), India is likely to suffer from an increased frequency of extreme temperature and precipitation events. Already in the past decades, a higher return frequency of extreme rain events and of disastrous flooding has been observed (e.g. Goswami et al. 2006; Guhathakurta et al. 2011). The Indian monsoon system has been identified as one tipping element of the global climate system meaning that a strong climate change might drastically change atmospheric circulation patterns (Lenton et al. 2008). In case of such a systemic shift, India's agricultural sector would be dramatically affected. Change in rainfall patterns would put millions of lives at stake. With a more erratic and extreme monsoon, the IPCC (2014) states that by 2030 India would face an agricultural loss of over US $7 billion, affecting the income of 10 % of the people. Without effective adaptation measures, this development would make it difficult for India to achieve a pathway of sustainable development consistent with the SDGs. Only with the successful implementation of climate resilience measures in the form adaptation strategies, 80 % of the losses could be averted, the report adds. Several missions and strategies (Green India Mission, REDD+ etc.) have been implemented at various levels under International initiatives. These initiatives are crucial in addressing the need for climate change adaptation and sustainable development at local level. UNFCCC also emphasises the need for national coordinated adaptation strategies and their support by regional centres; which endows a platform for a bottom-up approach to confer and adapt with climate change impacts at regional level. Strengthening livelihoods, developing sustainable land-use policy etc., have been increasingly seen as a critical strategy for supporting adaptation.

4 The Need for Integrated Scientific Approaches

From the authors' perspective scientists face two main challenges when they are aiming at providing valuable contributions to the climate change challenge. The first challenge is to better understand current and future vulnerabilities of human-environment systems under a changing climate. The second challenge includes the development of effective instruments to link their research efforts to the arena of societal, economic and political actors and decision makers.

Mastering the first challenge requires the development of a holistic understanding of the structure and the functioning of human-environment systems.

This can only be achieved when disciplinary boundaries between natural sciences and humanities are overcome. Consequently, integrative research tools can be designed that are suitable not only to describe the interactions of variables within each sub-system, but also the processes at the interface between them. Here, conceptual and mathematical models can play an important role as a tool to contextualise and organise knowledge and data. There are long traditions of environmental models covering different aspects such as biomass productivity or climate dynamics as well as of models for societal and economic systems. Early attempts to bring together both worlds in a consistent framework were the studies from the Club of Rome dating back to the 1970s (Meadows et al. 2004). Using coupled differential equations, the World-3 model (Forrester 1971) was applied to analyse global limits of societal and economic growth. It included sub-models for population development, food production and resource use, among others. Building on this pioneering work, Integrated Assessment Models (IAMs) were developed (Rotmans and van Asselt 2001). A prominent example is the IMAGE model (Stehfest et al. 2014) that couples relatively aggregated versions of process-based environmental models with econometric approaches and allows exploring global feedbacks between these sub-systems. Methodological challenges for model coupling include the often fundamentally different character of the individual models as well as questions of temporal and spatial scales. In respect to the climate change challenge, IAMs were playing a prominent role in the Fifth IPCC Assessment Report (IPCC 2014), where they were used to generate future trajectories of global greenhouse gas emissions based on societal development trends and to analyse potential mitigation and adaptation options (Van Vuuren et al. 2010). Closer to the spatial levels of actual societal decision-making are models that operate on a regional or local scale. This includes agent-based approaches of human decision-making coupled with environmental models (e.g. An 2012) and land system models that combine societal drivers of land-use change with environmental impact modelling (e.g. Brown et al. 2012). Here, applications range from the analysis of deforestation processes (Lapola et al. 2011) to the effects of agricultural intensification on pollination (Priess et al. 2007).

The second challenge might be even more demanding. In order to play an active role in defining and implementing applicable climate change mitigation and adaptation strategies, researchers need to reach out beyond their scientific communities and to establish transdisciplinary linkages, e.g. to societal actors and political decision makers (e.g. Nicolescu 2002; Lang et al. 2012). This touches all elements of scientific work: The definition of research questions, the selection of suitable analytical tools and the design of experiments, and last but not least, the communication of scientific findings and their transformation into practice. There is a wide range of the literature that deals with the design of transdisciplinary research projects in the context of climate change and sustainability studies and how scientific and non-scientific stakeholders can be involved in these projects (Hegger et al. 2012; Brandt et al. 2013; Mauser et al. 2013). The aforementioned models have proved to be valuable instruments also for transdisciplinary research, where they are often used in combination with scenarios (Scholz et al. 2006). Scenarios are defined as hypothetical but plausible sequences of future events considering the trends of driving forces, which are supposed to acknowledge the fundamental uncertainties regarding future development. They can be used to integrate knowledge, ideas and visions from different scientific and non-scientific stakeholders both in qualitative (e.g. narratives) and quantitative forms (numbers, maps). In this context, models can help to translate the qualitative assumptions for example about land management into quantitative data such as land-use maps (Castella et al. 2005). Consequently, they can provide valuable insights into human-environment interactions that can support the participating stakeholders in identifying desirable trajectories of future development.

5 About the Book

This book is the outcome of two International Conferences held at ISEC in Bangalore, India: the international conference on "Climate Change and Social-Ecological-Economical Interface-Building: Modelling Approach to Exploring Potential Adaptation Strategies for Bio-resource Conservation and Livelihood Development" held during 20–21 May 2015 and organized jointly by Centre for Ecological Economics and Natural Resources (CEENR), Institute for Social and Economic Change (ISEC) and Centre for Environmental System Research (CESR), University of Kassel, Germany, and the International Conference "Climate Change and Food Security—the Global and Indian Contexts", jointly convened between CEENR, ISEC and the School of Geosciences, University of Sydney, on 18–19 February 2015. This book contains four sections namely, Food Security, Biodiversity and Law (Section A); Methodological Issues; Socioeconomics and Vulnerability Assessments (Section B); Ecosystems and Production System Analysis (Section C); Vulnerability, Sectoral Assessment, Models and Interfaces (Section D). The selected papers that are presented in this book portray a large variety of international research efforts aiming at developing a deeper understanding of human-environment systems, but also at transferring scientific knowledge into political and societal solutions and responses to the climate change challenge.

Acknowledgements We take this opportunity to express our gratitude to the Alexander von Humboldt Foundation (AvH) Germany (through CESR, University of Kassel Germany) and University of Sydney for giving us the privilege to organize International conferences at ISEC, Bangalore, India which finally has come out in form of this volume. We extend our sincere thanks to the Director, ISEC, President, University of Kassel, Germany and President, University Sydney, Executive Director and Deputy Executive Director, CESR, University of Kassel for their kind encouragement and support. We would like to extend our thanks and appreciations to Ms. Vani Gopi (Production editor, Springer) and Mr. Govardhana K.M. (Production Editor, Scientific Publishing Services) for the kind help, support and assistance during the entire publication process and also to copy editor (Production team) and series editors Ulrich Förstner, Wim H. Rulkens and Wim Salomons for their careful reading of the book manuscript. The encouragement and support of Ms. Agata Oelschlaeger, Springer Book Production (Project Coordinator) is thankfully acknowledged. We are thankful to ISEC faculty and staff for their wholehearted support and cooperation in organising these conferences at ISEC.

References

An, L. (2012). Modeling human decisions in coupled human and natural systems: Review of agent-based models. *Ecological Modelling, 229*, 25–36.

Brandt, P., Ernst, A., Gralla, F., Luederitz, C., Lang, D. J., Newig, J., et al. (2013). A review of transdisciplinary research in sustainability science. *Ecological Economics, 92*, 1–15.

Brown, C., Ghile, Y., Laverty, M., Li, K. (2012). Decision scaling: Linking bottom-up vulnerability analysis with climate projections in the water sector.*Water Resource Research, 48*, W09537.

Castella, J. C., Trung, T. N., & Boissau, S. (2005). Participatory simulation of land-use changes in the northern mountains of Vietnam: The combined use of an agent-based model, a role-playing game, and a geographic information system. *Ecology and Society, 10*(1), 27.

Crutzen, P. J. (2006). *The "anthropocene"* (pp. 13–18). Berlin Heidelberg: Springer.

Foley, J. A., DeFries, R., Asner, G. P., Barford, C., Bonan, G., Carpenter, S. R., et al. (2005). Global consequences of land use. *Science, 309*(5734), 570–574.

Forrester, J. W. (1971). *World dynamics* (Vol. 59). Cambridge, MA: Wright-Allen Press.

Goswami, B. N., Venugopal, V., Sengupta, D., Madhusoodanan, M. S., & Xavier, P. K. (2006). Increasing trend of extreme rain events over India in a warming environment. *Science, 314* (5804), 1442–1445.

Guhathakurta, P., Sreejith, O.P., Menon, P.A. (2011). Impact of climate change on extreme rainfall events and flood risk in India. *Journal of Earth System Science, 120*, 359–373.

Hegger, D., Lamers, M., Van Zeijl-Rozema, A., & Dieperink, C. (2012). Conceptualising joint knowledge production in regional climate change adaptation projects: Success conditions and levers for action. *Environmental Science and Policy, 18*, 52–65.

IPCC, Intergovernmental Panel on Climate Change. (2014). *Climate change 2014, synthesis report, summary for policymakers*. http://www.ipcc.ch/pdf/assessment-report/ar5/syr/ AR5_ SYR_FINAL_ SPM.pdf

Lang, D. J., Wiek, A., Bergmann, M., Stauffacher, M., Martens, P., Moll, P., et al. (2012). Transdisciplinary research in sustainability science: Practice, principles, and challenges. *Sustainability Science, 7*(1), 25–43.

Lapola, D. M., Schaldach, R., Alcamo, J., Bondeau, A., Msangi, S., Priess, J. A., et al. (2011). Impacts of climate change and the end of deforestation on land use in the Brazilian Legal Amazon. *Earth Interactions, 15*(16), 1–29.

Lenton, T. M., Held, H., Kriegler, E., Hall, J. W., Lucht, W., Rahmstorf, S., et al. (2008). Tipping elements in the Earth's climate system. *Proceedings of the National Academy of Sciences, 105* (6), 1786–1793.

Mauser, W., Klepper, G., Rice, M., Schmalzbauer, B. S., Hackmann, H., Leemans, R., et al. (2013). Transdisciplinary global change research: The co-creation of knowledge for sustainability. *Current Opinion in Environmental Sustainability, 5*(3), 420–431.

Meadows, D., Randers, J., & Meadows, D. (2004). *Limits to growth: The 30-year update*. Chelsea Green Publishing.

Milder, J. C., Scherr, S. J., & Bracer, C. (2010). Trends and future potential of payment for ecosystem services to alleviate rural poverty in developing countries. *Ecology and Society, 15* (2), 4.

Millennium Ecosystem Assessment. (2005). *Ecosystems and human well-being*. Washington, DC: Island Press.

Misra, V. N. (2001). Prehistoric human colonization of India. *Journal of Biosciences, 26*(4), 491–531.

Nicolescu, B. (2002). *Manifesto of transdisciplinarity*. Suny Press.

Priess, J. A., Mimler, M., Klein, A. M., Schwarze, S., Tscharntke, T., & Steffan-Dewenter, I. (2007). Linking deforestation scenarios to pollination services and economic returns in coffee agroforestry systems. *Ecological Applications, 17*(2), 407–417.

Rockström, J., Steffen, W., Noone, K., Persson, Å., Chapin, F. S., Lambin, E. F., et al. (2009). A safe operating space for humanity. *Nature, 461*(7263), 472–475.

Rotmans, J., & van Asselt, M. B. (2001). Uncertainty management in integrated assessment modeling: Towards a pluralistic approach. *Environmental Monitoring and Assessment, 69*(2), 101–130.

Scholz, R. W., Lang, D. J., Wiek, A., Walter, A. I., & Stauffacher, M. (2006). Transdisciplinary case studies as a means of sustainability learning: Historical framework and theory. *International Journal of Sustainability in Higher Education, 7*(3), 226–251.

Stehfest, E., van Vuuren, D., Bouwman, L., & Kram, T. (2014). *Integrated assessment of global environmental change with IMAGE 3.0: Model description and policy applications*. Netherlands Environmental Assessment Agency (PBL).

Turner, B. L., Lambin, E. F., & Reenberg, A. (2007). The emergence of land change science for global environmental change and sustainability. *Proceedings of the National Academy of Sciences, 104*(52), 20666–20671.

UN, United Nations. (2015). *Transforming our world: The 2030 agenda for sustainable development*. [URL] http://www.un.org/pga/wp-content/uploads/sites/3/2015/08/120815_outcome-document-of-Summit-for-adoption-of-the-post-2015-development-agenda.pdf

Van Vuuren, D. P., Stehfest, E., den Elzen, M. G., van Vliet, J., & Isaac, M. (2010). Exploring IMAGE model scenarios that keep greenhouse gas radiative forcing below 3 W/m^2 in 2100. *Energy Economics, 32*(5), 1105–1120.

Part I
Food Security, Biodiversity and Law

The Impacts of Climate Change for Food and Nutrition Security: Issues for India

Bill Pritchard

Abstract This chapter reviews current knowledge on the potential effects of climate change on food and nutrition security in India. It finds that India's food systems are highly vulnerable to climate change on a number of fronts. A key finding emphasized in the chapter is that these effects will impact not just on agricultural production arrangements, but on the wider food system through the impacts of climate change on food prices, the composition of the rural economy, and the potential effects of severe weather and sea-level rise. Despite efforts by the Government of India, the nation remains inadequately prepared for the magnitude of effects, should the worst effects eventuate. There is a profound need for further research in India regarding the full food system implications of climate change, particularly with respect to the adaptation needs of vulnerable communities.

1 Introduction

Assessing how changes to the global climate system will impact upon the ways that humanity seeks to feed itself is an extremely difficult exercise. To do this requires consideration of a huge diversity of variables, both ecological and social, which interact in an uncertain and unpredictable fashion. Despite our best efforts, we can merely speculate upon many of these interactions. This problem persists notwithstanding the major advances in the science and social science of this issue during recent years, which form the basis of the landmark Chap. 7 ('Food security and Food Production Systems') of the Inter-Governmental Panel on Climate Change (IPCC) Fifth Assessment Report of Working Group II (AR5-WG II) (IPCC 2014).

The aim of this chapter is to revisit knowledge about the potential range of linkages between food security and climate change, and apply it to the contem-

B. Pritchard (✉)
School of Geosciences, University of Sydney, Sydney, Australia
e-mail: bill.pritchard@sydney.edu.au

B. Pritchard
International Visiting Faculty, University of Cologne, Cologne, Germany

S. Nautiyal et al. (eds.), *Climate Change Challenge (3C) and Social-Economic-Ecological Interface-Building*, Environmental Science,
DOI 10.1007/978-3-319-31014-5_2

porary Indian context. There is no suggestion that this chapter alone can give justice to this rich and complex topic. However, at the very least, it can establish key lines of enquiry in which more detailed, local studies can follow.

In the debate on food security and climate change, sometimes there is a tendency towards production-centric perspectives. This means that the impacts of a changing climate on crop and livestock production are given analytical priority over the effects of climate change on other food system aspects (such as social and economic access, and nutrition utilization). Without wanting to diminish the undeniably important effects in the production sphere, an important emphasis of this chapter is that the effects of climate change spread well beyond the direct effects on crop and animal systems. This emphasis corresponds to a similar line of argument made by the IPCC in AR5-WG II, which assessed the climate change-food security relationship *holistically*.

A holistic perspective is best pursued using the 'four pillar' concept of food security. This approach owes its genesis to the World Food Summit (WFS) of 1996 and subsequent efforts by the Food and Agriculture Organization (FAO). At the 1996 WFS, food security was seminally defined in the following terms:

> Food security, at the individual, household, national, regional and global levels [is achieved] when all people, at all times, have physical and economic access to sufficient, safe and nutritious food to meet their dietary needs and food preferences for an active and healthy life (FAO 1996)

The four pillars that underpin this aspiration are food availability, access, utilization and stability. Climate change has potential ramifications on each of these, both in the global and Indian contexts. The magnitude of its effects will depend on the pace of warming. Global temperatures are currently approximately 0.8 °C above the pre-industrial period. The scientific consensus is that accumulated stored energy from previous and current warming—most of which is in the oceans—will ensure the warming trend continues for some time. The pace and duration of this increase will be shaped by temperature forcing through continued expansion of greenhouse gas emissions, offset by mitigation measures. Ongoing global commitment to mitigation is difficult to assess. Moreover, it brings politically difficult issues to the fore, such as the extent to which developing countries should be held account for mitigation responsibilities. The intractability of these problems, given the continuing centrality of growth as a mantra for the world economy, suggests that the world may be heading towards a scenario of dangerous climate change. This would pose massive implications for the capacities of present food systems to maintain their essential logics. We return to this issue later in this chapter. In the following sections, the relationships between climate change on the one hand, and food availability, access and utilization on the other, are successively addressed. In each case, the issue is presented at the global scale prior to consideration of the Indian context.

2 Climate Change and the Availability of Food

The first widely-circulated attempt to assess the effects of climate change on global agriculture was the model published by Cline (2007). This work is notable because of its use of a range of climate scenarios and its discussion of the potential yield effects on major crop staples. In the years that followed, major efforts took place to build the assumptions that underlay Cline's work into more robust models that were sensitive to a wider array of variables.

Models subsequent to Cline have understood climate change not simply in terms of shifts to temperature gradients, but its embodiment of a range of different weather-related events. This is an important development, because in different regional agricultural-climate interfaces, it may be that *non-temperature* effects related climate change—such as shifts to the volume, timing and distribution of precipitation, or increased exposure to extreme weather events—may have critical impacts on the viability of crop and livestock production. Research in northern India, which we will discuss further below, has highlighted the dramatic effects of these factors in combination.

Knowledge about the relationship between climate change and agricultural production was expanded considerably by the World Bank-sponsored work of Müller et al. (2010). In addition to the modelling of climate per se, this model also incorporated potential effects of CO_2 fertilization (the effects of enhanced CO_2 in stimulating plant growth) in forecasting climate-agriculture interactions. Müller et al. were open-minded on the effects of CO_2 fertilization, painting a range of potential effects, noting that it was 'the most important factor' (p. 3) in uncertainty over climate change effects. This accorded with knowledge at the time, which proposed that increased CO_2 in the context of a changed climate may actually generate a net positive effect on agricultural yields. Since 2010, however, consensus has shifted such that any positive effects from CO_2 fertilization would be outweighed by net negative implications from climate change (Hemming et al. 2013). In AR5-WG II, fertilization effects were noted as being highly variable, based on plant type and agro-ecological context, and in general, rather less important than other factors in the determination of agricultural production responses to a changed climate. This position was summarized aptly by *The Guardian* at the time of the release of AR5-WG II, which observed: "While plants like carbon dioxide, they don't like heat waves, droughts, and floods" (Nuccitelli 2014).

By 2014, the accumulation of research based on experiments and field trials across the world allowed the IPCC to synthesize 66 yield-impact studies on the effects of climate change on major cereal crops for both tropical and temperate regions (IPCC 2014: 497–499). The focus on cereals is pertinent given that a few major cereal crops (rice, wheat and maize) contribute approximately 40 % of humanity's dietary energy. With temperature increases of 1–2 degrees over pre-industrial averages, yields in the tropics are forecast to begin to decline. In temperate zones, the negative effects of climate change kicked-in once temperatures were 3–5 degrees above pre-industrial averages. More recent evidence further affirms the general conclusions reached by the IPCC. Asseng et al. (2015) use an amalgam of 30 different models to argue that for global wheat cultivation, each increase in average temperature of 1 °C will result in a loss of global production of 6 %. Naturally, all reported forecasts deserve caution. However, the consensus of this sizable number of studies confirms that cereal production capabilities at a global level will be impaired by the ongoing effects of global warming.

Several important observations are necessary to be made in conjunction with the results reported by the IPCC. First, it is significant that the tropical zone bears the more immediate negative effects from climate change. The tropics currently contain slightly less than 50 % of the world's population but a considerably larger proportion of global under-nutrition. Moreover, within the tropics, there are generally high proportions of agriculturally-dependent households in populations. Within the large rural population of the tropics, own-consumption of foods by smallholder farms provides a major source of food access. Therefore, the projected negative effects on cereal yields in the tropics will have a direct and highly consequential impact on global under-nutrition and food insecurity. Second, the IPCC's focus on yield needs to be placed in broader agri-societal contexts. In most production areas of the tropics, land and water shortages mean that effects of climate in reducing cereal yields cannot easily be compensated by bringing more land into agricultural production, or applying additional water through expanded irrigation or groundwater schemes.

Contextualizing these issues in the Indian context highlights the highly problematic implications of climate change. The maintenance of national production levels for wheat and rice exerts massive significance for India's food security. Although cultivation of these crops is distributed across all States and Union Territories, the role of production surpluses from north-western India has vital importance for the national system as a whole. Large surplus volumes from Punjab, Haryana and western Uttar Pradesh are purchased by the Food Corporation of India to be stored and then transported nation-wide via the Public Distribution System (PDS) and ancillary food security programs. Yet the agro-environments for the rotation rice–wheat cultivation regime of north-western India are already facing severe degradation, and modelling by Ericksen et al. (2011) suggests that it will be a future hotspot for intensified climate-stress.

Ericksen et al. (2011) identify north-western India as a global hot spot of climate-stress for food availability because of its combination of key vulnerabilities. It has high exposure to climate change, because shifts in temperature spikes and more uncertain precipitation may imperil the maintenance of the reliable crop growing period for wheat–rice cultivation. Climate change will generate more extreme heat days, more intense rainfall patterns and a potential weather-induced shortening of the growing season that will curtail cropping patterns. As the IPCC noted: "half of the wheat-growing area of the Indo-Gangetic Plains could become significantly heat stressed by the 2050s" (IPCC 2014: 504). It has high sensitivity to climate change because of the dominance of this single cultivation regime across the landscape. And, it has poor coping capacity, which Ericksen define in terms of existing food insecurity measured by childhood stunting rates of >40 %.

Ericksen et al.'s model is significant because of its joint consideration of both climate-science and social factors. North-western India is identified as a global hot spot for climate-stress to food availability because of the combination, and therefore mutual reinforcing nature, of its vulnerabilities. Yet, Ericksen et al.'s model tells only part of the story about the significance of north-western India, because it does not address the significance of this region in making food staples available to the rest of the nation. Disruptions to agricultural production systems in north-western India would cause massive repercussions across the country. To put it simply, India's current food system depends crucially on what has been identified as the most climate-sensitive portion of the nation.

The problems of this dependence have not gone unnoticed by international research institutes and New Delhi policy-makers. Investments have taken place to stimulate higher rice yields and production levels in the eastern Gangetic Plains, notably in Bihar and Chhattisgarh. This agenda clearly also seeks to address the needs for development in these poverty-stricken states. Expansion plans for staple crop cultivation in the eastern Gangetic Plain would additionally seek to utilize the principles of climate-smart agriculture, whereby extension agencies and research institutes take into account climate-sensitivity when recommending seed varieties, agronomic practices and related livelihood arrangements.

These issues sketched above have vital long-term implications for India. Despite its impressive growth in agricultural production levels since the 1970s, under-nutrition is still widespread. Current estimates are that there are 195 million under-nourished persons in India, representing 15.2 % of the population (FAO et al. 2015: 46). At the same time, in recent years the storage of cereal grains by the state-owned Food Corporation of India has been at near-record highs, and India is a net exporter of cereals. Evidently therefore, being able to produce more than enough cereal grains to feed India's population has not translated into India's population being adequately fed. If climate change impacts adversely on India's agricultural production capabilities, especially in terms of cereal staples, the challenge of attaining food security for all of India's population will become increasingly complex. But as indicated earlier in this chapter, issues of food production are only one dimension of the climate change/food security dynamic. Of additional importance is the question of how climate change may affect people's social and economic access to food, and it is to this issue, this chapter now turns.

3 Climate Change and People's Access to Food

While the direct impacts on agricultural production is the obvious starting point for assessment of the climate change/food security relationship, it needs remembering that most of the world's population do not grow their own food, but gain access to food access through the market. Hence, of key significance in this debate is the extent to which climate change will impact upon food prices and hence, people's economic access to food. Oxfam typified this problem with the campaign statement:

'a hot world is a hungry world'. Additionally, as discussed below, increases in food prices can spark further processes of social and political instability that aggravate affect access to food, especially for poor population segments.

The IPCC in AR5-WG II propose that it is 'very likely' that temperature and precipitation changes due to climate change will increase food prices by 2050, without consideration of the effects of carbon fertilization (which, as suggested earlier, is probably of lesser significance overall to these macro-trends) (IPCC 2014: 512). Models vary (the IPCC indicates that model ranges vary from 3 to 84 % increase in prices) but with 'medium confidence' suggests that global food prices will have risen substantially by 2050.

The effects of higher food prices will be socially complex. After the food price shock of 2007–08, a stylized interpretation of food price inflation was that it was bad for (food buying) urban populations, but good for (food producing) rural populations. Yet whilst the numerous urban food riots across the world in 2007–08 certainly seems to bear out the validity of the former, the sanguine view of food prices enhancing rural populations' welfare has come under considerable scrutiny over the past few years. While it might appear that higher food prices should mean better returns for farmers, substantial proportions of the population of the rural global South have become de-agrarianized, as their livelihoods become detached from farming. For these households, food needs are met from market purchases, and higher food prices will have undeniably negative effects. It is therefore likely that even in rural populations climate-related increases in food prices would have negative effects on economic access to food (Pritchard 2014).

Reduced economic welfare and food insecurity arising from higher food prices would trigger a range of adaptive responses. It may lead to return migration to rural areas by some urban populations. But as noted, net food buying households in rural areas, which would commonly be expected to in the lower social strata, would also be negatively impacted by higher food prices. Hence, higher food prices could (perversely) encourage new streams of rural-to-urban migration among disadvantaged rural populations in greater degrees of distress. In short, it is difficult to foresee the complex interplay between economic welfare and migration that may arise from higher food prices.

These problems are certainly germane to India. There is a high proportion of net food buying households in rural India. Data is scarce, and often disputed, but one well-regarded study estimated that 74 % of rice smallholders in India are net food buyers (de Janvry and Sadoulet 2012: 21). This seems intuitively consistent with the fact that it has been estimated that an Indian farm household requires at least 4 ha of cultivated land to meet all their consumption requirements through farming (NSSO 2006; cited in Bhalla 2012: 19), and less than 5 % of Indian farmer households have holdings of this size. Nevertheless, as Headey (2014) has observed, these negative immediate effects could be turned around over the longer term if higher food prices flow into higher agricultural wages, and thereby bolster rural households' capacities to feed themselves through local labour markets. Using data from Bangladesh, Headey argues that rural wages are elastic over the longer term to increased food prices, however, it is unclear whether this conclusion would hold for India, or indeed, different regions of the country.

If climate-induced higher food prices do reduce access to food for large numbers of people, problems of hunger and anger clearly have the capacity to spill over into social unrest. One of the prominent lines of argument associated with the Arab Spring and the subsequent crisis in Syria was that it was triggered by drought and higher food prices. (The Arab Spring was set off by a disgruntled bread vendor in Tunisia.) This chapter is not the place for an extended critique of these issues, but notably, one study that has attempted to consider the link between agricultural production fluctuations and conflict in the Indian context is Wischnath and Buhaug's (2014) analysis of the association between year-on-year fluctuations in crop production and ongoing armed conflicts associated with the Naxalite insurrection. An extensive statistical examination by the authors finds that harvest losses are associated strongly with increased levels of political violence. The authors conclude:

> So, is a poor harvest a good indication of escalating violence in India? Based on the results presented above, the answer is a qualified yes. (Wischnath and Buhaug 2014: 13)

4 Climate Change and the Ability to Utilize Food for Long-Term Health

The foods we eat define our cultural identities, and are a source of enjoyment and pleasure. Yet viewed from a human development perspective, food can be thought of as a means to an end—a supplier of nutrients that assist us to have purposive, productive and healthy lives. As climate change alters environmental health conditions and the geographies of food production, connections between food, nutrition and health are recast. This is an important, though not always fully appreciated, manifestation of the climate change/food security interface.

Of key relevance is the fact that food security, when defined comprehensively, incorporates the nutritional needs of a full and active life. Indeed, although the term 'food security' is used in this chapter, the prevailing convention among many international organizations is to use the terminology 'food *and nutrition* security', to emphasize the latter (Committee on Food Security [CFS] 2012). Parallel with this change in terminology has been advocacy of 'nutrition-sensitive agriculture', which means the research and promotion of agricultural products and practices that facilitate good health. Clearly, these kinds of initiatives need also to take into account climate-sensitivity. An ideal set of research and intervention practices is that nutrition deficits are identified among food-insecure populations, appropriate crops are promoted to address these deficits, and research then takes place on how to ensure that the promoted crops are viable and sensitive to climatic change. An example of this is provided by the Partnership for Maternal Newborn and Child Health (PMNC) in Mozambique, where a very high prevalence of vitamin A deficiency was identified. This encouraged researchers to advocate and introduce new varieties of orange sweet potato (an effective provider of vitamin A), leading to a 63 % increase in vitamin A intakes for children aged 6–35 months, 169 % for children aged 3–5.5 years and 42 % among women (PMNC 2012), and to encourage research priorities on breeding sweet potato sensitive to heat spikes, and therefore, prepare for climate change.

However, recent research suggests the relationship between climate change and nutrition utilization has other manifestations too. A recent study published in Nature (Myers et al. 2014) found that for rice and wheat, increased CO_2 atmospheric levels were associated with reductions in zinc and iron content within existing cultivars. This has particular relevance in the context of hidden hunger, and where, according to data cited in the Nature article, there are 2.3 billion people were living in countries whose populations received at least 60 % of their dietary zinc and/or iron from rice or wheat. As the authors argue—

> The public health implications of global climate change are difficult to predict, and we expect many surprises. The finding that raising atmospheric [CO_2] lowers the nutritional value of C3 food crops [rice and wheat] is one such surprise

5 Food System Stability

The fourth pillar of food security is food system stability. Changes to climate systems will inevitably trigger an array of destabilizing processes on food systems. Some of these are already discussed above, such as shifts to the geographical distribution of crop and livestock activities, and the potential effects of climate-induced higher food prices on food access for non-farm populations. However, these destabilizations are potentially just the tip of the iceberg, no pun intended. In addition to these processes are the effects of climate-induced sea-level rise, and potential increases in the incidence of extreme weather events. These occurrences may bring into being vicious cycles of food-climate instability involving waves of distress and emergency migration, stretching political management capabilities.

Starting with the effects of extreme weather, a warmer climate, including warmer ocean temperatures, seems likely to generate a scenario of tropical storms with greater severity, although possibly less frequently (Knutson et al. 2010). Similarly, for India, Wang et al. (2012) find that since 1997, tropical storms in the pre-monsoon period over the Arabian Sea have become more intense and earlier in the season. Using a simulation model, Parth Sarthi et al. (2015) find that for the Bay of Bengal, there is a likelihood of greater storm intensity, earlier in the season. Tropical storms have several specific destabilizing effects for food systems. First and most obviously, there is a need for emergency evacuation and food supplies. Second, when tropical storms bring flooding and salt water invasion, fields are destroyed and may be not arable for an extended period. Third, the disruptive effects of storms on livelihoods, due to damage to agriculture and non-farm industries can lead to aggravated household financial crises that can have long-lasting effects.

The issue of tropical storms brings into focus the more general issue of the impacts of climate change on the predictability of the Indian monsoon. There is considerable debate over the longer-term implications of climate change on the monsoon. In terms of its strength, a recent study by Roxy et al. (2015) points to a gradual weakening of the monsoon as the sea–land thermal gradient reduces due to sea-level temperature rises. In contrast, Jayasankar et al. (2015) argue that modelled projections show a range of potential outcomes, but a cluster of models with highest reliability suggest increased frequency in the occurrence of extreme and high rainfall events, hence making more likely short-term flooding during the monsoon season. Extrapolating these findings in terms of agricultural production, it appears likely that farmers dependent on rain-fed systems will face intensified challenges because of shifts in rainfall patterns during the monsoon.

An insight into these issues is provided by Iwasaki and Shaw's (2008) the examination of the relationship between climate-related environmental change and household livelihoods in Chilika Lake, Odisha. The livelihoods of these fisherfolk communities are finely attuned to environmental conditions, and bore the brunt of major changes in the past two decades. These include the effects of intensified rainfall during the monsoon on erosion of hill slopes adjacent to the lake, causing aggravated siltation, the persistent impacts of regular tropical storms, and changes to ocean environments, including sea temperatures, with impacts for marine ecologies. Studies of similar conditions in the Sundarbans (both in India and Bangladesh) highlight a comparable medley of challenges, but also including the effects of land loss due to sea-level rise (Ayers and Forsyth 2009; Collins 2014).

Migration is one response to the threats and uncertainties accompanying climate change and linked food insecurity, however, as a huge volume of the literature has discussed, the precise link between climate change and migration can be conceptually difficult to untangle from wider migratory processes (Renaud et al. 2011). In India, huge waves of migration occur annually and semi-annually for all kinds of livelihood-related reasons. However, certainly in some parts of the country especially prone to physical change due to climate change—such as the threat facing the Sundarbans and parts of coastal Kerala to sea-level rise, or the effects of glacial melt in the Himalaya, resident populations might have little other option but to abandon their homes and communities due to chronic problems associated with climate change. As discussed above, populations on the move due to distress are especially vulnerable to food insecurities of various sorts, which is an important potential implication coming out of climate change.

6 Conclusion

This chapter has sought to provide an overarching perspective on the conceptual connections between climate change on the one hand, and food security on the other, paying particular attention to the Indian context. The important message from the chapter is the need to recognize the potentially wide-ranging and diverse connections between these two processes. On occasions, the climate change-food security nexus is portrayed in somewhat narrow terms, focusing strictly on the effects of higher temperature profiles on agricultural production. Yet as this chapter has sought to present, these issues, although important, are but one element of the relationship.

In closing, three further issues warrant mention. First is the issue of climate mitigation and the food system. This chapter has largely steered away from providing an extensive discussion of the potential for climate change scenarios to be redirected (positively) through mitigation policies and strategies. However, should global commitment to mitigation become stronger and binding on national governments, it is inevitable that policy-makers will look to the food system for emission reductions. This can and would take multiple forms including soil carbon sequestration, revegetation, the more prevalent use of non-fossil fuel-based agricultural inputs and technologies, and possibly, policies to encourage greater localism in food systems in order to reduce emissions from food miles. There is considerable research already underway on mitigation strategies within the food system, notably through Climate-Smart Agriculture programs. Mostly however, these remain voluntary and on the margins of the food system as a whole. Mainstreaming of Climate-Smart Agriculture would lead to major shifts in the composition and character of food production, distribution and consumption.

As part of any set of changes towards a food system with lesser emission propensities, the introduction of carbon taxes or emission trading arrangement may place cost pressures on those parts of agriculture with relatively higher emission levels, such as livestock for meat consumption purposes. As a country with a high rate of vegetarianism, the Indian food system does not have the high emission vulnerabilities relating to livestock consumption when compared to countries such as the US and, increasingly, China. However, higher costs of meat could place renewed nutritional importance on pulses, a food category that many Indians depend crucially reliant upon, but which has faced problems of supply in India over recent years. India increasingly depends on imports to meet its pulse requirements, and domestic production stocks have led to rapid rises in local prices. In a broader sense, pulses were displaced from much of Indian agriculture in conjunction with the adoption of Green Revolution rice and wheat hybrids with more intense cropping patterns, and perhaps, with a need to adapt the Indian food system to a more climate resilient future, there is a place for a renewed emphasis on pulses within Indian farming.

Second, there is the question of how the Indian food system, and farmers in particular, adapt to climate change. Scenarios of potential climate change migration, painted above, are contingent (in part at least) on farmers' adaptive capabilities. Higher temperatures and altered precipitation patterns will pose challenges for farmers, and their capacities to adapt will shape their futures. The enhancement of adaptive capabilities in Indian agriculture faces considerable hurdles. During the past decade, investment in agricultural extension has lagged requirements. It has been estimated that only 5 % of small and marginal farmers have access to agricultural extension services and that 40 % of agricultural extension posts in India remain unfilled (Agarwal 2011: 10). Adding further layers to this problem, forma education levels in much of rural India remain at low levels. Of course, there is a rich well of indigenous knowledge amongst Indian farmers, including knowledge about how to manage climatic adversity. Successful adaptation pathways will require improvements to the level of formal institutional and educational support for Indian agriculture, to dovetail with the knowledge already held by Indian farmers.

Third and finally, the discussion in this chapter presupposes a period of gradual climate change, however, there remains a possibility of the world moving into a period of dangerous climate change by the middle of this century, which would necessitate much more rapid and profound destabilizations of food systems. It is possible, for example, for India to be facing its own climate change struggles whilst simultaneously trying to manage a sure in climate refugees from low-lying neighbours such as Bangladesh and Maldives, and intensified geopolitical tensions with Pakistan and Nepal over the control of water in major river systems such as the Indus and Kosi. In a scenario of dangerous climate change, it is difficult to foresee what would happen to world food trade, but heightened protectionism could be one possibility, leading to greater difficulties in the procurement of key foods that India is import-reliant upon, such as pulses, as noted above. Sea-level rises would also, it needs noting, have some of their most important ramifications on coastal cities, which in the Indian context include Chennai and Mumbai, which are both prone to flooding.

To conclude, there are numerous uncertainties associated with climate change and its relationship to food security. With a changing climate worldwide, this relationship will be altered in numerous ways. Paying attention to this issue through prioritization in research and policy is a crucial contribution for humanity in the twenty-first century.

References

Agarwal, B. (2011). *Twelfth plan working group on disadvantaged farmers including women*. Final report submitted to the Planning Commission of India, online. http://indiagovernance.gov.in/files/working_group-report-farmers.pdf. Accessed January 29, 2013.

Asseng, S., Ewert, F., Martre, P., Rotter, R. P., Lobell, D. B., Cammarano, D., et al. (2015). Rising temperatures reduce global wheat production. *Nature Climate Change*, *5*(2), 143–147. http://dx.doi.org/10.1038/nclimate2470

Ayers, J., & Forsyth, T. (2009). Community-based adaptation to climate change. *Environment: Science and Policy for Sustainable Development, 51*(4), 22–31.

Bhalla, S. (2012). *Notes on land, long run food security and the agrarian crisis in India*. IHD Working Paper Series, 02/2012, Institute for Human Development, New Delhi.

CFS (Committee on World Food Security). (2012). *Coming to terms with terminology: Food security, nutrition security, food security and nutrition, food and nutrition security*. Document CFS 2012/39/4. Rome: FAO. http://www.fao.org/docrep/meeting/026/MD776E.pdf

Cline, W. (2007). Global warming and agriculture: Impact estimates by country. Washington, D. C.: Center for Global Development: Peterson Institute for International Economics.

Collins, J. (2014). A rising tide in Bangladesh: Livelihood adaptation to climate stress. *Australian Geographer, 45*(3), 289–307.

de Janvry, A., & Sadoulet, E. (2012). Subsistence farming as a safety net for food-price shocks. In M. J. Cohen & M. Smale (Eds.), *Global food-price shocks and poor people: Themes and case studies* (pp. 18–26). London: Routledge.

Ericksen, P., Thornton, P., Notenbaert, A., Cramer, L., Jones, P., & Herrero, M. (2011). Mapping hotspots of climate change and food insecurity in the global tropics. *CCAFS Report, 5.*

Food and Agriculture Organization (FAO). (1996). *Rome declaration on world food security and world food summit plan of action.* Rome: World Food Summit, November 13–17, 1996.

Food and Agriculture Organization (FAO), World Food Program (WFP), & International Fund for Agricultural Development (IFAD). (2015). *The state of food insecurity.* Rome: FAO.

Headey, D. (2014). *Food prices and poverty reduction in the long run.* IFPRI Discussion Paper 01331, International Food Policy Research Institute, Washington.

Hemming, D., Betts, R., & Collins, M. (2013). Sensitivity and uncertainty of modelled terrestrial net primary productivity to doubled CO_2 and associated climate change for a relatively large perturbed physics ensemble. *Agricultural and Forest Meteorology, 170*, 79–88.

IPCC. (2014). Climate change 2014: Impacts, adaptation, and vulnerability. Part A: Global and sectoral aspects. In C. B. Field, V. R. Barros, D. J. Dokken, K. J. Mach, M. D. Mastrandrea, T. E. Bilir, M. Chatterjee, K. L. Ebi, Y. O. Estrada, R. C. Genova, B. Girma, E. S. Kissel, A. N. Levy, S. MacCracken, P. R. Mastrandrea & L. L. White (Eds.), *Contribution of working group II to the fifth assessment report of the intergovernmental panel on climate change* (p. 1132). Cambridge, United Kingdom and New York, NY, USA: Cambridge University Press.

Iwasaki, S., & Shaw, R. (2008). Fishery resource management in Chilika lagoon: A study on coastal conservation in the Eastern Coast of India. *Journal of Coastal Conservation, 12*(1), 43–52.

Jayasankar, C. B., Surendran, S., & Rajendran, K. (2015). Robust signals of future projections of Indian summer monsoon rainfall by IPCC AR5 climate models: Role of seasonal cycle and interannual variability. *Geophysical Research Letters, 42*(9), 3513–3520. doi:10.1002/2015GL06365

Knutson, T. R., McBride, J. L., Chan, J., Emanuel, K., Holland, G., Landsea, C., et al. (2010). Tropical cyclones and climate change. *Nature Geoscience, 3*(3), 157–163.

Müller, C., Bondeau, A., Popp, A., Waha, K., & Fader, M. (2010). *Climate change impacts on agricultural yields.* Washington, DC: World Bank. © World Bank. https://openknowledge.worldbank.org/handle/10986/9065

Myers, S. S., Zanobetti, A., Kloog, I., Huybers, P., Leakey, A. D. B., Bloom, A. J., et al. (2014). Increasing CO_2 threatens human nutrition. *Nature, 510*(7503), 139–142. http://dx.doi.org/10.1038/nature13179

National Sample Survey Organisation (NSSO). (2006). *Income, expenditure and productive assets of farmer households.* NSSO Report, 497, Government of India, New Delhi.

Nuccitelli, D. (2014). *IPCC report warns of future climate change risks, but is spun by contrarians.* The Guardian, March 31, 2014. http://www.theguardian.com/environment/climate-consensus-97-per-cent/2014/mar/31/ipcc-warns-climate-change-risks

Parth Sarthi, P., Agrawal, A., & Rana, A. (2015). Possible future changes in cyclonic storms in the Bay of Bengal, India under warmer climate. *International Journal of Climatology, 35*(7), 1267–1277.

Partnership for Maternal Newborn and Child Health. (2012). *PMNCH knowledge summary #19 food security and climate change.* http://www.who.int/pmnch/knowledge/publications/summaries/knowledge_summaries_19_food_security_climate_change/en/

Pritchard, B. (2014). The problem of higher food prices for impoverished people in the rural global South. *Australian Geographer, 45*(4), 419–427.

Renaud, F. G., Dun, O., Warner, K., & Bogardi, J. (2011). A decision framework for environmentally induced migration. *International Migration, 49*(s1), e5–e29.

Roxy, M. K., Ritika, K., Terray, P., Murtugudde, R., Ashok, K., & Goswami, B. N. (2015). Drying of Indian subcontinent by rapid Indian Ocean warming and a weakening land-sea thermal gradient. *Nature Communications, 6*. doi:10.1038/ncomms8423

Wang, B., Xu, S., & Wu, L. (2012). Intensified Arabian Sea tropical storms. *Nature, 489*(7416), E1–E2.

Wischnath, G., & Buhaug, H. (2014). Rice or riots: On food production and conflict severity across India. *Political Geography, 43*, 6–15.

Climate Change Strategies and Developing Nations: Prospects and Priorities for India

P.J. Dilip Kumar

Abstract It is now accepted by the majority of people that climate change due to human activities is going to be a reality in the very near future, and that we may even currently be seeing some of the early manifestations in the form of increased fluctuations in weather and related phenomena, frequent extremes like excess rainfall leading to floods or deficiencies resulting in droughts, accelerated melting of glaciers and so on. It is no longer easy to ignore the warning signs that human pressure on the environment and natural resources is at unprecedented levels, and the damage to property and lives is also becoming unacceptably high because of the high populations, huge amount of built infrastructure, high financial stakes in extractive industries and so on.

Keywords Climate change · Greenhouse gases · Development strategy · Natural resources · Forest · Carbon sequestration · IPCC · REDD

All this leads to a heightened concern to find ways to reduce or slow down the impacts of climate change through mitigative and adaptive measures. As can be seen from successive international meetings on climate change, agreement has still not been reached on the balance between gross and per capita consumption of environmental resources and the concomitant obligations to the world at large of each country.

A developing country like India is faced with the twin problems of low per capita consumption levels (which demands a manyfold stepping up of use of natural

---

Paper prepared for the "International Conference on Climate Change and Social-Ecological-Economic Interface-Building", May 20–21, 2015, organized by the Centre for Ecological Economics and Natural Resources (CEENR), Institute for Social and Economic Change (ISEC), Nagarbhavi, Bengaluru-560072, India.

P.J. Dilip Kumar (✉)
Institute for Social and Economic Change (ISEC), Nagarbhavi,
Bangalore 560072, India
e-mail: pjdilip@gmail.com

S. Nautiyal et al. (eds.), *Climate Change Challenge (3C) and Social-Economic-Ecological Interface-Building*, Environmental Science,
DOI 10.1007/978-3-319-31014-5_3

resources, especially fossil fuels and minerals), and a high total contribution to global pollution (because of the huge population). While a developing country has to be mindful of the global effects of its increasing consumption, it also has to make a dispassionate calculation of what effect any sacrifices in current consumption will have on the global trends and global climate change scenario. It appears that any sacrifice by India at the current low levels of per capita consumption will have negligible impact on the global trends, whereas it will cause economic distress within the country, for example to the poor who are yet to achieve reasonable standards of consumption. In such a situation, as per the logic of collective action, the sensible choice would be to make the minimum contribution to the collective good while maximizing private (here read country) returns. In order to make wise choices that do not result in more poverty in the name of development, we may have to devise and apply more sensible measures of well-being, rather than blindly following the high-consumption paths of the developed world.

1 Introduction: The Climate Change Prognoses

The world's climate is like a global commons; a large number of countries, and billions of individuals, are doing things that are affecting the world climate in unexpected, and unpleasant, ways. This of course is only the latest addition to various other global commons that are known to be under unprecedented pressure, such as the air and water quality, the environment in general (especially the load of chemicals), the soil and vegetation (the erosion of land and forest resources), the natural habitats (forests, wetlands, grasslands, arid lands and wild areas in general), the world's stock of species (biodiversity) and so on. In more philanthropic spheres, there is the gradual erosion and impoverishment of traditional cultures and societies, with the consequent erosion of languages, local knowledge, art, culture, oral histories and literature, religions and so on. These effects are all traced back to the global adoption of western economic and technological models of development, coupled with a gradual homogenization of cultures and mores that have amounted to a second wave of domination over the less developed third world after the winding up of the imperial project of the last couple of centuries of the European powers.

Ever since the global environment has become a subject of study, discussion and action in the world forums, it is the Intergovernmental Panel on Climate Change (IPCC) that has been the main source of our understanding of climate change issues and our main guide to policy. It would be instructive, therefore, to take a brief look at the IPCC reports before going on to analysis of our options. The latest tranche is the Fifth Assessment Report, AR5, 2014, and naturally it has generated a plethora of documents by a large number of authors and drafters, that are available on the internet (thankfully they do not have to cut down too many trees to reach them to us nowadays, but on the other hand, the relatively negligible incremental cost of publishing encourages enormous output!). It therefore becomes difficult for any one individual to access all of these documents, let alone understand or analyze them;

we will, therefore, have to fall back on the executive summaries provided, and refer to published independent critiques to bring in some element of critical understanding.

Let us, then, take a look at one such document: *Climate Change 2014. Impacts, Adaptation, and Vulnerability. Summary for Policymakers*, (IPCC 2014a, b, c), which is the Working Group II contribution to AR5 (at https://ipcc-wg2.gov/AR5/images/uploads/WG2AR5_SPM_FINAL.pdf, accessed on 25 May 2015). The claim is made in this document that climate change effects are becoming distinguishable from other influences or causes in recent decades: changing precipitation, (accelerated) melting of snow and ice affecting water resources, shrinking of glaciers, permafrost warming and thawing, shifting geographical ranges of many terrestrial and aquatic species and a few recent species extinctions (op. cit., p. 4). Negative impacts on crop yields have been more common than positive, although some gains have also been studied in high-latitude regions (ibid.). What can be said, with a high level of confidence, is that "it is not yet clear whether the balance of impacts has been negative or positive" (p. 5). However, they do ascribe recent periods of "rapid food and cereal price increases" in key producing regions to climate extremes; as also "increased heat-related mortality and decreased cold-related mortality in some regions as a result of warming" (op. cit., p. 6). Marginalized, poor people are adjudged (not surprisingly) as more vulnerable to climate change, "and also to some adaptation and mitigation responses". Impacts are considerable even from recent climate-related extremes, such as heat waves, droughts, floods, cyclones and wildfires, showing a "significant vulnerability" and "significant lack of preparedness" to *current* climate variability, in some sectors (ibid.).

A possible source of relief for us in India is that the WG2 report ascribes only "minor" contribution of climate change (medium confidence level) to food production in the subcontinent (Figure SPM.2, p. 7). A previous report *Turn Down the Heat* of the World Bank (2013), which was released even as devastating floods were wreaking havoc in Uttarakhand state in June 2013, suggested that South Asia would be especially impacted by climate change as the world atmosphere warms up from the present level of 0.8 °C to say 4 °C above pre-industrial times by 2100, with sea levels rising by over 100 cm, monsoon rains becoming more variable, while floods (and droughts) would be more frequent and damaging (Balasubramanian and Dilip Kumar 2014). Kolkata is suggested to be among the 10 most exposed cities to flooding, and a significant reduction in crop yields is to be expected due to extreme heat by the 2040s. Reduced water availability due to changes in precipitation levels and falling ground water tables are likely to aggravate the situation. The worst effects, it was suggested in the report, could be avoided by holding global warming below 2 °C (above pre-industrial times) (ibid.).

We need not cover the climate change prognostications in any further detail here, because the material is available everywhere at whatever level of expertise is required. However, a few considerations come up immediately to the discerning observer. One is, of course, what is the rigour of these exercises: are these prognoses just guesses, or are they based on specific cause and effect models or formulae. Can we take them on face value, as actual predictions rather than just alternative scenarios? Second, what will be the likely actual rise in global temperatures due to human activities, especially the addition of carbon due to a greater burning of fossil fuels as the third world countries seek to increase their production levels? Third, what will be the efficacy of the mitigative (and adaptive) measures usually suggested (e.g. limiting the additions of each nation's future carbon contributions, investing in additional carbon-absorbing measures like new forests), and how would these measures relate to the quest for better standards of living and higher incomes in the poorer countries of the world? Finally, what should be the policy to be adopted by India (and other developing countries with large numbers of poor), that will balance the needs of growth and development in the immediate future, and the justifiable demands of the world at large, and of future generations of our own countries, to a fair and equitable share of the world's resources, which include the global climate commons?

Turning to the first issue, that of the dependability of the prognoses, the problem is that there is a wide range in the predictions according to different assumptions or scenarios. The IPCC reports ascribe different levels of confidence in a somewhat misleading way, because the words "high confidence" are often juxtaposed to these projections or to the impacts of mitigative and adaptive measures, whereas they usually refer to phrases in the earlier parts of the respective paragraphs that refer to other aspects, not the measures themselves. A simple example is this one on p. 9 of the WG2 summary (IPCC 2014a):

> **Responding to climate-related risks involves decision-making in a changing world, with continuing uncertainty about the severity and timing of climate change impacts and with limits to the effectiveness of adaptation (*high confidence*).** (emphasis in original)

This sentence suggests, to an unwary reader who is skimming the text, that there is a high confidence in the severity and timing of climate change impacts, and in high effectiveness of adaptation, whereas the text actually suggests the exact opposite: there is high confidence in the continuing ***uncertainty*** of impacts and in the ***limits*** to the effectiveness of adaptation. From the accompanying figures, it appears that there will be warming almost everywhere, except a few patches in the north Atlantic where the moderate cooling effect is, however, "not statistically significant". What we are given to understand is that the projections which range from say 1 to 4 °C (above pre-industrial times) by 2100, are based on a combination of diverse sources:

> …empirical observations, experimental results, process-based understanding, statistical approaches, and simulation and descriptive models. Future risks related to climate change vary substantially across plausible alternative development pathways, and the relative importance of development and climate change varies by sector, region, and time period (*high confidence*). Scenarios are useful tools for characterizing possible future socioeconomic pathways, climate change and its risks, and policy implications. … **Uncertainties about future vulnerability, exposure, and responses of interlinked human and natural systems are large (*high confidence*).** (Op. cit., p. 11, emphasis in original)

Once again, the casual reader may well go away with the impression that future vulnerability, exposure, etc., are large, whereas the sentence only states (with high confidence) that the *uncertainties* are large.

In a long and detailed critique of such dire environmental prognoses, Bjørn Lomborg (who describes himself in his Preface as "an old left-wing Greenpeace member"), also addresses the issues of global warming and climate change (Lomborg 2001, Chap. 24). On considering all the evidence, Lomborg concludes that there has been a rise in temperature of some 0.6 °C over the past century, and it is reasonable to ascribe this partly to an anthropogenic greenhouse effect, "although the impression of a dramatic divergence from previous centuries is almost surely misleading" (op. cit., p. 317). He points out the basic weakness of the prognoses, that still are unable to say with certainty whether a doubling of CO_2 concentrations will lead to a rise of 1.5 °C or the "dramatic" high estimate of 4.5 °C. The IPCC models do not deal with other effects (water vapour feedback and clouds, for example), that may reduce the impact. Lomborg feels that the large number of "scenarios" presented indicates that the IPCC has "explicitly rejected making predictions about the future, but instead gives us "computer-aided storytelling", basing the development of crucial variables on initial choice and depicting normative scenarios "as one would hope they would emerge" (quoting de Vries et al. 2000, p. 170). Indeed, Lomborg points out that some of the scenarios even suggest a net gain in incomes in both the developed and developing worlds. Not only is the quantum of temperature rise uncertain, but the consequences are also not as devastating as people have been led to assume: "Global warming will not decrease food production, it will probably not increase storminess or the frequency of hurricanes, …" and so on. However, it will have costs (to the order of $5 trillion), and it will "hit the developing countries hardest", "primarily because they are poor—giving them less adaptive capacity" (Lomborg 2001, pp. 317–318).

As can be expected, such sceptical ideas have not been received kindly, and there are numerous counters, e.g. the Danish biologist Kåre Fog's website www.lomborg-errors.dk, dedicated to debunking Lomborg's arguments. This paper does not claim to be a scientific assessment of these arguments and counterarguments, and concerned readers would do well to delve into the voluminous literature available. Fog states that it is not sufficient just to read Lomborg's book and think of the ideas in it, but one has to go back and check every piece of information to see whether it is "true" and if "the presentation is balanced" (Fog, op. cit., probably 2004, accessed 31-08-2015). Of course, we would have to state the same of the climate change prognoses as well, such as are contained in the IPCC documents. Obviously, every reader cannot undertake such verification, and since the effects are far out into the future, it becomes a matter of strategic choice how far the arguments are pursued (especially as regards national actions in response to the prognoses).

It would perhaps be fair to state here that Lomborg himself seems to be not so much a climate change denier today, as advancing arguments based on strategic priorities to use the available money, especially international development assistance, in areas that will offer the maximum gains in terms of human welfare: public health, for instance. Academics tend to dismiss this as pseudoscience, which provides for a lot of heated arguments in the media and on public platforms.

What about the likely impact of mitigative measures that we are all exhorted to take in the present so that our futures can be secured? Once again, we turn to an IPCC "Summary for Policymakers", which is all that the average reader is going to have time to look at, this time of the Working Group III to the fifth IPCC report (IPCC 2014b). This report, like the previous one cited, is prefaced by an equally punctilious and precise statement of the basis of the uncertainty evaluations. The report points out that "Effective mitigation will not be achieved if individual agents advance their own interests independently", which therefore demands "international cooperation" (op. cit., p. 5). This immediately brings up "Issues of equity, justice, and fairness", because past contributions to global CO_2 have been different (we point fingers at the industrialized countries), and current challenges (to reduce poverty and increase incomes) and capacity to take up mitigation and adaptation measures also differ (developing countries are at a disadvantage here). Further, "Climate policy intersects with other societal goals creating the possibility of co-benefits", which suggests a stronger basis for undertaking climate action, and so on (ibid.).

An important part of the WG3 report is the information on the trends in greenhouse gases (GHGs) and their drivers (op. cit., p. 6 et seq.). We are told that total anthropogenic GHG emissions *grew* by around 1.0 Gt CO_2 equivalent (a rate of 2.2 % per year) from 2000 to 2010, as against just 0.4 Gt per year (a rate of 1.3 %) from 1970 to 2000. It may be noted that these are just year-to-year *increases* in the annual GHG emissions; the actual emissions were of the order of 49 Gt in 2010 (the highest recorded so far). The major contribution to the total GHG emission *increase* is said to have come from CO_2 emissions from fossil fuel combustion and industrial processes: in 2010, these contributed around 65 %, followed by other major sources like CO_2 from forestry and other land use (11 %, with a very large uncertainty of the order of ±50 %), methane (16 %), nitrous oxide (6.2 %), fluorinated gases covered by the Kyoto Protocol (2.0 %) (op. cit., p. 7). Population and economic growth have been the major drivers of increases in CO_2 emissions, and have outpaced emission reductions from improvements in energy intensity (op. cit., p. 8).

What are the prognoses for global CO_2 levels? Without additional efforts at mitigation, the growth in emissions is expected to "result in global mean surface temperature increases in 2100 from 3.7 to 4.8 °C compared to pre-industrial levels" disregarding climate uncertainty, with a threefold increase in CO_2 concentration levels from around 430 ppm CO_2 eq in 2011 to 1300 ppm by 2100 (op. cit., p. 9). The WG3 summary document goes on to discuss mitigation measures and their possible costs, based on a collection of about 900 mitigation scenarios based on integrated models, which would leave the CO_2 load by 2100 between 430 and 720 ppm (op. cit., p. 10). As can be imagined, it is not easy to comprehend all these variations and their implications, but it appears that a concentration of around 250 ppm would be likely to keep temperature rise below 2 °C. Pledges made at Cancun are said to be "broadly consistent with cost-effective scenarios that are *likely* to keep temperature change below 3 °C" (p. 13). Lomborg (2001), however, has earlier cautioned that any action to reduce emissions in the immediate future will have heavy costs into the future, and we have to carefully balance the potential gains from accelerated development (and consequent higher welfare) in the intermediate run against the admittedly highly uncertain penalties that will come in the far future, on the unrealistic assumption that no improvements will take place in technology (even new sources of energy, for instance).

2 Logic of Collective Action and Optimal Country Strategy

Granted that we should be conscious of the limited capacity of the environment to absorb our emissions and repair the damaging effects of excessive resource extraction and unsustainable use, what should be our approach as a world community, and as an individual country? The global atmosphere (and climate) is like a commons, with hundreds of independent countries, thousands of corporate and other bodies, and billions of individual families. There is no mechanism to impose the required controls on carbon emissions apart from a voluntary agreement. But, here the "logic of collective action" comes into operation (Olson 1971). Even the nation-state in modern times is not capable of supporting itself solely on voluntary contributions, despite the strong emotional bonds that hold the citizens together, and the benefits that they undeniably get in the continuance of the state structure; the state has to impose compulsory taxes and collect them with a firm hand. This is because, when there is a large multitude of members, no one of them can be prevented from enjoying the benefits of the organization; hence there will be an underlying incentive for the individual to withhold his or her individual contribution as far as possible, as long as it is reasonably certain that the contributions of the myriad others will carry the organization forward. In fact, that would the *rational* thing to do, even if not ethical or fair.

Another metaphor used to describe such situations is the "tragedy of freedom in a commons" that Hardin (1968) talked about many decades back. The problem with regulating use of a commons is that private costs and benefits are mismatched with public ones: each person has to incur a private cost that may be substantial from the individual point of view, but the collective benefits are enjoyed by everybody, and moreover, the individual has little or no control on the actions of all the others, and hence on the outcome. If the others choose to break the agreement and maximize their private returns, our sincere individual may be left 'holding the short stick'. No one likes to be taken advantage of in this manner, so there is an in-built tendency for a mismatch between public professions and actual behaviour, if the individual feels that they can get away with it, or suspects that the majority of actors are intending to take advantage. In simple terms, when there are large numbers of actors, and where the benefits are collective, or public goods, and where no one person could be excluded from enjoying those benefits, it would be the narrowly rational choice for individuals to avoid, as far as possible, the incurring of costs on their side, while continuing to hold a cooperative public posture so as to preserve their position as beneficiary. The consequent loss of the cumulative benefits due to the individual's covert actions will likely be miniscule (after all, the individual's contribution is a miniscule proportion, and the enterprise will not break down even if there are a few defaulters), but the return to the defaulter will be disproportionately high. Of course, this sort of 'free-riding' will be penalized if it becomes chronic and the community gets tired of it, so the individual may have to pay his dues once in a while, just to maintain his membership in the community. The enterprise can still be viable if the majority adheres to the rules, but sometimes a few bad apples can spoil the whole lot.

Faced with this situation, one would have to resort either to some control from outside (the state makes and enforces strict laws, for example), or the resource is privatized so that both the costs and benefits are incident on the same entity (i.e. externalities are internalized). There is, however, a possible third alternative in-between state control and privatization (Dilip Kumar 1991), where the community and the state get together in a joint arrangement, an example of the 'nested' institutions that Elinor Ostrom recommended as the most conducive to successful management of common property (pool) resources (e.g. see Ostrom 1990). In the Indian forest scenario, this principle has been internalized in the Joint Forest Management regime wherein the state forests are managed in partnership with the communities, and conversely, communities also find it more effective to protect even their own forest resources if there is support from the state (e.g. Dilip Kumar 2013).

This problem of 'collective action' is recognized by the "Climate Change 2014 Synthesis Report" (IPCC 2014c) in the following words:

> **Effective mitigation will not be achieved if individual agents advance their own interests independently**. Climate change has the characteristics of a collective action problem at the global scale... Cooperative responses, including international cooperation, are therefore required...The evidence suggests that outcomes seen as equitable can lead to more effective cooperation. (IPCC 2014c, p. 76; emphasis in original).

In our climate change scenario, what would be the rational choice for a country like India, given this 'logic of collective action'? We need to be clear about two things: first, if India were to limit its GHG emissions (say at today's levels), what would be the impact on global climate change? India contributes a sizeable quantum of current emissions (because, like China, it has a huge population, although very low per capita levels), and it may be thought that a cap on aggregate emissions at the current level would have a discernible impact, but in the world situation this may be masked because other countries will continue to increase emissions, and moreover, in spite of such caps, India and other economies will still be adding sizeably every year to the carbon load in the atmosphere. Consider, for example, India's contribution to global CO_2 from the "energy and transformation industries" in 2007 according to NATCOM-2 (Government of India 2012): taking the figure as 749,617 Gg, or 749.6 mta (or 0.7496 Gt per annum), it constitutes a small (if not exactly negligible) fraction of the annual contribution of 49 Gt in 2010 (globally) referred to above. Even if India wants to double the energy production, say over a period of 10 years, this will add only around 0.075 Gt to the annual increase, compared to the annual increase of 1 Gt globally (see above). This is only 7.5 % of the increase in annual CO_2 accretions, but CO_2 levels will still be increasing year-to-year, unless some great new sink or method of fixing it could be found. In other words, India by itself will not be able to contribute meaningfully to the resolution of the GHG problem just by curbing energy production at current levels, or influence the actions of the rest of the world (this is not a situation where the famous moral force of Gandhism can be applied!). So, India will have to proceed as if the rest of the world is going to keep increasing its emissions in the near future. If India were to make any voluntary, unilateral sacrifices now, it would have neither the imagined benefits of the reduced carbon loads, nor the benefits of a stronger economy and greater world influence in the future. Indeed, the very fact that a huge population like India foregoes its present entitlements, will only prompt other countries to appropriate that share for themselves.

Therefore, it is not a rational strategy for a country like India to make any unilateral cuts or promises at the present juncture, and this has rightly been our position in the world climate conferences, with the only concession made that we will never exceed the current per capita emission levels of the industrial countries. This, of course, we will never even aspire to: as it has been repeated often enough, we will need another seven planet Earths if all were to consume at the western levels. We will, however, be aspiring to some intermediate level of consumption, which will require a manyfold increase in the production of energy and various other sectors. On the other hand, India has the world's largest population of poor, with low quality of life indicators; it has to vigorously pursue its course of positive affirmation to set right the historical wrongs to sections of the population, and increase investment in public health and medicare, education, communications, skills development, job creation, etc., and at the same time play its geopolitical role as one of the largest economies of the world. The experience of the last general election suggests that the masses also are looking for something beyond just subsidized food and other doles, and there is a huge middle class that has aspirations of making it in the modern economy.

This does not mean, however, that we should be indifferent to the issues of environmental degradation in our own sphere, quite regardless of the world climate conferences (just as, indeed, high carbon emitters like the developed west would be well advised to reduce per capita consumption levels, not for the sake of the global good, but in the interests of the health and well-being of their own citizens). Indeed, this has been the strategy adopted by our governments over the past decades: while insisting on our share of the global carbon cake, in terms of increasing per capita consumption (and emissions) levels, we have also steadily taken many steps to improving the sustainability of our development process and mitigating damage to our own environment. Of course, the complaint is that we are strong in legislation, but weak in enforcement, as the numerous articles in the popular media, and the spate of environmental cases in the courts, attest. This lag between intent and action, however, is to be expected during the catch-up phase, when all output levels are being racked up manyfold. In keeping with the form of the Kuznets curve, we may expect to see a lessening of this gap and an improvement in both the incidence of poverty and the laxity in enforcing environmental safeguards as GDP per capita progresses beyond a certain level (see Lomborg, op. cit., p. 177 for examples of this in the context of particle pollution and SO_2 pollution).

In the meanwhile, it will be the best strategy to identify points of greatest importance that need to be addressed to keep our environmental systems and natural resource base intact in anticipation of those future levels of relative prosperity. Some of these crucial sectors are our water resources, our forests and grasslands, our biodiversity and wildlife habitats, our marine resources and coastlands, our wetlands (especially mangroves), our catchment areas in the hills, and our urban air quality. While we need to accelerate our development process, we also need to ensure grater participation in the fruits of that development, by rapidly improving public education and public health, communications and transport facilities, raising skill levels and employability, minimizing displacement of people and ensuring proper rehabilitation where displacement is unavoidable, just as with diversion of natural habitats.

3 Forest Carbon Sequestration (CDM-AR, REDD) and Climate Change

These principles are well exemplified in our experience with the AR component (Afforestation and Reforestation) of the Clean Development Mechanism (CDM), and now REDD and REDD-Plus. The basic concept of REDD (Reduction of Emissions from Deforestation and Degradation) is simple enough (Dilip Kumar 2014a, b, c): worldwide, this is supposed to be the second largest source, contributing 12–20 % of GHG emissions, posing both a threat to world climate, and an opportunity, because growing trees absorb CO_2 and hence can be a significant low-cost, universally applicable mitigation device. Afforestation and reforestation (A&R) as a route to carbon mitigation were first introduced in UNFCCC CoP-3 (1997) under the Clean Development Mechanism (CDM). In 2005, at CoP-11, Reducing Emissions from Deforestation (RED) was introduced, and the concept later enhanced by adding forest Degradation (REDD). This was further expanded to REDD-Plus in the CoP-13 at Bali in 2007, by adding elements of forest conservation and sustainable forest management as devices for mitigating carbon emissions, giving rise to the expectation, among developing, tropical countries, that they would be paid for the carbon sequestered in trees and forests by sacrificing short-term benefits.

The CDM-AR component was at one time touted as the great new mechanism for carbon sequestration, and detailed and laboriously drafted guidelines and rules were put up on the UNFCCC websites (e.g. UNFCCC 2013). But, these guidelines and procedures were so involved that very few projects were actually approved: according to an online article on the Forest Carbon Asia website (Chokkalingam and Vanniarachchy 2014), by May 2011, there were only 22 registered CDM-AR projects (and 4 requesting registration), as against some 3000 CDM projects overall, mostly in energy, waste management, manufacturing, emissions, and agriculture. The main issue seems to have been that the process of developing standards and procedures for validating such projects (for instance, that it is an additionality, and will be long-lived) is complex and time consuming (which the authors have cited, rather unconvincingly, as an excusable process of learning that will help the successor scheme to get off the mark that much faster), and the same is bound to be true of the latest incarnation, REDD Plus, which introduces yet more components, like sustainable forest management, to sort out.

Apart from the intractable difficulties in actually proving that a particular afforestation or restoration project is an additionality, and that the forest crop will be sure to last indefinitely with its locked-in carbon (and not liquidated sometime in the near future), there is the fact that the world community has been rather niggardly in providing the funds required to implement the scheme: after a lot of meetings and discussions, some 280 million USD were pledged at the CoP-19 at Warsaw in November 2013 by certain developed countries for a "results-based" REDD programme in 48 partner countries, a miniscule amount compared to the many billions required to even secure the existing forests, let alone produce enough new trees to sequester the extra CO_2 the world is going to emit as per the IPCC scenarios. India is in any case probably not going to be a potential recipient of such payouts, and the relatively minor amounts being discussed are an obvious indication that much of the money will actually go to technical inputs like training, 'capacity building', consultancies and so on. For India, it is all the more meaningless to look to such sources for its national programmes, as there is already an accumulated Compensatory Afforestation (CAMPA) fund of over 5 billion USD waiting to be utilised.

Another potential problem with relatively stable and institutionally developed states like India is that it will be difficult to find any major additionalities over and above the existing national programmes (the 'business as usual' scenario). For India in particular, there are already the eight national Climate Missions devised under the previous government, of which the Green India Mission is one (see Prof. Ravindranath's presentation at the FAO online conference, 2015). It envisages the creation of 5 mha new growth and improvement of 5 mha of existing forest, on a landscape approach, over a period of 10 years, which will sequester carbon, along with a number of 'co-benefits' in the form of strengthening community participation, biodiversity conservation, and income augmentation. India also has a long-established forest service that administers the forest and wildlife areas (which cover some 65 mha out of the 300 mha of the area reported upon in the country), and many progressively stringent protective laws like the Forest Conservation Act (1980), Wildlife Protection Act (1972), National Forest Policy (1988), Environmental Protection Act (1986), National Environment Policy (2006), Forest Rights Act (2006), etc., along with a strong and active judiciary (and now the National Green Tribunal), which have succeeded in at least stabilizing, and even modestly increasing, the forest cover and wildlife habitats in recent decades (Dilip Kumar 2014c).

At the same time, there have been strenuous efforts to empower communities through decentralization of political and administrative roles through the Panchayati Raj institutions, and give them legal entitlements on a broad range of issues, to employment, food, education and of course the individual and community forest rights of traditional forest dwellers through the Forest Rights Act 2006. The forest service itself, although a rule-bound and rather authoritarian force of uniformed officers, that has traditionally seen its main role as protecting the forests *from* the people living around it, has developed mechanisms to work *with* the communities through the Joint Forest management Committees (JFMCs). These measures appear to have gone a long way in meeting the global objectives agreed at the UN Forum of Forests (UNFF), 2006, such as reversing loss of forest cover, enhancing economic, social and environmental benefits, and increasing the area under protection and sustainable forest management (SFM).

Any REDD scheme would therefore be, but a minor augmentation over and above these targets and instruments, and is likely to be nothing more than a token gesture of support to the international UNFCCC community and perhaps a display piece for the private sector. Indeed, according to the Second National Communication (NATCOM) of the environment ministry (Government of India 2012), there would seem to be little scope for REDD-type interventions in India, because forests are reported to be contributing very little to GHGs: just as an illustration, for 2007, of the total country levels of CO_2 emissions (1,476,357 Gg, or million tonnes) and removals (275,358 Gg), forest contributed only 87,840 Gg to emissions and 67,800 Gg to removals, leaving a minuscule net addition, nowhere near the 20 % contribution worldwide of GHG from forest degradation which is the basic assumption of the REDD strategy.

In any case, as already suggested by this author (Dilip Kumar 2014a, b, c), it does not appear that India is being viewed as a major destination for aid in the sphere of REDD or REDD Plus; indeed, as a major contributor to the carbon emissions as well as a major player in the climate mitigation and adaptation sphere (including afforestation and landscape restoration), India may have to get used to thinking and acting as a *provider*, rather than a recipient, of aid.

4 Sustainable Development as a Self-Imposed Precaution

Even though we may not agree to curb our carbon emissions at present levels at the urging of the developed countries, it is still to be recognized that each developing country should undertake conservation and sustainable utilization of its natural resources as a matter of common sense and abundant precaution in its *own* long term interests, and not necessarily because it feels under pressure in the world climate conferences.

In the short term, as already suggested, we should secure and sequester those natural habitats and biodiversity centres that are still left intact, notwithstanding the criticism of our oversensitive ecologists to the fact that these constitute 'islands' of biodiversity (Dilip Kumar 2014b). In the longer term, we need to have a clear pathway to higher income (GDP, growth) *and* improved quality of life of the mass of people (distribution, development). In our own interest, quite apart from the interests of controlling global GHG emissions, we need to become steadily more carbon-efficient, fuel-efficient and careful about how we use (or abuse) the environment and natural resources.

The philosopher Pascal is supposed to have said that even if we did not believe there was a God, just in case we were mistaken it wouldn't hurt to act as if He existed. In a similar spirit, whether one is an environmental warrior (like the IPCC), or a reformed sceptic (like Lomborg), we will have to probably agree with the IPCC finding that "**Warming of the climate system is unequivoval, and since the 1950s, many of the observed changes are unprecedented over decades to millennia. The atmosphere and oceans have warmed, the amounts of snow and ice have diminished, and sea level has risen**" (IPCC 2014c, SPM 1.1, p. 2, emphasis in original). Accordingly, it is advisable that we anticipate the possible climate change effects of the future, and start building up resilience in critical sectors. Indeed, we need not even go so far into the future of the climate change scenario: we just have to build up resilience to the *current* environmental factors and intermittent disasters, e.g. by proper water and soil (nutrients) management, regulating land use, controlling pollution, building defences against floods, droughts, and other climate-related hazards, diversifying rural occupations, increasing mobility, improving skills, minimizing losses, developing processing and storage facilities, developing different sectors of the economy, benefiting from the global trade and information network, moving increasingly to renewable energy and so on.

While gross carbon use (and GHG emissions) will inevitably be increasing manyfold given the growth targets of energy, fuels, infrastructure and other sectors, the country would have to take up these sustainability measures in any case to be viable even in the current situation. These measures would automatically contribute to better carbon management in the long term, and as our people move to slightly higher levels of income, reduced uncertainty and environmental hazards, lower levels of poverty, lower birth rates, it is expected that better management of the environment and natural resources would follow. In doing this, we need to ensure that there is a more equitable sharing of the benefits of development with the presently less well-off, especially identifiable minority groups and regional pockets of low performance. In order to make wise choices that do not result in more poverty in the name of development, we may have to devise and apply more sensible measures of well-being, rather than blindly following the high-consumption paths of the developed world; therein lies the governance challenge.

References

Balasubramanian, M., & Dilip Kumar, P. J. (2014). Climate change, Uttarakhand, and the World Bank's message. *Economic and Political Weekly, XLIX*(1), 65–68.

Chokkalingam, U., & Vanniarachchy, S. A. (2014). CDM AR did not fail and is not dead. In *Forest Carbon Asia*, June 7, 2014. At www.forestcarbonasia.org/articles/cdm-ar-did-not-fail-and-is-not-dead/. Accessed May 25, 2015.

de Vries, B., Bollen, J., Bouwman, L., den Elzen, M., Janssen, M., & Kreilman, K. (2000). Greenhouse gas emissions in an equity-, environment- and service-oriented world: An IMAGE-based scenario for the 21st century. *Technological Forecasting and Social Change, 63*, 137–174. (Quoted by Lomborg 2001, p. 317).

Dilip Kumar, P. J. (1991). Between common property and state monopoly: The institutional context of wasteland management in India. In *Managing the Village Commons, Proceedings of National Workshop*, December 15–16, 1991, Bhubaneswar, Orissa. IIFM: Bhopal.

Dilip Kumar, P. J. (2013). Village communities and their common property forests. *Economic and Political Weekly, XLVIII*(35), 33–36.

Dilip Kumar, P. J. (2014a). Climate change, forest carbon sequestration and REDD-plus. The context of India. *Economic and Political Weekly, XLIX*(22), 22–25.

Dilip Kumar, P. J. (2014b). Western ghats conservation: Experts' reports and a view from the ground. *Economic and Political Weekly, XLIX*(22), 224–229.

Dilip Kumar, P. J. (2014c). Forestry challenges in India-I. Strengths, achievements. www.forestmatters.blogspot.com/2014/12/02-challenges-facing-forestry-in-india.html

Government of India. (2012). India. In *Second national communication to the United Nations framework convention on climate change*. New Delhi.

Hardin, G. (1968). The tragedy of the commons. *Science, 162*, 1243–1248.

IPCC. (2014a). Summary for policymakers. In C. B. Field, et al. (Eds.), *Climate change 2014: Impacts, adaptation, and vulnerability. Part A: Global and sectoral aspects. Contribution of working group II to the fifth assessment report of the intergovernmental panel on climate change (IPCC AR5)* (pp. 1–32). United Kingdom and New York, USA: Cambridge University Press. https://ipcc-wg2.gov/AR5/images/uploads/WG2AR5_SPM_FINAL.pdf. Accessed on May 25, 2015.

IPCC. (2014b). Summary for policymakers. In O. Edenhofer, et al. (Eds.), *Climate change 2014: Mitigation of climate change. Contribution of working group III to the fifth assessment report of the intergovernmental panel on climate change (IPCC AR5)* (pp. 1–32). United Kingdom and New York, USA: Cambridge University Press. report.mitigation2014.org/spm/ipcc_wg3_ar5_summary-for-policymakers_approved.pdf. Accessed on May 25, 2015.

IPCC. (2014c). *Climate change 2014: Synthesis report*. Geneva: IPCC.

Lomborg, B. (2001). *The skeptical environmentalist. Measuring the real state of the world.* UK and New York: Cambridge University Press.

Olson, M. (1971). *The logic of collective action. Public goods and the theory of groups.* Cambridge (USA): Harvard University Press.

Ostrom, E. (1990). *Governing the commons. The evolution of institutions for collective action.* Cambridge: Cambridge Universities Press.

Ravindranath, N. H. (2015). Costs and benefits of the Green India mission for mitigation. In *Economics of climate change mitigation options in the forest sector.* Session: Afforestation and reforestation, Friday February 6, 2015. Rome: FAO. www.fao.org/forestry/cc-mitigation-economics/88445/en/

UNFCCC. (2013). Afforestation and reforestation projects under the clean development mechanism. A reference manual. In *United Nations framework convention on climate change*. Bonn. At www.unfccc.int/resource/docs/publications/cdm_afforestation_bro_web.pdf. Accessed May 25, 2015.

World Bank. (2013). *Turn down the heat: Climate extremes, regional impacts, and the case for resilience*. Washington, D.C.: World Bank. http://documents.worldbank.org/curated/en/2013/06/17862361/turn-down-heat-climate-extremes-regional-impacts-case-reslience-full-report. Accessed in January 2014.

Author Biography

Dr. P.J. Dilip Kumar, Ph.D. in Forest Economics (University of Wales), retired in November 2012 as the Director General of Forests, Ministry of Environment and Forests, Government of India. He is presently holding a Senior Fellowship of the ICSSR, New Delhi at the Institute for Social and Economic Change (ISEC), Nagarbhavi, Bangalore-560072. Email pjdilip@gmail.com, website www.forestmatters.blogspot.in.

Climate Change, Risk and Food Security: An Analysis of Wheat Crop in Pakistan

Muhammad Haroon Abdullah, Abdul Saboor, Irfan Ahmad Baig and Muhammad Arshad

Abstract Climate change is a growing threat for humanity, while agriculture is one of the most vulnerable sectors to this impending change. This study analyses the risk associated with wheat crop (staple food) due to climatic factors and its possible impacts on food security in Pakistan. As climate changes, it effects crop productivity and has heterogeneous impacts with respect to geographical locations. This concern led us to select geographical classification of wheat farming in Pakistan. To capture risk factor for wheat production, we employed Just and Pope (1978) production function using district-level disintegrated data of area, fertilizer, yield and climatic variables (temperature and precipitation) for wheat over the period of 1984–2013. We used 20 years moving averages of monthly mean temperature and precipitation and further classified them according to the growth stages of wheat. We estimated the weather shocks and combine effect of climatic variables by variation and interaction terms, respectively. Further, we translated the impacts of vulnerability of wheat grain on per-capita availability to depict the national food security level. Our empirical findings showed that impact of variation in temperature during growing stage is highly significant and contributes in lowering the wheat yield risk, while increase in precipitation during growing and flowering stages poses high risk to wheat crop. Interaction terms of climatic variables showed non-significant risk-decreasing effect during all the three growth stages of the crop. Further, we found that average per-capita wheat grain availability is 135 kg in Pakistan, significantly higher than self-sufficiency level but highly vulnerable. Our findings suggest that there is a need to develop robust adaptation strategies to combat the risk phenomenon and to sustain and improve wheat grain production for the masses in future. Specifically, developing climate resistant cultivars and

M.H. Abdullah · A. Saboor · I.A. Baig
Department of Economics and Agricultural Economics,
PMAS-Arid Agriculture University, Rawalpindi, Pakistan

M. Arshad (✉)
Institute of Socio-Economics, Leibniz Center for Agricultural Landscape Research (ZALF), Müncheberg, Germany
e-mail: Muhammad.Arshad@zalf.de; Arshad.Muhammad125@gmail.com

S. Nautiyal et al. (eds.), *Climate Change Challenge (3C) and Social-Economic-Ecological Interface-Building*, Environmental Science,
DOI 10.1007/978-3-319-31014-5_4

educating farmers about balance use of inputs (e.g. fertilizer and irrigation) coupled with engendering awareness about climate change could lead to secure the wheat grain production in the country.

1 Introduction

Climate change, an antediluvian phenomenon, is not only the matter of scientific conjecture but also fast becoming reality. It has become an earth-shattering issue after the onset of nineteenth century referring to altering the weather conditions and other climatic variables mainly derived from human activities, directly or indirectly, coupled with natural variability (UNFCCC 2014). Change in climate has been projected to have dire consequences on number of sectors including water and energy; agriculture and food production; human health and biodiversity; and the economy itself severely suffers from shocks resulting from extreme weather events (IPCC 2014). Changing climatic conditions affects the crop productivity, which may lead to threaten the food and livelihood security of rural communities (Wilson and Buckle Henning 2016). Further, variability in climatic parameters affects the socioeconomic and natural systems in multiple ways (Malone 2009).

Change in climate has several socioeconomic dimensions; each carries different mechanisms and consequences (Abumhadi et al. 2012). While agriculture sector is expected to hit severely by climate change, which serves as victim and abettor, however, assessments across regions show different consequences (Rosenzweig et al. 2013). The Intergovernmental Panel on Climate Change (IPCC 2007) estimated that 31 % of total emissions in 2004 came from agriculture and forestry. Changes in climate and natural calamities are the foremost cause of hunger and affect all dimensions of food security including access to food, availability and stability of supplies, and nutrition across the world (Habiba et al. 2016). Climate change has been predicted to alter food production and to reform the food distribution system and markets (Nelson et al. 2010). However, the risks triggered by climate are inequitable across countries and anticipated to vary according to the topographic locations (Dickinson et al. 2015); ecological circumstances (Edger et al. 2016); and the level of economic development of each area/country (Myers-Smith et al. 2015), which defines the capacity of local peoples to deal with such antagonistic consequences. Despite having miniscule contribution in greenhouse gases, Pakistan is ranked very high in climate risk index (Khan 2015) due to diverse topography, meagre economy and inadequate infrastructure.

Pakistan is a developing country, geographically located in diverse ecological zones between 24–37°N latitudes and 62–75°E longitudes (Salma et al. 2012). It is ranked as sixth most populace country in the world, with substantial population growth rate and ever-increasing demand of food production. The country has diverse topography with variety of ecological systems from wet mountains to dry deserts. Its economy is predominantly dependent upon agriculture, which provides food, fibre and livelihood to its population. The abrupt variability in climatic parameters leads to extreme events having adverse consequences for agriculture production and ultimately food security. Unfortunately, the frequency and intensity of extreme weather events in the country has increased from the beginning of current century ranking it among (top ten) vulnerable countries in the world (GOP 2015). Pakistan's economy is mainly based on agriculture sector having a large share in country's GDP (21 %) and it provides employment to 43.5 % of the country's labour force (GOP 2015). Despite agriculture sector is of critical importance to growth, rural household income and food security, it has been facing continuous decline in production due to variabilities in climate (Khan 2015). It is an alarming situation that 61 % of the country's population is food insecure, 80 out of 131 districts are facing food insecurity (SDPI 2003), while 22 districts of Baluchistan Province are severely food insecure (SDPI 2009). Furthermore, almost 50 % of the population does not have access to adequate food supply to fulfil their daily dietary needs for healthy life (WFP 2009). In 2014, Sustainable Development Policy Institute (SDPI) stated in policy brief that 103 districts are food deficient and only 10 districts were sufficient in food availability and food availability is the first pillar to food security. Unavailability of food in different regions may increase in future due to decline in agriculture production, rise in input prices and possible adverse effects of climate change.

It is not an exaggeration to say that wheat crop is the core one staple food crop, which delivers major energy requirement of human diet across the world. Global wheat production is more than 600 t annually and ranked as third largest growing crop (Asseng et al. 2011) and one-third of all grains trade accounts for wheat globally (Sharma et al. 2015). Pakistan is placed at sixth position in wheat production across the world, while contributes 3.5 % into the global wheat production. The yield of wheat in Pakistan is 32 % of its potential, which is quite low, yet the country stands at 59th position in terms of wheat yield per unit of land globally (Zulfiqar and Hussain 2014). Wheat serves as staple crop in Pakistan, providing 48 % caloric needs (SDPI 2003), though its share is even higher for rural households. It is grown at 9039 thousand hectares in the country, with a share of 10 % in agricultural value addition and contributes 2.1 % to the national GDP (GOP 2015). Wheat straw is also used as forage to feed livestock, which has large share in agriculture value addition, i.e. 56.3 %. It is an established fact that self-sufficiency in wheat means self-sufficiency in food for Pakistan (Ahmad 2009).

Given the importance of wheat crop, we find it imperative to quantify the variations in climatic indicators, which are threatening the food production, security and livelihood sustainability of the masses in Pakistan. International Food Policy Research Institute (IFPRI) estimated up to 30 % decline in irrigated wheat yields and almost 90 % increase in price level in 2050 in developing countries, Pakistan is one of them. To the best of our knowledge, current study is a very first attempt for Pakistan to assess food security linked with climate change and risk in wheat productivity. Zulfiqar and Hussain (2014) estimated the wheat production gap to evaluate the food security without taking climate change in account, while

Ashfaq et al. (2011) found climate change as major factor in determining wheat productivity. Ahmad et al. (2014) and Janjua et al. (2014) assessed the impact of climate change on mean wheat yield but did not estimated the associated risk and linked it to food security. However, present study is an organized attempt to assess the impact of climate change on wheat production and the associated risk and per-capita food security level is also estimated with reference to wheat grain. Current study is an idiosyncratic in nature and could be beneficial for all concerned stakeholders, particularly for policy makers to define the future direction. We define the specific objectives of our study as follows:

- Assessing the variability in climatic parameters across different zones in Pakistan.
- Evaluating the risk in wheat production due to climate change during each growth stage across different zones for Pakistan.
- Translating the consequences of wheat production variability on food security in Pakistan.
- Suggesting policy measures to the stakeholders in the country based on our empirical findings.

2 Methodological Framework

This section comprised subsections including classification of study area, selection of variables, data collection from the source departments in Pakistan, model specification and application of appropriate techniques to analyse the data and estimation.

2.1 Classification of Study Area

Pakistan has been classified into five zones according to latitudinal level (Salma et al. 2012). Poudel and Kotani (2013) revealed significant difference in climatic variability across latitudes; hence, current study undertook such a classification. We selected one district (administrative unit) from each zone and the selection is based on three considerations namely (a) availability of metrological observation centre since early 1960s, (b) having large share in wheat production and (c) declared as district in 1984 or earlier. We used data of five major wheat-producing districts of Pakistan over the period of 1984–2013. Mardan, Sialkot, Sukkur, Bahawalpur and Banazir Abad are the districts selected from Zone A, B, C, D and E, respectively, as depicted in Fig. 1.

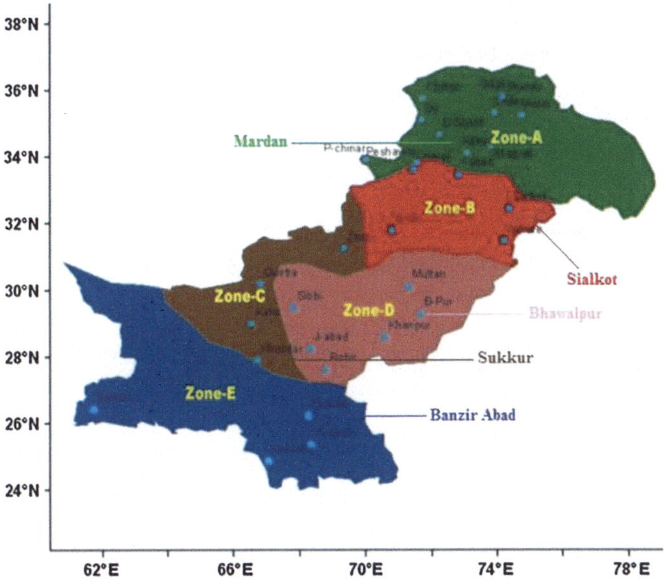

Fig. 1 Climatic zones according to the latitude of the study area and selected districts. *Source* Salma et al. (2012)

2.2 *Variables and Data*

The district-level time series data for the period of 1984–2013 on yield, area under cultivation and production of wheat were collected from Federal Bureau of Statistics, Pakistan. District-level share of fertilizer (NPK nutrients) for wheat crop was obtained from National Fertilizer Development Centre (NFDC) Islamabad for the same time period, whereas the monthly data of metrological variables of wheat-growing season (November to April) is collected from the Pakistan Meteorological Department (PMD) Islamabad for the period of 1964–2013. The collected monthly data was categorized according to growth stages of wheat crop as suggested by Ahmad et al. (2014) to avoid the problem of autocorrelation and endogenity. Defined growth stages of wheat crop are presented in Table 1. Yield of wheat crop was taken as a dependent variable in the model, while stage-wise climatic variables, area under cultivation and amount of fertilizer were used as explanatory variables as shown in Table 2. Moreover “Time trend” a dummy variable was used to capture technology improvement and to avoid trend in explanatory variables as suggested by Poudel and Kotani (2013).

Table 1 Growth stages for wheat crop

Sowing stage	Flowering stage	Harvesting stage
November–December	January–February	March–April

Author’s own

Table 2 Dependent and explanatory variables

Variables	Abbreviates	Definition	Source of data	Unit of measurement
Dependent variable				
Wheat yield	Y	District-wise wheat yield	Pakistan Bureau of Statistics (1984–2013)	Tones per hectare
Explanatory non-climatic variables				
Area under cultivation	A_{Wheat}	District-wise per year use of land under wheat	Pakistan Bureau of Statistics (1984–2013)	Thousand hectares
Fertilizer	Fr	District-level fertilizer data	Nation Fertilizer and Development Center (1984–2013)	Kg/hectare
Explanatory climatic variables				
Temperature	Temp.	District-wise mean monthly temperature for November–April	Pakistan Metrological Department (1964–2013)	Degree celsius (°C)
Precipitation	Precp.	District-wise mean monthly precipitation for November–April	Pakistan Metrological Department (1964–2013)	Millimetre (mm)

Author's own

In the recent climate change impact assessment on crop productivity literature, different researchers have used different definitions of climatic variables, as Shakoor et al. (2011) utilized simple annual average temperature and precipitation as normal climatic parameters. Contrarily, Lobell and Field (2007) and Janjua et al. (2010) used average of just growing season as normal climate, whereas Luo et al. (2005) modelled integrated averages of both growing and non-growing seasons, while Tubiello et al. (2002), Chang (2002) and You et al. (2009) employed 20 years monthly averages of temperature and precipitation as normal climate. We modelled 20 years moving averages of the mean monthly temperature and precipitation for wheat-growing season (November–April) as normal climatic parameters (i.e. temperature and precipitation), respectively. Normal climate variables can not only be used as future climate predictors but also as a reference value to assess climate anomalies (WMO 2011). Hence, we estimated the variation in both temperature and precipitation parameters as a difference of normal temperature and precipitation from their monthly corresponding values as proposed by Chang (2002). Furthermore, we also modelled interaction terms of both temperature and precipitation following Ahmad and Ahmad (1998) and Ludwig et al. (2009), as they showed that different climatic variables interact with each other and do not act independently.

2.3 Theoretical Spectrum

Based on methodologies, we can divide previous climate change impact assessment studies into three categories, viz., production function approach, Ricardian approach and crop modelling. Although crop modelling approach arrives to the most reliable estimates, however, it is costly and difficult to implement in developing countries (Kurukulasuriya and Ajwad 2007). Out of three, Ricardian approach seems most suitable in case of Pakistan. However, due to feeble documentation of agricultural farmland values and imperfect land markets in Pakistan, it would not be optimal to employ the Ricardian approach (Gbetibouo and Hassan 2005; Guiteras 2009). Moreover, Cline (1996) pointed out that constant price assumption in Ricardian method will lead to biased estimates in the welfare calculations. Given the above-mentioned facts, we employed a widely used method, i.e. production function approach, to assess the impact of climatic parameters on wheat yield in Pakistan (Isik and Devadoss 2006; Poudel and Kotani 2013; De Salvo et al. 2013).

2.4 Panel Data Model Specification

As already stated we employed panel data estimation model as it is the most reliable model when there is a problem of omitted variables.

The generalized form of panel data yield model is given in Eq. (1):

$$\text{Yield} = C_{\text{norm}}, C_{\text{var}}, C_{\text{interction}}, \text{Area}, F, T \tag{1}$$

The specific form of Cobb Douglas function for wheat crop used in this study is given below in Eq. (2):

$$Y_{\text{it}} = e^{\beta_a C_{\text{norm}} + \beta_b C_{\text{var}} + \beta_c C_{\text{interction}}}.\text{Area}^{\beta_d}.F^{\beta_e}.e^{\beta_f T}.e^{\mu_{\text{it}}} \tag{2}$$

After taking log on both sides Eq. (2) becomes

$$\ln Y_{\text{it}} = \beta_a C_{\text{norm}} + \beta_b C_{\text{var}} + \beta_c C_{\text{inter}} + \beta_d \ln \text{Area} + \beta_e \ln F. + \beta_f \text{Time} + \mu_{\text{it}} \tag{3}$$

where $C_{\text{norm},}$ C_{var} and $C_{\text{interction}}$ are vectors of climatic variables, variation and interaction terms, respectively. Area is the area under wheat crop, F is the fertilizer use, T is the time trend, and μ is the error term. Each climatic vector further consists of three components of temperature and precipitation according to growth stages.

These variables are defined in more detail in the following Eq. (4), which is the expanded version of Eq. (3):

$$\begin{aligned}\text{Ln}\ (W.\,\text{Yield}) = {} & \beta_{0t} + \beta_{1t}(\text{TEMP}_{\text{Sowing}}) + \beta_{2t}(\text{TEMP}_{\text{Flowering}}) + \beta_{3t}(\text{TEMP}_{\text{Harvesting}}) \\ & + \beta_{4t}(\text{PRECP}_{\text{Sowing}}) + \beta_{5t}(\text{PRECP}_{\text{Flowering}}) + \beta_{6t}(\text{PRECP}_{\text{Harvesting}}) \\ & + \beta_{7t}(\text{V.\,TEMP}_{\text{Sowing}}) + \beta_{8t}(\text{V.\,TEMP}_{\text{Flowering}}) + \beta_{9t}(\text{V.\,TEMP}_{\text{Harvesting}}) \\ & + \beta_{10t}(\text{V.\,PRECP}_{\text{Sowing}}) + \beta_{11t}(\text{V.\,PRECP}_{\text{Flowering}}) \\ & + \beta_{12t}(\text{V.\,PRECP}_{\text{Harvesting}}) + \beta_{13t}(T \times P_{\text{Sowing}}) + \beta_{14t}(T \times P_{\text{Flowering}}) \\ & + \beta_{15t}(T \times P_{\text{Harvesting}}) + \beta_{16t}\text{Ln}(A_{\text{wheat}}) + \beta_{17t}\text{Ln(Fr)} + \beta_{18t}(\text{Time}) \\ & + \beta_{20t}(\text{B.A}) + \beta_{20t}(\text{Skr}) + \beta_{21t}(\text{Bwp}) + \beta_{22t}(\text{Mar}) + \mu_t\end{aligned} \tag{4}$$

Variables in Eq. (4) are explained in Table 3.

Table 3 Description of variables

Variables	Description
$\text{TEMP}_{\text{Sowing}}$	20 years moving average of temperature during November- December
$\text{TEMP}_{\text{Flowering}}$	20 years moving average of temperature during January–February
$\text{TEMP}_{\text{Harvesting}}$	20 years moving average of temperature during March–April
$\text{PRECP}_{\text{Sowing}}$	20 years moving average of precipitation during November–December
$\text{PRECP}_{\text{Flowering}}$	20 years moving average of precipitation during January–February
$\text{PRECP}_{\text{Harvesting}}$	20 years moving average of precipitation during March–April
$\text{V. TEMP}_{\text{Sowing}}$	Variation in climatic temperature during November–December
$\text{V. TEMP}_{\text{Flowering}}$	Variation in climatic temperature during January–February
$\text{V. TEMP}_{\text{Harvesting}}$	Variation in climatic temperature during March–April
$\text{V. PRECP}_{\text{Sowing}}$	Variation in climatic precipitation during November–December
$\text{V. PRECP}_{\text{Flowering}}$	Variation in climatic precipitation during January–February
$\text{V. PRECP}_{\text{Harvesting}}$	Variation in climatic precipitation during March–April
$(P \times T)_{\text{Sowing}}$	Interaction terms of climatic temperature and precipitation during November–December
$(P \times T)_{\text{Flowering}}$	Interaction terms of climatic temperature and precipitation during January–February
$(P \times T)_{\text{Harvesting}}$	Interaction terms of climatic temperature and precipitation during March–April
Ln (A_{wheat})	Natural log of area under wheat crop
Ln (Fr)	Natural log of fertilizer used for wheat crop
Time	Technological Improvement with Time (dummy variable)
B.A	Banazir Abad (dummy variable)
Skr	Sukkur (dummy variable)
Bwp	Bahawalpur (dummy variable)
Mar	Mardan (dummy variable)
μ	Heteroscedastic error term

Author's own

2.5 Stochastic Production Function

Risk in wheat production due to climatic parameters was calculated using stochastic production function approach proposed by Just and Pope (1978) to estimate the effect of climatic variables on mean as well as variance of wheat yield; the model takes the following form:

$$Y = f(X,\beta) + \mu = f(X,\beta) + h(X,\alpha)^{0.5}\varepsilon \tag{6}$$

where "*Y*" is the crop yield, "*X*" is the vector of explanatory variables, $f(X,\beta)$ is the mean production function relating "*X*" to average yield with β as the estimated parameters of associated vector variables and μ is a heteroscedastic disturbance term with mean zero. $h(X,\alpha)$ is the risk production function (Asche and Tveterås 1999) which relates *X* to the standard deviation of yield with α as the corresponding vector of estimated parameters, and ε is a random error with zero mean and variance σ^2. Both increasing as well as decreasing risk effects of inputs on yield were estimated by this function. This method has previously been employed by Cabas et al. (2010), Barnwal and Kotani (2010), Holst et al. (2011) and Poudel and Kotani (2013).

To estimate the production risk, two stages of feasible generalized least square (FGLS) procedure were employed by following the basic procedure of Just and Pope (1978) as used by McCarl et al. (2008) and Barnwal and Kotani (2010). The following steps are involved while estimating the risk model. First, the regression model on Y at $f(X,\beta)$ was estimated using ordinary least squares (OLS). District-specific effect was captured by incorporating district-specific dummies. Residuals obtained from the OLS were used as an input for the second step, i.e. regressing the logarithm of squared residuals against independent variables, which is "yield risk". All the estimates obtained are reported in the result section.

2.6 Per-Capita Availability

We presented the level of food security relative to wheat production by its per-capita availability, as assessed by Zulfiqar and Hussain (2014) and Tariq et al. (2014) dividing total production with total population of the country/region. We used the population statistics of Pakistan from Population Censes, (1998) and Economic Survey 2014–15. We generated growth rate of population by employing exponential population growth formula. General form of exponential growth formula is defined as

$$P(t) = \mathrm{P(o)}e^{rt} \tag{7}$$

where $P(t)$ depicts the population at time (t), P(o) is the initial population, t is the number of years and r is the growth rate. Using calculated growth rate, population of several years were estimated to compare with wheat crop production for the respective years to evaluate per-capita availability level for analyzing the food security conditions at national level.

3 Results and Discussion

This section is divided into three subsections according to defined study objectives. First subsection describes the results of variability in climatic parameters across latitudinal zones of Pakistan. Second subsection presents the results of regression analysis explaining the risk in wheat yield due to change in climatic parameters. The last subsection showed the self-sufficiency in wheat crop to assess wheat availability at local level.

3.1 Variation in Climatic Parameters

Temperature and precipitation are two important climatic variables in crop production. Variability in these variables referred to variability in climatic parameters was evaluated across different latitudinal zones of Pakistan. For a keen insight, both seasonal variation and variation during different growth stages of wheat for different time periods were estimated. Figures 2 and 3 present the average variations between climatic temperature and precipitation for selected districts, respectively; both variations were drawn for three decades (1984–2013), two decades (1994–2013), single decade (2004–2013) and for the five-year (2009–2013) periods. A decreasing trend in temperature for almost all districts could be observed during wheat-growing season, while Banazir Abad and Sukkur suffered large variation

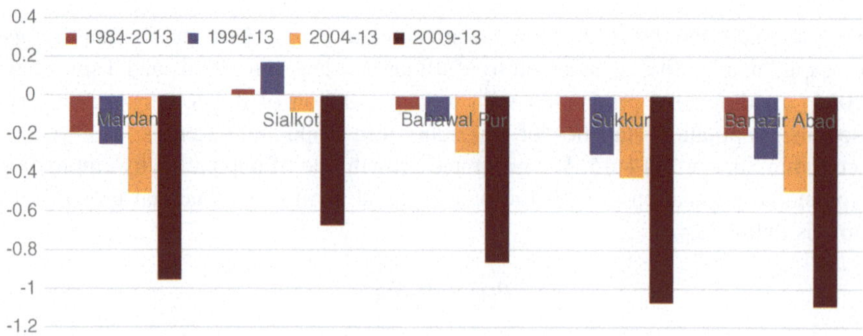

Fig. 2 Average variation in climatic temperature (°C). Author's own

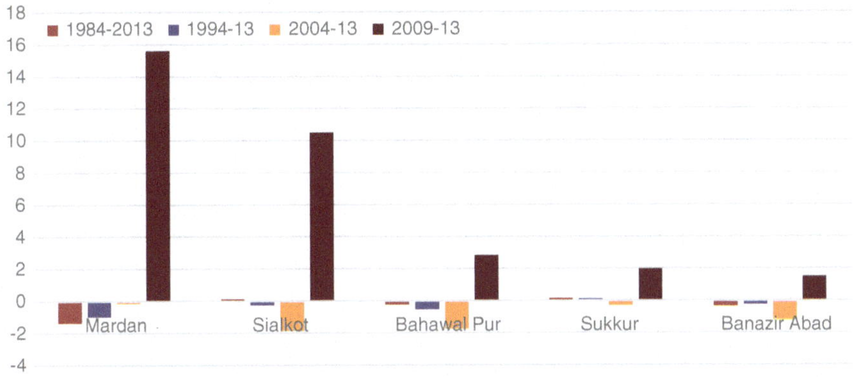

Fig. 3 Average variation in climatic precipitation (mm). Author's own

(>−1 °C) during past 5 years. On the other hand, Mardan and Sialkot district faced outsized (>10 mm) variation in past 5 years in precipitation, as Sialkot and Mardan districts are located in northern side of Pakistan for which Salma et al. (2012) and Asmat and Athar (2015) mentioned that the northern Pakistan has higher absolute precipitation coupled with higher variability in it. Moreover, variation in both climatic parameters is becoming stout especially in past 5 years (2009–2013), pointing towards increased frequency of extreme events.

The hypothesis of increasing variation in climatic temperature is verified by Figs. 4 and 5, in which variations are depicted according to growth stages of wheat crop for the period of (1984–2013) and (2009–2013), respectively. The figures show that threatening condition of intrusive variations in temperature is becoming intense during flowering and harvesting stage of the wheat-cropping season. Our results are consistent with the findings of Gornall et al. (2010) that the high variation in temperature may cause decline in crop production. Variation in precipitation according to growth stages is presented in Figs. 6 and 7 for the same time periods. Large variation for district Mardan and Sialkot drives towards loss in crop

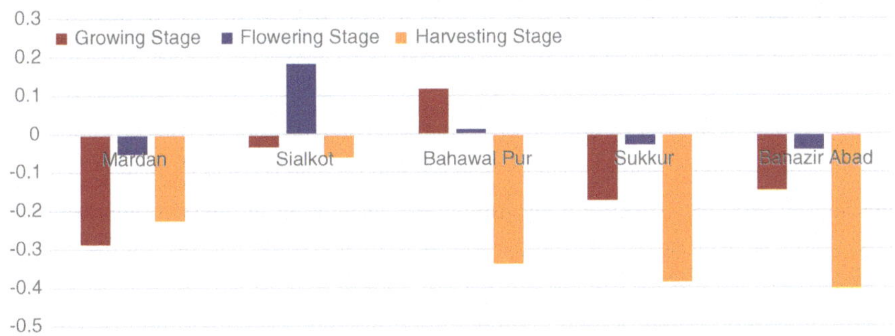

Fig. 4 Average variation in climatic temperature (°C) during 1984–2013. Author's own

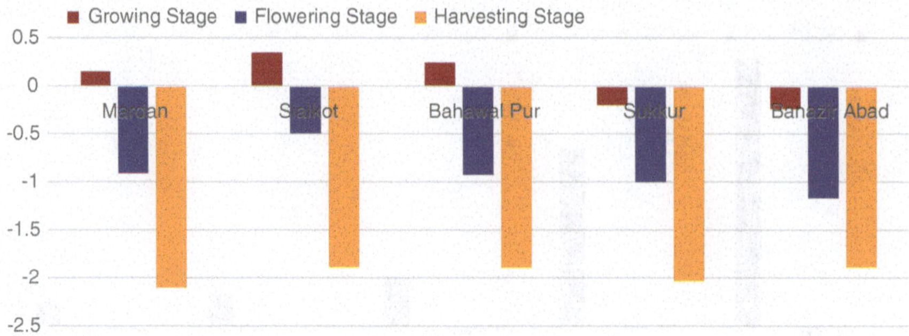

Fig. 5 Average variation in climatic temperature (°C) during 2009–2013. Author's own

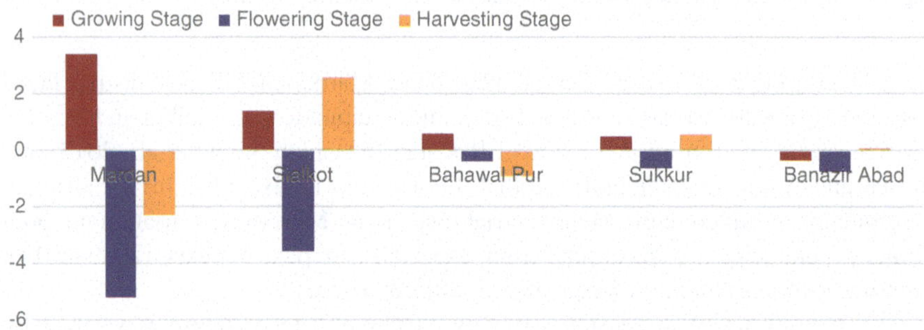

Fig. 6 Average variation in climatic precipitation (mm) during 1984–2013. Author's own

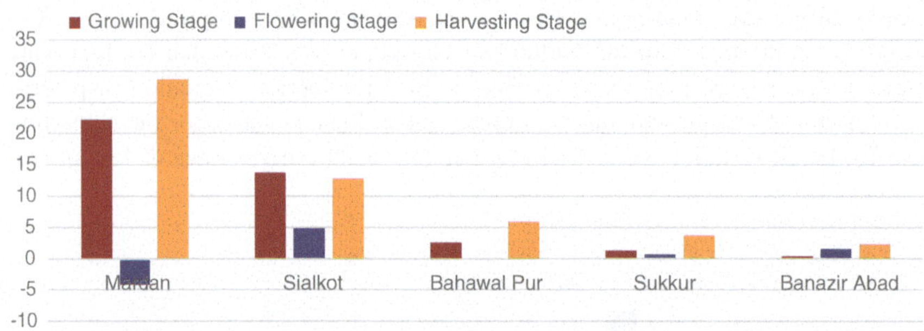

Fig. 7 Average variation in climatic precipitation (mm) during 2009–2013. Author's own

production, especially during harvesting stage. Ahmad et al. (2015) report similar impacts on yield, assessed by single-pole pattern for winter–spring rainfall in the northern region of Pakistan, whereas Ray et al. (2015) reported that more than 60 % variation in crop yield is explained by the climatic variability. Average temperature and precipitation parameters are presented in Appendixes 1 and 2.

3.2 Risk in Wheat Production

First, to satisfy stationary condition we performed unit root test for time series data. Both variations in precipitation and temperature were found stationary at level, I(0), while all other climatic and non-climatic variables became stationary at first difference, I(1). The results of unit root test are presented in Appendix 3. Second, Wald test was employed to assess about model significance as suggested by Kohansla and Aliabadi (2014) and Ahmad et al. (2014). Wald test results are presented in Appendix 4; however, insignificant *P* value rejected the null hypothesis and significant intercept was found. Hence, the model used in this study was verified as significant. Third, as stochastic frontier analysis and production risk both capture the variation in output, we tested the heteroscedasticity as suggested by Wood and Mendelsohn (2015). The Breusch–Pagan test was applied and is presented in Appendix 5. Null hypothesis for Breusch–Pagan was that the model holds homoscedasticity or constant variance and is rejected due to insignificant *P* value. The test revealed the presence of heteroscedasticity (Production Risk). Finally, two-step stochastic production function was estimated and the estimates of risk in wheat crop production are presented in Table 4.

The estimation results for risk in yield revealed that the relationship between risk and time (technology) was significant and negative sign showed a decrease in risk due to improvement in technology every year, as improved seed varieties and advance cultivation technology are encouraging the wheat production and declining the level of risk. Similar significant and risk-decreasing trend by technology was reported by Holst et al. (2011) for wheat production in China. Area under cultivation and amount of fertilizer applied were not significant but the positive coefficient exposed increasing risk by increasing both area and fertilizer application. This may be because of limited management skills and inadequate information/knowledge of balance use of fertilizer to the farmers in the study area, as the same was observed by Shakoor et al. (2015) for rice crop, and Zulfiqar and Ashfaq (2014) for wheat in Pakistan. Holst et al. (2011) reported decreasing but non-significant risk due to increase in area in China. Conversely, Poudel and Kotani (2013) reported a significant decreasing trend in risk for wheat in Nepal.

Moreover, no significant production risk was found due to temperature during all the three growth stages, though coefficient is positive for sowing stage and flowering stage and negative for harvesting stage. It means that increasing temperature during flowering and sowing stage raised the level of risk while rise in temperature during harvesting stage caused decline in wheat yield risk. The optimum temperature during sowing stage is 12–25 °C; increase in temperature during flowering stage leads to shorten length of growing season and instigates decline in wheat yield. Increasing temperature has positive impact on mean yield of wheat crop during harvesting stage in Pakistan, as Ashfaq et al. (2011) found that high temperature during harvesting stage helps in the formation of healthy seed. On the other hand, precipitation during sowing and flowering stage showed a positive and significant effect on wheat yield risk. Similarly, significant increasing risk due to

Table 4 Risk for wheat production in Pakistan

Dependent variable: log of squared residual from OLS				
Variables	Coefficient	Std. error	*t*-stat	*P* value
Time	−0.0854787	0.0498817	−1.71	**0.089****
Area	0.5792516	4.177157	0.14	0.890
Fertilizer	0.593977	0.8242305	0.72	0.472
$\text{Temp.}_{\text{Sowing}}$	0.6882323	1.109605	0.62	0.536
$\text{Temp.}_{\text{Flowering}}$	2.131086	2.2742	0.94	0.351
$\text{Temp.}_{\text{Harvesting}}$	−1.286809	1.326634	−0.97	0.334
$\text{Precp.}_{\text{Sowing}}$	0.203595	0.1235375	1.65	**0.102****
$\text{Precp.}_{\text{Flowering}}$	0.2365704	0.1234259	1.92	**0.058****
$\text{Precp.}_{\text{Harvesting}}$	0.0456467	0.0698164	0.65	0.514
$\text{V. Temp.}_{\text{Sowing}}$	−0.9123751	0.2724321	−3.35	**0.001***
$\text{V. Temp.}_{\text{Flowering}}$	0.2195584	0.3407196	0.64	0.520
$\text{V. Temp}_{\text{Harvesting}}$	0.1048872	0.2466652	0.43	0.671
$\text{V. Precp.}_{\text{Sowing}}$	0.0038577	0.0074054	0.52	0.603
$\text{V. Precp.}_{\text{Flowering}}$	0.0164532	0.0068063	2.42	**0.017***
$\text{V. Precp.}_{\text{Harvesting}}$	0.0019829	0.0046666	0.42	0.672
$(T \times P)_{\text{Sowing}}$	−0.0172617	0.0218308	−0.79	0.431
$(T \times P)_{\text{Flowering}}$	−0.0327468	0.0284407	−1.15	0.252
$(T \times P)_{\text{Harvesting}}$	−0.0135489	0.0112734	−1.20	0.232
Banazir Abad	−118.187	99.23305	−1.19	0.236
Bahawalpur	−77.48033	72.99428	−1.60	0.290
Sukkur	−117.7447	94.37629	−1.25	0.214
Mardan	−59.63737	66.27053	−0.90	0.370
$F(22, 127) = 1.76$		Prob > F = 0.0275		
No. of Obs. = 150		Adjusted-R sqr = 0.101		
Level of significance = 5 %(*)		Level of significance = 10 %(**)		

Author's own

change in precipitation was reported by Holst et al. (2011) and Poudel and Kotani (2013). Climatic variation disrupts the hydrological cycle, and rain become erratic and unpredictable (Treydte et al. 2006); hence, long-term increase in precipitation increases the risk level for wheat crop particularly during harvesting stage.

Variation in temperature from normal during growing stage has significant and decreasing risk effect, while variation in temperature during flowering and harvesting stage did not show any significant effect on risk. Negative trend in the variability of temperature was observed during harvesting stage, which negatively affects the crop productivity. The significant variation from optimal temperature during any biological stage initiated decline in wheat production (Ashfaq et al. 2011). Holst et al. (2011) reported decrease in risk due to deviation in temperature in China. Excessive rainfall during flowering and harvesting stage also drives loss in wheat production by adding risk in it. Moreover, Babar and Amin (2014) reported the same decreasing trend of mean wheat yield due to change in precipitation pattern. Our results support the findings of (Ashfaq et al. 2011) that extensive variation in rainfall is seen for past 3 years in Pakistan and it caused high risk to wheat crop.

Interaction terms of both normal temperature and precipitation are non-significant with decreasing (negative) risk coefficients during all the three stages of wheat crop. It means that both long-term temperature and precipitation have conjointly diminishing risk effect for wheat crop during whole growing season. In above risk analysis, Sialkot is considered as base district not included in regression analysis to avoid dummy variable trap. However, coefficients of other district dummies were negative and non-significant showing that wheat in Sialkot region was much vulnerable compared to other zones. This is due to high variation in precipitation patterns especially during harvesting stage. Mardan district was vulnerable at top second level having highest variation in climatic parameters but has least area under cultivation. Overall, Zone-A had small share in area under wheat crop cultivation, and Banazir Abad and Sukkur were the least vulnerable districts to climatic variations as our findings suggest.

3.3 *Per-Capita Availability*

As wheat crop serves as staple food and largest grain source in the country, hence, it is one of the most essential crops for national food security (Ahmad et al. 2014). Nuclear Institute for Food and Agriculture (NIFA) Pakistan estimated that annual average per-capita availability was quite high in the country, whereas international number is 126 kg/person. At national level, wheat production increased from 4.6 mt in 1965 to 7.3 mt in 1970. Pakistan was self-sufficient in wheat production by 1968. Yields were over 21 mt by 2000 (GOP 2015). The remarkable rise in wheat production was due to Green Revolution in which genetically reformed varieties were introduced in the country. National per-capita availability level is shown in Fig. 8.

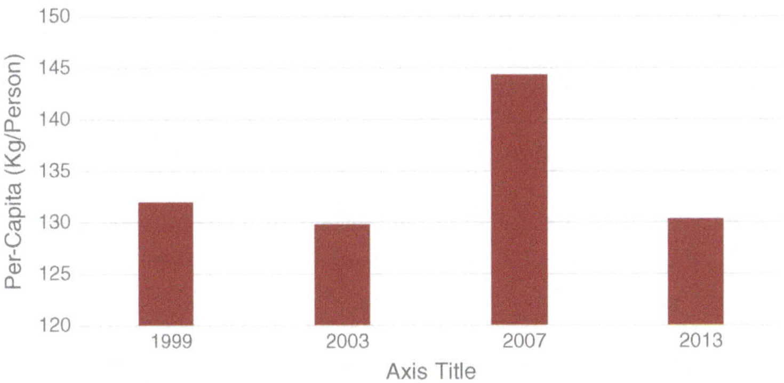

Fig. 8 Per-capita availability of wheat crop in Pakistan. Author's own

The average annual per-capita availability was 130 kg/person in 2013. A substantial rise in per-capita availability during 2007 was due to wheat sowing campaign launched by Provincial Governments to educate farmers about the upcoming challenges to wheat crop and strategies to deal with them. Agriculture extension services also played significant role to educate farmers in the adaptation of heat-resistant varieties and modern technologies (Islam 2015).

Fluctuation in per-capita wheat grain availability in Pakistan in the recent years was mainly due to loss in production and high population growth rate, although support price was also a reason to some extent (Ahmad 2009). Moreover, ever-increasing cost of inputs coupled with erratic and intensive extreme climatic events caused low wheat productivity in the country. Sustainable production of wheat crop ensures the food security, while escalating variability in climate parameters threatens the production (availability) and price stability as well. According to Ashfaq et al. (2011) climate change was a major factor in determining wheat productivity in Pakistan, where wheat is contributing nearly 50 % of total calories consumed by an individual per day (Zulfiqar and Hussain 2014), and this number is higher for people residing in rural areas of Pakistan. According to Pakistan Agriculture Research Council (PARC), per-capita annual consumption was 125 kg in Pakistan. In order to meet country's wheat grain requirements, there is a dire need to exploit the frontier of knowledge, technology coupled with improved farm management practices and robust adaptive strategies to cope up with climate anomalies. The development of heat-resistant wheat cultivars, improving the existing production technologies, may lead to decrease the risk in wheat production and ensure the availability of wheat to meet the national requirements.

4 Conclusions and Policy Recommendations

Current study is an organized attempt to estimate the variability in climatic parameters and identify the associated risks in wheat production, which may ultimately disrupt the food security pattern in Pakistan. We followed latitudinal classification of Pakistan and exposed significant variation in climatic parameters according to latitude across the zones. We used district-level data for the analysis, having large area under cultivation, from each of the five zones. We found negative variation in temperature while positive variation in precipitation during wheat-growing season for all districts (Zones). Our results suggest that the range of variation may be outsized by each passing year in the country. Large negative variation in temperature was found in District Sukkur (Zone-C) and Banazir Abad (Zone-E) among all districts. On the other hand, district Mardan (Zone-A) faced the highest positive variation for precipitation, while district Sialkot (Zone-B) was at top second in terms of variation in precipitation. As per growth stages of wheat season, negative variation was found in temperature during flowering and harvesting stage, whereas very high positive variation in precipitation during harvesting stage for district Mardan and Sialkot in Zone-A and Zone-B, respectively.

Two-step Just and Pope Production Function (1978) was employed to identify the risk factor in wheat production. The results suggested that improved technology (time) has significantly contributed in diminishing wheat risk. While vertical expansion would be better than horizontal expansion, positive elasticity of risk was found due to increase in area under cultivation. Almost all the climatic variables used in our model influenced wheat yield significantly. Increased precipitation during growing and flowering stage posed risk-increasing effect for wheat yield. However, variation in temperature during growing stage caused decline in yield risk significantly. However, an increase in risk elasticity was revealed during flowering and harvesting stage due to variation in temperature. Moreover, variation in precipitation pattern during flowering stage triggered the risk significantly in wheat yield.

We undertook district Sialkot (Zone-B) as base district for risk analysis and found it highly vulnerable compared to other districts/zones, due to decline in temperature coupled with high variation in precipitation pattern especially during flowering and harvesting stage, while having large share in area under wheat cultivation. Mardan district was found vulnerable at top second level, while Banazir Abad and Sukkur were less vulnerable than other districts because of less variation in precipitation pattern. Per-capita availability of wheat is quite high, i.e. 131 kg per person in 2013 but fluctuating with the production level because of rapid fluctuations in climatic parameters.

We inferred the following conclusions based on our empirical findings. As the climatic variables have heterogeneous impact across the latitudes (zones) in the country, highly significant for precipitation, improvement in production technology of wheat and region-specific management skills can boost the wheat yield levels which may ultimately ensure the livelihood and food security of rural masses in the country. Furthermore, better management practices and appropriate use of inputs (e.g. fertilizer and irrigation) by farmers would lead to a significant rise in wheat per-capita availability.

Moreover, our findings suggest that the extension services and training programmes at farm level in order to cope with the variation in temperature and rainfall could play pivotal role in maintaining and boosting the wheat yield levels. Vertical expansion of extension services focusing on climate resilient crop production and management practices may be preferred over horizontal expansion. Well-designed adaptation strategies including development of new verities given the pattern of region-specific climatic conditions would be important coping strategy to mitigate the future risk in wheat crop production. Moreover, prioritizing the investment in research and development of wheat crop farming could lead to a much better situation and ensure food security at national level.

Appendix 1

Average climatic parameters for the period of 1984–2013

Zones	Districts	Climatic temperature (°C)			Climatic precipitation (mm)		
		Sowing stage	Flowering stage	Harvesting stage	Sowing stage	Flowering stage	Harvesting stage
Zone-A	Mardan	11.15	9.79	21.72	34.68	25.45	36.52
Zone-B	Sialkot	12.15	11.78	25.33	27.00	16.5	17.63
Zone-C	Bahawalpur	16.36	15.59	28.31	6.95	3.64	3.62
Zone-D	Sukkur	18.93	17.79	29.84	3.59	1.10	2.40
Zone-E	Banazir Abad	19.70	18.57	30.06	2.11	1.06	1.93

Author's own

Appendix 2

Average climatic parameters for the period of 2009–2013

Zones	Districts	Climatic temperature (°C)			Climatic precipitation (mm)		
		Sowing stage	Flowering stage	Harvesting stage	Sowing stage	Flowering stage	Harvesting stage
Zone-A	Mardan	11.55	9.89	21.65	33.86	25.71	37.28
Zone-B	Sialkot	12.11	12.00	25.30	24.93	16.15	17.93
Zone-C	Bahawalpur	16.18	15.68	28.19	5.24	3.63	3.39
Zone-D	Sukkur	19.16	17.86	29.73	3.12	1.14	2.03
Zone-E	Banazir Abad	19.88	18.64	29.95	2.04	1.25	1.44

Author's own

Appendix 3

Augmented Dickey Fuller unit root test

Variables	Unit root test at level		Unit root test at 1st difference		Decision	Order of integration
	t-stat	Prob.	*t*-stat	Prob.		
$TEMP_{Sowing}$	7.2401	1.0000	−3.1362	0.0026*	Non-stationary at level but stationary at first difference	I(1)
$TEMP_{Flowering}$	7.0394	1.0000	−7.0866	0.0000*	Non-stationary at level but stationary at first difference	I(1)

(continued)

(continued)

Variables	Unit root test at level		Unit root test at 1st difference		Decision	Order of integration
	t-stat	Prob.	*t*-stat	Prob.		
$TEMP_{Harvesting}$	0.1682	0.7289	−2.1974	0.0289**	Non-stationary at level but stationary at first difference	I(1)
$PRECP_{Sowing}$	5.4085	1.0000	−2.8188	0.0066*	Non-stationary at level but stationary at first difference	I(1)
$PRECP_{Flowering}$	−0.0864	0.6466	−3.9536	0.0003*	Non-stationary at level but stationary at first difference	I(1)
$PRECP_{Harvesting}$	0.7726	0.8759	−4.7179	0.0000*	Non-stationary at level but stationary at first difference	I(0)
V. $TEMP_{Sowing}$	−3.1745	0.0021*	–	–	Stationary at level	I(0)
V. $TEMP_{Flowering}$	−2.2543	0.0015*	–	–	Stationary at level	I(0)
V. $TEMP_{Harvesting}$	−4.9154	0.0000*	–	–	Stationary at level	I(0)
V. $PRECP_{Sowing}$	−6.8745	0.0000*	–	–	Stationary at level	I(0)
V. $PRECP_{Flowering}$	−7.8527	0.0000*	–	–	Stationary at level	I(0)
V. $PRECP_{Harvesting}$	−2.9239	0.0041*	–	–	Stationary at level	I(0)
$(P \times T)_{Sowing}$	5.6934	1.0000	−6.5432	0.0000*	Non-stationary at level but stationary at first difference	I(1)
$(P \times T)_{Flowering}$	4.8658	1.0000	−5.8567	0.0000*	Non-stationary at level but stationary at first difference	I(1)
$(P \times T)_{Harvesting}$	5.0837	1.0000	−4.9743	0.0000*	Non-stationary at level but stationary at first difference	I(1)
Ln (A_{wheat})	6.9731	1.0000	−5.9388	0.0000*	Non-stationary at level but stationary at first difference	I(1)
Ln (Fr)	−0.0649	0.6278	−3.4893	0.0009*	Non-stationary at level but stationary at first difference	I(1)

Author's own
Level of significance = 95 %(**); 100 %(*)

Appendix 4

Wald test

Null hypothesis	*F* statistics	*P* value
No significant intercept	22.75	0.000

Author's own

Appendix 5

Breusch–Pagan test

Null hypothesis	*F* statistics	*P* value
Constant variance	19.85	0.000

Author's own

References

Abumhadi, N., Todorovska, E., Assenov, B., Tsonev, S., Vulcheva, D., Vulchev, D., et al. (2012). Agricultural research in 21st century: Challenges facing the food security under the impacts of climate change. *Bulgarian Journal of Agricultural Science, 18*(6), 801–818.

Ahmad, F. (2009). Food security in Pakistan. *Pakistan Journal of Agriculture Sciences, 46*(2), 83–89.

Ahmad, M., & Ahmad, A. (1998). An analysis of the sources of wheat output growth in the "Barani" area of the Punjab. *The Pakistan Development Review*, 231–249.

Ahmad, M., Siftain, H., & Iqbal, M. (2014). *Impact of climate change on wheat productivity in Pakistan: A district level analysis. Climate change.* Working Paper Series 1, Pakistan Institute of Development Economics, Islamabad.

Ahmad, I., Ambreen, R., Sun, Z., & Deng, W. (2015). Winter-spring precipitation variability in Pakistan. *American Journal of Climate Change, 4*(1), 115. doi:10.4236/ajcc.2015.41010

Asche, F., & Tveterås, R. (1999). Modeling production risk with a two-step procedure. *Journal of Agricultural and Resource Economics*, 424–439.

Ashfaq, M., Zulfiqar, F., Sarwar, I., Quddus, M. A., & Baig, I. A. (2011). Impact of climate change on wheat productivity in mixed cropping system of Punjab. *Soil and Environment, 30*(2), 110–114.

Asmat, U., & Athar, H. (2015). Run-based multi-model interannual variability assessment of precipitation and temperature over Pakistan using two IPCC AR4-based AOGCMs. *Theoretical and Applied Climatology*, 1–16.

Asseng, S., Foster, I. A. N., & Turner, N. C. (2011). The impact of temperature variability on wheat yields. *Global Change Biology, 17*(2), 997–1012.

Babar, S., & Amin, A. (2014). Climate change impact on Rabi and Kharif crops of Khyber-Pakhtunkhwa. *PUTAJ Humanities and Social Sciences*, 21(1).

Barnwal, P., & Kotani, K. (2010). *Impact of variation in climatic factors on crop yield: A case of rice crop in Andhra Pradesh.* India: Economics and Management Series. 17.

Cabas, J., Weersink, A., & Olale, E. (2010). Crop yield response to economic, site and climatic variables. *Climatic Change, 101*(3–4), 599–616.

Chang, C. C. (2002). The potential impact of climate change on Taiwan's agriculture. *Agricultural Economics, 27*(1), 51–64.

Cline, W. R. (1996). The impact of global warming of agriculture: Comment. *The American Economic Review, 86*(5), 1309–1311.

De Salvo, M., Begalli, D., & Signorello, G. (2013). Measuring the effect of climate change on agriculture: A literature review of analytical models. *Journal of Development and Agricultural,* 5(12), 499–509.

Dickinson, D. M., Prentice, I. C., & Mace, G. M. (2015). *Climate change and challenges for conservation.* Imperial Collage London, Grantham Institute, Briefing Paper No. 13.

Edgar, G. J., Bates, A. E., Bird, T. J., Jones, A. H., Kininmonth, S., Stuart-Smith, R. D., et al. (2016). New approaches to marine conservation through the scaling up of ecological data. *Annual Review of Marine Science, 8*, 435–461.

Gbetibouo, G. A., & Hassan, R. M. (2005). Measuring the economic impact of climate change on major South African field crops: A Ricardian approach. *Global and Planetary Change, 47*(2), 143–152.

GOP. (2015). *Pakistan economic survey 2014–15*, Chap. 2, Agriculture.

Gornall, J., Betts, R., Burke, E., Clark, R., Camp, J., Willett, K., et al. (2010). Implications of climate change for agricultural productivity in the early twenty-first century. *Philosophical Transactions of the Royal Society of London B: Biological Sciences, 365*(1554), 2973–2989.

Guiteras, R. (2009). *The impact of climate change on Indian agriculture.* College Park, Maryland: Manuscript, Department of Economics, University of Maryland.

Habiba, U., Abedin, M. A., & Shaw, R. (2016). Food security, climate change adaptation, and disaster risk. In *Sustainable development and disaster risk reduction* (pp. 87–101). Springer Japan.

Holst, R., Yu, X., & Grün, C. (2011). *Climate change, risk and grain production in China* (No. 68). Poverty, Equity and Growth-Discussion Papers. Courant Research Centre.

IPCC. (2007). Climate change, synthesis report, a report of the intergovernmental panel on climate change (p. 104). In Core Writing Team, R. K. Pachauri & A. Reisinger (Eds.), *Contribution of working groups I, II and III to the fourth assessment report of the intergovernmental panel on climate change.* Geneva, Switzerland: IPCC.

IPCC. (2014). Climate change 2014: Synthesis report. In Core Writing Team, R. K. Pachauri & L. A. Meyer (Eds.), *Contribution of working groups I, II and III to the fifth assessment report of the intergovernmental panel on climate change* (p. 151). Geneva, Switzerland: IPCC.

Isik, M., & Devadoss, S. (2006). An analysis of the impact of climate change on crop yields and yield variability. *Applied Economics, 38*(7), 835–844.

Islam, S. M. S. (2015). *Print media coverage on climate change issues in Bangladesh.*

Janjua, P. Z., Samad, G., Khan, N. U., & Nasir, M. (2010). Impact of climate change on wheat production: A case study of Pakistan (with comments). *The Pakistan Development Review*, 799–822.

Janjua, P. Z., Samad, G., & Khan, N. (2014). Climate change and wheat production in Pakistan: An autoregressive distributed lag approach. *NJAS-Wageningen Journal of Life Sciences, 68*, 13–19.

Just, R. E., & Pope, R. D. (1978). Stochastic specification of production functions and economic implications. *Journal of Econometrics, 7*(1), 67–86.

Khan, M. A. (2015). *Climate change risk and reduction approaches in Pakistan* (pp. 195–216). Japan: Springer.

Kohansla, M. R., & Aliabadi, M. M. F. (2014). Impact of technological change on wheat production risk in Northwest of Iran. *International Journal of Agricultural Science and Technology.*

Kurukulasuriya, P., & Ajwad, M. I. (2007). Application of the Ricardian technique to estimate the impact of climate change on smallholder farming in Sri Lanka. *Climatic Change, 81*(1), 39–59.

Lobell, D. B., & Field, C. B. (2007). Global scale climate–crop yield relationships and the impacts of recent warming. *Environmental Research Letters, 2*(1), 014002.
Ludwig, F., Milroy, S. P., & Asseng, S. (2009). Impacts of recent climate change on wheat production systems in Western Australia. *Climatic Change, 92*(3–4), 495–517.
Luo, Q., Bellotti, W., Williams, M., & Bryan, B. (2005). Potential impact of climate change on wheat yield in South Australia. *Agricultural and Forest Meteorology, 132*(3), 273–285.
Malone, E. L. (2009). Vulnerability and resilience in the face of climate change: Current research and needs for population information. *Battelle Memorial Institute, PNWD-4087*.
McCarl, B. A., Villavicencio, X., & Wu, X. (2008). Climate change and future analysis: Is stationarity dying? *American Journal of Agricultural Economics, 90*(5), 1241–1247.
Myers-Smith, I. H., Elmendorf, S. C., Beck, P. S., Wilmking, M., Hallinger, M., Blok, D., et al. (2015). Climate sensitivity of shrub growth across the tundra biome. *Nature Climate Change, 5* (9), 887–891.
Nelson, G. C., Rosegrant, M. W., Palazzo, A., Gray, I., Ingersoll, C., Robertson, R., et al. (2010). *Food security, farming, and climate change to 2050: Scenarios, results, policy options* (Vol. 172). International Food Policy Research Institute.
Poudel, S., & Kotani, K. (2013). Climatic impacts on crop yield and its variability in Nepal: Do they vary across seasons and altitudes? *Climatic Change, 116*(2), 327–355.
Ray, D. K., Gerber, J. S., MacDonald, G. K., & West, P. C. (2015). Climate variation explains a third of global crop yield variability. *Nature communications, 6*.
Rosenzweig, C., Elliott, J., & Deryng, D. (2013). Assessing agricultural risks of climate change in the 21st century in a global gridded crop model intercomparison. *PNAS, 111*(9), 1–6.
Salma, S., Rehman, S., & Shah, M. A. (2012). Rainfall trends in different climate zones of Pakistan. *Pakistan Journal of Meteorology* (Pakistan).
SDPI. (2003). *Food insecurity in rural Pakistan*. Pakistan: Sustainable Development Policy Institute. http://www.sdpi.org/publications/publication_details-287-29.html
SDPI. (2009). *Food security in Pakistan*. Pakistan: Sustainable Development Policy Institute. https://sdpi.org/research_programme/researchproject-22-22-50.html
Shakoor, U., Saboor, A., Ali, I., & Mohsin, A. Q. (2011). Impact of climate change on agriculture: Empirical evidence from arid region. *Pakistan Journal of Agricultural Science, 48*(4), 327–333.
Shakoor, U., Saboor, A., Baig, I., Afzal, A., & Rahman, A. (2015). Climate variability impacts on rice crop production in Pakistan. *Pakistan Journal of Agricultural Research, 28*(1).
Sharma, I., Tyagi, B. S., Singh, G., Venkatesh, K., & Gupta, O. P. (2015). Enhancing wheat production-A global perspective. *Indian Journal of Agricultural Sciences, 85*(1), 3–13.
Tariq, A., Tabasam, N., Bakhsh, K., Ashfaq, M., & Hassan, S. (2014). Food security in the context of climate change in Pakistan. *Pakistan Journal of Commerce and Social Sciences, 8*(2), 540–550.
Treydte, K. S., Schleser, G. H., Helle, G., Frank, D. C., Winiger, M., Haug, G. H., et al. (2006). The twentieth century was the wettest period in northern Pakistan over the past millennium. *Nature, 440*(7088), 1179–1182.
Tubiello, F. N., Rosenzweig, C., Goldberg, R. A., Jagtap, S., & Jones, J. W. (2002). Effects of climate change on US crop production: Simulation results using two different GCM scenarios. Part I: Wheat, potato, maize, and citrus. *Climate Research, 20*(3), 259–270.
UNFCC (United Nations Framework Convention on Climate Change). (2014). http://unfccc.int/essential_background/convention/background/items/2536.php. Retrieved on March 3, 2014.
WFP. (2009). *Food security in Pakistan*. Pakistan: World Food Program. https://www.wfp.org/countries/Pakistan/News/Hunger-in-the-news?page
Wilson, I., & Buckle Henning, P. (2016, January). A call to action for the systems sciences community. In *Proceedings of the 59th Annual Meeting of the ISSS-2015 Berlin, Germany* (Vol. 1, No. 1).
WMO. (2011). *Climate normals*. Commission for Climatology Management Group Meeting, Denver, USA. CCl/MG/2011/Doc.10.

Wood, S. A., & Mendelsohn, R. O. (2015). The impact of climate change on agricultural net revenue: A case study in the Fouta Djallon, West Africa. *Environment and Development Economics, 20*(1), 20–36.

You, L., Rosegrant, M. W., Wood, S., & Sun, D. (2009). Impact of growing season temperature on wheat productivity in China. *Agricultural and Forest Meteorology, 149*(6), 1009–1014.

Zulfiqar, F., & Hussain, A. (2014). Forecasting wheat production gaps to assess the state of future food security in Pakistan. *Journal of Food and Nutritional Disorders, 3*(3), 2.

Methodological Issues in Social Science Research for Bioresource Conservation and Livelihood Development Under Global Climate Change

B.K. Narayana Swamy and Y. Nagaraju

Abstract Think of the time when the first *Homo sapien* was born. What did we know about the world or the universe or the galaxy at that time? Not even what a three or four years old child of today knows. Probably, this world was restricted to basic needs and surroundings. During those days, managing bioresources conservation and livelihood development was very easy. Slowly but steadily, we started to know the difference between doubt and belief. To satisfy our doubts, therefore, it is necessary that a method should be found to manage bioresources conservation and livelihood development by which our beliefs may be determined by nothing human, but by some external permanency by something upon which our thinking has no effect. Science is the systematic observation of natural events and conditions in order to discover facts about them and to formulate laws and principles based on these facts, theories to make unequivocal predictions that can be tested by anybody. The results can be reproduced by others; the systematic observation of natural events and conditions in order to discover facts about them and to formulate laws and principles based on these facts therefore, managing bioresource conservation and livelihood development unlike other three methods of knowing, the knowledge generated through method of social science is amenable to testing. The sociological scientific approach has been developed with a set of assumptions that cannot be proved. This will also help in understanding the superiority of scientific method over other methods to manage bioresources conservation and livelihood development. The nature is orderly and regular. The state of knowledge is a relative term to manage bioresources conservation and livelihood development what we know today may be disproved tomorrow. The man/woman has been restless, inquiring, and soul searching because he/she knows the power of knowledge. The process of research to manage bioresource conservation and livelihood development starts

B.K. Narayana Swamy (✉)
Department of Agricultural Extension, Sociology, Psychology and Distance Education Unit Policy, University of Agricultural Sciences, Bangalore, India
e-mail: narayanaswamy.konappa@gmail.com

Y. Nagaraju
Department of Economics, Center for Study of Social Exclusive and Inclusive Policy, Bharathidasan University, Khajmali Campus, Tiruchirappalli, Tamil Nadu

S. Nautiyal et al. (eds.), *Climate Change Challenge (3C) and Social-Economic-Ecological Interface-Building*, Environmental Science,
DOI 10.1007/978-3-319-31014-5_5

with problem identification, setting objectives, formulating hypothesis, finalizing research design, determining measurement processes, collecting data, analyzing data, and finally making generalization. When earlier social researchers first started to use research method as means of knowing, i.e., knowing with objectivity without any subjectivity, they started with anthropological type of research where in, they were interested to know what is where to manage bioresources conservation and livelihood development. Exploratory Research what is where? Descriptive Research what is what? Explanatory Research what causes which effect? Experimental Research what is the extent of effect under controlled causal condition? However, **Narayana's innovation attributes Lotus Model** explains bioresources conservation and livelihood development research must have relative advantage, compatibility, practicability and complexity. One of the goals is description and other goals being prediction and explanation. Demarcating the research field into manageable parts by dividing the main problem into subproblems is of the utmost importance. SMART can help you critically evaluate the objectives you have set. For every objective, ensure it meets the following criteria: Specific, Measurable, Achievable, Realistic, Time limited. In India, most of the researchers do not report the hypothesis formulated in their studies on managing bioresources. It raises a serious doubt whether the researcher formulate hypothesis for their research or not. One of the important purposes that this blueprint serves is to help draw causal inferences. The idea of causality is central to most of the scientific investigations. To conclude there is need for capacity building among professional, i.e., **Narayana's Wheel Model** explains seven orchestrated, concerted, comprehensible like innovativeness, decision-making ability, achievement motivation, information seeking ability, risk taking ability, coordinating ability, and leadership ability for capacity building.

1 Introduction

There are four methods of knowing, i.e., method of tenacity, method of authority, method of intuition, and method of science. Science is the systematic observation of natural events and conditions in order to discover facts about them and to formulate laws and principles based on these facts. Social Science implies two things, a body of knowledge as well as a system of methods. The aim of Social science itself is explaining natural phenomena. The purpose of research in Social sciences is to discover answers to questions or problems through the application of scientific procedures. How to manage bioresources conservation and livelihood development? It is a systematic, controlled, empirical and critical investigation of hypothetical propositions about the presumed relations among natural phenomenon to manage bioresources: conservation and livelihood development. Basically, there are four types of research, viz, exploratory, descriptive, explanatory, and experimental. In fact, in the era of information explosion where information is available at the tip of the finger, it is difficult to visualize the importance of generating this information.

Think of the time when the first *Homo sapien* was born. What all we knew about the world or the universe or the galaxy at that time? Not even what a three or four years old child of today knows. Probably, this world was restricted to basic needs and surroundings. During those days, managing bioresource conservation and livelihood development was very easy. Slowly but steadily, we started to know. Difference between doubt and belief explained that doubt is an uneasy and dissatisfied state from which we struggle to free ourselves and pass into the state of belief which is a calm and satisfactory state which we do not wish to avoid, or to change to a belief in anything else. However, methodological issues of research advancements in social science to manage bioresources conservation and livelihood development there are four methods of fixing belief or say knowing something. The first is the method of tenacity. It means holding or tending to hold persistently to something, such as a point of view. We believe what we have always believed to be truth. The second method takes care of fixing the belief in the community, which may be called the method of authority. This method has, from the earliest times, been one of the chief means of upholding correct theological and political doctrines, and of preserving their universal character. However, no institution can regulate opinions of every individual. There are logical thinkers and they fix their belief through a superior and intellectual method of knowing, i.e., the method of intuition. This method involves deductive reasoning, i.e., proceeding from a known or assumed cause to a necessarily related effect. It is derived without reference to particular facts or experience and made before or without examination not supported by factual study. To satisfy our doubts, therefore, it is necessary that a method should be found to manage bioresource conservation and livelihood development by which our beliefs may be determined by nothing human, but by some external permanency by something upon which our thinking has no effect. It must be something which affects, or might affect, every man and though these affections are necessarily as various as are individual conditions, yet the method must be such that the ultimate conclusion of every man shall be the same. Such is the method of science, the fourth method of fixing belief to manage bioresource conservation and livelihood development through methodological issues of research advancements in social science.

2 Scientific Research for Bioresource Conservation and Livelihood Development

It is amazing that many educated persons do not know what social science is although the word social science is so familiar; its meaning is not widespread. Science is the systematic observation of natural events and conditions in order to discover facts about them and to formulate laws and principles based on these facts, theories to make unequivocal predictions that can be tested by anybody. The results can be reproduced by others the systematic observation of natural events and conditions in order to discover facts about them and to formulate laws and principles based on these facts. Therefore, managing bioresources conservation and livelihood development unlike other three methods of knowing, the knowledge generated through method of social science is amenable to testing. In social science, knowledge is acquired using scientific method, i.e., body of techniques for investigating phenomena and acquiring new knowledge, as well as for correcting and integrating previous knowledge. It is based on gathering observable, empirical, measurable evidence, subject to the principles of reasoning. Scientific researchers propose specific hypotheses as explanations of natural phenomena, and design experimental studies that test these predictions for accuracy. These steps are repeated in order to make increasingly dependable predictions of future results. Scientific advance to manage bioresource conservation and livelihood development theories that encompass wider domains of inquiry serve to bind more specific hypotheses together in a coherent structure. This in turn aids in the formation of new hypotheses, as well as in placing groups of specific hypotheses into a broader context of understanding. Among other facets shared by the various fields of inquiry is the conviction that the process must be objective so that the scientist does not bias the interpretation of sociological advancement to manage bioresources conservation and livelihood development methodological issues in results or change the results outright. Another basic expectation is that of making complete documentation of data and methodology available for careful scrutiny by other scientists and researchers, thereby allowing other researchers the opportunity to verify results by attempted reproduction of these results. This also allows statistical measures of the reliability of the results to be established. The scientific method also may involve attempts, to achieve control over the factors involved in the area of inquiry, which may in turn be manipulated to test new hypotheses in order to gain further knowledge of managing bioresources conservation and livelihood development.

3 Social Science Research for Bioresource Conservation and Livelihood Development

The sociological scientific approach has been developed with a set of methodological issues and assumptions that cannot be proved. This will also help in understanding the superiority of scientific method over other methods to manage bioresources conservation and livelihood development. The nature is orderly and regular. The basic assumption of scientific approach is that there exists a definite regularity and order in the nature. Example could be that of atmospheric changes that lead to distinct pattern of climate, i.e., winter, summer, and rainy season. We can know nature. Human mind is capable of knowing nature and also itself. It can understand, comprehend, analyze, interpret, and infer. Knowledge is superior to ignorance. The state of knowledge is a relative term to manage bioresources conservation and livelihood development. What we know today may be disproved tomorrow. The man/woman has been restless, inquiring and soul searching because he/she knows the power of knowledge. All natural phenomena have natural causes. Science assumes the existence of cause in nature for every effect. It rules out the possibility of metaphysical explanation until the science cannot account for the causation of the natural phenomena. Nothing is self-evident. Science is not common sense. The truth must be objectively verifiable to manage bioresources conservation and livelihood development. Knowledge is derived from the acquisition of experience. If science has to tell anything about the real world, then it should be empirical, that it must rely on perception, experience and observation. Broadly speaking, the aim of science itself is explaining natural phenomena. Similarly, the aim of social science is to explain social phenomena that occur in natural setting. When these explanations are applied to solve some societal problems like manage bioresources conservation and livelihood development. The branch of science is called applied social science. Scientific research is systematic, controlled, empirical and critical investigation of hypothetical propositions about presumed relations among natural or the community at large. That is, the scientific explanations are used to predict and control the social phenomena for the greater interest of the society. Strictly laid down conditions, a standard methodology is required to conduct the research. Fundamentally, the research is a process. It is cyclic in nature. Theory is the product of this process which also becomes input for future researches. The process of research to manage bioresources conservation and livelihood development starts with problem identification, setting objectives, formulating hypothesis, finalizing research design, determining measurement processes, collecting data, analyzing data, and finally making generalization. These generalizations help develop theory to manage bioresources conservation and livelihood development.

4 Appropriate Technology Development for Bioresource Conservation and Livelihood Development

The several research and evaluation studies conducted in India revealed that utilization of for bioresource conservation and livelihood development is far from satisfactory in different parts of the world under global climate change. However, **Narayana's Innovation Attributes Lotus Model** (Fig. 1) explains four attributes with seven sub components under each for appropriateness of technologies to adopt like Relative advantage, Compatibility, Practibility and Complexity in the emerging global order. (A) **Relative Advantage**: is the degree to which an innovation is superior to the idea it supersedes. It can be explained with seven sub items like. (1) **Market feasibility**: as the extent of market demand for the product derived out of innovation and also the extent of scope for marketing product. (2) **Market stability**: as the consistency of market price and demand of product derived out of Innovation. (3) **Cost**: is of two types, initial cost and continuing cost. Initial cost

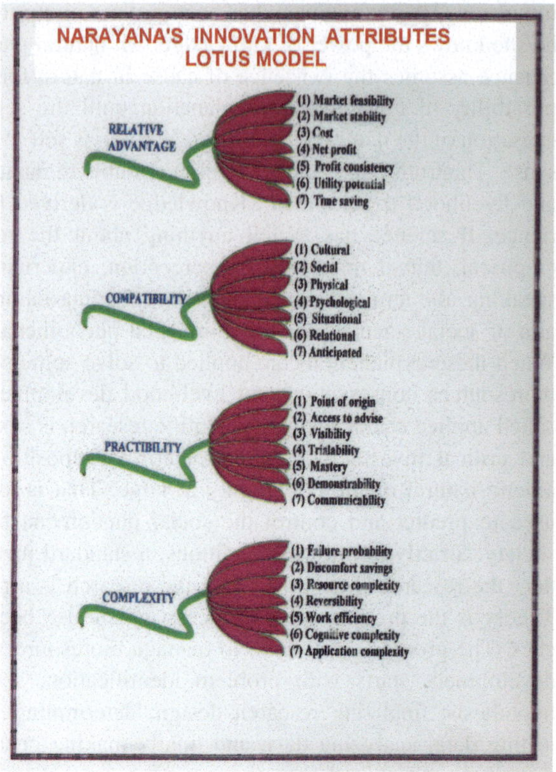

Fig. 1 Appropriate attributes of modern technologies are shown in the lotus

represents the capital investment required for adoption of innovation. Further, the cash or inputs required for subsequent years use of innovation is termed as continuing cost. (4) **Net Profit**: as the quantum of monetary benefit obtained by an individual through adoption of innovation. (5) **Profit consistency**: denotes the regularity of net returns obtained by an individual or group of individuals over a period of time by adoption of an innovation. (6) **Utility potential**: as the degree to which the multiple use potential of an innovation to an individual or group of individuals through adoption of innovations. (7) **Time saving**: indicates the best efficiency of an innovation in terms of saving time in different aspects. (B) **Compatibility**: an innovation is consistent with past experiences, existing values, and future plans of the adopters of technologies. Compatibility is divided into seven sub groups like (1) **Cultural compatibility**: an innovation is consistent with the values and norms of the society. (2) **Social compatibility**: denotes prestige gain or esteem by individual in the society through adoption of an innovation. (3) **Physical compatibility**: an innovation is consistent and should fit into the needs and interests of the adopters. (4) **Psychological compatibility**: Is innovation usefulness as perceived by the members of social system. (5) **Situational compatibility**: Denotes consistency and harmony of the innovation with previous practices followed by adopters. (6) **Relational compatibility**: an innovation can be adopted independently by the adopters just like other practices. (7) **Anticipated compatibility**: an innovation should be consistent with the future ideas of the adopter over a period of time. (C) **Practibility**: an innovation can be easily communicated, tested, demonstrated and practiced. (1) **Point of origin**: indicates the credibility of the source from where the innovation originated. (2) **Access to advice**: for implementation of innovation, its extent of availability of original and detailed information for guidance and clearing doubts that arise while implementing it. (3) **Visibility**: the results of an innovation are visible. (4) **Trialability**: is the degree to which new idea can be tried on a small scale. (5) **Mastery**: is the practice of an innovation could be learned or mastered in a short period of time. (6) **Demonstrability**: an innovation can be demonstrated to members of social system easily. (7) **Communicability**: is the information about the new idea can be diffused to members of the social system easily and speedily. (D) **Complexity**: an innovation is relatively difficult to understand and use. (1) **Failure probability**: an innovation chances of failure and uncertainty of results after its adoption. (2) **Discomfort saving**: represents avoidance of physical discomfort may be derived by adoption of an innovation. (3) **Resource complexity**: difficult in getting the necessary inputs and other resources for the application of an innovation. (4) **Reversibility**: degree of ease with which the innovation can be replaced in case of its failure. (5) **Work efficiency**: the adoption of new idea saves labor or increases the available labor efficiency. (6) **Cognitive complexity**: an extent of relative difficult in understanding an innovation. (7) **Application complexity**: relative difficulty of an innovations use and application on the farm. The appropriate technology developed must be used for bioresource conservation and livelihood development. This is the demand for HRD.

5 Different Types of Research for Bioresource Conservation and Livelihood Development

When we say types of research, we mean different ways of conducting research to manage bioresources conservation and livelihood development but having same ultimate goal. When earlier social researchers first started to use research method as means of knowing, i.e., knowing with objectivity without any subjectivity, they started with anthropological type of research where in, they were interested to know what is where to manage bioresources conservation and livelihood development. That means exploring the study area. This type of research is called exploratory research. Exploratory research is often conducted because a problem has not been clearly defined as yet, or its real scope is as yet unclear. It allows the researcher to familiarize him/herself with the problem or concept to be studied, and perhaps generate hypotheses to be tested. The results of exploratory research are not usually useful for decision-making by themselves, but can provide significant insight into a given situation. Once we were clear about what is where, the next question was what is what to manage bioresources conservation and livelihood development. In other words, describing different things. This type of research is called descriptive research. There are three main types of descriptive methods like observational methods, case study methods and survey methods. Observational method, under this method, animal and human behavior are closely observed in natural or laboratory conditions which are also called naturalistic observation and laboratory observation, respectively. Naturalistic observation is Exploratory Research What is where? Descriptive Research what is what? Explanatory Research what causes which effect? Experimental Research what is the extent of effect under controlled causal condition? One of the goals of advancements to manage bioresources conservation and livelihood development is description and other goals being prediction and explanation. Descriptive research methods are pretty much as they sound, i.e., they describe situations, having greater ecological validity than laboratory observation. Laboratory observations are usually under controlled conditions, less time-consuming and cheaper than naturalistic observations. However, both naturalistic and laboratory observations are important for managing bioresources conservation and livelihood development. Case Study method research involves an in-depth study of an individual or group of individuals. Case studies often lead to testable hypotheses and allow us to study rare phenomena. Case studies cannot be used to determine cause and effect and they have limited use for making accurate predictions to manage bioresources conservation and livelihood development. Survey method research, participants answer questions administered through interviews or questionnaires. After participants answer the questions, researchers describe the responses given. In order for the survey to be both reliable and valid it is important that the questions are constructed properly. Questions should be framed so they are clear and easy to comprehend. It is important for you to understand that descriptive research methods can only describe a set of observations or the data collected. It cannot draw conclusions from that data about which way the relationship goes Does A cause B, or does B cause A? This leads to another type of research called explanatory research. In case of explanatory type of research to manage bioresources conservation and livelihood development we are interested to know what causes which kind of effect. This brings in the topic of Principle of Cause and Effect relationship. The cause and effect principle is based on four basic characteristics. These four characteristics are like. (1) Causes and effects are the same things. (2) Causes and effects are part of an infinite continuum of causes. (3) Each effect has at least two causes in the form of actions and conditions. (4) An effect exists only if its causes exist at the same point in time and space. You might be surprised to know that cause and effect are same things. Actually, it is our perception based on which you may call something cause or effect to manage bioresources conservation and livelihood development.

6 Formulation of Research Problem for Bioresource Conservation and Livelihood Development

Advancements to manage bioresource conservation and livelihood development research originate from a need that arises. You should make a clear distinction between the problem and the purpose. The problem is the aspect that the researcher worries about, thinks about, and wants to find a solution for managing bioresources conservation and livelihood development. The purpose is to solve the problem, i.e., find answers to the question(s). If there is no clear problem formulation, the purpose and methods are meaningless. The researcher should (1) Outline the general context of the problem. (2) Highlight key theories, concepts and ideas current in this area. (3) What appear are the underlying assumptions of this area? (4) Why are these identified issues important? (5) What needs to be solved? (6) Read the subject to get to know the background and to identify unanswered questions or controversies, and/or to identify the most significant issues for further exploration. The criteria of problem statement to manage bioresources conservation and livelihood development indicate (a) Problem should express a relation between two or more variables. (b) The problem should be stated clearly and unambiguously in question form. (c) Problem statement should be such as to imply possibilities of empirical testing. However, the research problem should be stated in such a way that it would lead to analytical thinking on the part of the researcher with the aim of possible concluding solutions to the stated problem. Research problems can be stated in the form of either questions or statements. The research problem should always be formulated grammatically correct and as completely as possible. You should bear in mind the wording (expressions) you use. Avoid meaningless words. There should be no doubt in the mind of the reader what your intentions are. Demarcating the research field into manageable parts by dividing the main problem into sub-problems is of the utmost importance. SMART can help you critically evaluate the objectives you have set. For every objective, ensure it meets the following criteria Specific, Measurable, Achievable, Realistic, and Time limited. Research problems are conveyed with a set of concepts. In order to move from conceptual to empirical level, we convert concepts into variables by mapping concepts into a set of values. Our concepts will eventually appear as variables in hypothesis to be tested.

7 Formulation of Hypothesis for Bioresource Conservation and Livelihood Development

In India, most of the researchers do not report the hypothesis formulated in their studies on managing bioresources conservation and livelihood development. It raises a serious doubt whether the researcher formulate hypothesis for their research or not. In fact, managing bioresources conservation and livelihood development setting up and testing hypotheses is an essential part of statistical inference. In order to formulate such a test, usually some theory has been put forward, because it is to be used as a basis for argument but has not been proved, for example, claiming that educated people would be quick to adopt certain technology. In each problem considered, the question of interest is simplified into two competing claims/hypotheses between which we have a choice the null hypothesis, denoted by H0, against the alternative hypothesis, denoted by H1. These two competing claims/hypotheses are not, however, treated on an equal basis special consideration is given to the null hypothesis. Which research design should be used and why? Once the extension researcher determines the objectives of the study, explicates its hypothesis and defines variables, he/she confronts with the problems like whom to study? What to observe? When will observations be made? How to collect data? The research design serves as a blueprint of the study to manage bioresources conservation and livelihood development and provides solutions to these problems. One of the important purposes that this blueprint serves is to help draw causal inferences. The idea of causality is central to most of the scientific investigations. In practice, the demonstration of causality involves three distinct operations demonstrating covariance, eliminating spurious relations and establishing the time order of the occurrence. The classical research design consists of three components like comparison, manipulation and control to manage bioresource conservation and livelihood development.

8 Capacity Building Among Professionals for Bioresource Conservation and Livelihood Development

Under the global climate change capacity building among professionals for bioresource conservation and livelihood development is too complex phenomenon to be explained by a single factor. However, **Narayana's Wheel Model** explains combination of **seven components** for capacity building among professionals, viz., innovativeness, decision-making ability, achievement motivation, information seeking ability, risk taking ability, coordinating ability, and leadership ability. The combined contribution of the above seven factors to an individual behavior is being expressed in terms of capacity building among professionals, so far attention given is limited. (1) **Innovativeness**: Considered as socio-psychological orientation closely associated with change, adopting new ideas and practices. An individual adopts new ideas relatively earlier than others in his organization. However, innovativeness in professionals is very essential for bioresource conservation and livelihood development. (2) **Decision-making ability**: Considered as the nature of decision-making either individually or consulting with others while performing activities. It is the degree to which an individual justifies his/her selection of most efficient means from among the available alternatives on the basis of scientific criteria for achieving maximum profits. Hence, decision-making ability is very important among professionals for bioresource conservation and livelihood development. (3) **Achievement motivation**: Every man/woman has a desire to achieve certain things in their life. Achievement motivation is considered as the extent to which an individual is oriented toward maximizing profits. Achievement motivation as a social value that emphasizes a desire for excellence in order for an individual to attain a sense of personal accomplishment. So, achievement motivation increases efficiency of professionals and important for bioresource conservation and livelihood development (4) **Information seeking ability**: It refers to the frequency of contact by professional with various information sources. This is the pattern by which an individual gets information either on his/her own seeking or as a consequence of his/her being a part of the network. This component is important for bioresource conservation and livelihood development. (5) **Risk taking ability**: Some take more risk, some others take moderate risk and many hesitate to take risk. Risk taking ability considered an individual orientation toward risk and uncertainty in adopting new ideas and courage to face the problems for bioresource conservation and livelihood development demands ability to take risk by professionals. (6) **Coordinating ability**: In order to complete the required work in stipulated period, one has to harmonize and synchronize the various activities for better profit. It is an individual coordinates action in a time dimension. This ability helps to increase the efficiency of professional for bioresource conservation and livelihood development. (7) **Leadership ability**: To get things done properly, a professional has to initiate the action, motivate the followers and decision should be taken. It is an individual initiates or motivates the action of the other fellows. Hence, leadership ability is an important component in professional for bioresource conservation and livelihood development (Fig. 2).

However, it is generally assumed that opinion leadership is unique and specific to each situation. Nevertheless, the increasing technological developments in

Fig. 2 Narayana's wheel model for capacity building

bioresource conservation and livelihood development, have underlined the importance of farm leaders in spreading and stabilizing the adoption of new technologies among small farmers. This has given rise to the need for understanding certain basic leadership roles to be played by farm opinion leaders as group leader. The opinion leader is a small farmer who is sought after by other small farmer for information and advice and who is instrumental in influencing the decision-making behavior of other small farmers on sustainable crop production and allied enterprises. Hence, there is need to develop opinion leadership among small farmers. The farm opinion leader provides the needed assistance to extension agencies, in planning and implementing bioresource conservation and livelihood development programs. The **Narayana's Star model** (shown in Fig. 3) explains **seven qualities** of farm opinion leaders like (1) **Learner and Legitimizer**: The farm opinion leader learns more about recent technologies in sustainable crop husbandry and allied fields, judge pros and cons, accept, and adopts them. The farm opinion leader, by virtue of his/her social position in their community as the authority to give sanction or approval for the introduction of new sustainable technologies in the village keeping in view the village norms and values. (2) **Planner and Organizer**: The farm opinion leader is able to plan, visualize in his imagination, ways by which needs of his/her group can be satisfied. He/she has greater social insight into the structure and functioning of group and actually plans with the groups. He/she help farmers to make plans for getting maximum returns from farming. The farm opinion leader initiates common action in the villages. He/she arranges development meetings, mobilizes farmers and organizes group and community action in the village. (3) **Listener with self confidence**: The farm opinion leader should be a good listener to develop self confidence in motivating farmers. (4) **Linker of programs**: The farm opinion leader acts as a liaison person between small farmers and technical staff, input and market agencies, development officers, media persons and other agencies and thus facilitates effective contact between them to achieve the purpose of his/her group. (5) **Motivator with enthusiasm and eagerness**: The farm leader inspires and motivates the small farmers and attempts to sustain their interest in pursuing their plan even under critical situation in sustainable crop production and allied enterprises. (6) **Harmonizing spokesman**: The farm opinion leader promotes harmony among the members inline with the basic purpose of the group. He/she places more emphasis on the uniformities among the members rather than upon individual differences, and minimizes conflicts. The farm opinion leader is the responsibility of speaking-farmers group and representing the group's interest and position. (7) **Educator with group ideals**: The farm opinion leader shares his knowledge and experience with his/her followers to raise their level of understanding. He/she promotes training of small farmers for assuming leadership responsibility and group functioning. The farm opinion leader adopts group norms and lives by them. As per the expectation of his/her group he/she embodies the group ideals in relation to farming professions (Fig. 3).

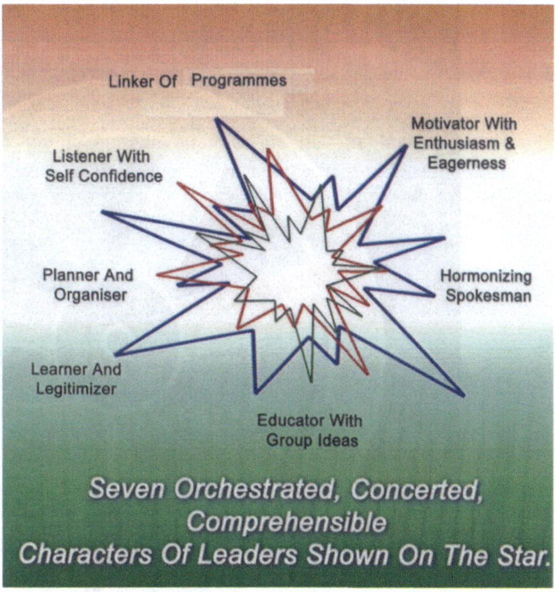

Fig. 3 Narayana's star model of leadership

9 Skills Development Among Professionals for Bioresource Conservation and Livelihood Development

Professional for convergence and collaboration need a broad range of skills in order to contribute to a modern economy and take their place in the technological society of the twenty first century. Training and development of Professionals are increasingly under pressure to be innovative, skilled and cost effective in nurturing a learning culture that will result in the development of professional. Skill development remains a significant area, of work and a range of initiatives focused on professional skill development. The program of work outlined to build convergence and collaboration in recent years to motivate rural youth for sustainable agriculture development with learning has the necessary knowledge, skill and abilities to support. However, **Narayana's Wheel Model** describes seven orchestrated, concerted, comprehensible skills required among professional for bioresource conservation and livelihood development like technical skill, managerial skill, human skill, conceptual skill, design skill, creative skill, and communicative skill (Fig. 4).

(1) *Technical skills of professionals*
Knowledge and proficiencies required in the accomplishment of specific task by the strong technical skills can save time, increase income, and enable to extract the most bang-per-buck from technology transactions to motivate bioresource conservation and livelihood development. There are seven ways

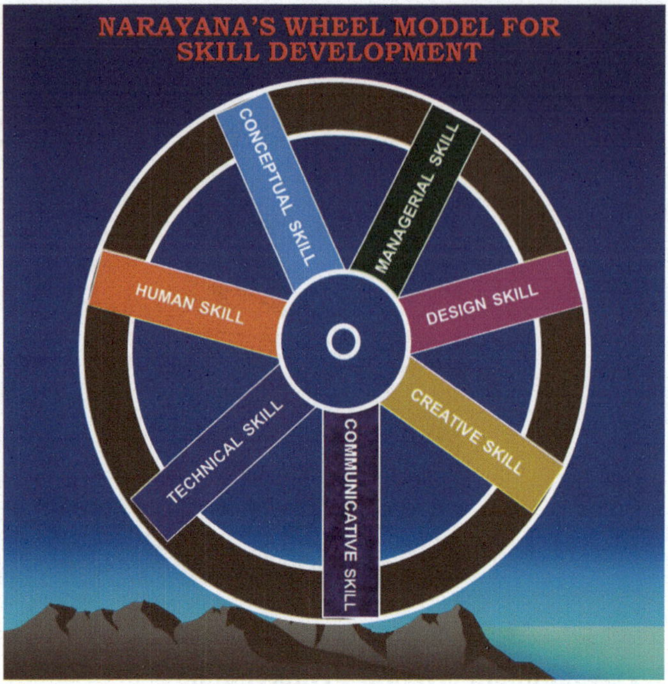

Fig. 4 Seven orchestrated, concerted, comprehensible shown in the wheel

to improve technical skills, regardless of current skill level: (1) **Reading technical books** in relation to bioresource conservation and livelihood development is one of the best ways to improve technical skills. Today, it is far better to shop online because one can more easily find the true gems and avoid the lemons. Although technical books can be expensive and are often padded with lengthy code listings, the good ones make up for it with clearly organized, well-edited, well-indexed content. Further, the books in their second edition or later is a great choice because they have already been through at least one round of testing in the marketplace. (2) **Read online tutorials** advantage over books is that they are accessible, timely, and of course free. The disadvantage is that they usually are not professionally edited, which can leave them lacking in completeness and clarity. However, they often sport other features like abundant interlinking, user comments, and interactive demonstrations. Sometimes the comments are better than the original information, since they can contain lots of additional tips and suggestions. (3) Spend enough time with technical people, some of their knowledge will rub off on you, a great way to accelerate the development of your technical skill is to join a local computer club or users group to be one of favorite outlets for learning. **Take classes** if group learning is your interest, look for courses and other classroom and workshop offerings in your area. (4) **Create your**

own web site setting a goal is a great way to learn practical skills. When one have a compelling reason to learn, goals will accelerate learning, with a focus on practical application. (5) **Build your own PC** to develop better hardware skills; a great project is to build your own PC from scratch. You will save money, learn a lot about how your computer works, and end up with a nicely customized machine that you can easily upgrade. After all the components arrived, it may take about a day to assemble everything and install the necessary software. (6) **Embrace a variety of software** to improve productivity with breadth of experience, so use many different software programs (online or offline) to improve overall ability to get things done through software. (7) **Learn to** Programming is the art of instructing a computer to perform a task. The key to accomplishing this is learning to think like a computer. Programming is one of the most mentally challenging tasks a human being can perform, but nothing compares to the satisfaction of engineering a piece of code to solve a specific problem of bioresource conservation and livelihood development.

(2) *Managerial Skills of professionals*

The managerial skills indicate the quality of professionals involved in motivating. The work need of the different convergence and professional require the different skills in order to handle to make it successful. There are different types of skills which the professional need in order to exercise among different bioresource conservation and livelihood development. So professional have to deal with the lot of problems which require special skill in order to solve them. When the professional counter a problem they require some special skill in order to deal with the specific problem. Hence, seven managerial skills professional need to develop as a leader in working like (1) Observation is an important aspect that often gets neglected due to the demands on a leader's time and schedule. Observation and regular visits to the work environment are a priority and should be scheduled into the calendar. Observing professionals at work, the procedures, interaction and work flow is foundational for implementing adjustments to improve results. To have credibility, a professional need to be seen and be known to be up to date with what is happening in the bioresource conservation and livelihood development. (2) Monitoring of Professional performance needs to be in mutually accepted ways. Policies and procedures need to be clear. Conferencing should be on a regular basis and not just when there is a problem. Assessments and evaluations should not be merely all formality or viewed a necessary paperwork to be done and filed away. Individual and group conferencing should be undertaken not only to monitor performance, but with the expectation of ongoing professional development and support for bioresource conservation and livelihood development. There should be frequent encouragement and clear criteria for ongoing goals both for the group and individual. (3) The planning of professional development programs should be tailor made. The professional should have the ability for selection of right bioresource conservation and livelihood development at right time in right ways for better performance. (4) A good professional need to implement program properly and evaluate bioresource conservation and livelihood development programs, provide training and development strategies to strengthen the weaker skills in the team. (5) Good professional for bioresource conservation and livelihood development need to demonstrate working knowledge and expertise come from a place of strong knowledge and experience of the production and process leading to results. If a leader does not possess all the expertise and knowledge personally, then regular consultations with experts involved in the departments should be held. This is important in order to maintain an accurate and informed overall picture. (6) Good Decision-making of a professional is characterized by the ability to make good decisions. A professional considers all the different factors before making a decision. Clear firm decisions, combined with the willingness and flexibility to adapt and adjust decisions when necessary, create confidence in bioresource conservation and livelihood development. (7) The professional need to conduct and evaluate ongoing review and research on bioresource conservation and livelihood development is vital in order to keep on the cutting edge. While managing the present to ensure ongoing excellence in product and performance, a good professional is also able to look toward the future.

(3) *Human skills of professionals*
Human skills involve the ability to work well with bioresource conservation and livelihood development both individually and in group. Because a professional deal directly with this skill, it is crucial for Professional with good human skills is able to get the best out of. They know how to communicate, motivate, lead, and inspire enthusiasm and trust. These skills are equally important at all levels among professional.

(4) *Conceptual skills of professionals*
Conceptual skills are skills that utilize the ability of professionals to form concepts. Such skills include thinking creatively, formulating abstractions, analyzing complex situations, and solving problems. These skills are considered an integral requirement of professional for motivating bioresource conservation and livelihood development. In general, conceptual skills involve abstraction. It is easier to think abstractly about a concrete, physical item, such as a car or house, than an abstract idea, such as justice or happiness. Conceptual skills can also require analysis, such as looking at patterns in events or objects. The ability to understand interrelationship of ideas to totality. Professional must have conceptual skills to think and to conceptualize about abstract and complex situations to see development as a whole, understand the relationships among various subunits, and visualize how the bioresource conservation and livelihood development fits into its broader environment.

(5) *Communication skills of professionals*
Communication is a process whereby meaning is defined and shared between professionals. Communication requires a sender, a message, and an intended

recipient, although the receiver professionals need not be present or aware of the sender's intent to communicate at the time of communication. Thus, communication can occur across vast distances in time and space. However, communicating professional share an area of communicative commonality. The communication process is complete once the receiver of bioresource conservation and livelihood development has understood the sender. Speaking, writing, listening, reading, and thinking are the important communication skills. Speaking is skill consisting of two concepts, verbal and nonverbal. Verbal skills like pronunciation, articulation, use of language, etc. Non-verbal skills are eye contact, dress, and personality, how you stand, how you observe audience, etc. Speaking optimum word must be 120–140 words per minute. What to speak about it should be clear and serve the purpose. Materials may be collected from different source. But it should have clarity, brevity, simplicity, logical sequence and style. Listening is vital for the professional to pay attention, understand and retention. The professional must direct his/her conciseness to perceive and understand the message but understanding the message demand ability. However, if one listens with definite purpose keeping main idea in view will understand better. Good hints for attention is listen with face, might directed toward speech and listen to speech as such with any physical disturbance. Reading is also important skill like listening. However, it is essential for readers to set a goal, start with small sequential reading step, immerse in active reading, obtain feedback yourself, start self phasing and end with self evaluation. Thinking is very essential for both encoding and decoding.

(6) *Design skills of professionals*
Design skill as an excellent resource for inspiration to professional to motivate bioresource conservation and livelihood development for sustainable agriculture development. I am not listening too many designs blogs here of course you might as well find inspiration in non-design blogs, just be curious. There are tons of books with tons of logos, letterheads, web designs, or other designs featured. Appropriate design for motivating is important.

(7) *Creative Skills of professionals*
Creative skills are skills that let you expand on what there already is. One can continue on an idea and are not limited in ability to create. But things like driving, one can really improve upon that skill very much. It is the professional responsible for development of bioresource conservation and livelihood development for all creative practices. Creative skill is the competency that gets closer to realizing the vision of what it wants to create. Trying to nail it down further, to mind is an activity of exclusion that may not be useful when creating any skill and any tool at disposal will not work all the time and guess that is the nature of creativity. All the seven skills are very important for professional to motivate bioresource conservation and livelihood development.

Impact of Ruminants on Global Warming: Indian and Global Context

Partha Sarathi Swain, George Dominic, K.V.S. Bhakthavatsalam and Megolhubino Terhuja

Abstract According to the Food and Agriculture Organization of the United Nations (FAO), agriculture is responsible for 18 % of the total release of greenhouse gases worldwide and among this emissions from livestock constitute nearly 80 % of all agricultural emissions. Indian poor peoples' economy is based on his livestock which is mostly ruminants as a source of meat, milk, wool, etc. Ruminants produce a huge quantity of the greenhouse gases which significantly contributes to global warming namely, carbon dioxide (CO_2), methane (CH_4) and nitrous oxide (N_2O), etc. On an average, a cow produces around 70–120 kg CH_4 per year which is 23 times more potent factor to climate change than that of CO_2. So, 100 kg CH_4 is as potent as 2300 kg CO_2 as a greenhouse gas. So, as a whole, ruminants in world emit about two billion metric tonnes of CO_2 equivalents of methane per year and 2.8 billion metric tonnes of CO_2 per year due to grazing and grazing land related issues like deforestation. 70 % forests in the Amazon is been converted to grazing land. Climate change affects the animal husbandry in India by affecting the dairy, meat and wool production and thus has severe consequences to Indian economy. For an instance, heat stress can reduce the feed intake causing reduction in growth, fertility and reproduction. Water shortage and drought causes reduction in productivity of animal. Again, flood and irregular rain cause life threatening loses to the farmers. Along with this, change in the environment predisposes to vector-borne

P.S. Swain (✉) · G. Dominic
Dairy Cattle Nutrition Division, National Dairy Research Institute,
Karnal 132001, Haryana, India
e-mail: parthavet@yahoo.com

K.V.S. Bhakthavatsalam
Dairy Economics Division, National Dairy Research Institute,
Karnal 132001, Haryana, India

M. Terhuja
Veterinary Virology, Indian Veterinary Research Institute,
Izatnagar 243122, Uttar Pradesh, India

S. Nautiyal et al. (eds.), *Climate Change Challenge (3C) and Social-Economic-Ecological Interface-Building*, Environmental Science,
DOI 10.1007/978-3-319-31014-5_6

diseases and parasites. Responses to climate change include adaptation to reduce the vulnerability to climatic changes and mitigation to reduce this change. Climate change can be controlled by either by adaptation to the ecosystem or mitigation (to reduce the level of emission of gases) measures but both are necessary to get an effective result.

Keywords Climate change · Global warming · Methane · Ruminants

1 Global Warming

The evolution of the various living forms on this earth is supported due to the vigour of the earth's climate. It is observed that the temperature of our environment has increased continuously which is impacting a change in the climatic variation and its occurrence. It is observed that in the last one hundred years there has been around 1 °C increase in Earth's temperature and is expected to increase by 2–3 °C in the coming century. This raise in temperature is due to trapping of the Sun's heat by Earth's atmosphere and thus Earth is getting warmer. For now, this change is small but if appropriate measures are not taken, it is expected to get quicker. The impact of the climate change is eminent as the polar ice is melting and the sea level is rising with increasing in this atmospheric temperature, which will lead to flooding along coasts. The tide gauge observations in the last four decades across the coast of India also indicate a rise in sea level at the rate of 1.06–1.25 mm/year. In India, the meteorological records indicate rise in the mean annual surface air temperature by 0.4 °C with not much variations in absolute rainfall. However, the rates of change in temperatures and precipitation have been found to vary across the region. The intensity and frequency of heavy precipitation events have increased in the last 50 years. Further, some preliminary assessments point towards a warmer climate in the future over India, with temperatures projected to rise by 2–4 °C by 2050s. No change in total quantity of rainfall is expected, however, spatial pattern of the rainfall is likely to change, with rise in number and intensity of extreme rainfall events. Raise in temperature will lead to heat stroke and drought which may cause destruction to forests and food crops like ecosystems. So global warming is not a local rather a global issue. As the temperature all around the globe is rising, it is said as global warming. This will affect plants and animal ecosystems. This global warming is caused due to many factors. The changes in the climate of the atmosphere caused due to humans in the recent decades are a major concern for the scientific fraternity, through, fossil fuel combustion, widespread deforestation and burning of biomass. These anthropogenic activities have resulted in an increased emission of radioactively active gases, e.g., carbon dioxide (CO_2), methane (CH_4) and nitrous *oxide* **(N_2O),** popularly known as the 'greenhouse gases'. Human activities causing global warming include burning fuels, increased industrial emissions, cutting down forest, nuclear explosions, changes in land-use and land management practices that have led to drastic changes in the atmospheric composition of the Earth and by taking part in other activities that release certain heat trapping gases, i.e. green house gases (GHG) into atmosphere. Along with this, agriculture is also playing a vital role in this global warming and in agriculture, ruminants are among the worst offenders as they produce some gases that aid in global warming. Intergovernmental Panel on Climate Change (IPPC) included six gases namely, carbon dioxide (CO_2), methane (CH_4), nitrous oxide (N_2O), hydrofluorocarbons (HFCs), perfluorocarbons (PFCs) and sulphur hexafluoride (SF_6), which have high global warming potential.

The Earth would have been 4 °C colder than its normal temperature, if the greenhouse gases were absent in the Earth's temperature. So these GHGs may be conferred as a necessary evil. CO_2, CH_4 and N_2O are present in the atmosphere and are also produced as a result of agricultural and livestock and also from the activities of the human beings. The global warming potential of CH_4, N_2O is 23 and 296 times higher with respect to carbon dioxide in global warming, respectively (IPCC 2001). Anthropogenic gases like water vapour, CO_2, CH_4 and Ozone contribute 36–72, 9–26, 4–9 and 3–7 %, respectively, to the global warming (Kiehl et al. 1997).

2 Agriculture and Global Warming

Agriculture sector emitted 334.41 million tonnes of CO_2 equivalent, of which 13.76 million tonnes is CH_4 and 0.15 million tonnes is N_2O. Enteric fermentation constituted 63 % of the total CO_2 equivalent emissions from this sector, 21 % of the emissions were from rice cultivation. Crop soils emitted 13 % of the total CO_2 equivalent emission from agriculture. Rest 2.7 % of the emissions is attributed to livestock manure management and burning of crop residue. India emitted 3.3 million tonnes of CH_4 in 2007 from 43.62 million ha cultivated for this purpose. Of the total rice area cultivated, 52.6 % was irrigated, 32.4 % was rain-fed lowland, 12 % was rain-fed upland and 3 % was deep water rice (Huke at el. 1997). The annual amount of CH_4 emitted from a given area of rice is a function of the crop duration, water regimes and organic soil amendments. The CH_4 emissions from rice cultivation have been estimated by multiplying the seasonal emission factors by the annual harvested areas. The total annual emissions are equal to the sum of emissions from each subunit of harvested area using the following equation. Global warming, in turn, leads to regional changes in climate related parameters such as rainfall, soil moisture and sea level. The extensive and frequent occurrence of climatic extremes such as droughts, heat and floods in the last decade in many parts of the world may be the fallout of this. The sea level has risen by 10–20 cm with regional variations (IPCC 2001). Similarly, snow cover is also believed to be gradually decreasing.

2.1 Agriculture Soils

Agricultural soils are a source of N_2O, mainly due to application of nitrogenous fertilizers in the soils. Burning of crop residue leads to the emission of a number of gases and pollutants. Amongst them, CO_2 is considered to be C neutral, and therefore not included in the estimations. Only CH_4 and N_2O are considered for this report. The total CO_2 eq. emitted from these two sources were 50.00 million tonnes.

Nitrous Oxide is produced naturally in soils through the processes of nitrification and denitrification. Nitrification is the aerobic microbial oxidation of ammonium to nitrate, and denitrification is the anaerobic microbial reduction of nitrate to nitrogen gas (N_2). Nitrous oxide is a gaseous intermediate in the reaction sequence of denitrification and a by-product of nitrification that leaks from microbial cells into the soil and ultimately into the atmosphere. One of the main controlling factors in this reaction is the availability of inorganic nitrogen (N) in the soil. This methodology, therefore, estimates N_2O emissions using human-induced net N additions to soils (e.g. synthetic or organic fertilizers, deposited manure, crop residues and sewage sludge), or of mineralization of N in soil organic matter following drainage/management of organic soils, or cultivation/land-use change on mineral soils (e.g. forest land/grassland/settlements converted to cropland).

2.2 Burning of Crop Residue

Crop residue is burnt in the fields in many Indian states such as Uttar Pradesh, Punjab, West Bengal, Haryana, Bihar, Madhya Pradesh, Himachal Pradesh, Maharashtra, Gujarat, Chhattisgarh, Jharkhand, Tamil Nadu, Uttaranchal and Karnataka producing CO, CH_4, N_2O, NO_x, NMHCs, SO_2 and many other gases. In this report only the CH_4 and N_2O emissions have been reported.

2.3 Methane Emission from Rice Fields

Methanogenesis, the process responsible for methane formation, occurs in all anaerobic environments in which organic matter undergoes decomposition. Rice is generally grown in waterlogged condition, which creates an anoxic environment and is conducive to methane production by the strictly anaerobic methanogenic bacteria. Methanogens use organic compounds as electron donors for energy and synthesis of cellular constituents and, in turn, reduce C to CH_4.

3 Greenhouse Gases in Ruminant Context

Livestock, mainly cows and buffaloes are identified as one of the primary contributors of greenhouse gases in India. These animals play a major role in the emission of methane—a gas with a much more lethal impact on global warming than the usual suspect carbon dioxide. Livestock is known to release a huge amount of methane through belching and flatulence, out of which flatulence accounts for a smaller quantity.

A worrying trend outlined in the study is that India's methane emission from livestock is the highest in the world.

India's greenhouse gas emissions from agriculture fell to about 18 % of the total between 1994 and 2007. But, livestock sector in 2007 produced methane with a warming potential equivalent to that of 334 million tonnes of carbon dioxide.

4 Greenhouse Effect (GHE)

Atmospheric scientists first used the term 'greenhouse effect' in the early 1800s. At that time, it was used to describe the naturally occurring functions of trace gases in the atmosphere and did not have any negative connotations. It was not until the mid-1950s that the term greenhouse effect was coupled with concern over climate change. In recent decades, we often hear about the greenhouse effect in somewhat negative terms. The negative concerns are related to the possible impacts of an enhanced greenhouse effect. It is important to remember that without the greenhouse effect, life on earth as we know it would not be possible. GHG allows solar radiation to pass through the Earth's atmosphere but they absorb some amount of solar radiation which is not radiated back and in turn causes the warming of the Earth surface. In fact, these GHGs are important for our environment to keep our climate warmer, but excess production of these gases is causing more rises in the temperature than normal. This phenomenon is called as GHE. In this effect, the short wavelength (visible light) from the sun passes through the atmosphere and is absorbed, but the longer wavelength (infrared) reradiation from the heated objects are unable to pass through that medium. The trapping of the long wavelength

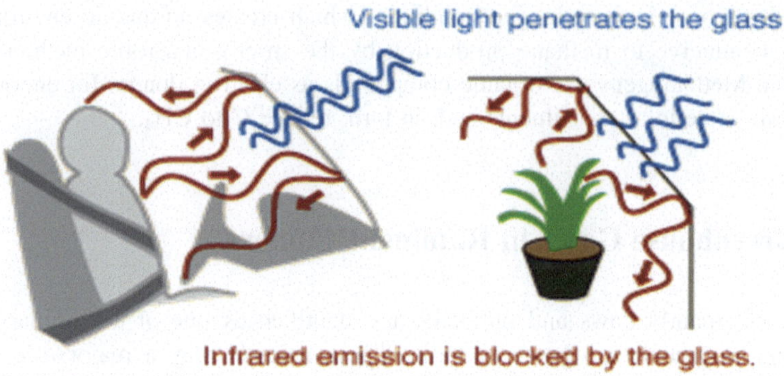

Source http://hyperphysics.phy-astr.gsu.edu/hbase/thermo/grnhse.html

radiation leads to more heating of atmosphere and resulting in rise in Earth's temperature. The GHE has been widely used to describe the trapping of excess heat due to the rising concentration of carbon dioxide in Earth surface. The carbon dioxide strongly absorbs infrared and does not radiate it back into the space. A major part of the efficiency of the heating of an actual greenhouse is the trapping of the air so that the energy is not lost by convection. Keeping the hot air from escaping out the top part of the practical "greenhouse effect", but it is common usage to refer to the infrared trapping as the greenhouse effect in the atmospheric applications where the air trapping is not applicable.

Noteworthy greenhouse gases are methane, nitrous oxide, carbon dioxide, hydrofluorocarbons (HFCs), perfluorocarbons (PFCs) and sulphur hexafluoride (SF_6). These gases are thought to affect the climate directly and indirectly, even though they constitute only a small fraction of the blanket of gases that make up the atmosphere. The greenhouse effect (20.00 %) is attributed directly to carbon dioxide and 5 % to all other greenhouse gases according to 2010 models cited by NASA. The remaining 75 % of the greenhouse effect is thought to be due to water vapour and clouds, which are naturally occurring. However, even though carbon dioxide and the other greenhouse gases are such a small percentage of the total gas in the atmosphere, they affect when, where and how clouds form, so greenhouse gases have some relevance when it comes to 100 % of the greenhouse effect. Carbon dioxide is thought to modulate the overall climate, like an atmospheric thermostat.

4.1 Ruminants and Global Warming

Ruminants are the herbivores that consume plants and digest them through the process of enteric fermentation in their four chambered stomach. Ruminants include cattle, sheep, goats, buffalo, deer, elk, giraffes and camels. Unlike to the mono-gastric species (human, dog, cat, etc.), ruminant stomach is having four chambers.

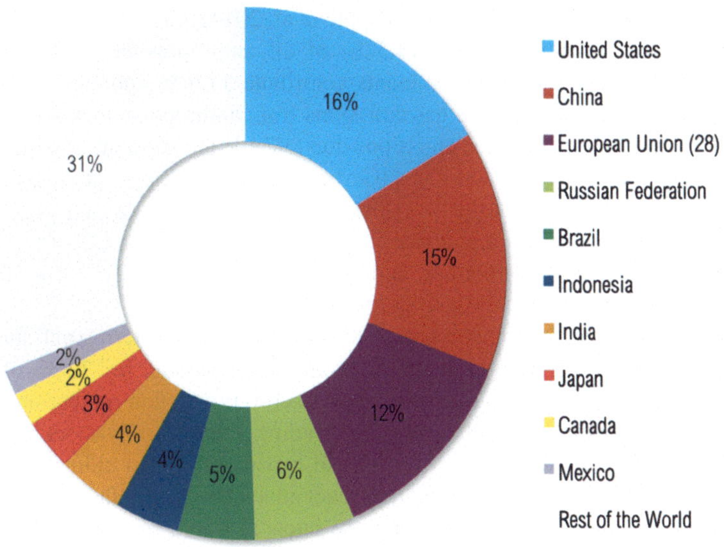

Cumulative greenhouse gases emissions 1990–2011 (% of world total). *Source* World Resources Institute

Methane emission rates from agricultural sources

Sources related to agriculture	Methane emission rate (million tonnes/year)
Enteric fermentation	80
Paddy rice production	60–100
Biomass burning	40
Animal wastes	25
Total	205–245

Source Watson et al. (1992)

Of the four compartments the rumen is the largest compartment and the main digestive centre. The rumen is unique in its flora and fauna which is the fermentation vat. Rumen microbes ferment the fibrous food taken by the animals and produce volatile fatty acids (VFA) which serve as energy source. Methane is produced as a by-product of rumen microbial fermentation. Energy lost as methane from cattle ranges from 2 to approx. 12 % of GE intake (Johnson and Johnson 1995). Compared to ruminants, non-ruminants (pigs and poultry) have a single-chambered stomach which has negligible methane emission.

There are 1.4 billon cattle, 1.1 billion sheep, 0.9 billion goats and 0.2 billon buffalo which sums up to 3.6 billion domestic ruminants on Earth in 2011 (FAOSTAT 2013). On an average, 25 million domestic ruminants are added to this count every year in the past 50 years (Ripple et al. 2014). Globally, livestock sector is responsible for approximately 14.5 % of all anthropogenic GHG emissions (Gerber et al. 2013) and in it, ruminants contribute 11.6 % and cattle alone contributes 9.4 % of all greenhouse gas emissions from anthropogenic sources (Ripple et al. 2014). Among ruminants, emissions due to cattle are substantially higher than those from buffalo or sheep and goats (Gerber et al. 2013; Ripple et al. 2014). CH_4 contributes about 44 % of the livestock sector's emissions which comes from enteric fermentation, manure and rice feed. The other gases are CO_2 (27 %) from land-use change and fossil fuel use, and nitrous oxide (N_2O) (29 %) from fertilizer applied to feed-crop fields and manure (Gerber et al. 2013).

The total grazing land encompasses almost 26 % of the terrestrial surface of earth (Steinfeld et al. 2006). Livestock production accounts for 70 % of global agricultural land and the area dedicated to feed-crop production represents 33 % of total arable land (Steinfeld et al. 2006). The feeding of crops to livestock is in direct competition with producing crops for human consumption and climate mitigation (Smith et al. 2013). Ruminant rearing can adversely affect the ecosystem by deforestation of the grazing lands, land-use intensification, soil erosion, and competition with due to grazing prominent mostly in tropical region where animals are left loose for grazing (McAlpine et al. 2009). Control over this kind of deforestation may allow regrowth of forests and grassland vegetation and thus improves our environment.

The United Nations Food and Agriculture Organization (FAO) has pointed out that the growth of the meat industries is also having a great impact in fastening the global warming scenario (Steinfeld et al. 2006). The greenhouse gas production from meat industry on average is 19–48 times higher than that of high-protein foods obtained from plants which include both direct and indirect environmental effects like enteric fermentation, manure, feed, fertilizer, processing, transportation and land-use change, etc. (Ripple et al. 2014). So, decrease in meat consumption from ruminant origin, i.e. beef or more consumption of pork and chicken may also play an important strategy in controlling the climate change.

Change in the climate may impact in the quality and quantity of feeds and forages. This includes herbage growth rate, composition of pasture, dry matter of the forages which is mostly governed by drought or excess rainfall which increase or decrease the DM content of the available forages, respectively (Hopkins and Del Prado 2007). The continuous change in this environment temperature may hamper the normal physiology of the animals. At higher temperature, there will be decline in feed intake and in reproductive performances. Drop in the feed intake may be reflected in a decline in milk yield also (Thornton et al. 2009). Global warming also causes melting of polar snow and thus the coastal area is submerged in the seas. The rise in the temperature may hasten the hatching process which may lead to larger population of the pathogen (Harvell et al. 2002). Along with all these direct effects

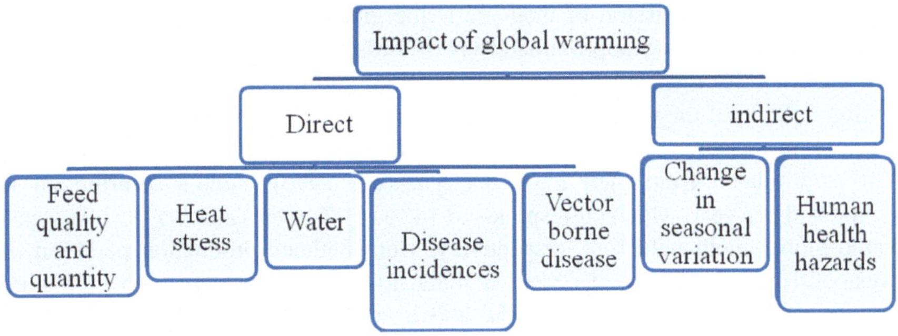

Possible impacts of global warming (Thornton et al. 2009)

on livestock, human population is also affected due to this global warming. This includes emergence of newer diseases and mortality and morbidity due to heat stroke.

5 Ruminants in Indian Context

Being an agricultural country, Indian economy is basically based on agriculture. India is equipped with a dense population of livestock and poultry. Indian livestock encompasses 16.1, 56.5 and 16.5 % of world's cattle, buffalo and goat population, respectively, and thus occupies first position in milk, third in egg and fifth in broiler production. Livestock sector in India provides employment, social as well as nutritional security to the millions of resource poor farmers who are engaged in farm operations in the rural areas. The animals are mainly fed with poorly digestible and low quality feeds and fodders which are mostly the agricultural by-products. Animals grazing on poor quality feeds and fodders produce more methane per unit livestock product (Kirschgessner et al. 1995). Ruminal fermentation of feeds is the largest source of methane from enteric fermentation and the molar percentage of the different VFA produced during fermentation influences the production of methane in the rumen. Acetate and butyrate promote methane production while propionate formation can be considered as a competitive pathway for hydrogen use in the rumen. So, such a high density of livestock raised mostly on crop residues and agricultural wastes is a cause of major concern as this is going to add to the total pool of greenhouse gases. More over 71 % of cattle, 63 % of buffaloes, 66 % of small ruminants, 70 % of pigs and 74 % of poultry are maintained by poor and marginal farmers (Sampath and Bhatta 2010) who are not able to provide good quality foods and fodders. This enhances the production of these GHGs. Again most of the animals in India are either non-producers or under-producers. But, due to social taboo and restrictions these animals cannot be culled and they are the source of the daily livelihood of the poor and marginal farmers.

An agricultural emission of methane is the greatest source (45 %) of methane emissions, out of which approximately two-thirds come from enteric fermentation and one third from livestock manure (Moss et al. 2000). Enteric methane accounts for about 75 % of the total on-farm methane emissions (EPA 2010). The domestic livestock produces about 80 Tg enteric methane per year and 25 Tg methane per year is produced from their manure (Cynoweth 1996). India's contribution is around and per year, which corresponds to 14 % (11 Tg/year) and 4.5 % (1 Tg/year) of the global methane influx, respectively. But, methane production per head of livestock (methane emission factor) is much lower than the very high producing ones of America and Europe. So, the above index might not be alarming if we consider the quantum of livestock population in India. But, India often is blamed for higher emission of methane per unit of livestock production even though the methane emission factor is lesser than the other countries. This may be due to our socioeconomic compulsions to maintain a large population of unproductive or low productive animals.

6 Effects of Global Warming

Global warming is the process of increasing in the Earth's atmospheric temperature due to many anthropogenic and GHG. The effect of this global warming is well felt in the past few decades as the routine cyclicity of the seasonal variation is not proper. According to FAO (2007) if global average temperature increases by 1.5–2.5 °C, approximately 20–30 % of plant and animal species are expected to be at the risk of extinction. The heat in summer and the cold in winter also not same like a decade before. The summery of climate change can be briefed as follows:

1. Increase in Earth's temperature, i.e. increase in number of hot days and hot nights and heat waves during the last fifty years and there are less frequent cold days, cold nights and frost.
2. For the next two decades a warming of about 0.2 °C per decade is projected by International Panel on climate change.
3. The ice on the mountains and polar region melts leading to rise in global sea level. Global average sea level rose by 1.0 mm per year over 1961–2003.
4. There is change in the seasonal variation. That is why it is termed as climate change.
5. Changes in ecosystem due to altered temperature and rainfall which affects the quality and quantity of vegetation, fruit, cereal grain, aromatic and medicinal plants.
6. Insects and pathogens are highly susceptible to the atmospheric temperature and humidity. So, this also gives rise to many new diseases and thus affecting plants, animals and human beings.
7. The propionate enhancers should be used in the ration that will decrease A:P ration and in turn reduces methane.

8. In addition, there are a large number of proposed additives that could be added in the ruminant ration to lower methane emissions. Ex: yeasts, essential oils, fibre degrading enzymes and compounds such as garlic or oregano, etc.

7 Mechanism of Methane Production by Ruminants

CH_4 is produced in the rumen of the ruminants by microbial fermentation of feed and fodders. Rumen flora and fauna comprising of bacteria, protozoa and fungi ferment the feed nutrients like proteins, carbohydrates (starch and plant cell wall) into amino acids and sugars and thereafter they are converted into VFA, hydrogen (H_2) and carbon dioxide (CO_2). VFA (acetate, propionate, butyrate, etc.) are then absorbed and utilized by the host animal as energy source. Acetate and butyrate are substrate for CH_4 production. So, the feed like more fibrous feeds produce more acetate thus increases CH_4 production, but on the contrary high concentrates produce more propionate and thus reduces enteric CH_4 production.

7.1 Mitigation Strategies

There are many possibilities of mitigation of methane emission by the livestock, e.g. providing balanced feeds, feeding less fibrous feeds, use of chemicals, ciliate protozoa removal (defaunation), culling of unproductive animals, microbiological and biotechnological inhibition of ruminal methanogens. All the above are having some demerits as chemicals have toxic effects on fermentation of feed in the rumen; providing high nutrition diet to the low producers are not economic and culling unproductive and lower productive animals are having social and religious restrictions. Reducing methane production can be of direct economic benefits as it enhances the efficiency of energy utilization of the feed by the animal.

7.2 Reduction of Methane Mitigation from Ruminants

There are numerous attempts made to reduce the CH_4 production and release from the ruminants. All are found successful to some extent and none is having without limitations. Some are having economic barrier, some are toxic in nature after certain limit and some are having social and religious taboo. Establishing a balance is the main aim of all the researchers. However, some of the methods are summarized below.

1. CH_4 mitigation is an intrinsic property of feed that vary with feed ingredients. So, if possible, the farmers should select feeds with low methane emission

potential during ration formulation that will reduce the methane mitigation from those animals.

2. Frequency of feeding also affects the methane production in the ruminants. Feeding the animals more times will cause less methane release as passage rate is increased. But it is having labour intensive and cost related limitations.
3. Alternate hydrogen sinks:

   A. Fumaric acid is having the potential to alter the production of volatile fatty acids. When fed, fumaric acid increases total VFA concentration, with increase in propionate proportion and decreased acetate: propionate ratio (Beauchemin and McGinn 2006) and thus reduces methane production. To inhibit methanogenesis, higher doses of fumaric acid is needed which may cause a fall, if ruminal pH and thus hamper the normal rumen ecosystem.
   B. Use of nitrate can reduce the quantity of the methane production in ruminants (Sakthivel et al. 2010). *Selenomonas ruminantium*, and *Wolinella succinogenes* are nitrate-reducing bacteria found in the goat rumen (Stewart et al. 1997). These bacteria may act as alternate hydrogen sink in the rumen as they can convert nitrate to ammonia and nitrite is an intermediate. The demerit of this method is that reduction of nitrate is 2–3 times faster than nitrite. So nitrite may accumulate in rumen. Therefore, reduction of nitrite in the rumen should be enhanced to avoid the toxic effects of nitrite in rumen.

4. Plants and their parts may reduce the methane production in rumen due to the presence of some secondary metabolites like tannin and saponin for an instance; some of the saponin rich plants includes *Glycine max, Cicer arietinum L, Medicago sativa, Lupinus Species, Sapindus mukorossi* (Fenwick and Oakenfull 1983) *whereas Terminalia chebula, Populus deltoides, Mangifera indica*, and *Psidium guajava* leaves are rich sources of tannin.
5. Feeding balanced ration, dietary energy and protein could be utilized in a more efficient manner in lactating ruminants (Kannan et al. 2010) which will reduce the methane production in the animals. Again proving, low fibrous and ingestible feeds may also reduce the rumen methane production. More fibrous feeds and imbalanced feeds lead to change rumen fermentation pattern and produce more acetate and butyrate which in turn leads to production of more hydrogen and carbon dioxide and thus produces more methane.
6. Ionophores like monensin are found effective in reducing the rumen methane production. This selectively reduces gram positive ruminococci which increases the propionate production in the rumen and thus resulting in the decrease in the CH_4 production (Newbold et al. 1988).
7. Vegetable oil addition to the ration can also reduce the methane production by the ruminants as they are energy rich source and also acts as rumen modifier. Some of the components of oils have antimicrobial activity and antifibrolytic activity. Oil if incorporated more than 7 % of the diet will hamper the fibre degradation and dry matter degradability will be greatly reduced. Beauchemin

and McGinn (2006) reported that adding canola oil at the rate of 4.6 % of DM intake inhibited methane emission by 32 %. This is achieved due to less fibre degradation and also low feed intake by the animals.

8. Use of Anti-Methanogen Vaccine: One of the most promising approaches in the recent years is the use of a vaccine against specific rumen microbes like methanogenic archea to reduce the methane production. The use of vaccines in animals has been conventional for disease prevention or control for many decades, but novel uses such as production increases and methane abatement have only been proposed recently. Vaccination of ruminants against rumen methanogens has the potential to reduce methane emissions by decreasing the number or activity of methanogens in the rumen. The anti-methanogen vaccine works by triggering an animal's immune system to produce antibodies against the methanogens. Animals would be vaccinated to stimulate production of specific salivary antibodies against rumen methanogen and delivery of antibodies to the rumen should result in impairment of methanogen function and their ability to produce methane (Wedlock et al. 2013). Several researches have been done in ruminants especially in sheep as a ruminant model in Australia and New Zealand and found methane reduction of about 23 %, even though this value may be affected by the type of diet, climate and other factors. Williams et al. (2009) reported a vaccine which evoked a substantial production of serum antibody against the methanogens especially the *Methano brevibacter* strains. This vaccine found to have affected the diversity and composition of the methanogen population.

8 Conclusions

It can be concluded that by improving the animal productivity and reducing the numbers of non-productive animals, the environmental impacts of animal production can be minimized, as low producing animals are mainly involved in maximum emission of methane. Reducing the numbers of ruminants will be a difficult and complex task, both politically and socially. Again in the world scenario, India is equipped with a dense population of low producing animals and these animals provide food security to many poor and land less labours. So, in such a condition implementing culling of low producing animals looks very difficult. Again providing high nutritious food by these poor farmers also does not look feasible. Hence, implementing a programme that will reduce the methane production and indirectly global warming is very difficult in Indian context. But providing high nutritious feeds and development of new and effective mitigation strategies to reduce enteric rumen fermentation is today's call for all the scientists.

References

Beauchemin, K. A., & McGinn, S. M. (2006). Methane emissions from beef cattle: Effects of fumaric acid, essential oil, and canola oil. *Journal of Animal Science, 84*, 1489–1496.

Cynoweth, D. P. (1996). Environmental impact of biomethanogenesis. *Environmental Monitoring and Assessment, 42*(1–2), 3–18.

EPA. (2010). Greenhouse gas emissions reporting from the petroleum and natural gas industry. *Background Technical Support Document*. Accessed January 3, 2011 from http://www.epa.gov/climatechange/emissions/downloads10/Subpart-W_TSD.pdf

Food and Agriculture Organisation of the United Nations (FAO). (2007). Adaptation to climate change in agriculture, forestry and fisheries: Perspective, framework and priorities. Rome: FAO. ftp://ftp.fao.org/docrep/fao/009/j9271e/j9271e.pdf

FAOSTAT (FAO). (2013). Food and Agriculture Organization of the United Nations (FAO), Rome. Accessed August 12, 2013 from http://go.nature.com/Z23f7E

Fenwick, D. E., & Oakenfull, D. (1983). Saponin content of food plants and some prepared foods. *Journal of the Science of Food and Agriculture, 34*, 186–191. doi:10.1002/jsfa.2740340212

Gerber, P. J., Steinfeld, H., Henderson, B., Mottet, A., Opio, C., Dijkman, J., et al. (2013). *Tackling climate change through livestock: A global assessment of emissions and mitigation opportunities*. Rome: Food and Agriculture Organization of the United Nations (FAO).

Harvell, C. D., Mitchell, C. E., Ward, J. R., Altizer, S., Dobson, A. P., Ostfeld, R. S., et al. (2002). Climate warming and disease risks for terrestrial and marine biota. *Science, 296*(5576), 2158–2162.

Hopkins, A., & Del Prado, A. (2007). Implications of climate change for grassland in Europe: Impacts, adaptations and mitigation options: A review. *Grass and Forage Science, 62*(2), 118–126.

Huke, R. E., & Huke, E. H. (1997). *Rice area by type of culture: South, Southeast, and East Asia*. Los Baños, Philippines: IRRI. p. 59.

Intergovernmental Panel on Climate Change (IPCC). (2001). Climate change 2001: The scientific basis. In J. T. Houghton et al. (Ed.), *Third Assessment Report from Working Group 1*. UK: Cambridge University Press.

Johnson, K. A., & Johnson, D. E. (1995). Methane emissions from cattle. *Journal of Animal Science, 73*, 2483–2492.

Kannan, A., Garg, M. R., & Singh, P. (2010). Effect of ration balancing on methane emission and milk production in lactating animals under field conditions in Raebareli district of Uttar Pradesh. *Indian Journal of Animal Nutrition, 27*, 103–108.

Kiehl, J. T., & Kevin, E. T. (1997). Earth's annual global mean energy budget. *Bulletin of the American Meteorological Society, 78*(2), 197–208.

Kirchgessner, M., Windisch, W., & Muller, H. L. (1995). Nutritional factors affecting methane production by ruminants. In W. V. Engelhardt, S. Leonhard-Marek, G. Breves, & D. Giesecke (Eds.), *Ruminant physiology: Digestion, metabolism, growth and reproduction* (pp. 333–348). Stuttgart: Ferdinand Enke Verlag.

McAlpine, C. A., Etter, A., Fearnside, P. M., Seabrook, L., & Laurance, W. F. (2009). Increasing world consumption of beef as a driver of regional and global change: A call for policy action based on evidence from Queensland (Australia), Colombia and Brazil. *Global Environmental Change, 19*(1), 21–33.

Moss, A., Jouany, J. P., Newbold, J. (2000). Methane production by ruminants: its contribution to global warming. *Annales de zootechnie, 49*(3), 231–253. doi:10.1051/animres:2000119. <hal-00889894>

Newbold, C. J., Wallace, R. J., Watt, N. D., & Richardson, A. J. (1988). The effect of the novel ionophore tetranosin (ICI 139603) on ruminal microorganisms. *Applied and Environmental Microbiology, 54*, 544–547.

Ripple, W. J., Smith, P., Haberl, H., Montzka, S. A., McAlpine, C., & Boucher, D. H. (2014). Ruminants, climate change and climate policy. *Nature Climate Change, 4*(1), 2–5.

Sakthivel, P. C., Kamra, D. N., Agarwal, N., & Chaudhary, L. C. (2010). Effect of sodium nitrate and nitrate reducing bacteria on in vitro methane production and feed fermentation with buffalo rumen liquor. In *Proceedings of Fourth Greenhouse Gases and Animal Agriculture Conference*, Banff, Canada, p. 178.

Sampath, K. T., & Bhatta, R. (2010). Feeding strategies with special reference to climatic changes for high yielding buffaloes. In *International Buffalo Conference on Optimizing buffalo productivity through conventional and novel technologies*. February 1–4, NASC Complex, New Delhi, pp. 47–54.

Smith, P., Haberl, H., Popp, A., Erb, K. H., Lauk, C., Harper, R., et al. (2013). How much land-based greenhouse gas mitigation can be achieved without compromising food security and environmental goals? *Global Change Biology, 19*(8), 2285–2302.

Steinfeld, H., Gerber, P., Wassenaar, T., Castel, V., Rosales, M., & de Haan, C. (2006). *Livestock's long shadow: Environmental issues and options*. Rome: Food and Agriculture Organization of the United Nations.

Stewart, C. S., Flint, H. J., & Bryant, M. P. (1997). The rumen bacteria. In: P. N. Hobson & C. S. Stewart (Eds.), *The rumen microbial ecosystem* (2nd ed., pp. 10–72). London: Blackie Academic and Professional.

Thornton, P. K., Van de Steeg, J., Notenbaert, A., & Herrero, M. (2009). The impacts of climate change on livestock and livestock systems in developing countries: A review of what we know and what we need to know. *Agricultural Systems, 101*(3), 113–127.

Watson, R. T., Meira Filho, L. G., Sanhueza E., & Janetos, T. (1992). Sources and sinks. In J. T. Houghton, B. A. Callander & S. K. Varney (Eds.), *Climate change* (pp. 25–46). Cambridge: Cambridge University Press.

Wedlock, D. N., Janssen, P. H., Leahy, S. C., Shu, D., & Buddle, B. M. (2013). Progress in the development of vaccines against rumen methanogens. *Animal, 7*(2), 244–252.

Williams, Y. J., Popovski, S., Rea, S. M., Skillman, L. C., Toovey, A. F., Northwood, K. S., et al. (2009). A vaccine against rumen methanogens can alter the composition of archaeal populations. *Applied and Environmental Microbiology, 75*(7), 1860–1866.

Effects of Demographic Change on Environmental Degradation: Evidence and Implications for India

C.M. Lakshmana

Abstract In the recent decade, India has seen remarkable changes in the age structure due to demographic transition. This has further influenced the society and economy in a positive side and however, changing demographic trends and consumption pattern in India has been causing eco-degradation and environmental pollution, and therefore an issue of changing demographics and its effects on environmental degradation is a serious issue in the present pattern of population dynamics. Though there is a currently fair understanding of the ways in which the India's population is changing as well as of the social and economic implications of these changes, little attention has so far been paid to its possible environmental impacts. In this background, this particular paper tries to understand how the changing demographic trends have affected environment in India, and seeks to identify measures both to mitigate environmental damage and increase resource use efficiency.

1 Introduction

Currently the annual average world population growth has been increasing at 1.1 %. It is slightly higher when compared to India. Though, India has 2.4 % of the world's geographical area, it is home to nearly 17 % of the world's population. The demographic trends such as increase in the total population, changing age structure, change in household size, distribution, and size of urban population across various segments have been seriously caused to environment decay, which directly degraded the human welfare, health, and well-being. However, in India, demographic trends are not uniform across the country; the backward states such as Bihar, Madhya Pradesh, Chhattisgarh, Uttar Pradesh, Orissa, and Rajasthan have witnessed above the average growth rate of population while other states are

C.M. Lakshmana (✉)
Institute for Social and Ecnomic Change, Dr. VKRV Rao Road, Nagarabhavi, Bangalore 560072, India
e-mail: lakshmana@isec.ac.in

S. Nautiyal et al. (eds.), *Climate Change Challenge (3C) and Social-Economic-Ecological Interface-Building*, Environmental Science,
DOI 10.1007/978-3-319-31014-5_7

experiencing the growth rate of population relatively low. But, on the other hand, the level of socio-economic development in the above said states is also low when compared to the southern states and western states of India. In this background, measuring environmental pollution in India based on the physical characteristic of a region is deemed appropriate. This paper highlights the linkage of changing demographic trends and its potential environmental impact in India.

2 India's Demographic Change

Rapid decline of fertility in most part of India in the recent decades has significant impact on the changing age structure of population. However, primarily, three important factors have been emerged from the recent demographic change in India. They are: (a) people are living longer (b) there will be more number of households and the average number of people per household will be lower (c) the rate of population growth varies across regions in the country. And further, the demographic character is changing rapidly and the country is now entering the final stage of demographic transition. But due to regional inequality in the progress of socio-economic development especially in some states like Bihar, Rajasthan, Madhya Pradesh, Rajasthan, and Assam were not reaching the replacement level fertility. However, it has evidence in most of the state in southern, western, and northern states of Punjab and Haryana. However, India exhibits one of the highest demographic heterogeneities ever experienced anywhere in the world at the regional and state level. Despite the recent decline in the birth rate in the country, India has recorded a growth rate of 1.6 % per year during 2001–2011 census periods, adding around 181 million people to the total, but the annual addition to the total population has remained nearly the same. However, there is a general consensus that demographic change in India has led to opening up new economic opportunities and has had a greater impact on economy and society (Bloom and Williamson 1997). But, at the same time the demographic change such as the size of population, household numbers, age structure of population, urbanization, etc., are found to be important proximate causes for adverse impact on environment. Additionally, our current patterns of consumption such as the amount of energy and water use in homes, waste generation, changing landscape, and biodiversity loss are going to have a greater impact on the environment than all the above demographic factors.

3 The Changing Age Structure

The increase in life expectancy—an important driver contributing to an increase in population—and the consequent growth in the number of older people are likely to continue in India. This calls for timely action to ward off the social and economic consequences of aging population. For instance, the most advanced country of the United Kingdom has already taken precautionary measures to mitigate the potential impact of environment to anticipate by these demographic trends, particularly the increasing number of older people in total. In this regard, the Royal Commission on Environmental Pollution in the UK has found that, if people have more years of active life that will potentially have a greater environmental impact. Older people and particularly those over the age of 75, on average, contribute disproportionately to carbon emissions from heating their homes, etc. This is because they spend a greater amount of time at home and feel the cold more intensely. Further, there is growing body of evidence to show that the potential environmental impact of residues of pharmaceuticals released into the environment by excretion or disposal particularly by the elderly is greater as they use more medicines than younger people. And now a similar situation is evident in India. For example, as per 2001 census, India had 49.10 million people in 65+ age group which increased to 90 million (7.43 %) in 2011. And by the year 2050, their number is expected to reach 315 million, constituting about 20 % of the total population. Hence, it is considered that the increasing number of older people in the country will be a potential source of environmental effects in the future.

In this regard, an effort has been done to understand the role of fertility and changing age structure population across southern states of India focusing the elderly population in Table 1. The TFR decline was evident in all four southern states of India during the last two decades. Interestingly, a decline of 1.6 in Karnataka has resulted in an increase of about 3 % in the proportion of aging population during 30 years. Surprisingly, the absolute TFR decline was very low in Kerala (0.9), but it has resulted in significant increase in the proportion of aging population in the state with 4.8 %. Tamil Nadu also witnessed a comparatively higher increase in the proportion of aging population (4.7 %) during the period.

Coming to the extent of increase in the share of aging population in the total, across the southern states in the decade 1991–2001, it can be seen that a slight TFR decline of −0.1 % in Kerala and Tamil Nadu had caused a relatively higher in the proportion of aging population. During the same period, the TFR decline was relatively high at −0.4 in both Andhra Pradesh and Karnataka, but increase in the proportion of aging population in the total was not significant, i.e., below 1 % in Karnataka and below 0.5 % in Andhra Pradesh.

In the following decade of 1991–2001, the decline in TFR and resultant change in age structure of the population was substantial; proportion of aged population in the total population increased in all the southern states. Percentage increase in the proportion was relatively higher in both Kerala and Tamil Nadu (1.68 and 2.15 % respectively). Karnataka and Andhra Pradesh also had registered considerable increase in the proportion of aging population in total population in the decade 1991–2001, but the increase in proportion was below one per cent in Karnataka, and about one percent in Andhra Pradesh. Further, if we look at the major changes in the proportion of aging population in the total attributable to TFR decline, it can be seen that TFR decline was almost identical in both the decades of 1991–2001 and 2001–2011 in the respective states. However the effects of TFR decline on the

Table 1 Decadal change of TFR and age structure transition in southern states of India (60+ population)

State	TFR decline				Total increase of age 60+ population			
	In 30 years	Decadal change			In 30 years	Decadal change		
	(1981–2011)	1981–1991	1991–2001	2001–2011	(1981–2011)	1981–1991	1991–2001	2001–2011
Andhra Pradesh	−2.1	−1.3	−0.4	−0.4	2.7	0.38	1.02	1.3
Karnataka	−1.6	−0.8	−0.4	−0.4	3	0.79	0.81	1.4
Kerala	−1.1	−0.8	−0.1	−0.2	4.8	1.32	1.68	1.8
Tamil Nadu	−1.7	−1.3	−0.1	−0.3	4.7	0.95	2.15	1.6

Source (1) Census of India, series-1, paper 5 of 1984, registrar general, government of India
(2) Technical group of population projections, registrar general, and census commissioner of India, 2001–2026

increase of aging population was quite impressive in the recent decade not only in India but also in the southern states. For example, a TFR decline of −0.04 had resulted in an increase in the proportion of aging population by about 1.4 % in Andhra Pradesh and Karnataka. In contrast, a TFR decline of just −0.02 % resulted in a significant increase in the proportion of aging population in Kerala (1.8 %) and Tamil Nadu (1.6 %).

What is clear from the above discussion on change in age structure of population and fertility transition across the southern region in over 30 years is that the total increase in the proportion of aging population in total was comparatively higher in Kerala and Tamil Nadu. It is relatively low in Karnataka and Andhra Pradesh. But finally, the above description is evident that there has been a positive decrease of child population under age 6 and increase of elderly population in the total population is relatively higher across southern states with a continues decline of TFR. This has certainly influenced the environmental degradation having more elderly and working population in future. It means, more is the working population and elderly more is the utilization of energy use and generating huge waste at the household level, which caused land degradation, water and air pollution, which further caused human welfare, health, and well-being.

4 The Changing Household Structure and Energy Use

It is widely assumed that, in general, smaller size households make higher energy use and results in higher CO_2 emissions per head. The data collected and conclusions reached as part of the integrated impact assessment of the London Housing Strategy by The Royal Commission on Environmental Pollution (2011) in the UK has found that larger households consume proportionately fewer resources than the smaller ones, with a three-person household using only about twice (rather than three times) the water used in a one-person household. Hence, household size is one of the factors which contribute the amount of water consumed per head. A similar trend was also observed in India (UN 1999; NCAER 2011). Electricity use, gas use, and waste generation show quantitatively the same trends not only in the UK but also in other countries like US, China, and India (Lakshmana 2013). Although, the changing demographic trends have led to a reduction in the size of family, and nuclear family[1] is the common trend in India.

Reduction in family size naturally increases household income. This means, less number of dependents, and more money available for savings, investments, and consumption. Besides, demographic change has led to the preponderance of young and working-age population in the country. Population in 15–29 age group is

[1]A nuclear family is a family unit that consists of father, mother, and children.

growing at a faster rate than the other age-groups (Census of India 2011). This has led to two important developments: (1) the proportion of those joining the labor force and earn wage/salary has increased many fold; and (2) joint family system[2] is losing ground and nuclear families are gaining ground (Census of India 2011). Therefore, increased per capita expenditure at the household level has been leading to acquisition of household assets like cars, refrigerators, etc., which consume enormous quantities of energy on one hand, and release carbon and other pollutants on the other. Now the burgeoning middle class in Asia, particularly China and India is consuming energy that is more or less equal to the US energy consumption (UN 1999; NCAER, New Delhi 2011). The demand for new housing and related development as a result of demographic change will increasingly come up against environmental constraints in all parts of the country.

As per 2011 census, India has 24.66 million households as against 19.19 million of the previous census of 2001. The overall increase in growth rate of total households during the period of 10 years is 29 % of which urban areas account for over 50 %. Percentage of nuclear families in India increased from 61 % in 2001 to 68 % in 2011. The total reduction in fertility rate of 0.71 % (3.39 % in 1991 and 2.68 % in 2005) has resulted in an increase of about 8 % in the number of nuclear families over 15 years. The reduction in fertility rate is found faster in rural areas than in urban areas. Hence, on the one hand, a reduction of fertility has led to a decline in population growth, but at the same time it has also led to an increase in the nuclear families with fewer members per household. Increasing household income has resulted in the acquisition of electronic goods and personal motor vehicles by individuals. And, the socio-economic transformation in the context of technological change has been accompanied by huge acquisition of television sets, computer, telephone/mobile phone and automobiles not only by urbanites but also by rural people. Interestingly, the percentage of households that do not have such household goods decreased from 34.5 % in 2001 to 17.8 % in 2011. According to the latest (2013) report by the Planning Commission, Government of India, this is exactly equal to the proportion of population below poverty line (BPL). Hence, it confirms that drastic decline of fertility in most part of the country (except perhaps in the backward states) has consequent increase in disposable per capita income and empowered individuals to acquire household assets like automobiles and electronic goods extensively. Therefore, there is a need to address the emerging issues of population and development in general and seek answers to the question whether demographic change in terms of fertility decline is a dividend or a disaster in terms of its adverse affect on environment.

[2]A family includes in one household near relatives in addition to a nuclear family or a family that includes not only parents and children but also other relatives (such as grandparents, aunts and uncles: (An Encyclopaedia, Britannica Company).

5 The Effects of Urbanization

Generally the urban population growth in India was slow during pre-independence; however, there was a gradual increase of urban population during the liberalization period. And as per the latest census of 2011, about 31 % of population in the total is urban. Interestingly this population is mostly accumulating in class I cities. And the class three tier cities are losing the ground. However, during the post-liberalization period, India has witnessed a rural influx into urban areas. This has put tremendous pressure on fertile land and resulted in resource depletion and environmental pollution. The most striking feature of India's urbanization is that it is 'large-city oriented' and almost 70 % of India's urban population is now concentrated in class I cities (Lakshmana 2008). The poor quality of India's urban centers and the consequent haphazard kind of urbanization has been worsened by the burden of this rural influx. The magnitude of environmental degradation in such urban sprawls is alarming. However, increased urban population over the past 30 years is of greater significance in the western region compared with other regions in India. Besides, the west region has the highest number of most populous cities with only 16.76 % of the total land area, and double the proportion of population, i.e., 31.33 % of the total (Table 1). Similarly, the northern region with 14 % of land area had 18 % of population in the same time-span. This means that the most populous cities are located in the western and northern regions, and therefore one could conclude that urban growth and urbanization has led to increased use of natural resources; as a result, environmental pollution could be expected to be higher in the western and northern cities. While the determinants of population growth such as birth rate, death rate, etc., have declined over time, migration seems to have neutralized the expected advantages, as evidenced by the fact that there has not been any decline in the rate of resource use and consequent environmental decay (Table 2).

Table 2 Most populous cities and their population by region in India

Region	No. of cities	Population (in millions)	% age	Population (in millions)	% age	Proportion of urban area to the total urban area	Man-land ratios per km^2
		2001		2011			
North	28	24.61	17.90	31.65	18.03	14.00	2893
Central	46	27.28	19.85	34.62	19.72	23.00	1926
East	32	16.36	11.90	19.93	11.35	10.14	2516
North–East	3	1.25	0.91	1.56	0.89	3.18	629
West	36	39.91	29.04	55.01	31.33	16.76	4200
South	47	30.22	21.99	35.13	20.01	32.92	1365
India	192	137.45	100.00	175.57	100	100.00	2246

Source Compiled by the author from census data

6 Summary and Conclusion

Currently, annual average growth of world population has been growing at the rate 1.1 %. However, India has witnessed for higher than the world population with an increase of 1.7 %. No doubt the population and health policies have certainly helped in reduction of fertility. But even though, the size of population is second highest after China. Increasing working population is caused to have more money and they are going to spend on electronic gadgets and motor vehicles. This has directly influenced to air pollution and the use of high energy. The current patterns of consumption in India have a greater impact on the environment than all the above demographic factors. Therefore, demographic trends not only imply significant social and economic issues but also potential environmental impacts. Therefore, meaningful efforts need to be done to increase resource use efficiency to protect the environment. At the same time, the rapid change in age structure and resultant increase in the number of nuclear families demands for new housing, and the related development will increase and despite inherent environmental constrains in most regions of the country. Hence, remedial measures need to be taken to reduce consumption level and waste generation by individuals and households in order to protect the environment. Until and unless the government gives adequate attention to the implications of demographic change for the environment, the country will not be able to face the challenge of environmental decay and the related human issues in the coming decades. Efforts have been done in India in order to reduce population growth. But the size and quality of population were the major threat to environment in the country at different levels due to regional inequality in terms of geographic, social, and economic variables. However, changing demographics in the country certainly is a ray of hope for the economic development; however it should not be the disaster to the sustainable environment in India.

References

Bloom, D. E., & Williamson, J. G. (1997). *Demographic transition and economic miracles in emerging Asia*. Working paper 6268, Cambridge, M.A.: NBER.

James, K. S. (2011). India's demographic change: Opportunities and challenges. *Science Journal, 333*, 576–580.

Lakshmana, C. M. (2008). Effects of population growth on environmental degradation: With reference to India. *Demography India, 37*(1), 63–79.

National council of applied economic research. (2011). *Report on middle class population*. New Delhi: National Council of Applied Economic Research.

Registrar general and census commissioner of India: Tables on houses, amenities and assets of census 2001 and 2011.

The Royal Commission on Environmental Pollution. (2011). *The 29th report of demographic change and the environment*. London: United Kingdom.

United Nations. (1999). *The landmark report of the world commission on environment and development*. USA: United Nations Development of Economic and Social Affairs (DESA).

Biodiversity Status and Climate Change Scenario in Northeast India

P. Saikia, A. Kumar and M.L. Khan

Abstract Conservation of biodiversity and impacts of climate change are perhaps the most critical challenges faced by all sectors of the society. The patterns and processes related to climate change and biodiversity are so complex that corrective measures are often taken with an imperfect scientific knowledge base. This article addresses the biodiversity status and climate change scenario in Northeast India. The normalized difference vegetation index (NDVI) is used to explain the vegetation pattern of Northeast India and majority of area (58 %) are under tropical climatic zone, followed by subtropical (23 %), alpine (10 %), and temperate (9 %) zones. Although the Northeastern India is very susceptible to climate change because of its ecological fragility, the forest cover is reported increasing in the past few years because of the successful implementation of the different plantation and community development schemes. In order to avoid rapid and irreversible change in biodiversity, conservation strategies needed to focus on supporting the species' natural capacity to adapt to climate change.

Keywords Biodiversity · Climate change · Mitigation · NDVI · Northeast India

1 Introduction

Climate change, global warming, and biodiversity depletion are considered as the most critical challenges for sustainable development worldwide. Environmental conditions in combination with other factors play a key role in spatial distribution of biodiversity. Climate change is recognized as a significant manmade global

P. Saikia · A. Kumar (✉)
School of Natural Resource Management, Central University of Jharkhand, Brambe, Ranchi 835205, Jharkhand, India
e-mail: amit.kumar@cuj.ac.in; amit.iirs@gmail.com

M.L. Khan
Department of Botany, Dr. Hari Singh Gour Central University, Sagar 470003, Madhya Pradesh, India

S. Nautiyal et al. (eds.), *Climate Change Challenge (3C) and Social-Economic-Ecological Interface-Building*, Environmental Science,
DOI 10.1007/978-3-319-31014-5_8

environmental challenge, which has enormous impacts on biodiversity patterns in the past and is seen as having significant contemporary impacts (Sahney et al. 2010). It is predicted that it will remain one of the major drivers of biodiversity patterns in the future also (Sala et al. 2000). A thickening layer of carbon dioxide pollution, mostly from power plants and automobiles that traps heat in the lower atmosphere, is considered as the main reason of climate change. It has a direct impact on biodiversity (Chapin et al. 2000) influencing the reduction in species' diversity (Franco et al. 2006) which ultimately affect the ability of biological systems to support human needs (Vitousek et al. 1997). The industrialization, urbanization, and agricultural intensification has led to a significant change in land use and associated land cover (Kumar et al. 2011), which intensified the pressures on habitats and landscapes and biodiversity in general (Stanners and Bordeaux 1995). The steady decline of habitats and landscapes demonstrates the need for protection, which has also been addressed in Convention on Biological Diversity (CBD) in order to identify and monitor ecosystems, habitats, species, communities, genomes, and genes. Global climatic changes also affect agriculture through their direct and indirect effects on crops, soils, livestock, and pests (Pathak et al. 2012).

Forests have been influencing the gas composition in the atmosphere, which in turn influenced temperatures and weather patterns on Earth (Zachos et al. 2001; Sigman and Boyle 2000). The two important measures of climate change are the variation in precipitation and temperature (Sukumar 2000). Changes in rainfall pattern are likely to lead either to severe water scarcity or flooding and on the other hand, rising temperature with increasing number of days having high temperature accelerates the extinction rate of many species and also cause shifts in crop growing seasons, which ultimately affects food security. Plant growth, flowering, animal reproduction, and migration also depend in part on temperature. Changing environments are expected to lead to changes in species distribution (Lynch and Lande 1993) and life cycle events, which have been recorded for many plant species (Parmesan and Yohe 2003). Phenology is being used as an indicator of species sensitivity to climate change (Bharali and Khan 2012). Plants used to respond to climate change in four possible ways: (a) phenotypic plasticity enabling species survival, with alterations in eco-physiological processes in the changed climate, (b) evolutionary adaptation to new climate, (c) emigration to favorable habitats, and (d) extinction (Bawa and Dayanandan 1998; Saxena and Purohit 1993). Global climate change along with continued habitat loss and fragmentation is now being recognized as a major threat to future biodiversity (Bharali and Khan 2012). Ongoing distributional changes may not necessarily allow the species to persist throughout its range. There is strong evidence that plant species are shifting their ranges in altitude and latitude as a response to changing climatic conditions (Parmesan and Yohe 2003). Human-induced climate change impacts biodiversity and on the other hand, biodiversity reduces the impacts of climate change on the environment. Genetic evidence for *Fagus sylvatica* suggests that populations may show some capacity for an in situ adaptive response to climate change (Jump et al. 2006).

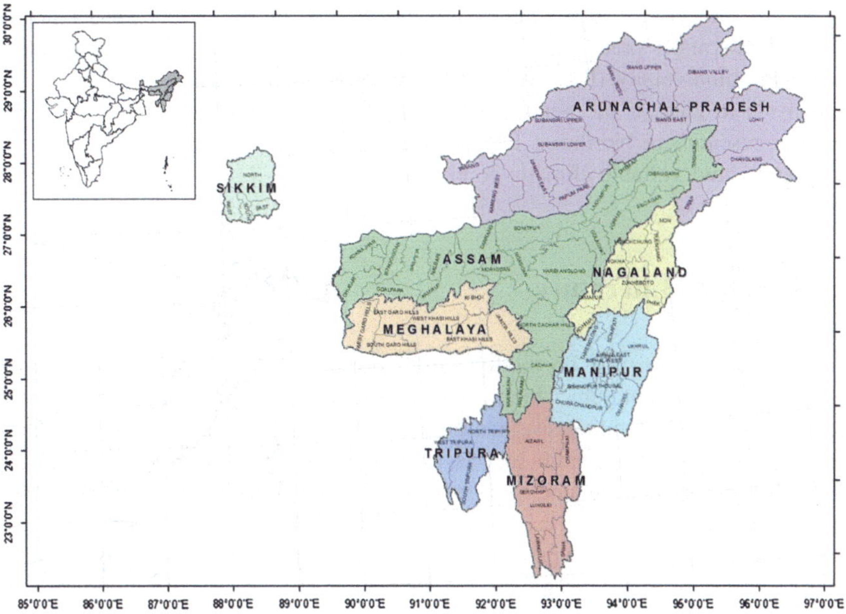

Fig. 1 Map showing the locations of eight Northeastern states of India

The Northeast region of India is consisting of eight states covering a geographic area of 26.2 million hectare (Fig. 1). The Brahmaputra and Barak are the two main river basins of the region. Northeast India has diverse vegetation types encompassing from tropical, subtropical, temperate, sub-montane, montane, subalpine to alpine. The local tribal populations are highly dependent on forest's resources, which provides livelihood to more than 225 tribal groups native to the region. The timber trade, tourism and wildlife resorts, and shifting cultivation in the hills are closely woven with the region's forest wealth. The total forest cover was 54 % of the total area in 1993 (FSI 1995) and increased to 66 % in 2005 (FSI 2008) although doubts have been expressed over official data (Bose 2005). Official reports (FSI 2008) state that forest cover varies from 80.9 % (of the total geographical area) in Arunachal Pradesh to 35 % in Assam, with the other states placed from 76 % in Manipur to 88.6 % in Mizoram. The region is characterized by diverse climatic regimes and is highly dependent on the southwest monsoon. It has two globally recognized biodiversity hotspots and an eco-region renowned for its high species diversity and endemism. According to biodiversity records, Northeast India supports nearly 50 % of the total flowering plants recorded in India, out of which 31.58 % are endemic. It is also recognized as one of the 'centers of origin of cultivated plants.' It is the original locations of over 50 important tropical and subtropical fruits, cereals, and rice (Vavilov 1926; Dhawan 1964; Hore 2005). In addition, out of an estimated 800 species used for food in India, about 300 species occur in Northeast India alone (Rao and Murti 1990). The natural resources of the region are subjected to degradation and loss due to deforestation, unsustainable shifting cultivation practices, fragmentation, and degradation which ultimately impact the biodiversity (Ravindranath et al. 2011). In this article an attempt has been made to address the geoinformatics application in climate change and biodiversity research in Northeast India, a region distinguished globally for its rich biodiversity and variable climatic conditions.

2 Review of Literature

The scattered information on biodiversity and climate change of Northeast India exists in available literature which needs a systematic compilation. The impact of global warming on biodiversity has emerged as an active area in contemporary conservation biology research and is extremely important for Northeast India, where community dependence on forests is very high (Ravindranath et al. 2006). Earth is facing biodiversity crises with very high rates of biodiversity loss and biodiversity change (Duraiappah et al. 2005). Conservation of biodiversity is one of the key issues in terms of global environmental change and its monitoring is necessity as it is strongly related to human well-being. Roy et al. (2012) used remote sensing and GIS data for identification, assessment, and monitoring biodiversity and they considered it as better and contemporary tool because it ensures uniformity in biodiversity data collection. The need to conservation of biodiversity and natural resources in Northeastern states with special reference to Manipur was addressed by Singh et al. (2009). Recent State of Forest Reports point to continued forest losses in Northeast India, including its protected landscapes (FSI 1995, 2005, 2008). Saikia (2009) studied the forest cover changes in Northeast India using normalized difference vegetation index (NDVI) and explained the behavior of NDVI among different land use and land cover changes with reference to rainfall and temperature.

The use of continuous thematic fraction layers, derived from linear unmixing, provides a good basis for monitoring land cover changes. However, gradual and small changes in habitats and their quality are not easily detected from space by satellite imagery, and therefore, additional information from field surveys is needed (Mücher 2011). Singh (1989) used Landsat data for detecting the changes in the forests of Northeastern region of India and considered it as an effective tool for forest cover mapping. On the other hand, Chakraborty (2009) used moderate resolution imaging spectroradiometer (MODIS) to study the change in forest cover of Barak basin, Northeastern part of India. The climate change vulnerability profiles for Northeast India were studied by Ravindranath et al. (2011) and they reported that majority of the districts are subjected to climate-induced vulnerability at present and in the near future. Another important study was done by Bharali and Khan (2012) on impact of climate change on biodiversity and its mitigation in the state of Arunachal Pradesh, which indicates that climate change not only threatens the biodiversity, but also affects the socioeconomic condition of the indigenous people of the state.

The possible causes of climate change in India with satellite measurements were done by Ganguly (2011) and reported population growth accompanied by uncontrolled urbanization and rapid industrialization as the main causes of high levels of pollution and imbalances in the regional climate. Recent research shows that climate change will be even more pronounced in high-elevation mountain ranges, which are warming faster than adjacent lowlands. Hydrological and ecological changes of this magnitude would result in a loss of unique biodiversity, as well as a loss of many of the environmental goods and services provided by these mountains, especially water supply, basin regulation, and associated hydropower potential (The World Bank 2008). Integrated Model to Assess the Global Environment (IMAGE) is a helpful tool for investigating climate change, loss of biodiversity, water scarcity, accelerated nitrogen cycle, and their causes and inter-linkages in a comprehensive framework (Kram and Stehfest 2011). With respect to the altitudinal spectrum, climate change is affecting mountain ecosystems and biodiversity. Glacial retreat in the eastern Himalayas is occurring at an alarming rate, which are likely to result in substantial impacts on water flows to Northeastern (Assam) valleys. At lower mountain altitudes, changes observed include loss of water regulation, increased likelihood of flash fires, and changes in ecosystem composition and resilience. Moreover, as temperatures increase, there is a substantive risk of recurring glacial overflows caused by ice melting, placing large downstream populations and infrastructure at imminent risk. Warming is also affecting the moorlands, high-altitude ecosystems with unique and abundantly diverse flora and fauna that are also a storage area for water and carbon in the soil (The World Bank 2008).

3 Application of Geoinformatics in Biodiversity and Climate Change Research

The rate of species extinction will overtake the rate of biodiversity inventorization and characterization (Chapin et al. 2000). Maximum inaccessible areas of the natural habitats have been yet to be inventorized as most of the biodiversity documentations in our country are concentrated in the areas accessible to the researchers which led to a gap in the biodiversity exploration. Traditional identification and monitoring of biodiversity has many lacunae, some of them serious enough to question the authenticity of the results (Roy et al. 2012). Field sampling provides detailed information but it is quite expensive and time-consuming process. Strategic ground sampling, expert knowledge, and interpretation of remote sensing data can form a reliable, repeatable, and cost-effective analytical framework for accurately assessing the rate of biodiversity change. Remote sensing is well recognized for the integral role it plays in assessment and monitoring of biodiversity and climate-related indicators. Although field surveys provide higher levels of accuracy, remote sensing techniques make it possible to increase the speed and frequency (Arockiaraj et al. 2015). The recent advancement in the fields of remote sensing and geographic information system has enabled accurate and uniform documentation of biodiversity helping in identifying the gap in biodiversity exploration (Roy et al. 2012). Using remote sensing and GIS, vegetation-type mapping, habitat mapping, ecological niche modeling, spatial disturbance regimes and biological richness mapping, etc. can be done which are of special importance in the field of biodiversity research. Researches focusing on remote sensing data maximize the opportunities for ensuring long-term preservation of endangered species, better understanding of how fauna make use of maximum space occupied by vegetation, and to predict the future impacts of climate change and land management on species distributions. Satellite remote sensing has been proved to be the most cost-effective means of mapping and monitoring environmental changes in terms of vegetation and other ecological issues (Deka et al. 2013). Biodiversity monitoring by remote sensing enables us to scale up the understanding and knowledge on biodiversity. Remote sensing brings tremendous possibilities to estimate the climatic and anthropogenic impacts on biodiversity and moreover, to predict its temporal change in the future through ecosystem modeling (Suzuki et al. 2010). Remote sensing will provide us plenty of biodiversity-related information as ecosystem types, their distribution patterns, and information about habitat structure for organisms. Biodiversity can be assessed and interpreted at each level of ecological organization using various approaches at several spatial and temporal scales (Noss and Cooperrider 1994). Monitoring the extent and quality of biodiversity is also required in a more comprehensive fashion across the countryside, ranging from regional to global scales (Mücher 2011). He suggested that operational remote sensing enables land cover characterization at various scales but the classification accuracies are still insufficient at continental and global scales for monitoring purposes.

Climate is one of the most important factors controlling the growth, abundance, survival, and distribution of species as well as regulating natural ecosystems (Faisal 2008). Climate change will affect all natural ecosystems, but the impacts will be more prominent on the already stressed ecosystems of the Northeastern region (ICIMOD 2010). It poses a major challenge to conservation efforts worldwide from habitat shifts (Parmesan 2006) to the threat of new invasive species (Hellmann et al. 2008). Changes in climate have potential direct and indirect affects on individuals, populations, species, ecosystems, and the geographic location of ecological systems which ultimately cause extinction of wildlife, change in phenology, and hatching and immigration of species, disrupted plant communities, species, and ecosystems (Trisurat et al. 2011). Variety of remotely sensed data is being used to monitor and quantify numerous climate change indicators at different scales (temperature/Land Surface Temperature, precipitation, water content in atmosphere, vegetation cover, changing phonology/crop growing patterns, aerosol concentration, etc.). A series of satellite and airborne sensors have been developed to collect thermal behavior of land data from the earth surface, such as HCMM, Landsat TM/ETM+, AVHRR, MODIS, ASTER, and TIMS (Quattrochi and Luvall 1999; Weng et al. 2004). LST modulates the air temperature of the lower layer of urban atmosphere, and is a primary factor in determining surface radiation and energy exchange, the internal climate of buildings, and human comfort in the cities (Voogt and Oke 2003). Important elements such as land cover and land use processes, the global current and historical carbon cycle, the global nitrogen cycle, management of nutrients in agricultural systems, and climate variability including interaction with land use are addressed by Kram and Stehfest (2011) in the integrated modeling of global environmental change.

The normalized difference vegetation index ($NDVI = (\rho_{NIR} - \rho_{RED})/(\rho_{NIR} + \rho_{RED})$) is the most commonly used vegetation index. It quantifies the contrast between red surface reflectance (ρ_{RED}), which decreases with increasing chlorophyll content, and near-infrared surface reflectance (ρ_{NIR}), which increases with growing leaf area index and crown coverage. Atmospheric noise in the NDVI caused by clouds, dust, and aerosols is generally considered negatively biased. This is because additive path radiance causes an increase in red reflectance, while lower atmospheric transmission reduces near-infrared reflectance (Guyot et al. 1989). NDVI has been widely used across diverse biomes as a proxy for rainfall (Barbosa et al. 2006), as a substitute for vegetation growth and health (Schmidt and Karnieli 2000) and in ecological studies (Pettorelli et al. 2005).

4 Vegetation Assessment in Northeast India Using NDVI

In the present study, the IRS P6 (RESOURCESAT-1) AWiFS product is utilized to quantify NDVI of the region during 2010. The IRS P6 AWiFS-based NDVI of the Northeast India exhibits that the majority of area (55 %) having higher NDVI (≥7.0) largely located in eastern Himalayas (Arunachal Pradesh, Nagaland, Manipur, Mizoram, Tripura) and Meghalaya followed by 33 % (NDVI 0.4–0.7) located in the adjoining areas of eastern Himalayas (including Assam). In contrast, only 8 % area is under low positive NDVI (0.1–0.4) mainly located in Assam (Figs. 2a and 3a).

The Cartosat-I digital elevation model (DEM) of Northeast India exhibits that the majority of the total area (48 %) exist in 0–500 m elevation spatially found mainly in Assam, Tripura, western Meghalaya, south Arunachal Pradesh, and western Mizoram, followed by 500–1000 m (17 %) and 1000–1500 m (12 %), whereas only 8.5 % area is above 3000 m elevation spatially found only in northern Arunachal Pradesh and northern Sikkim (Figs. 2b and 3b). Different climatic zones were classified on the basis of elevation, i.e., tropical (0–800 m), subtropical (800–1800 m), temperate (1800–2800 m), and alpine (>2800 m) zones. The NDVI distribution of Northeastern India in four different climatic zones exhibited that

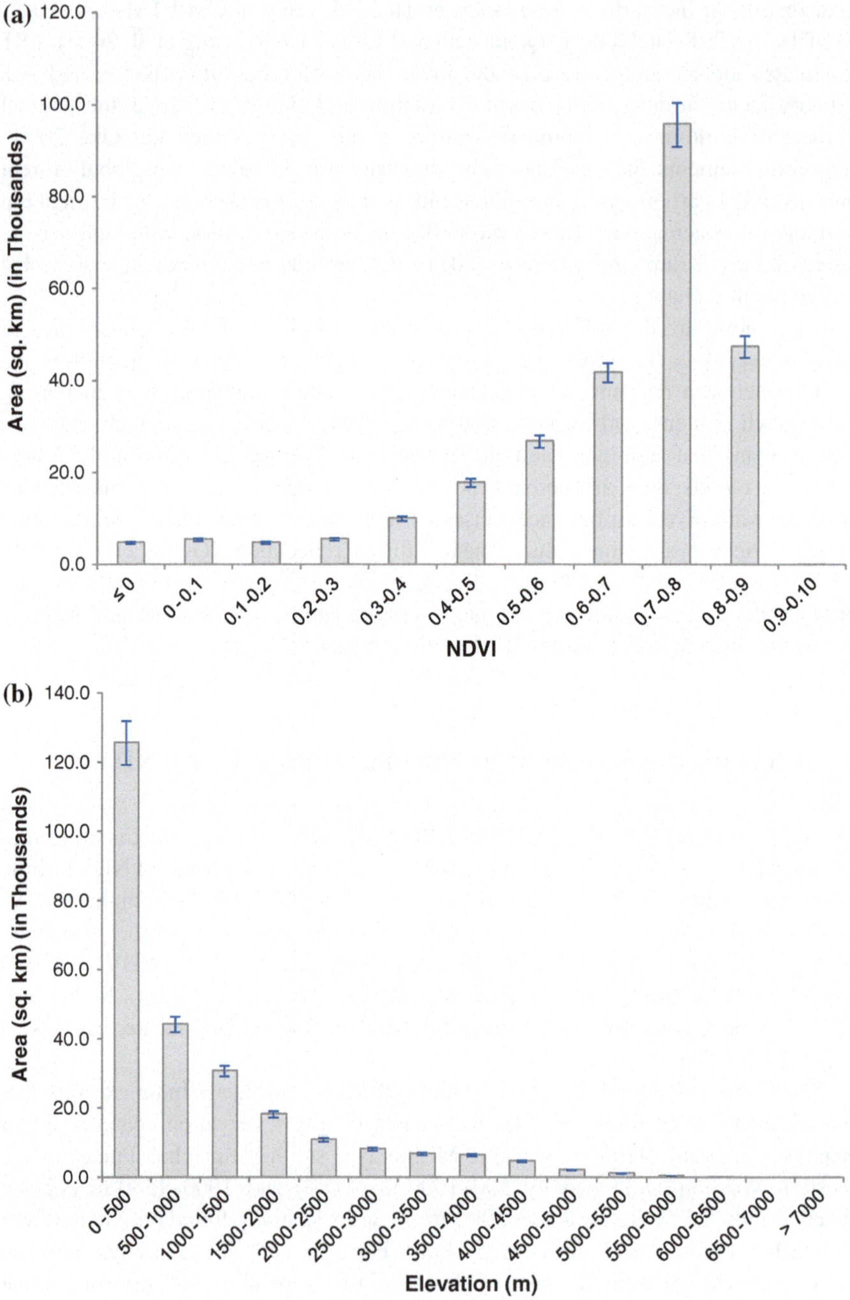

Fig. 2 **a** NDVI distribution and **b** Elevation zones of Northeast India

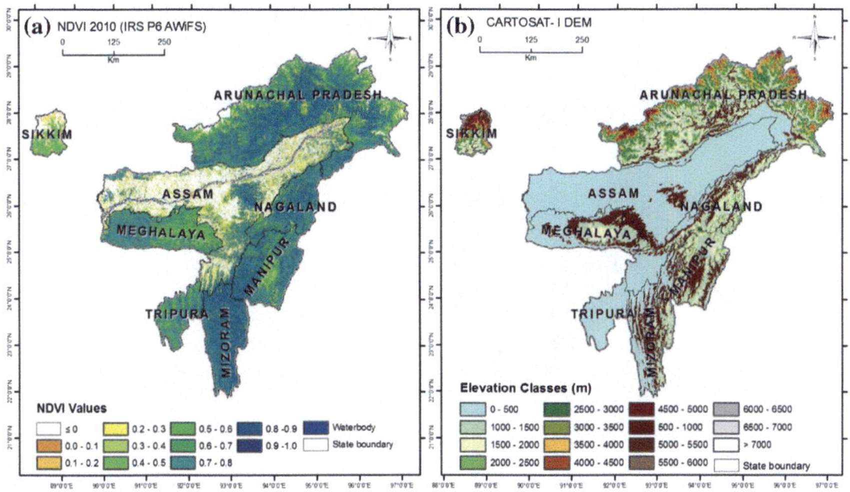

Fig. 3 Spatial distribution of (**a**) NDVI and (**b**) Cartosat-I Digital Elevation Model (DEM) of Northeast India

majority of area (58 %) are under tropical climatic zone, followed by subtropical (23 %), alpine (10 %), and temperate (9 %) zones. NDVI distribution in different climatic zones is shown in Fig. 4.

5 Challenges to Conserve Biodiversity and Climate Change Mitigation of Northeast Region

The exceptional biodiversity in Northeastern region is mainly due to the multiple biogeographic origins and its location at the juncture of two continental plates. It is in an ecotone represented by flora and fauna of both Asian continental shelf and the Indian plate having the Gondwana origin (Roy et al. 2012). The climatic variability associated with the vast, complex, and steep topography is another reason which offers variety of biodiversity in diverse habitats of the region. The region is subject to tremendous threat of landslide, flood, earthquake, as well as human disturbances including roads, agricultural development, forestry activities, deforestation, and energy development which may lead to the loss of biodiversity of different endangered and endemic species of flora and fauna. The threats to biodiversity arising from climate change are likely to be very acute in Northeast region because of ecological fragility, economic marginality, and richness of threatened and endemic species. The vulnerability is very high mainly for ecologically delicate species like lichens, orchids, and insectivorous plants which occupy highly specific and narrow niches.

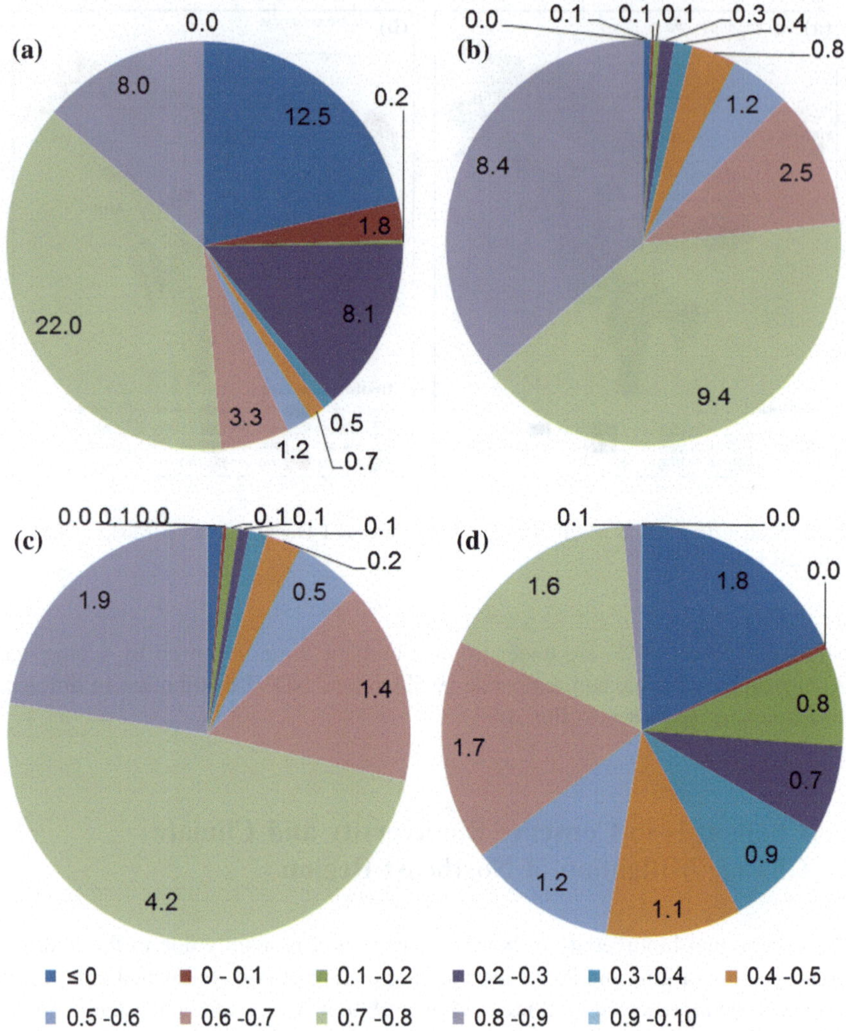

Fig. 4 NDVI distribution in different elevation-based climatic zones **a** tropical, **b** subtropical, **c** temperate and **d** climatic zones in Northeast India

A number of adaptations for conservation of biodiversity have been proposed by several workers, which include assisted colonization (Hoegh-Guldberg et al. 2008), use of invasive species to restore ecosystem services (Hershner and Havens 2008), creation and expansion of protected areas, corridors, and networks (Hannah et al. 2007), and efficient use of resources. Provision of affordable energy; keeping air pollution and climate change under control; management of water systems in support of agriculture, industry, and human settlements; increasing agricultural production; protecting soil, groundwater, and surface water quality; and slowing down and eventually halting further loss of biodiversity are considered as the key elements of sustainable development (Kram and Stehfest 2011).

6 Future Research Prospect

1. There is a need for further research on climate change impact assessment using multiple models as well as multiple approaches for the development of vulnerability profile of the region.
2. Application of geoinformatics in the field of biodiversity requires some new approaches like identification of potential risk zones and habitats to take proper precautions for appropriate conservation measures.
3. The climate change assessment using geoinformatics must consider their consequences with ecosystem structure and functioning so that it could provide more valuable information toward its adaptation and mitigation.
4. A qualitative as well as quantitative approach to assess the impacts of climate change on human health.

7 Conclusion

Biodiversity is intrinsically linked to climate and large-scale changes impacts on the habitats of species within ecosystem. Climatic changes are likely to intense the problems of future food security by exerting pressure on biodiversity. A number of rare and endemic species of flora and fauna found in Northeast India may soon be extinct due to climate change unless conservation efforts are accelerated. For avoiding rapid and irreversible change in biodiversity, conservation strategies needed to focus on supporting the species' natural capacity to adapt to climate change.

Acknowledgements The author would like to thank NRSC Open EO Data Archive (NOEODA) from where the (IRS P6) AWiFS satellite images and CARTOSAT-I DEM datasets were downloaded and used in the present study.

References

Arockiaraj, S., Kumar, A., Hoda, N., & Jeyaseelan, A. T. (2015). Identification and quantification of tree species in open mixed forests using high resolution quickbird satellite imagery. *Journal of Tropical Forestry and Environment, 5*(02), 40–53.

Barbosa, H. A., Huete, A. R., & Baethgen, W. E. (2006). A 20-year study of NDVI variability over the Northeast region of Brazil. *Journal of Arid Environments, 67*, 288–307.

Bawa, K. S., & Dayanandan, S. (1998). Global climate change and tropical forest genetic resources. *Climate Change, 39*, 473–485.
Bharali, S., & Khan, M. L. (2012). Climate change and its impact on biodiversity; some management options for mitigation in Arunachal Pradesh. *Current Science, 101*(7), 855–860.
Bose, R. (2005). In a sorry state. *Down to Earth, 14*(7).
Chakraborty, K. (2009). Vegetation change detection in Barak Basin. *Current Science, 96*, 1236–1242.
Chapin III, F. S., Zavaleta, E. S., Eviner, V. T., Naylor, R. L., Vitousek, P. M., Reynolds, H. L., et al. (2000). Consequences of changing biodiversity. *Nature, 405*, 234–242.
Deka, J., Tripathi, O. P., & Khan, M. L. (2013). Implementation of forest canopy density model to monitor tropical deforestation. *Journal of the Indian Society of Remote Sensing, 41*(2), 469–475.
Dhawan, N. L. (1964). Primitive maize in Sikkim. *Maize Genetics Co-op Newsletter, 38*, 69–70.
Duraiappah, A., Naeem, S., Agardi, T., Ash, N., Cooper, D., & Díaz, S. (Eds.). (2005). *Ecosystems and human well-being: Biodiversity synthesis* (p. 100). Washington, DC: Island Press.
Faisal, A. M. (2008). Climate change and phenology. *New Age*.
Franco, A. M. A., Hill, J. K., Kitschke, C., Collingham, Y. C., Roy, D. B., Fox, R., et al. (2006). Impacts of climate warming and habitat loss on extinctions at species low-latitude range boundaries. *Global Change Biology, 12*, 1545–1553.
FSI. (1995). *State of forest report (1993) forest survey of India*. Dehradun: Ministry of Environment and Forests.
FSI. (2005). *State of forest report (2003) forest survey of India*. Dehradun: Ministry of Environment and Forests.
FSI. (2008). *State of forest report (2005) forest survey of India*. Dehradun: Ministry of Environment and Forests.
Ganguly, N. D. (2011). Investigating the possible causes of climate change in India with satellite measurements. *International Journal of Remote Sensing, 32*(3), 687–700.
Guyot, G., Guyon, D., & Riom, J. (1989). Factors affecting the spectral response of forest canopies: A review. *GeoCarto International, 3*, 3–18.
Hannah, L., Midgley, G., Andelman, S., Araújo, M., Hughes, G., Martinez-Meyer, E., et al. (2007). Protected area needs in a changing climate. *Frontiers in Ecology and the Environment, 5*, 131–138.
Hellmann, J. J., Byers, J. E., Bierwagen, B. G., & Dukes, J. S. (2008). Five potential consequences of climate change for invasive species. *Conservation Biology, 22*(3), 534–543.
Hershner, C., & Havens, K. J. (2008). Managing invasive aquatic plants in a changing system: Strategic consideration of ecosystems services. *Conservation Biology 22*(3), 544–550.
Hoegh-Guldberg, O., Hughes, L., McIntyre, S., Lindenmayer, D. B., Parmesan, C., Possingham, H. P., et al. (2008). Assisted colonization and rapid climate change. *Science 321*(5887), 345–346.
Hore, D. K. (2005). Rice diversity collection, conservation and management in northeastern India. *Genetic Resources and Crop Evolution, 52*, 1129–1140.
ICIMOD, (2010). *Climate change impact and vulnerability in the Eastern Himalayas*. Synthesis Report.
Jump, A. S., Hunt, J. M., Martinez-Izquierdo, J. A., & Peñuelas, J. (2006). Natural selection and climate change: Temperature-linked spatial and temporal trends in gene frequency in *Fagussylvatica. Molecular Ecology, 15*, 3469–3480.
Kram, T., & Stehfest, E. (2011). *Integrated modeling of global environmental change (IMAGE)* (pp. 104–118). Pennsylvania: IGI Global (Chapter 5).
Kumar, A., Pandey, A. C., Hoda, N., & Jeyaseelan, A. T. (2011). Evaluating the long-term urban expansion of Ranchi township, India using geospatial technology. *Journal of Indian Society of Remote Sensing, 39*(2), 213–224.
Lynch, M., & Lande, R. (1993). Evolution and extinction in response to environmental change. In R. B. Huey, P. M. Kareiva, J. G. Kingsolver (Eds.), *Biotic interactions and global change* (pp. 234–250). Sunderland, Mass: Sinauer Associates.

Mücher, C. A. (2011). *Monitoring biodiversity using remote sensing and field surveys* (pp. 78–102). Pennsylvania: IGI Global (Chapter 4).

Noss, R. F., & Cooperrider, A. Y. (1994). *Saving nature's legacy*. Washington, D.C.: Island Press.

Parmesan, C. (2006). Ecological and evolutionary responses to recent climate change. *Annual Review of Ecology Evolution and Systematics, 37*, 637–669.

Parmesan, C., & Yohe, G. (2003). A globally coherent fingerprint of climate change impacts across natural systems. *Nature, 421*(6918), 37–42.

Pathak, H., Aggarwal, P. K., & Singh, S. D. (2012). *Climate change impact, adaptation and mitigation in agriculture: Methodology for assessment and application*. New Delhi: Indian Agricultural Research Institute.

Pettorelli, N., Vik, J. O., Mysterud, A., Gaillard, J., Tucker, C. J., & Stenseth, N. C. (2005). Using the satellite-derived NDVI to assess ecological responses to environmental change. *Trends in Ecology and Evolution, 20*, 503–510.

Quattrochi, D. A., & Luvall, J. C. (1999). Thermal infrared remote sensing for analysis of landscape ecological processes: Methods and applications. *Landscape Ecology, 14*(6), 577–598.

Rao, R. R., & Murti, S. K. (1990). North East India a major center for plants diversity in India. *Indian Journal of Forestry, 13*(3), 214–222.

Ravindranath, N. H., Joshi, N. V., & Sukumar, R. (2006). Impact of climate change on forests in India. *CurrSci 90*(3), 354–361.

Ravindranath, N. H., Murthy, I. K., & Swarnim, S. (2011). Options for forest management for coping with climate change in South Asia. In Lal R., et al. (Eds.), *Climate change and food security in South Asia.* Berlin: Springer.

Roy, P. S., Roy, A., & Karnataka, H. (2012). Contemporary tools for identification, assessment and monitoring biodiversity. *Tropical Ecology, 53*(3), 261–272.

Sahney, S., Benton, M. J., & Falcon-Lang, H. J. (2010). Rainforest collapse triggered Pennsylvanian tetrapod diversification in Euramerica. *Geology, 38*(12), 1079–1082.

Saikia, A. (2009). NDVI variability in North East India. *Scottish Geographical Journal, 125*(2), 195–213.

Sala, O. E., Chapin, F. S., & Armesto, J. J. (2000). Global biodiversity scenarios for the year 2100. *Science, 287*(5459), 1770–1774.

Saxena, K. G., & Purohit, A. N. (1993). Greenhouse effect and Himalayan ecosystems. In P. Narain (Ed.) *Proceedings of the conference on First Agricultural Science Congress-1992*. New delhi: National Academy of agricultural Sciences, Indian Agricultural Research Institute, pp. 83–93.

Schmidt, H., & Karnieli, A. (2000). Remote sensing of the seasonal variability of vegetation in a semi-arid environment. *Journal of Arid Environments, 45*, 43–59.

Sigman, D. M., & Boyle, E. A. (2000). Glacial/interglacial variations in atmospheric carbon dioxide, *Nature, 407*.

Singh, A. (1989). Review article-digital change detection techniques using remotely-sensed data. *International Journal of Remote Sensing, 10*, 989–1003.

Singh, E. J., Singh, N. K. S., & Singh, N. R. (2009). Biodiversity conservation and natural resources in Northeast India—With special reference to Manipur. *NeBIO, 1*(1), 42–47.

Stanners, D., & Bordeaux, P. (Eds.). (1995). *Europe's environment; The Dobris assessment. European Environment Agency, Copenhagen* (676 pp). Luxembourg: Office for Official Publications of the European Communities.

Sukumar, R. (2000). Climate and ecosystem change: What does it mean for biodiversity conservation in India? *Journal of the Indian Institute of Science, 80*, 609–618.

Suzuki, R., Muraoka, H., & Ishii, R. (2010). Remote sensing working group for JBON (2010) Enhance the link between remote sensing and in-situ observation network for biodiversity monitoring. *International Archives of the Photogrammetry, Remote Sensing and Spatial Information Science 38*(8), 178–181.

The World Bank. (2008). *Biodiversity, climate change and adaptation nature-based solutions from the World Bank portfolio*, Washington DC.

Trisurat, Y., Alkemade, R., & Shrestha, R. P. (2011). *Conclusion and recommendations* (pp. 403–413). PA: IGI Global (Chapter 20).

Vavilov, N. I. (1926). Studies on the origin of cultivated plants. *Trudy ByuroPrikl. Bot. 16*, 139–248 (Bull. Appl. Bot. Pl. Breeding, *16*, 1–245).

Vitousek, P. M., Mooney, H. A., Lubchenco, J., & Melillo, J. M. (1997). Human domination of earth's ecosystems. *Science, 277*, 494–499.

Voogt, J. A., & Oke, T. R. (2003). Thermal remote sensing of urban climates. *Remote Sensing of Environment, 86*, 370–384.

Weng, Q., Lu, D., & Schubring, J. (2004). Estimation of land surface temperature-vegetation abundance relationship for urban heat island studies. *Remote Sensing of Environment, 89*(4), 467–483.

Zachos, J. C., Shackleton, N. J., Revenaugh, J. S., Palike, H., & Flower, B. P. (2001). Climatic responses to orbital forcing across the Oligocene-Miocene boundary. *Science, 292*, 274–278.

A Critical Analysis of Law Relating to Biodiversity Conservation and Forest Ecosystem Management in India

Suresh V. Nadagoudar

Biodiversity is extremely complex and dynamic. India is rich in biodiversity but facing lots of challenge in protecting its environment. Main problems with regard to the environment in India are from the use and misuse of biological resources. The development of environment jurisprudence in India is obliged to the Apex Court and the numerous legislations and policies. The approach of the 'Green Bench' has been very appreciable. There is a need for effective implementation of law so that the goals for conservation and sustainable management for ecosystem can be achieved together. This paper describes a critical analysis of law regarding biodiversity conservation and forest ecosystem management in India.

1 Introduction

Biodiversity is a term which is complex by nature as it varies, which no feature on the Earth varies. It includes all living being and microbes which are physically and chemically, whether in the form of solid part on earth or in the form of liquid such as water, ice and water vapour or in the form of gas state. Its very existence is everywhere in the biosphere and occurs on land and water, in every organisms from bacteria to complex plants. Yet it is invaluable.

Though we know that it is invaluable, we are commodifying the same by assessing its value. Preserving and protecting this biosphere has become utmost responsible for all the human beings on earth as without them it is impossible to live on earth. This is where we speak about environment protection.

S.V. Nadagoudar (✉)
Faculty of Law, University Law College, Bangalore, India
e-mail: svnadagoudar@rediffmail.com

S.V. Nadagoudar
P. G. Department of Studies and Research in Law, University Law College, Bangalore University, Jnanabharati, Bangalore 560056, India

S. Nautiyal et al. (eds.), *Climate Change Challenge (3C) and Social-Economic-Ecological Interface-Building*, Environmental Science,
DOI 10.1007/978-3-319-31014-5_9

It is the duty of all nations to protect the environment. India like other developing country is daunted with many challenges of having a developed economy on one hand and on the other preserving and protecting environment. Major environmental problems have resulted in India from the use (and more often misuse) of natural resources base. [1]Biodiversity is hence crucial not only for socio-economic and industrial development of a nation but for maintaining environment stability such as forest, water resources and flora fauna. Forest biodiversity has its significance due to its reduction on the earth. Hence the author likes to concentrate more on legislations and decisions of the Apex Court in this regard.

2 Biodiversity and International Scenario

For the development of civilized society biodiversity cannot be ignored and has been recognized not only in earlier civilization but also by the today's scientific world. Until the early 1980s, the focus was still on wild species of plants and animal protection both at national and international level. The shift came when developing countries' government raised questions on protection of wild life and plant. To answer this, a hot debate took place which resulted in the Convention on Biological Diversity (CBD) which was signed in 1992 at the 1992 UN Conference on Environment and Development (UNCED) in Rio de Janeiro and ratified in 1993.[2] This convention is a comprehensive and a binding agreement covering the use and conservation of biodiversity. The convention aims at implementing strategies for sustainable use and protection of biodiversity, and provides a platform by organizing conferences where parties have a dialogue on issues related to biodiversity. In other words, its objective is to develop national strategies for the conservation and sustainable use[3] of biological diversity. This is the key to sustainable development. The convention reminds decision-makers that natural resources are not infinite and sets out a philosophy of sustainable use.

The convention is a multilateral treaty. Its objects are conservation of biological diversity, sustainable use of its components and fair and equitable sharing of benefits arising from genetic resources. It requires countries to have their own strategies on all the sectors which impact (positive or negative) on biodiversity. The countries are required to bring biodiversity into their mainstream of planning and activities. And in line with this countries have prepared policies and legislations to conserve natural resources.[4] Under this convention, forest biological diversity is a broad term which refers to all the life forms found in the forest areas including plants, animals and

[1]Dr. Madhuri Parikh, 'The Forest Conservation in India and the Role of Indian Supreme Court: A Critical Analysis', IOSR Journal of Humanities And Social Science, vol. 13, Issue 4, pp. 55–56.

[2]Entered into force on 29th December, 1993.

[3]The organizing principle for sustainability is sustainable development, which includes the four interconnected domains: ecology, economics, politics and culture.

[4]Under Article 6 of National Biodiversity Strategies and Action Plans.

micro-organisms that inhabit in the forest and surrounding areas and their associated genetic diversity. Convention on Biological Diversity (CBD) defines forest biodiversity as, "Forest biological diversity means the variability among forest living organisms and the ecological processes of which they are part; this includes diversity in forests within species, between species and of ecosystems and landscapes."

3 Evolution of Forest Law in India

India is home to the largest indigenous population in the world. Forestry constitutes the second largest land use in India after agriculture covering about 641,130 km^2, or 22 % of the total land base. These statistics are also very important, driving home the quantum of forest dependence in our country. An estimated 200 million people depend on forests for at least part of their livelihoods.[5] Forest dwellers, including a high proportion of tribals, are among the poorest and most vulnerable groups in society. Historically, forests in India during the pre-British period were managed by communities living in and around the forests and by people dependent on them for their sustenance and livelihood. The Forest Charter of 1855 was the first attempt by the British Indian government in the direction of forest governance. The Charter of 1855 was followed by the *Indian Forest Act, 1865*,[6] which was amended in 1878 and then again in 1927. It was the *Indian Forest Act, 1878*[7] which radically changed the nature of common property and made it state property. Limited access was provided on reserved forests. After independence, food security for the millions of starving and hungry Indians, industrialization and development activities such as irrigation projects and large hydroelectric power projects were some of the important issues that needed immediate attention. The main objective of forest management those days therefore was to serve the purpose of industry and agriculture. Hence, protection of forest as a biodiversity tool was never thought of.

In the year 1972, at the request of the states, the federal government (6) passed the *Wildlife Protection Act, 1972*.[8] Though not directly related to forests, the Act did have a significant impact on their management and therefore also on people living in forests and dependent on forests for their sustenance and livelihood. Biodiversity is a new term introduced in twentieth century, where a balance has to be made between the protection of forest on one hand and protecting the people who are dependent on the livelihood on forest.

Forests hold the majority of the India's terrestrial species: tropical, temperate and boreal forests offer a diverse set of habitats for plants, animals and micro-

[5] 'State of Forests Report 2009' available at http://www.fsi.nic.in/sfr_2009/executive_summary.pdf.

[6] *Act* No. I of *1865*.

[7] Act No. VII of 1878.

[8] Act No. 53 of 1972.

organisms. However, these biologically rich systems are increasingly threatened, largely as a result of anthropogenic activities.

4 Legislative Control on Conversation of Forest

In ancient India, the *Vedas, Puranas, Upanishads*, and other scriptures of the Hindu religion gave a detailed description of trees, plants and wildlife and their importance to the people. The *Rig Veda* highlighted the potentialities of nature in controlling the climate, increasing fertility and improvement of human life emphasizing on intimate kinship with nature. *Atharva Veda* considered trees as abode of various gods and goddesses. *Yajur Veda* emphasized that the relationship with nature and the animals should not be that of dominion and subjugation but of mutual respect and kindness. Kautalya's Arthashastra, written in Mauryan period, emphasized the need for forest administration. Samrat Ashoka went further, and his Pillar Edicts expressed his view about the welfare of environment and biodiversity. To show that a nature should balance all living things on the earth, Hindu Gods have come in an incarnation where one half is human and the other half is either a bird or an animal.[9] Trees are worshiped as a sacred object.

In the modern world, the same is discussed in the name of biodiversity and sustainable development. Leaning towards sustainability[10] is also social challenge that entitles international and national law urban planning and transport local and individual lifestyles and ethical consumerism. According to the Food and Agricultural Organizations of the United Nations,[11] forest management and conservation are continuously evolving to manage and conserve forests effectively and to address the causes of deforestation and biodiversity loss. India has managed to protect its forest through much legislation and lately after signing biodiversity convention has taken steps to make policies also in this regard. To discuss a few among them are:

1. Indian Forest Act, 1927.
2. Wildlife (Protection) Act, 1972.
3. Environment Protection Act, 1986.
4. Biological Diversity Act, 2002.
5. Forest Rights Act (FRA), 2006.
6. Central and State Policy.

Hence, each law has to be analyzed in view of protection of forest and biodiversity.

[9]Ex. Narasimha avatar, Hanuman, Garuda, etc.

[10]The name sustainability is derived from the Latin *sustinere* (*tenere*, to hold; *sub*, up). *Sustain* can mean "maintain", "support", or "endure".

[11]Available at www.fao.org/docrep/012/i0680e/i0680e.pdf, visited on 21st April, 2015.

4.1 *Indian Forest Act, 1927*

This was the first legislation drafted during the British Regime for the protection of forest.[12] The preamble to the *Indian Forest Act, 1927*[13] states that the object of the law is to consolidate the law relating to forests, the transit of forest produce and the duty that can be levied on timber and other forest produce. The striking feature in the law is the absence of any definition of forest or forest land. This Act provides definitions for the forest produce and includes timber, charcoal, caoutchouc, catechu, wood-oil, resin, natural varnish, bark, lac, mahua flowers, mahua seeds, kuth and myrabolams, and trees and leaves, flowers and fruits, and all other parts or produce not hereinbefore mentioned, of trees, wild animals and skins, tusks, horns, bones, silk, cocoons, honey and wax, and all other parts or produce of animals, and eat, surface soil, rock and minerals. It also defined reserved forest,[14] protected forest[15] and village forest.[16]

4.2 *Wildlife (Protection) Act, 1972*

The Wildlife Act, 1972[17] was passed to protect the wildlife and their habitats. This was in relation to the International conventions[18] signed by India. What worried was that habitat destruction due to agriculture, industries, urbanization and other human activities had led to the erosion of the country's wildlife. Thus, the Act regulated the setting up and control of game parks to be referred to as National Parks and declared many species as protected animals and also provided for stringent punishments for poachers or other persons who killed wild animals. Effectively the Act banned hunting for pleasure or sport. The Act has six schedules which cover the entire gamut of wild life. There is a constitution of wild life

[12]Though the Act was drafted in 1865 it was amended twice and the final Act came in 1927. Before this there were only charters.

[13]Act XVI of 1927.

[14]Under section 20 of the Forest Act, 1927, reserved forest means 'an area mass of land duly notified under the provisions of India Forest Act or the State Forest Acts having full degree of protection'. In Reserved Forests all activities are prohibited unless permitted.

[15]Under section 29 of the Act, protected forest means, 'an area or mass of land notified under the provisions of India Forest Act or the State Forest Acts having limited degree of protection'. In Protected Forests all activities are permitted unless prohibited.

[16]Village forest is defined under section 28 of the Act under which the Government may assign to any village community the rights over a land which may not be a part of a reserved forest for use of the community.

[17]Act No. 56 of 1972.

[18]Convention on International Trade in Endangered Species of Wild Fauna and Flora (CITES), Convention on the Conservation of Migratory Species of Wild Animals, etc.

advisory board under the legislation.[19] Project tiger and project elephant are one of the main features of this legislation. The law provides for creation of protected areas for protection of wild life and also provides for punishment for hunting of specified fauna. Wildlife Crime Control Bureau has been established for control of illegal trade in wildlife, including endangered species. Apart from this, Central Bureau of Investigation (CBI) has been empowered under the Act, to apprehend and prosecute wildlife offenders. The major amendment to the law has been in 2002 where stringent punishment has been imposed for violation of the law.

4.3 Environment Protection Act, 1986

The paradigm shift and the efforts of the legislature and executives efforts have been notable over the past two decades towards including the Principles of Environmental Protection in the Legal Jurisprudence in India, most notably the 46th Amendment to the Constitution of India in 1976 laid down Environmental Protection as a part of the Constitutional Mandate and the enactment of the *Environment Protection Act of 1986.* [20] The Act was drafted to regulate the social relations with regard to and protection of the environment for the present and future generations and protection of human health. According to Rawls, "The destruction of a whole species can be a great evil". [21] Destroying species is like tearing pages out of an unread book, written in a language humans hardly know how to read, about the place where they live.[22]

The salient features of the Act is to confer powers on the Central Government to take all necessary measures for protecting quality of environment, co-ordinate actions of States, officers and other authorities under this Act, Plan and execute a nationwide programme for prevention, control and abatement of environmental pollution and protection of environment in toto. The objects of the Act are providing protection and improve the environment. It empowers the Central Government to establish authorities [under section 3(3)] charged with the mandate of preventing environmental pollution in all its forms and to tackle specific environmental problems that are peculiar to different parts of the country. The Act was last amended in 1991.

---

[19]Section 6 of the Wildlife Act, 1972.

[20]Act No. 29 of 1986.

[21]John Rawls, '*A Theory of Justice*', Harvard University Press, U.S., 1971, p. 512.

[22]Holmes Rolston Ill, 'Duties to Endangered Species', *BioScience*, vol. 35, p. 718.

4.4 The Biological Diversity Act, 2002

The *National Biological Diversity Act*[23] which was passed by the Parliament in 2002, after a process of consultation of stakeholders, provides provisions for regulated access to biological resources by *bonafide* end-users for various purposes including scientific research, commercial activities and sustainable use of non-timber forest produce. The Act provides for conservation of biological diversity, sustainable use of its components and equitable sharing of benefits arising out of the use of biological resources. This Act is implemented through three functional bodies viz., NBA at the national level, State Biodiversity Boards[24] (SBBs) in different states and Biodiversity Management Committees (BMCs) at the level of local community (Panchayat). At the national level, NBA is responsible for decisions pertaining to access and benefit sharing (ABS), approval for access to and transfer of biological resources, results or technology of scientific research to foreign citizens, companies or non-resident Indians and several other matters related to conservation of India's biodiversity.[25]

The Act insists upon appropriate benefit sharing under mutually agreed terms related to access and transfer of biological resources or knowledge occurring in or obtained from India for various purposes. Biodiversity heritage sites have been recognized under the law which should include both wild and domesticated biodiversity and human cultural relations with such biodiversity. It is the duty of the State government under the Act to relocate people who have been depending on their living with the production from forest as forest are now protected area.

4.5 Forest Rights Act (FRA), 2006

The Scheduled Tribes and Other Traditional Forest Dwellers (Recognition of Forest Rights) Act, 2006,[26] is a key piece of forest legislation passed in India. The law is concerned with the rights of forest-dwelling communities to land and other resources who are denied over decades certain benefits and rights as a result of the continuance of colonial forest laws in India. Under this law, the rights of people living in or depending on the area to be declared as a forest or protected area are to be "settled" by a "forest settlement officer." The issue is of crucial importance considering that forest landscapes cover over 23 % of the country and affect the livelihoods of around 200 million citizens, or 20 % of the population in our democratic polity. The basic requirement under the Act is that the officer who is appointed should enquire into the

---

[23]Act no. 18 of 2003.

[24]Authorities established under section 8 of National Biological Diversity Act, 2002.

[25]K. Venkataraman, 'Intellectual Property Right, Traditional Knowledge and Biodiversity of India', *Journal of Intellectual Property Rights*, vol. 13, July 3013, pp. 326–335 at p. 326.

[26]Act No. 2 of 2007.

claims of people to land, minor forest produce, etc., and, in the case of claims found to be valid, to allow them to continue or to extinguish them by paying compensation. The Act has the potential to recognize the diversity of use, access, and conservation practices and traditional knowledge of forest communities that have contributed to the conservation of forests and biodiversity.

4.6 Central and State Policy on Environment and Forest

The mandates of National Forest Policy and National Environment Policy were to recognize the importance of biodiversity and increasing forest productivity by restoring degraded areas. Over the period, a variety of policy measures has been developed. Many of these measures provide opportunities for strengthening documentation and data collection; empowering local communities by recognizing responsibilities, ownerships, rights, and concessions; and creating suitable institutions. India launched its National Forest Policy in 1988. This led to a programme named Joint Forest Management, which proposed that specific villages in association with the forest department will manage specific forest blocks. In particular, the protection of the forests would be the responsibility of the people. The basic objectives that should govern the National Forest Policy are the following: maintenance of environmental stability through preservation and, where necessary, restoration of the ecological balance that has been adversely disturbed by serious depletion of the forests of the country, conserving the natural heritage of the country by preserving the remaining natural forests with the vast variety of flora and fauna, which represent the remarkable biological diversity and genetic resources of the country, checking soil erosion and denudation in the catchment areas of rivers, lakes, reservoirs in the "interest of soil and water conservation, for mitigating floods and droughts and for the retardation of siltation of reservoirs, checking the extension of sand-dunes in the desert areas of Rajasthan and along the coastal tracts, increasing substantially the forest/tree cover in the country through massive afforestation and social forestry programmes, especially on all denuded, degraded and unproductive lands, meeting the requirements of fuel-wood, fodder, minor forest produce and small timber of the rural and tribal populations, increasing the productivity of forests to meet essential national needs, encouraging efficient utilisation of forest produce and maximising substitution of wood and creating a massive people's movement with the involvement of women, for achieving these objectives and to minimise pressure on existing forests".

The principal aim of Forest Policy must be to ensure environmental stability and maintenance of ecological balance including atmospheric equilibrium which are vital for sustenance of all life forms, human, animal and plant.

In 2002, India set up a National Forest Commission to review and assess India's policy and law, its effect on India's forests, its impact of local forest communities, and to make recommendations to achieve sustainable forest and ecological security in India. Its major recommendation consisted destruction of local forest cover;

fodder must reach forest dependent communities on reliable roads and other infrastructure, in all seasons year round, declare ecologically sensitive areas and government should reform regulations and laws that ban felling of trees and transit of wood within India.

National Environment Policy, 2006 was to a response to our national commitment to a clean environment, mandated in the Constitution in Articles 48 A and 51 A (g) and Article 21. National Biodiversity Action in Plan was prepared in 1999 which is based on the principle of National Environment Policy as human beings are the centre of concerns for sustainable development and they are entitled to a healthy and productive life in harmony with nature and cross cutting with biodiversity.

5 Apex Court of India and Protection of Environmental Law Under Sustainable Development

Our Apex Court has been a major contributory towards the environmental jurisprudence in India through a two-pronged approach that is by interpreting the Constitution and laying down dicta to protect the environment and also through innovating the processes of enforcing these protections such that they do not remain empty promises. It has through its verdict made clear that the future of the earth is entirely linked with the sustainable development. The countries have to adopt a visionary approach in consonance with the needs of the man and the earth. There is human threat to air, water and land. One of the first steps taken by the Supreme Court of India was the incorporation of the right to a pollution free environment to water and air for full enjoyment of 'life' in the list of rights guaranteed to an Indian citizen under the expandable vision of Article 21 of the Constitution. This was done by taking the balancing interest approach to the interpretation of the Constitution in the *Subhash Kumar v. State of Bihar.*[27]

Another innovation has been the development of the "Absolute Liability" Principle in the case of *M. C. Mehta v. Union of India*[28] where Justice Bhagwati laid down a stricter principle of law than the principle of strict liability in the sense that all the exceptions to the *Rylands v. Fletcher*[29] rule were not held applicable in this particular principle applicable to enterprises engaged in hazardous activities and the size of the industry determined the amount of compensation payable by it.

The concept of sustainable development has been introduced in the Indian judicial scenario by the judges of the Supreme Court including such international principles in the context of the development that was necessary in the view of the developing economy that India was and to a certain extent still is. In the *Vellore*

[27](1991) 1 SCC 598.

[28](1987) 1 SCC 395.

[29](1868) L.R. 3 H.L. 330.

Citizens Forum v. Union of India[30] whereby the concept of sustainable development was applied for the first time in an Indian the Supreme Court had observed in its judgment that ecological protection and economical development should not necessarily be seen as radically opposite to each other, rather the answer to the balance should lie in sustainable development.

The attempt of the Supreme Court to assign a meaning to the term 'forest' as per the dictionary meaning has seen a spate of interventions in the Court due to its wide ambit. Forest as per the above definition, may include private, common pasture, or cultivable land. In *T N Godavarman vs Union of India*[31] case in the Supreme Court, also known as the "forest case", is an example of the judiciary overstepping its constitutional mandate.

6 Conclusion

The development of environment jurisprudence in India is obliged to the Apex Court and the numerous legislations and policies. The approach of the 'Green Bench' has been very appreciable. If the law, as they are properly implemented, it is not a distance dream to achieve clean India movement and also in row go with sustainable development. But unfortunately this is not happening. Forest is the main concern. Day by day the area which is called as protected is encroached by people by providing untenable reasons. Project tiger, project elephant are ongoing, yet the number of these species is declining. It is a sorry affair to say that they do not enter into our dwelling place, but we have entered into their living area.

7 Suggestions

In view of the above discussion, the author wishes to make certain pertinent suggestions for protection of forest taking into consideration sustainable development and biodiversity:

1. Attempts made by the courts have to be appreciated, but the efforts of the courts can only achieve marginal success unless there is social, political and economic change in the Government as well as of people towards adhering to a model of sustainable development. Hence it is recommended to educate people to respect and conserve forest resource which in turn help the environment to be friendly.
2. The laws drafted should be properly implemented by the authorities and should not heed to the needs of the wealthy and influential people.
3. Adopt lifestyle which shall conserve the forest to its best.

---

[30]AIR 1996 SC 2718.

[31](2011) 7 SCC 338.

Part II
Methodological Issues; Socioeconomic and Vulnerability Assessments

Knowledge-Based Climate Economy: Integrated Sciences, Accelerated Convergence and Knowledge Resources Dynamics

Giridhari Lal Pandit

Submitted on **23rd** June and revised on **20th** September 2015, the article is based on the *key note* lecture the author delivered at the 'International Conference: Climate change and Social-Ecological-Economical Interface-Building: Modelling Approach to Exploring Potential Adaptation Strategies for Bio-Resource Conservation and Livelihood Development' (20–21st May 2015) held at the ISEC, Bangalore. The author thanks, *first*, an anonymous referee for receiving comments, triggering this revised version, and, *second*, all those who were present during the key note lecture, particularly the Director, ISEC: Bangalore, Prof. M.R. Narayana, distinguished colleagues Prof. Ruediger Schaldach, Prof. K.V. Raju, Prof. Sunil Nautiyal as well as the distinguished participants, including the many young scientists, who had all come to the International Conference from far and wide—from India, Germany and elsewhere. The article is a part of a larger research project (i) articulating and analyzing strategic multidisciplinarity as a method of knowledge resources dynamics; (ii) exploring the new possibilities/horizons of disciplinary interface-building on the one hand and enhancing knowledge environments through clustering of research institutions, on the other; *and* (iii) exploring the negative feedback that thought experiments and scenario-building models can bring to bear on policy development and implementation within a KBCE.

The concept of *knowledge resources dynamics* is for the first time introduced by the author in Pandit (2015b, c) and in Pandit and Meusburger (2015).

G.L. Pandit (✉)
Centre for Ecological Economics and Natural Resources (CEENR), Institute for Social and Economic Change (ISEC), Dr. V.K.R.V. Rao Road, Nagarabhavi, Bangalore 560 072, India
e-mail: giripandit@yahoo.co.in

G.L. Pandit
Methodology of Science and Ecological Economics Research Programme, University of Delhi South Campus, New Delhi 110021, India

G.L. Pandit
Faculty of Physics and Astronomy, Institut fuer Theoretische Physik, University of Heidelberg, Heidelberg, Germany

G.L. Pandit
Institutional Human Ethics Committee, Institute of Genomics and Integrative Biology, Council of Scientific and Industrial Research (Government of India), Delhi University North Campus, Mall Road, Delhi 110007, India

G.L. Pandit
Mitglied: Internationale Ernst Cassirer-Gesellschaft E. V., Hamburg, Germany

S. Nautiyal et al. (eds.), *Climate Change Challenge (3C) and Social-Economic-Ecological Interface-Building*, Environmental Science, DOI 10.1007/978-3-319-31014-5_10

Abstract This article explores the question *how* rationally and wisely humanity might in the coming decades hammer out solutions to global problems. In particular, how, it might *rethink* the challenges the global knowledge-based climate economy (KBCE) is currently confronted with, and, how crucial it is for the sciences to heed the methodology and strategy of knowledge resources dynamics (Pandit and Meusburger "University and knowledge environment: A study in knowledge resources dynamics", 2015), a newly introduced research tool, in order to accelerate the scientific response to these challenges. If equitable growth and reduced climate change risks are to go together hand-in-hand, the building of the new KBCE with a political-economical framework for economic growth that is also able to tackle the climate change risks is imperative. We can no more avoid discussing how convergence, and more importantly, the knowledge resources dynamics might interface with the KBCE, to develop efficient sustainable solutions to the problems confronting society and humanity at the beginning of the twenty-first century.

Keywords Knowledge resources dynamics (KRDs) · Knowledge-based climate economy (KBCE) · Global climate model (GCM) · Global value chains (GVCs) · Green house gases emissions (GHGs emissions) · Risk society · Ecology of knowledge · Environmental nesting · Strategic interface-building and multidisciplinarity · Rethink-tanks · Carbon sink capacity · Culture of wisdom

A *rethinking* is imperative as science is increasingly being seen as part of the complex problems that we are belatedly addressing now (Pandit 2013). However, it can be argued that science is also a *part* of the solution, if not *the* solution, that humanity must hammer out for the next half a century and beyond. In other words, the global challenges that confront the humanity at many frontiers today should be seen, *first*, as a warning against the fragmented knowledge of the universe that science produces without being able to prevent its abuse, and, *second*, as offering science an opportunity to 'self-heal' through dynamic interface-building. Thus, they should be seen as offering an opportunity to science, society, political economy and state to move forwards, fostering path-breaking interface-building linkages between biodiversity and ecosystem resilience, on the one hand, and between research, discovery, innovation and entrepreneurship, on the other. Just think of the *first two* revolutions in the life sciences (Alan 1978) that *started* with the discovery of the structure of DNA in 1953 and genomics in the 1980s that lead to the mapping of the human genetic blueprint in 2000s,[1] *in turn* blueprinting a host of innovations. There

[1]When Watson and Crick (1953; Allen 1978, pp. 187–249) discovered the double helical structure of the DNA, deducing how in cell division genetic information is transmitted, it revolutionized the life sciences, extending vertically Darwin's theory of evolution and Mendel's discovery of the transmission of the genes. This *first* revolution in life sciences was followed, 10 years later, by the discovery of the steps in the flow of information from the DNA to proteins through the intermediate of RNA, by the French biologists Jacob, Lwoff and Monod. The Nobel Prize in Physiology or Medicine 1965, that was awarded jointly to François Jacob, André Lwoff and

is now a high expectation of a *third* revolution in the life sciences that will have to build on the first two revolutions, blueprinting the future innovations (Sharp 19 Dec. 2014, p. 1470; http://www.nap.edu/catalog/12764.html, vii).[2]

It is important, however, to remember that any technological solutions triggered by the scenario of life sciences *convergence*, that may prove effective and efficient in meeting the global challenges in the *short-term*, will be only a *part* of the solution, invariably *accompanied* by much unintended baggage of *unintended consequences*. The other part of the solution must, therefore, come from the *long-term* perspective of a *culture of wisdom*, the *endless frontier* that includes accelerated knowledge resources dynamics, enhanced public understanding of science, large-scale lifestyle changes in society, the drivers of moral progress *and* ecology-society-economy-technology-entrepreneurship strategic interface-building at the most challenging frontiers.

1 Climate Change Agenda: Knowledge-Based Climate Economy, Risk Society, Rethink-Tanks

> 'How can a fundamental understanding of living systems reduce uncertainty about the future of life on earth, improve human health and welfare, and lead to the wise stewardship of our planet? (http://www.nap.edu/catalog/12764.html, vii)'

In the ringing of the climate change warning bells, so loudly and clearly being heard for the past few decades, may be seen an opportunity for the various sciences and engineering to increasingly *converge* on a common frontier. As regards the question, whether it is possible for the sciences to go even beyond *convergence*, I want to suggest in what follows, *first*, that it is imperative to go even beyond *re-integration on way to convergence*. No doubt, convergence on common frontiers is fast emerging as a promising scenario for strategic action in the life sciences' horizon. It promises to benefit not just the climate modelling techniques within ecological economics and

(Footnote 1 continued)

Jacques Monod "*for their discoveries concerning genetic control of enzyme and virus synthesis*", together with the elucidation of the genetic code, established how information in DNA is transferred within the cell to synthesize cellular proteins and thus other components. In the next decade, scientists wanting to look still deeper into the secrets of the gene investigated biology at the molecular level, triggering three advances in the mid-1970s that changed society: recombinant DNA, DNA sequencing, and chemical synthesis of DNA. The *second* revolution in the life sciences, the sequencing of the human genome that began in the late 1980s and early 1990s, was completed in the early 2000s, with the draft sequence in 2001 and the complete sequence in 2003. This *second* revolution has led to several innovations, among them the insights into the genetic causes of cancer and other chronic diseases and the development of inexpensive and rapid technology to sequence DNA (Sharp, 19 Dec. 2014; http://www.nap.edu/catalog/12764.html).

[2]A New Biology for the twenty-first century: Ensuring the United States Leads the Coming Biology Revolution http://www.nap.edu/catalog/12764.html Committee on a New Biology for the twenty-first century: Board on Life Sciences Division on Earth and Life Studies.

ecosystem resilience research institutions but the *rethinking* of the linkages between discovery, innovation and entrepreneurship. *Second*, it is equally imperative to build a knowledge-based climate economy (KBCE) within a framework for economic growth that is also able to manage climate change risks. In a nutshell, it is necessary to go beyond convergence in order to be able to choose and shape KBCE where equitable growth and reduced climate change risk go together hand-in-hand. *Third*, what the 'ecosystems' of research institutions and the reformed universities of tomorrow need most urgently is a strategic multidisciplinary interface-building where they can lay the foundations for meeting the global challenges—from the shaping of climate change agenda to managing the climate change mitigation and adaptation strategies. And, *fourth*, the multidisciplinary interface-building can be best advanced and understood through knowledge resources dynamics, a new framework that builds on the assumption that in dealing with global challenges we have got to understand the huge complexities of KBCE (turn to Fig. 1, p. 149).

Within a growing KBCE[3] that is driven by improvements in efficiency in resource use, strategic investments in infrastructure and strong linkages between discovery, innovation and entrepreneurship it is possible to work with two kinds of strategies, namely, mitigation and adaptation in order to manage climate change. The *mitigation strategies* mainly address the issues of how to reduce the climate risk by reducing the GHGs emissions without compromising economic growth, equity and poverty reduction. Essentially addressing the issues of resilience to climate change, *adaptation strategies* are generally described as processes of making improvements in resilience with which the whole communities and populations on the one hand and ecosystems on the other can adapt to global warming driven climate change. Among the key questions that arise in this context the following deserve a special mention:

> The first question concerns high-resolution regional climate change projections: How accurate are regional projections of climate change derived from downscaling of GCM projections of future climate change? What can realistically be discovered about climate change at spatial scales smaller than GCM grid spacing (Hall 19 Dec. 2014, p. 1461)?
>
> The second question concerns the context-dependent knowledge resources dynamics: How can KRDs interface with the development and growth models within the KBCE projected over next 50 years, where strategies of equitable growth and reduced climate risk may interplay in harmony, so as to innovate and develop efficient sustainable solutions to the problems confronting society and humanity in the twenty first century? Will there be fundamental limitations to their interface-building strategies, that originate in the knowledge resources themselves?

[3]Launched in September 2003, The Global Commission on the Economy and Climate (www.newclimateeconomy.net) is a major international initiative to examine the economic benefits and costs of acting on climate change, with the New Climate Economy as its flagship project. Set up to examine whether it is possible to achieve lasting economic growth while also tackling the risks of climate change, its reports seek to inform economic policy and decision-makers *how* actions to promote economic prosperity—jobs, competitiveness and poverty reduction—can go together with the mitigation and adaptation strategies of managing the climate change risks.

Beyond these, there arise far deeper questions that need to be addressed seriously even before answering the above two questions. In terms of the proliferating think-tanks, that shape the management of climate change and economic growth agenda, the last 25 years have seen lots of global initiatives, the United Nations Millennium Ecosystem Assessment (MEA: www.maweb.org, www.ecosystemvaluation.org, www.ecosystemservicesproject.org, www.ecosystemmarketplace.com, www.naturalcapitalproject.org), launched in 2000, being one of them (MEA Report 2006).

I think that in order to sail through the twenty-first century, what the world needs most urgently are *rethink-tanks*, if I may use this term. The *rethink-tanks* are imperative for taking a fresh and critical look at the global challenges that remain largely unarticulated and unresolved at the vastly deeper levels of their complexity and dynamics, the proliferating international initiatives notwithstanding. Since the finiteness of the planet's natural resources puts stringent limits to expansion of economic activity, i.e. economic growth and societal prosperity, the *rethink-tanks* will have to rethink, *first of all*, prosperity and freedom themselves. *Second*, the rethink-tanks will have to rethink the strategic management of *finite* natural resources in terms of promoting *trade-offs* between different ecosystem services, between prosperity and green growth or between prosperity and freedom (Pandit 2015b). *Third*, the research communities interested in building and testing the models of sustainable development will have to *rethink* growth and development scenarios of the future by giving up some of the cherished assumptions regarding the nature of economic activity and economic growth themselves. We must rethink the very nature of interaction between development and natural environment within a KBCE. *Fourth*, and finally, it is imperative to *rethink* climate change mitigation strategies not only in terms of drastic reductions in economic activities causing anthropogenic GHGs emissions, such as the burning of fossil fuels, but in terms of investments in enhancing the carbon-sink-capacity of the planet's ecosystems such as the forests and Oceans (Pandit 2015b).

But how do we go about this task of Earth stewardship? In a world that is going to be populated with over nine billion people in the next fifty years, demography is, evidently, a most crucial subject for deliberation not only for the educationists, educational institutions and educational systems but for economists and ecologists as well. We must keep in mind that overpopulation and overconsumption are the top two drivers of the current global biodiversity crisis. Yet demography remains a most neglected subject in the discussions that focus on the humanity's future in the context of the global challenges like global warming driven climate change. In particular, world's overpopulated democracies are faced with the following formidable problem: For any number of reasons, they cannot feed everybody well. The number of children born every single minute or every single day far exceeds their capacity to assure them heath care, nourishment, quality of life and well-being, from birth to the cradle and from the cradle to the school (Pandit 2014b, 2015a). And that has the adverse consequence that the state and society must reduce 'freedom' of large numbers of the population. That, in its turn, has the further consequence that there will be much less democracy left against what is being promised by the constitution and by the state and its mechanisms everywhere. As I have argued elsewhere (Pandit 2014b), even when there is a single individual who is not free in a country or in a society because she must remain hungry and undernourished, then no one is free in that country. I think that it is high time that the developing countries of the South Asian region and other regions adopt policies that can control the population explosion *without coercion* so as to reduce the *risks* that today's *risk societies* are additionally prone to due to poverty, population explosion, and overconsumption and rising inequality.

Moreover, by the end of the twenty-first century, scientific knowledge resources will *certainly* top the kinds of resource that humanity will need to keep up its commitment to the goals of (i) sustainable development for one and all, irrespective of the myth and reality of human rights in our deeply troubled world and (ii) environmental justice for nature's diverse ecosystems in distress that host biodiversity. In all likelihood, then, faced with a choice to opt for either of the *two* kinds of scenario roughly described below, humanity may still have time left to make its choice.

(1) The first kind of scenario relates to the kind of society that can be called *risk society*, based on market-mediated relations and driven by technological totalitarianism, where individuals and societies are made *totally* dependent on technology for *everything*, and where the top species will no longer be the humans, but machines; *and*
(2) The second kind of scenario captures a world of science *and* technology-based knowledge society driven by wisdom inquiry or by wisdom enterprises, where humanity is not yet totally dependent on technology and where individuals and societies are not yet overwhelmed by the market-mediated relations.

2 Emergence of Ecology of Subjective and Objective Knowledge

> 'What began as esoteric explorations of the workings of the physical world – the nature of electromagnetism and the atomic structure of matter, for example – became, in the hands of inventors and innovators, telecommunications, new drugs, medical imaging and devices, nuclear power, the computer chip, and the Internet. There are now more connected mobile devices than there are people on the planet'- Susan Hockfield, MIT President Emerita and Professor of Neuroscience (Davos 22. Jan. 2015).

By way of a word of caution, in this article I will avoid any direct discussion of the complexities of 'knowledge' and 'information' as *tradable commodities* within a knowledge-based economy. This is not to deny their role in economics of prosperity and poverty reduction in terms of value creation (Machlup 1962, 1980; Allen 1990; Pandit 2015b), to quote Clemenz (2007, p. 1):

(i) In advanced economies the information sector contributes around 50 % to gross domestic product, and it has been the fastest growing sector over the last twenty years.
(ii) Research, development and innovations are main contributors to economic growth and explain up to 50 % of the increase of labour productivity.
(iii) Accounting for information poses a major challenge to economic theory in almost all of its branches.

Yet I think that it is not helpful at all if we use the terms 'knowledge' and 'information' synonymously, as Clemenz, and many other scientists, are prone to do, spelling out the following research agenda:

> 'It has to be acknowledged that there is no such thing as a commodity "knowledge" which can be treated as homogeneous at least to the same degree as cars or computers. "Knowledge" is created in various forms, under various conditions and for various purposes, and it looks impossible to determine its properties and its handling in markets or other institutions without specifying the context. So in some sense we are back at the old and unsolved epistemological question what knowledge is. However, economic considerations along the lines presented above should be helpful. We should look at the type of knowledge we are considering, and Machlup's classification is probably a good starting point. We should look at the technical conditions under which knowledge is produced, e.g. the extent to which it can be modularized, and we should look at the ways in which it can be utilized, commercially and otherwise. And finally, we should keep in mind that economics provides a useful way to look at things, but it needs to be supplemented by the insights of other disciplines as well (Clemenz 2007, p. 9)'.

Let me make a digression by drawing attention to the classical Plato-Descartes-tradition (Pandit 1982, 1989, 1991, 2010a, 2012a). According to this tradition, *all* knowledge can be represented, even if metaphorically, by a single tree symbolizing *its* unity within the dualism of the knowing *subject* and the *objec*t of knowledge. In the sense of subjective knowledge, all knowledge claims share the same standard form within the Cartesian formalization as follows:

> 'S knows/claims to know that p', where 'S' symbolizes the knowing subject, whose personal identity, gender, interests, human situation, variable contexts and lifeworlds are dogmatically assumed to be without a bearing on how we define 'knowledge'; and 'p' symbolizes the individual proposition stating a relatively simple matter of fact.

With the universally identifiable knowing subject as its frame of reference, epistemology in the classical Plato-Descartes-tradition takes off as an inquiry that aims at analyzing the structure and properties of *all* knowledge claims that are *assumed* to be reducible to this standard form. Not only does this tradition make the *knowing subject* the owner of all kinds of knowledge, *invariably* it implies a denial of the existence of *objective knowledge* (Popper 1972; Pandit 1982, 1991). However, in the twentieth century, the methodologist of science Popper (1934) became the first to challenge this classical tradition that *reduces* all knowledge to the subjective states in the mind. Popper had the advantage of learning from Einstein's special and general theory of relativity (Popper 1934). These theories not only *displaced* the Newtonian universe, they *replaced* an entire corpse of scientific *beliefs* corresponding to the well-established nineteenth century view that Newtonian physics is the final and unchallengeable truth. Popper argued that there is a far more important form of knowledge, namely, objective knowledge exemplified by the *objectively formulated falsifiable* theories in physics and astronomy (Popper 1972; Pandit and Dosch 2013; Pandit 1982, 1991). Most important of all, a physical theory is scientific *only if* it is falsifiable (Popper 1934). In the next step that Popper took he argued for a radical alternative to the classical Plato-Descartes epistemology, namely, an *epistemology without a knowing subject* that studies objective knowledge (Popper 1972). The objective knowledge exists in the World**3**, argued Popper. Thus, he distinguished the World**3** of objective knowledge, *first*, from the subjective World**2** of minds, consciousness and beliefs, and *second*, from the World**1** of physical objects, processes and laws that, in fact, embeds the Earth's biodiverse world of wonderful ecosystems (Popper 1972).

How does ecology of knowledge, as I propose here to introduce it, emerge as a new discipline on the intellectual horizon? It emerges as an inquiry into the diversity of contexts/forms of *knowledge* whether understood in a subjective or an objective sense. In either case, knowledge must be distinguishable from 'information' and 'interpersonal communication' that are far wider in scope. Within the framework of these finer distinctions, it is now possible not only to turn Descartes' tree of knowledge *around* but to bring the knowing subject with varied interests and human situations *back in*. I have here in my mind the possibility of adopting the standpoint of ecology of knowledge. One immediate consequence of doing so is that morality and moral self-development assume priority over our roles as members of epistemic communities, i.e. as producers of knowledge and creators of knowledge environments. This, in its turn, makes us more responsible as problem-posing and problem-solving agents in whatever we want to create or produce in order to secure a world in which we can live with freedom and dignity. Moreover, instead of waiting for one and the same 'tree' of knowledge until it bears fruit, we can now think of different 'trees', representing diverse forms of knowledge and interpersonal communication. Some of these forms can be discovered in interpersonal relations as these develop in the child's knowledge of his/her mother as also in the mother's knowledge of her child, *in* friendship between persons and in our knowledge of the world shaped by our personal experiences. Think of what children can know about their mothers or what mother's can know about their children, and think of the adult's knowledge of the world. In short, in what follows, I want to propose that ecology of knowledge includes in its study a whole diversity of forms of knowledge that can be studied as part of knowledge resources dynamics (Pandit and Meusburger 2015). What is true of knowledge applies to interpersonal communication generally. By arguing and requiring that all scientific knowledge must be identified in terms of its developmental context (Pandit 1989, 2010a, 2012a; Pandit and Dosch 2013), we can expect the structure of knowledge to vary according to the variable contexts of its development. In the context of my argument in this article, I shall redefine 'knowledge' as a context-dependent resource in the following sense (Pandit and Dosch 2013).

3 Debating the Nature of Objective Knowledge, the Resource of All Resources

If we embrace the view that the goals of knowledge inquiry are better served only if we regard knowledge as a context-dependent resource rather than context-independent unchangeable truth, then one can strongly argue for its wise production and use for purposes of helping humanity to cope with many of the challenges that it faces now and will face in the future (Pandit and Dosch 2013).

If we decide to define knowledge as a resource that can be produced, transferred and utilized to generate more resources of various kinds, it may be asked what makes knowledge resources different from other kinds of resource including information, and the human and natural resources. It is reasonable to answer this question by saying that *knowledge is the resource of all resources*. This can be articulated by showing how it takes care of the following important constraints (i–ix):

(i) as a resource, knowledge is not only context-dependent but subject to change according to its own dynamics (Popper 1972, 1975; Pandit and Dosch 2013; Pandit 1982, 1991, 2010a, b, c, 2012a, b);
(ii) not every kind of resource is knowledge: not *all* kinds of resources that are accessible to mankind can be called as knowledge;
(iii) information, a resource, can become part of knowledge *only if* it is incorporated as an element in a theory-problem driven knowledge system;
(iv) knowledge gives us humans *orientation* when we are dealing with one another and with nature and natural resources in our environment (Pandit 1991; Meusburger 2015);
(v) knowledge change is a function of knowledge resources dynamics, scientific revolutions, interface-building among multiple sciences at the emergent frontiers and enhanced knowledge environments;
(vi) as a rule, it will be possible for society and humanity to design and adopt healthy and wise policies with regard to the natural and human resources *only if* the various epistemic communities—namely, researchers, scientists, policy makers—find it possible to produce and employ knowledge resources *wisely* in appropriate practical contexts and at the emergent frontiers;
(vii) knowledge environments, i.e. universities, institutes of research, organizations and firms are necessary for accelerating the knowledge resources dynamics;
(viii) as a matter of policy, a KBCE is expected to use knowledge as *the resource of all resources*; and
(ix) what sustainability research is to policy in the context of our dependence on natural and human resources, *wisdom inquiry* is to policy in the contexts of production, transfer and utilization of knowledge taken as the *resource of all resources.*

No surprise if in today's information society, it is becoming increasingly urgent to ask the following question: How is knowledge creation and utilization different from information processing, say within a firm or an organization? Earlier, I have already suggested that these *two* cannot be taken as synonymous, particularly in the context of globalization, in the context of our knowledge of markets for knowledge (Clemenz 2007), and in the context of science. Markets for knowledge deal with knowledge as a public good, as a resource with a dynamics that is highly context-dependent. Context-specific knowledge is not as homogeneous as the philosophical schools of epistemology debating the nature of knowledge over past thousands of years have assumed it to be. It must be made clear that it is the assumption of context-independence of knowledge, on which most epistemologies of the past have been founded, which is responsible for the *divide* between philosophy and the sciences on those very issues that are today at the forefront, becoming so crucial to grasping the very idea of knowledge resources dynamics, a new research tool I am introducing here (Pandit and Meusburger 2015).

As if knowledge was something that could be acquired or produced independent of the cultural, social and economic interactive contexts, it is the philosophers who are 'famous' for asking the following celebrated question: *What* is knowledge? Different philosophical traditions have for thousands of years argued for context-independent boundaries of knowledge, while attempting to define it by distinguishing it sharply and decisively from (systems of) belief or bodies of information. With very few exceptions, this is also true of the dominant traditions of contemporary schools of philosophy across the globe. This kind of situation makes the just posed question "How is information different from knowledge?" more complicated and challenging. Should we, then, say that this is not a correct way to asking questions concerning knowledge? In the context of our search for knowledge, both information and communication tend to function analogous to infrastructure, as the roads, railways, airways and their sign posts do across the globe, across different geographical areas and regions, facilitating local and regional interaction. However, like infrastructure, the flow of information always comes at a price and requires enabling environments in which local interactions among interested people can take place. More importantly, unlike information and belief, scientific theories and problems—taken as the developmental structures/resources of scientific knowledge (Pandit 1982, 1989, 1991, 2010a, b, c, 2012a, b; Pandit and Dosch 2013)—do not necessarily have to *reside* in somebody's head or mind. On the other hand, a belief without an owner, like information without a receiver, makes no sense at all.

Most important of all, therefore, it is imperative not only to distinguish between explicit and tacit knowledge but to articulate the concept of knowledge in two major dimensions and contexts. *First*, in the context of scientific discovery of the laws of the physical universe, we can speak of context-independence of knowledge in the following sense. Once knowledge is produced within a theory-problem interactive system, its existence does not depend on the individual knowing subject or an epistemic community, e.g. in an institute of research or a university. The books, libraries and discussions embody knowledge in this sense, quite independent of the individual users of books and libraries. Popper was the first among the methodologists of science to capture this kind of knowledge in his formalization by introducing his conception of objectivistic epistemology *without* the knowing subject (Popper 1972; Pandit 2012a; Pandit and Dosch 2013). As we have noted above, Popper argued for the existence of the World3 of objective knowledge, but we will not discuss the details here. *Second*, in the fast changing social-ecological-political-economical-environmental contexts, where newly emerging frontiers—biodiversity, land use, urbanization and energy systems—pose a challenge to public understanding of science and to established knowledge and practise, knowledge has to be strongly context-dependent. This must be understood in the newly emerging context of KBCE that is replacing the one based on industrial society. In the KBCE, the emphasis is on the value of intangible dimensions of products and services, since wealth and economic growth are "driven primarily by intangible (intellectual) assets" (Lev 2000, p. 1). No surprise, if the intangibles and intellectual capital (RICARDIS 2006) are building up an interface between the world of science, corporate and business worlds, technocrats, governments, regulators and enterprises (Drucker 1993; Hussi 2004).

With the emergence of the KBCE, context-specific knowledge increasingly plays an important dynamic role of a valuable resource. In this sense, knowledge resources cannot only be generated, acquired, transferred, combined and used in process/product innovation to fulfil the societal need. They can also be deployed to find solutions to pressing global problems. It is no surprise if in today's globalized world the crucial managerial question to ask is as follows:

> Given its knowledge resources, how an organization – a firm, an industry, an enterprise, a state or an economy - can go up the regional/global value-chains (GVCs), creating more and more value and generating economic growth?

From such a strategic perspective, the context-specific knowledge resources that have the potential to create value are defined as intellectual capital. Thus, it is primarily the intangible knowledge resources that give an organization, a firm or an economy, a sustainable competitive advantage. As a corollary thereof, 'identifying, measuring and managing intellectual capital is crucial for corporate innovation and competitiveness'.

As a next step, therefore, strategic knowledge resources management for creating value and staying competitive in the changing contexts of KBCE assumes significance. To cite an example, India's participation in the GVCs, while rising, is evidently much lower than that of many other countries (Gasiorek 2015) including smaller countries. While there are challenges and opportunity for India to enhance her participation, many experts think that there exist no 'ecosystems' in India for her to go up the GVCs in particular segments, if we may use the term 'ecosystem' rather very vaguely as referring to industrial 'ecosystems' and 'competitiveness', i.e. to supportive basic production chains that could be used as a trigger for enhancing India's going up the GVCs sector to sector.

Going by their role in the recent period, GVCs are best conceptualized as main drivers of globalization. Individual countries can enhance their participation in the

GVCs that are in turn seen as drivers of globalization and enhanced participation in global KBCE. If a country's participation in GVCs is lower in particular sectors than that of others, smaller or big, then that country's participation in globalization, and by implication in global KBCE, is also lower. Thus, participation in GVCs can be used as a measure of individual participation in globalization and global KBCE.

4 Knowledge Resources in the European Union's Lisbon Strategy

> 'Physicists gave engineers the electron and they created the IT revolution. Biologists gave engineers the gene and together they will create the future'. - Susan Hockfield, MIT President Emerita and Professor of Neuroscience

At the Lisbon Summit 2000 the European Union set herself the goal of transforming the EU by 2010 into 'the most competitive and dynamic knowledge based economy in the world capable of sustainable economic growth with more and better jobs and greater social cohesion'. It is remarkable that at the dawn of the twenty-first century, just 15 years ago, this declaration set the scenario-building for knowledge-based economy converging on the common frontier of sustainable future for the EU. One could also say that it set the scenario-building for knowledge-based global economy where the less developed economies of the world stand to benefit from trading with the EU. Knowledge-based economy is an economy in which knowledge is the most important input factor, irrespective of the challenging frontiers. In a knowledge-based economy, information has value in the interaction with human capital. This includes local interactions that build up enabling environments for people to use information to solve complex problems and adapt to change. The European Council of March 2005 relaunched the Lisbon strategy by refocusing on growth and employment in Europe. The main goal has been to strengthen competitiveness by investing, above all, in knowledge, innovation and human capital, in order to raise the potential for economic growth. Knowledge, accumulated through investment in R&D, innovation and education, is a key driver for long-term growth. Policies aimed at increasing the investment in knowledge and strengthening the innovation capacity of the EU economy are at the heart of the Lisbon strategy for growth and employment (RICARDIS, p. 18).

In contrast to the conventional economic theory, the new economic theory for the knowledge economy is developed in and for the knowledge era. It is especially characterized by the law of increasing returns.

> 'The economy is rapidly becoming a global marketplace characterized by fierce competition, increasing consumer demands and the need for value added products and services. The only way for enterprises to survive in this Knowledge-based Economy is to differentiate themselves by continuous innovation, in order to improve their processes, products, services, networks and reputation. Enterprises innovate in many different ways, ranging from *technological product innovation*, based on new knowledge generated by in-house basic research and via innovation by applied research, to *marketing innovation*, based on existing

models and concepts. Innovation may involve the generation of knowledge that is new to the world through R&D activities and it can be based on existing knowledge that is only new to the enterprise. R&D can be done in-house, or alternatively the results of R&D can be in-sourced from specialized Research and Technology Organizations or Universities (RICARDIS, p. 18)'.[4]

As was noted earlier, as and when the economists engage in a discourse on knowledge economy, they generally tend to take knowledge resources to be the same as information resources (Clemens 2007). However, they need to be reminded of the fundamental question that needs to be taken seriously but is hardly ever asked.

What is it that *can* sustain all that which sustains humanity and the planet's bio-diverse world?

As to the humanity, in final analysis, there are three things that are important. *First*, it is the ecology of knowledge, interpersonal interactions and freedom to trade that sustain a society, an economy and the world as we keep making and remaking it. *Second*, as to the life in general, it is the universal interconnectedness across nature's ecosystems, and across the universe, that sustains earth's biodiversity across these systems. And *third*, at a deeper level, it is the complex dynamic processes of *environmental nesting* (Pandit 1995, 1999, 2000, 2001a, b, 2006, 2007a, b, 2012b, 2013; Schwarzburger 1998; Hebeisen 1997, p. 154), going on across and deep within nature's ecosystems, in which humanity is embedded, that sustain the biodiverse world including us, the *Homo sapiens*. Environmental *nesting* among species involves both ecological and evolutionary complexities that shape their interactions with other species in the food web. Knowledge of both kinds of complexity is necessary for predicting how the species will respond to environmental change including land use change and climate change (Harmon et al. 2009, p. 1347).

Knowledge resources are indispensable for policy designing, for wise investment in sustainable development, for organizing economy and society in the most sustainable mode. They are necessary for developing innovative multidisciplinary research frontiers and for developing technology, and for building up interfaces across the disciplinary walls and boundaries. Context-dependent knowledge resources are necessary to making it possible for society and humanity to adopt healthy and wise policies with regard to natural resources. If creation of knowledge resources and information processing were the same, as many people think, there would be no need for investment in science, in basic research or in R&D. Therefore, the question that we should be asking is this: What is the structure and dynamics of (scientific) knowledge (Pandit 1982)? More than information in the ordinary sense,

---

[4]Knowledge plays a vital role in all these various types of innovation, which is why the Lisbon European Council in March 2000 has set the objective of making Europe the most dynamic knowledge-based economy in the world. The central role of knowledge was restated by the high level group on the Lisbon strategy, which recommended the realization of the knowledge-based economy as the top priority for the EU (RICARDIS, p. 18).

knowledge requires enabling environments that may be called knowledge environments. In order that the dynamics of knowledge resources (knowledge resources dynamics or KRDs) take off—i.e. in order to be developed, transferred and utilized as a valuable resource for sustainable development and for generating further resources in varied contexts, beyond knowledge itself, there must exist the enabling knowledge environments of local interaction in the sense of Sharp (19 Dec. 2014, p. 1469).

5 The Convergence Model on the Life Sciences Horizon

> 'Universities and research institutions are engines of innovation because they are the major source of discovery, new technology, and scientifically trained people. Technology companies arise on the boundaries of universities for many reasons: Translation of the science is accelerated by the recruitment of recently trained person and by consultation with academic scientists involved in discovery and invention; the literal proximity of basic and applied scientists with entrepreneurs fosters conversations and collaborations, accelerating the translation process (Sharp 19 Dec. 2014, p. 1469)'. This is illustrated by the Kendall Square Cluster around MIT in Cambridge, Massachusetts, and other clusters around the campuses in California, Ohio, and many other campuses: The 'immediate proximity of research institutions, high-tech companies, and venture capital highlights the fact that even in this age of the internet and rapid transport, local interactions are still vital. Leaders across the world recognize this and are establishing research institutes to stimulate local innovation (Sharp 19 Dec. 2014, p. 1469)'.

No doubt, the scientific advances of the past centuries, and the technologies, innovations and market-mediated luxurious urban lifestyles based on them have contributed to prosperity, economic growth and development, regionally and globally. But, in their turn, all of these have contributed to anthropogenic global warming and climate change. No surprise, if the growth economists working with consumerist assumptions tend to believe that it is the growth triggered by the ever increasing linkages of discovery, technology and innovation that will facilitate the *great escape* of hundreds of millions of people from the pull of poverty. Yet, the past scientific and technological advancements remain at the centre of controversy regarding the problems of living the global community is addressing *now* to secure the next steps for a dignified living and well-being for one and all in the future. The global problems of today build up the single greatest challenging frontier for humanity as these past advancements and the luxurious urban lifestyles based on them have contributed greatly to these challenges. Just think of the fast disappearing Earth's green cover as society races for accelerated urbanization with flourishing luxurious lifestyles, the anthropogenic global warming-driven climate change, the challenging climate change mitigation, sustainable food production and sustainable development, the explosive growth of global population on a finite planet with finite resources, inequity in education, the problem of improvement of health care, renewable energy frontier, poverty reduction, biodiversity loss and loss of ecosystem resilience. No surprise, if post-Industrial Revolution, science and technology are, quite rightly, increasingly being recognized as an essential part of the complex problems of climate change due to the accelerated domestication of the planet's ecosystems (Pandit 2013; Marean 2015) and increasing concentration of GHGs (mainly carbon dioxide, methane and nitrous oxide) in its atmosphere.

However, the good news is that science already is, or is soon going to be, in a state of self-repair, a state of self-healing, where it has to work through convergence on the life sciences horizon, also being described as the *third revolution*. Accelerated convergence on common frontiers may facilitate this process of self-healing, bringing newer perspectives to bear on possible solutions to global problems. Advocating 'A New Biology for the 21st Century' *as* a convergence of life sciences with engineering, physical, mathematical and computational sciences, the Nobel prize winning molecular biologist Phillip A. Sharp (Sharp 19 Dec. 2014, p. 1470)', spoke of continued investments in basic research at their interface as an *imperative* of future change. In the convergence model is seen a promise of meeting a host of future challenges (Sharp 19 Dec. 2014, p. 1470; http://www.nap.edu/catalog/12764.html, vii), envisaging a '… re-integration of the many sub-disciplines of biology, and the integration into biology of physicists, chemists, computer scientists, engineers, and mathematicians to create a research community with the capacity to tackle a broad range of scientific and societal problems'. These challenges include the challenge of *beneficially* impacting innovation and entrepreneurship and bringing a *newly reintegrated science* to bear on society. It is not surprising if in this context Sharp urged the scientists to 'continue to talk about the link between research and innovation, and economic and global need (Sharp 19 Dec. 2014)'.

No doubt, there is great merit in the convergence model. Convergence in the life sciences' horizon is seen as full of promise as regards the frontier of modelling regional and global environmental changes, namely, the changes taking place in the socio-ecosystems such as the Arctic Ocean. Let us take a concrete example. Think of the application of DNA sequencing to analyzing the range of microorganisms in healthy habitats and environments, on the one hand, and in Earth's ecosystems in distress, on the other. We are here thinking of ecosystem complexity as a function of organisms, of species diversity and of *environmental nesting* within the ecosystem. Since these are the very foundation of the food chain within and across the ecosystems, any 'shifts in their population will eventually drive change through the surrounding environment (Sharp 19 Dec. 2014, p. 1471)'. To take another very simple example, there would have been no Aral Sea catastrophe in Central Asia, had a knowledge-based policy and governance been in place as part of the development strategy that had aimed at regional economic prosperity but that landed the local population in the catastrophic collapse of the Aral Sea ecosystem itself (Pandit 2013). In general, this is true of all the Earth's ecosystems that are undergoing environmental change due to correlated anthropogenic global warming and habitat-specific developmental activity. This becomes very clear when we look at today's rapidly changing ecosystems such as the Arctic Ocean. The Arctic Ocean is a large, complex and highly dynamic socio-ecological system that is undergoing environmental change beyond the environmental threshold, transforming it 'from a perpetually ice-covered region to a seasonally ice-free area within the next few decades' (Berkman and Young 2009, p. 339).[5] The newly emerging convergence model in the life sciences horizon is expected to re-integrate the life sciences with the physical sciences and engineering, *promising* to help in developing an effective strategy of knowledge-based governance for the ecosystems in distress, such as the Arctic Ocean. Assuming that the real challenge is as simple as that, we must not fail to ask the question: *Where will the humanity go from here*?

6 Creating Knowledge-Environments for Accelerated Knowledge Resources Dynamics

But is convergence enough? If we consider the underlying complexities of global problems, the answer is in the negative. Beyond convergence, there is apparently no last step but an *endless frontier* that we may identify with knowledge resources dynamics. Creating the knowledge environments for knowledge resources dynamics is an imperative, at least in the author's view. The question what kind of resource is knowledge, briefly considered above, raises a further question: Is it a public good in so far as public goods are neither excludable nor rival (Mankiw 2004)? According to Mankiw (2004, p. 225) the answer is that '... people cannot be prevented from using a public good, and one person's use of a public good does not reduce another person's ability to use it'. This becomes even more evident if we look at the universities and research institutions as knowledge environments that foster local interactions most conducive to the creation of knowledge resources. Yet there are contextual complexities inherent in the concept of knowledge that makes any task of articulating it independent of contextual diversities an impossible task. In agreement with Clemenz (2007, p. 9) we can say that 'Knowledge is created in various forms, under various conditions and for various purposes, and it looks impossible to determine its properties and its handling in markets or other institutions without specifying the context'.

In today's KBCE, no doubt, the knowledge-discovery-innovation linkages play a crucial role as an input in the engines of economic growth. For humanity's survival, it is imperative to understand first how the accelerated domestication of nature (Pandit 2013), accelerated urbanization, increasing commodification of ecosystem services by 'creative destruction' (Joseph Schumpeter) and destruction of the planet's carbon sinks adversely change the Earth's ecosystems beyond recognition, aggravating global warming, climate change, biodiversity loss and land degradation. At a deeper level, in the context of the destruction that the Earth's ecosystems have suffered, the loss of knowledge, loss of a public good, has far reaching multiple implications. Destruction of ecosystem resilience, e.g.

[5]See Arctic Climate Impact Assessment (2004).

threatens the loss of nature's valuable resources. But it also threatens the loss of knowledge as a dynamic intangible resource, since the destruction of ecosystem resilience destroys the very possibility of studying the complexities and *nested dependencies* that sustain ecosystem resilience. Nothing short of a land change science, as part of multidisciplinary environmental-change-and-sustainability-research, is needed to understand the coupled human-environment systems, taken in their fragility as well as dynamics. There is also an urgent need to articulate the concept of knowledge as a dynamic resource. While analyzing the strategic multidisciplinarity as part of knowledge resources dynamics, as part of the enriched knowledge environments in which knowledge is generated and knowledge change and innovation are engineered, it is necessary to explore the new possibilities/horizons of disciplinary interface-building on the one hand and knowledge-enhancing *local interactions* across research institutions on the other.

Moreover, there is an urgent need for new research tools within the framework of knowledge resources dynamics. First, such tools should help us in studying how national economies have accelerated the process of moving from their industrial economic base, shifting towards a knowledge base, in which wealth creation is associated with the ability to develop and manage dynamic knowledge resources. Second, they should help us in studying the challenges of development arising within the framework of knowledge-based national economies and national and international institutions that have produced various Intellectual Capital (IC) frameworks and guidelines to guide in the management, measurement and reporting of IC. And third, they should help us in understanding better the dynamic linkages between discoveries, knowledge generation, knowledge change, innovation and entrepreneurship (turn to the knowledge resources dynamics Fig. 1).

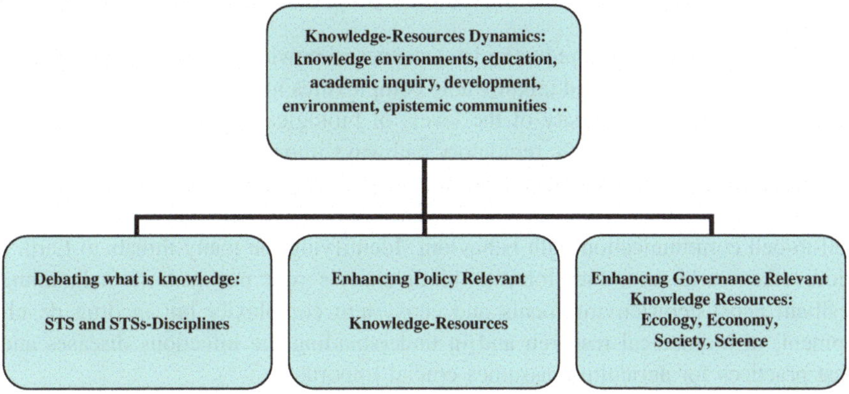

Fig. 1 *Knowledge resources dynamics* as a new multidisciplinary field of research and application (Pandit and Meusburger 2015)

7 Emerging Frontiers of Problem and Theory-Driven Science: Modelling Biodiversity-Ecosystem Linkages

> 'Earth is presently undergoing a major loss of biodiversity that is often referred to as sixth extinction by analogy to the five great extinctions preceding it… This loss began thousands of years ago as humans became sufficiently organized to dominate and change their environment. With the growth of this organization the loss of biodiversity accelerated to reach an extraordinary level at the beginning of the twenty-first century. Humanity is now so dominant that the destruction of small biomes such as wetlands through drainage and mangrove swamps by dredging is commonplace. However, some of the biomes now threatened are so large that their destruction may cause substantial local, and possibly global, climate change as exemplified by the huge Indonesian rainforest fires of 1997 (Lincoln 2006, p. 20)'.

Knowledge resources dynamics, a newly introduced research tool (Pandit and Meusburger 2015), focuses on the strategic interface-building among many sciences to accelerate the scientific response to the challenges humanity is confronted with at the newly emerging frontiers. At best, knowledge resources dynamics takes off horizontally. It is about how best the different disciplinary sciences, on the one hand, and the relevant branches in the humanities, on the other hand, might create a wider knowledge environment for enhancing their capacity to find long-term wise solutions to global problems and challenges at the newly emerging frontiers. The local interactions, among the various scientific groups, scientific-minded humanities groups and epistemic communities as well as among the institutions of scientific research, can serve as an engine of knowledge resources dynamics. In this sense, as a methodological framework, knowledge resources dynamics is more about building innovative strategic interfaces horizontally, that includes frontier-crossing and breakdown of disciplinary boundaries. The knowledge resources dynamics may not always be *necessarily* driven by the discoveries within a particular science. On the contrary, it can serve humanity best if it is driven by ethics and wisdom inquiry, as I will indicate below.

Unlike the individual reductionist sciences, knowledge resources dynamics recognizes the unknown and unknowable complexities in Earth's biodiverse world. Just think across the diversity of the levels of biological complexity: e.g. *how* in molecular biology the genetic regulatory pathways impact the health of organisms; *how* in ecology environmental changes impact the gene pool of an ecosystem across the species diversity (Marean 2015); and *how* neuroscientists *correlate* the cell-to-cell communication with behaviour. Identifying the many threats to Earth's biodiverse world and, therefore, to biodiversity's role not just in maintaining resilient ecosystems/environments and ecosystem complexity but in drug development, in biomedical research and in understanding the infectious diseases and best practices for agriculture assumes crucial importance.

One might ask *why* have the STS, including the philosophy and methodology of science, failed in their most important yet most neglected task of enhancing the *public understanding* of the challenges of stabilizing climate change and decarbonising the global economy. The centrality of public understanding of science to the challenging task of deployment of scalable low carbon technologies by 2050–2100 cannot be denied. This task itself calls for innovative energy-technology research and development (Galiana and Green 3 Dec. 2009, pp. 570–571). In order to *rethink* knowledge, science, discovery, innovation and entrepreneurship and to rebuild their optimal linkages for meeting the twenty-first century global and regional challenges, the rationalities of convergence and knowledge resources dynamics warrant a fresh look at the STS including the philosophy and methodology of science. Rebuilding the just alluded linkages in their role in addressing the global challenges is bound to result in the breakdown of traditional disciplinary boundaries across the sciences (Bigirimana 2011; Srinivasan 2010; Pandit 1989, 2014a, b; Pandit and Dosch 2013). In this context, it is imperative to accelerate the convergence of different sciences on *common frontiers* for achieving the goals of innovation-driven KBCE, on the one hand, and management of knowledge resources within the institutions that are engaged in the task of addressing the challenging frontiers at the society-political economy-industry interface, on the other. But how do the two models, convergence and knowledge resources dynamics, then compete with each other? Or, are they in some sense complimentary to each other?

In deciding how rational and wise it is to move, *first*, from the life sciences' *first* and *second* revolutions to convergence and, *then*, from there to the knowledge resources dynamics-driven KBCE, the steps indicated here, if considered at many levels of complexity, warrant a closer and critical attention. Where regional climate change projections are derivable from the GCM projections of future climate change profiles, using the appropriate techniques (Hall 19 Dec. 2014), this is all the more necessary at a more empirical level. One of the biggest challenges that the reintegrated sciences must address is the rethinking of ecosystem resilience in the context of KBCE. Quite explicitly, the KBCE accepts climate change as the single dominant driver of social-ecological-economical change. As a result, it is imperative to take into account global and regional climate change impact on natural resources, biodiversity and ecosystem services (Côté and Darling 2010, pp. 1–5).

In its Preface, the 2009 National Academy of Sciences **112** pages Report entitled "A New Biology for the 21st Century: Ensuring the United States Leads the Coming Biology Revolution (http://www.nap.edu/catalog/12764.html)" declares that

> 'Biological research is in the midst of a revolutionary change due to the integration of powerful technologies along with new concepts and methods derived from inclusion of physical sciences, mathematics, computational sciences, and engineering. As never before, advances in biological sciences hold tremendous promise for surmounting many of the major challenges confronting the United States and the world. Historically, major advances in science have provided solutions to economic and social challenges. At the same time, those challenges have inspired science to focus its attention on critical needs. Scientific efforts based on meeting societal needs have laid the foundation for countless new products, industries, even entire economic sectors that were unimagined when the work began (http://www.nap.edu/catalog/12764.html, vii)'.
>
> 'The lessons of history led the Committee on a New Biology for the 21st Century to recommend that a New Biology Initiative be put in place and charged with finding solutions

to major societal needs: sustainable food production, protection of the environment, renewable energy, and improvement in human health. These challenges represent both the mechanism for accelerating the emergence of a New Biology and its first fruits. Responding to its Statement of Task, the committee found the answer to the question: "How can a fundamental understanding of living systems reduce uncertainty about the future of life on earth, improve human health and welfare, and lead to the wise stewardship of our planet?" in calling for a national initiative to apply the potential of the New Biology to addressing these societal challenges (http://www.nap.edu/catalog/12764.html, vii)".'

The Report makes a number of recommendations while posing a number of crucial research questions, notably the following:

'How can a fundamental understanding of living systems reduce uncertainty about the future of life on earth, improve human health and welfare, and lead to the wise stewardship of our planet? Can the consequences of environmental, stochastic or genetic changes be understood in terms of the related properties of robustness and fragility inherent in all biological systems?'

Envisaging a *problem-driven* New Biology for the twenty-first century, it indicates an answer to the above question as follows (P. 3): 'The essence of the New Biology, as defined by the committee, is integration—reintegration of the many sub-disciplines of biology, and the integration into biology of physicists, chemists, computer scientists, engineers and mathematicians to create a research community with the capacity to tackle a broad range of scientific and societal problems. Integrating knowledge from many disciplines will permit deeper understanding of biological systems, which will both lead to biology-based solutions to societal problems and also feed back to enrich the individual scientific disciplines that contribute new insights. The New Biology is not intended to replace the research that is going on now; that research, much of it fundamental and curiosity-driven by individual scientists, is the foundation on which the New Biology rests and on which it will continue to rely. Instead, the New Biology represents an additional, complementary approach to biological research (http://www.nap.edu/catalog/12764.html)'.

In Chap. 2, entitled 'How the New Biology Can Address Societal Challenges (http://www.nap.edu/catalog/12764.html, pp. 17–38)' the Committee articulates four kinds of challenge, viz., generating food plants capable of adapting and growing under environmental change, modelling ecosystem function and biodiversity under unpredictable change, expanding sustainable alternatives to fossil fuels *and* monitoring individual health in order to provide predictive surveillance and care. Making the future less and less *uncertain* by addressing the above stated four kinds of challenge, by following a technology-led climate policy and by participating in an energy-technology race with R&D of low carbon technologies should be the focus of convergence and knowledge resources dynamics *equally*. The convergence model and knowledge resources dynamics accordingly require a rebuilding of the knowledge environments in the universities and research institutions, on the one hand, and a rethinking of the science-technology-industry interface, on the other. The four kinds of challenge could be respectively tackled by (http://www.nap.edu/catalog/12764.html, pp. 17–38):

(1) Developing the knowledge-based new technologies that will make it possible to adapt all sorts of crop plants for efficient production under different conditions and generate food plants to adapt and grow sustainably under local conditions in changing environments. In the long run, it is then possible to feed people around the world with abundant, healthful food, adapted to grow efficiently in many different and ever changing local environments.
(2) Integrating the knowledge base of ecology with those of organismal biology, evolutionary and comparative biology, climatology, hydrology, soil science, and environmental, civil and systems engineering, through the unifying languages of mathematics, modelling and computational science, so that it is possible to monitor ecosystem function, identify ecosystems at risk, and develop effective interventions to protect and restore ecosystem function, minimizing harmful impacts of human activities and climate change.
(3) Making efficient use of plant materials—biomass—to make bio fuels and optimizing both the plant system that serves as the source of cellulose and the industrial process that turns the cellulose into a useful product.
(4) Accelerating the fundamental understanding of the web of interacting networks of staggering complexity between an individual's genome sequence and the endpoint of that individual's health, so as to develop the tools and technologies that will in turn lead to more efficient approaches to developing therapeutics and enabling individualized, predictive medicine.

Going a step beyond convergence, the knowledge resources dynamics accelerates error elimination in science by trying to falsify scientific theories by means of experimentation (the CERN-LHC 2012 being the latest example from physics and astronomy), though this is more restricted to the natural sciences. As a methodological framework, knowledge resources dynamics may be described as strategic multidisciplinary research and interface-building within a knowledge environment, as the ultimate endless frontier in service of humanity. Knowledge resources dynamics acknowledges the ambivalence of scientific discovery. It acknowledges the dangers of justifying the arrogance of scientific practice and of public policy and decision-making in terms of this ambivalence. Often, science has gone astray and failed to fully recognize the complexity and universal interconnectedness across nature's fragile ecosystems (Pandit 2012b, 2016). The good news is that, being part of the global problems that humanity is confronted with today, science seeks convergence to *repair* itself. Science is now in a state of self-repair and self-healing, if I may so formulate it. If you prefer, we may simply call it convergence or the third revolution on the life sciences horizon. Knowledge resources dynamics qualitatively improves on this state of self-healing undertaken by science itself, allowing a dynamic role to ethics and wisdom inquiry. In this sense, knowledge resources dynamics supports convergence while taking humanity far beyond it.

8 Conclusion: A Warning Against Paradigms[6]

Convergence is sometimes described as the third revolution in the making on the life sciences horizon. Briefly, this has important implications, more relevantly for the public understanding of science and scientific change. *First of all*, it becomes amply clear from the life sciences horizon that the scientific revolutions are rational in so far as they must build upon the previous revolutions by a process of knowledge resources dynamics. In the life sciences, the rationality of the third revolution can be seen as strongly building upon the rationalities of the first and second revolutions starting from 1953 and 1980s, respectively. *Second*, the epistemic communities, as diverse as scientists, methodologists of science and innovators, among others, can accelerate the processes of knowledge resources dynamics by working together at the challenging frontiers of development, including the frontiers of problem and theory-development in the sciences. *Third*, the sciences themselves can go into a state of self-repair, a state of self-healing, only if they are not in *fetters*, confined within 'paradigms' with 'gate keepers', and only if they are able to learn from their past mistakes. As a corollary to this, notice that there are no so-called 'paradigms' (Kuhn 1962, 1970) or their 'gate keepers' to be found playing a key role either in science or in scientific revolutions. What is remarkable and most important is that at no stage of the third revolution of convergence in the life sciences horizon, and in turn in the physical sciences, is there any indication of 'Kuhn's paradigms' coming into play, peripherally or centrally. Therefore, at least in the interests of public understanding of science, we must say a final good bye to the myth of the 'paradigms-driven science' that Kuhn created in the name of the sociology of science (Kuhn 1962, 1970). *Fourth*, it cannot be denied that there are many among the global science communities and the STS communities who have been psychologically as well as sociologically so much overwhelmed by Kuhn's (1962, 1970) book 'The Structure of Scientific Revolutions' that they are able to pretend to understand what 'paradigms' stand for but in truth do not understand anything about 'paradigms', nor about the multiple senses in which one could use this notoriously ambiguous term (Masterman 1970, Wittgenstein 1953). Perhaps, it is no exaggeration to warn that 'paradigms' are best suited for arrogant *dictators* and *totalitarian regimes* (Wynn 2015). *Fifth*, and finally, to avoid a takeover by technological totalitarianism, it is imperative that the sciences join hands to accelerate knowledge resources dynamics. It is urgent to create a culture of wisdom and a knowledge environment for knowledge resources dynamics to bear fruit.

As is clear from the current state of development of the life sciences, the convergence model is essentially discovery-driven. No surprise, if it is expected to take off vertically. It is about how one scientific revolution within a scientific discipline successfully builds upon the previous revolutions within that discipline, creating a potentially larger knowledge environment for several disciplinary sciences, even without a previous

[6]The main arguments are developed in Pandit (1982, 1991, 2010a, 2012a, 2014a, 2015c); and Pandit and Dosch (2013).

history of interface- building, to come together to address the pressing global problems at a common frontier. In this sense, convergence is essentially a *vertical model*, of which convergence in the life sciences-horizon is a good example.

Moreover, the convergence model is a very good model for various initiatives but it is silent on the role of *ethics and wisdom inquiry*. Therefore, it should be made clear that convergence alone is not enough in meeting the global challenges within a KBCE. Ethical principles and wisdom inquiry must also play their dynamic roles in building a new world with a reduced climate change risk. Concerted efforts are needed in this context to bring ethics and wisdom inquiry into play in the reintegrated science scenario, more so in terms of knowledge resources dynamics.

Already nurtured by the *wisdom of culture*, namely, the world's diverse cultures, many of them unfortunately disappearing or already disappeared like the world's many dying species and languages, the humanity must now be guided to dynamically move *forwards* to a *culture of wisdom*, if it is to negotiate its future prospects under the global challenges it is confronted with now. Hopefully, the *culture of wisdom*, envisaged here, would include the newly emerging wisdom inquiry at its very core (Pandit 1995, 2007b, 2008, 2010b, c, 2012b). It is important, particularly for the developing countries, to redesign the trajectories of economic growth and human development, including scientific and technological development, within the *culture of wisdom*, fostering a harmonious, dynamic and life-enhancing interaction between culture, ecology and biodiversity.

But what is the *wisdom of culture* exactly? To put it very briefly, in the past philosophical traditions it had been simply *assumed*, without stating it explicitly, that in their search for knowledge the different sciences would take into consideration the *wisdom of culture* that finds a mention in the oldest definition of philosophy as *love of wisdom*. After several thousand years, particularly in the context of modern science, one is prompted to ask *where* is the wisdom of culture to be found in the sciences that have an in-built efficiency to produce highly *fragmented knowledge* of the universe *that* we take for granted as final truth. Fragmented knowledge allows its *ambivalence* easily to be misused and abused, depending on what kinds of interests are at play, once knowledge and discovery are moved from universities and research institutions into society and industry. In short, reductionism of fragmented knowledge flourishes everywhere (Whitehead 1923; Waddington 1977, xi–xii, pp. 23–24; Pandit 2007b, 2010c; Wynn 2015).

What about the *culture of wisdom*? While thinking of today's global challenges we must *rethink* our approach to education, global population explosion, humanity's ecological overshoot, destruction of the planet's carbon sinks, urban lifestyle-based overconsumption, overproduction of goods and services, anthropogenic global warming-driven climate change, the challenge of designing carbon free green technologies, and the challenge of green growth within the KBCE. All of these are already ample evidence, *first*, how science and technology, instead of being an exception, are a part of the larger problem and, *second*, how the uses and abuses of science thrive on the forces of power and arrogance. However, a consensus among the academics and scientists across the disciplinary boundaries is growing in favour of the view that concerted efforts at rebuilding a *culture of wisdom* are urgently needed so that the integrating sciences can learn a lesson how to foster healthy change, driven by wisdom inquiry (Maxwell 2010; Pandit 2007a, b, 2008, 2010b, c, 2012b).

In the preceding discussion, I have indicated how important it is to move from the convergence model to the rationality of knowledge resources dynamics that might help humanity to hammer out rational and wise solutions to global challenges. Should such a development take place in the next 20–50 years, the *wisdom of culture* that had originated in ancient philosophical texts, compositions and world-views might have a chance to come back to life in the *culture of wisdom*, as it were, completing the wisdom circle for the sciences. In final analysis, however, it seems that in moving from convergence to the rationality of knowledge resources dynamics, we are still moving from the wisdom of culture to the culture of wisdom. By 'culture of wisdom' I am referring to a knowledge environment in the universities and research institutions that is nurtured by *local interactions*, strongly fostering linkaÊges between ecology, culture, discovery, innovation, ethics, wisdom inquiry and entrepreneurship in the service of humanity, human well-being, biodiversity, ecosystem resilience and environmental justice.

While sounding optimistic, such an approach receives considerable support from the unexpected quarters, namely, ethics and wisdom inquiry (Maxwell 2010; Pandit 2007a, b, 2008, 2010b, c, 2012b) that have emerged on the humanity's intellectual horizon for steering it through the twenty-first century towards its future goals. To put it in a nutshell, wisdom inquiry tests the sciences not by their power to produce knowledge *but* by their promise for interface-building so as to hammer out rational and wise solutions to problems of living in a world that is witnessing accelerated risk and environmental injustice. The rationality of sustainable development teaches all of us how humanity might learn to work with, and not against, our planet; how it might learn to cultivate and inculcate a new culture of wisdom and the values of intergenerational equity.

References

A New Biology for the 21st Century, National Academy of Sciences. (2009). http://www.nap.edu/catalog/12764.html

Allen, G. (1978). *Life Sciences in the twentieth century*. London/New York/Melbourne: The Cambridge History of Sciences Series, Cambridge University Press. 1978.

Allen, B. (1990). Information as an economic commodity. *American Economic Review, 80*(2), 268–273.

Arctic Climate Impact Assessment. (2004). *Impacts of a warming arctic: Arctic climate impact assessment*. Cambridge: Cambridge University Press.

Berkman, P. A. & Young, O. R. (April 17, 2009). Governance and environmental change in the Arctic ocean. *Science, 324*, 339–340.

Bigirimana, S. (2011). Patterns of human knowing in the information society: A philosophical study of the epistemological implications of the information revolution. A Thesis Presented to the Department of Philosophy Ruprecht-Karls-University, Heidelberg in Partial Fulfilment of the Requirements for the Degree of Doctor of Philosophy, May 2011, Supervisor: Prof. Dr. Peter McLaughlin (citing Pandit 1991).

Clemenz, G. (August 2007). Our knowledge of markets for knowledge (pp. 1–19). In International Wittgenstein symposium, Kirschberg: Lower Austria, (Unpublished).

Côté, I. M., & Darling, E. S. (2010). Rethinking ecosystem resilience in the face of climate change. *PLoS Biology, 8*(7), e1000438. doi:10.1371/journal.pbio.1000438

Drucker, P. F. (1993). *Postcapitalist Society* (Vol. 346, Issue 6216). New York: Herper Collins Publishers.

Gasiorek, M. (March 2015). *India's engagement in global value chains: An overview*. A paper presented at the ICRIER, University of Sussex and InterAnalysis Workshop on: "Enhancing India's Participation in Global Value Chains", New Delhi.

Galiana, I., & Green, C. (2009). Let the global technology race begin. *Nature, 462*(2009), 570–571.

Global Commission on the Economy and Climate. (January 2015). *New Climate Economy Technical Note: Abatement Reduction Potential*.

Hall, A. (December 19, 2014). Projecting regional change—How accurate are regional projections of climate change derived from downscaling global cimate model results? *Science, 346* (6216), 1461–1462.

Harmon, J. P., Moran, N. A., & Ives, A. R. (2009). Species response to environmental change: Impacts of food web interactions and evolution. *Science, 323*, 1347. doi:10.1126/Science. 1167396

Hebeisen, M. W. (1997). Umweltschutz als Staatsaufgabe und als Gegenstand der Rechtspolitik. In Adrian Holderegger (Hrsg.) Oekologische Ethik als Orientierungs Wissenschaft: Von der illusion zur Realitaet. Schweiz: Universitaetsverlag Freiburg Schweiz, Pauludruckerei Freiburg. ISBN 3-7278-1132-3 (p. 154, citing Pandit 1995).

Hussi, T. (2004). Reconfiguring knowledge management—Combining intellectual capital, intangible assets and knowledge creation. *Journal of Knowledge Management, 8*(2), 36–52.

Kuhn, T. S. (1970). *The structure of scientific revolutions* (1st ed., 1962). Chicago: Chicago University Press.

Lev, B. (2000). Intangibles: Management, measurement and reporting. http://www.baruch-lev.com

Lincoln, S. F. (2006). *Challenged earth: An overview of humanity's stewardship of earth*. London: Imperial College Press.

Machlup, F. (1962). *The production and distribution of knowledge in the United States*.

Machlup, F. (1980). *Knowledge and knowledge production*. Princeton: Princeton University Press.

Mankiw, N. G. (2004). *Principles of microeconomics*. Southwestern: Thompson.

Marean, C.W. (August 2015). How we conquered the planet. *Scientific American*, 23–29.

Masterman, M. (1970). The nature of a paradigm. In I. Lakatos, & A. Musgrave (Eds.), *Criticism and the growth of knowledge* (pp. 59–89). Cambridge: Cambridge University Press (ISBN 0-521-09623-5).

Maxwell, N. (2010). Reply to comments on science and the putsuit of wisdom. *Philosophia, 38*, 667–690. doi:10.1007/s11406-010-9260-y

Meusburger, P. (2015). Relations between knowledge and power: An overview of research questions and concepts. In Meusburger, P. et al. (Eds.), *Geographies of knowledge and power, knowledge and space* (Vol. 7, pp. 19–74). doi:10.1007/978-94-017-9960-7_8

Pandit, G. L. (1982). The structure and growth of scientific knowledge—A study in the methodology of epistemic appraisal. In *Boston studies in the philosophy of science* (Vol. 73, xxii + 231 pp, with an Editorial Preface by Robert S. Cohen & Marx W. Wartofsky). Dordrecht/Boston/London: D. Reidel Publishing Company, The Netherlands: Springer.

Pandit, G. L. (1989). Scientific change: The possibility of a unified approach. In Paul Weingartner & Gerhard Schurz (Eds.), Berichte Des 13. Internationalen Wittgenstein-Symposiums: Grenzfragen Zwischen Philosophie und Naturwissenschaft. Wien 1989, 168–179.

Pandit, G. L. (1991). Methodological variance: Essays in epistemological ontology and the methodology of science. In *Boston Studies in the Philosophy of Science* (Vol. 131, xxiv + 422 pp). Dordrecht/Boston/London: Kluwer Academic Publishers, The Netherlands: Springer.

Pandit, G. L. (1995). Von der Oekologie des Bewusstseins zum Umweltrealismus - Die Wiederentdeckung menschlicher und nicht-menschlicher Interessenssphaeren. Wiener Vorlesungen, Band 41. Picus Verlag: Wien (Vienna), Austria.

Pandit, G. L. (September 1999). Paradoxes of progress: How can humanity improve environmental thinking and environmental action? In *Politics and the Life Sciences, 18*, 241–242.

Pandit, G. L. (2000). Environmental actions: Can a part control the whole? In *Problems of Action and Observation*, a Special Issue of *Systemica* 12 (2000, 351–368), Journal of the Dutch Systems Group. Proceedings of the International Conference held in Amsterdam, the Netherlands, in April 1997. BKS + ad.

Pandit, G. L. (2001a). Participate and reconnect: The problem of improving the human condition. In Gerard de Zeeuw, Martha Vahl and Ed Mennuti (Eds.) *Problems of Participation and Connection*, a Special Issue of *Systemica* 13 (337–341), the Journal of the Dutch Systems Group. Proceedings of the International Conference held in Amsterdam, the Netherlands, 5–9 April 1999. Lincoln Research Centre.

Pandit, G. L. (2001b). Leibniz and the Changing Images of Universal Interconnectedness. In Hans Poser et al. (Hrsg.) *VII. Internationaler Leibniz-Kongress "NIHIL SINE RATIONE – Mensch, Natur und Technik im Wirken von G. W. Leibniz*", Vortraege 2. Teil. Technische Universitaet Berlin (942–947).

Pandit, G. L. (2006). Leibniz´s Weltbild of Universal Interconnectedness. In Herbert Breger, Jürgen Herbst und Sven Erdner (Hrsg.) *Einheit in der Vielheit*. Vorträge 2. Teil, VIII. Internationaler Leibniz-Kongress Vorträge (769–779).

Pandit, G. L. (2007a). Ethical tasks of media advocacy in the 21st century. In Philosophie der Informationsgesellschaft/Philosophy of the Information Society, Beitraege des 30. Internationalen Wittgenstein Symposium (August 5–11, 2007), Band XV, edited by Herbert Hrachovec, Alois Pichler and Joseph Wang, pp. 169–173. Kichberg am Wechsel: Austrian Ludwig Wittgenstein Society. Available at http://wab.uib.no/agora-alws/

Pandit, G. L. (2007b). Rethinking science—Studies with a return to natural philosophy. *Friends of Wisdom News Letter* (No. 1, November 2007: 28–32), A Review Essay based on Nicholas Maxwell: *Is Science Neurotic?* Imperial College Press: 2004, pp. xx + 240. For details, visit FOW archive at: www.jiscmail.ac.uk/lists/friendsofwisdom.html.

Pandit, G. L. (2008). Universities with a room for wisdom inquiry. In *Friends of Wisdom News Letter*, No. **3**, July 2008, pp. 4–9.

Pandit, G. L. (2009). Ethics in public domain: Biomedical research and beyond. In Alok Srivastava and Ipsita Roy (Eds.), *Bio-Nano-Geo sciences: The future challenge*, Humboldt-Kolleg-Palampur (India), Anne Books: New Delhi/Chennai/Mumbai 2009, pp. 187–208. A pre-print is accessible in Indo-German Science Circle Website of the Embassy of Federal Republic of Germany, New Delhi. www.science-cooperation.org/www.science-circle.org

Pandit, G. L. (2010a). The structure and growth of scientific knowledge—A study in the methodology of epistemic appraisal (Springer Paperback). In *Boston studies in the philosophy of science* (Vol. 73, xxii + 231 pp, with an Editorial Preface by Robert S. Cohen and Marx W. Wartofsky). The Netherlands: Springer (New Paperback ISBN 10: 9048183758 ISBN 13: 9789048183753)

Pandit, G. L. (2010b). "Aim Oriented Empiricism: How We Might Improve the Aims and Methods of Science, Making it More Rational, Responsible and Far Reaching", Friends of Wisdom Newsletter, No. 6, January 2010, pp. 30–37, a Review Essay based on Nicholas Maxwell: What is Wrong with Science? Second Edition, Pentire Press: London (first published by Bran's Head Books 1976).

Pandit, G. L. (2010c). How simple is it for science to acquire wisdom according to its choicest aims? *Philosophia, 38*, 649–666. A Special No. of Philosophical Quarterly of Israel, doi:10.1007/s11406-010-9252-y

Pandit, G. L. (2012a). Methodological variance: Essays in epistemological ontology and the methodolohy of science (Springer Paperback). *Boston studies in the philosophy of science* (Vol. 131, xxiv + 422 pp). Dordrecht/Boston/London: Springer Verlag: Netherlands. Dedicated to Karl R. Popper (1902–1994), this work was completed by the author during his first research stay in the Federal Republic of Germany (Berlin/Heidelberg, 1986–1987). ISBN 10: ISBN 10: 9401054002 ISBN 13: 9789401054003.

Pandit, G. L. (2012b). A science of universal interconnectedness: Rationality of wisdom inquiry. In IX. International Leibniz-Kongress: Nature and Subject, Nachtragsband: Herausgegeben von Herbert Breger, Juergen Herbst und Sven Erdner, Hannover: September 26–1.Oktober 2011, S. 214–225. Herstellung: Druckerei Hartmann GmbH, Weidendamm 18, 30167 Hannover 2012.

Pandit, G. L. (2013). Ecosystem-resilience: A long journey to nature policy. In S. Nautiyal, K. S. Rao, H. Kaechele, K. V. Raju, & R. Schaldach (Eds.), *Knowledge systems of societies for adaptation and mitigation of impacts of climate change* (Environmental Science and Engineering, pp. 57–86). Berlin, Heidelberg, Germany: Springer.

Pandit, G. L. (2014a). Epistemology and methodology of science: Contours, contexts, theories, interfaces, frontiers… University of Heidelberg, Faculty of Physics and Astronomy (in preparation).

Pandit, G. L. (2014b). Rebuilding democracy, rethinking education. In V. Byczkiewicz (Ed.), *Democracy and education: Collected perspectives* (pp. 161–222). Los Angeles, CA: Trebol Press.

Pandit, G. L. (May 2015a). Breaches in the reach of empowerment: When can we put a final end to collective crimes? In V. Byczkiewicz (Ed.), Democracy and education (Vol. 2). Los Angeles, CA: Trebol Press (forthcoming).

Pandit, G. L. (2015b). Turning the growth-discourse and trajectories around (in preparation).

Pandit, G. L. (2015c). Beyond the STS-myth and jargon. University of Heidelberg, Faculty of Physics and Astronomy. http://gsfp.physi.uni-heidelberg.de/graddays/index.php?m=2&s=5

Pandit, G. L. (2016). Scenarios of global interconnectedness: Turning Leibniz's research programme around, X. International Leibniz kongress—ad felicitatem ninstrum alienamve—for our happiness or the happiness of others. Gottfried Wilhelm Leibniz University Hannover, Germany, under the Patronage of the Prime Minister of Lower saxony, Stephen Weil (2016)

Pandit, G. L. & Dosch, H. G. (December 4, 2013). *The frontiers of theory development in physics: A methodological study in its dynamical complexity*. Los Angeles, CA: CreateSpace Independent Publishing Platform, Trebol Press.

Pandit, G. L. & Meusburger, P. (2015). University and knowledge environment: A study in knowledge resources dynamics (forthcoming).

Popper, K. R. (1934). *Logik der Forschung*, Wien. English translation, *Logic of scientific discovery*, 1959, London: Hutchinson/New York: Basic Books.

Popper, K. R. (1972). *Objective knowledge*. Oxford: Oxford University Press.

Popper, K. R. (1975). The rationality of scientific revolutions. In R. Harre (Ed.), *Problems of scientific revolution* (pp. 72–101). Oxford: Oxford University Press.

RICARDIS. (2006). Reporting Intellectual Capital to Augment Research, Development and Innovation in SMEs Report to the Commission of the High Level Expert Group on RICARDIS. European Commission, pp. 1–160.

Schwarzburger, H. (1998). "Kein blindes Technik-Vertrauen - Indischer Philosoph fordert Umweltrealismus und neue Ethik" - Bildung und Wissenschaft - *Der Tagesspiegel* (Berlin): Sonntag, **7**. Juni 1998, S.44 (Review of Pandit 1995).

Sharp, P. A. (2014). Meeting global challenges: Discovery and innovation through convergence. *Science, 346*(6216), 1468–1471.

Srinivasan, K. (2010). Rebordering the borders created by multidisciplinary sciences: A study. Presented at XIV World Congress World Council of Comparative Education Societies (WCCES) "Bordering, Re-Bordering and New Possibilities in Education and Society", Istanbul 2010 14–18 June, Page 8: Online at http://mpra.ub.uni-muenchen.de/25090/ MPRA Paper No. 25090, posted 20. September 2010/09:09 (citing Pandit 1982).

Waddington, C. H. (1977). *Tools of Thought*. London: Jonathan Cape.

Watson, J. D., & Crick, F. H. C. (1953). A structure for deoxyribose nucleic acid. *Nature, 171*, 737–738.
Whitehead, A. N. (1923). *Science and the modern world.* Cambridge: Cambridge University Press.
Wittgenstein, L. (1953). *Philosophical Investigations (Philosophische Untersuchungen).* Oxford: Basil Blackwell.
Wynn, G. (2015). The power of words and the tides of history: Reflections on man and nature and silent spring. In P. Meusburger et al. (Eds.), *Geographies of knowledge and power, knowledge and space* (Vol. 7, pp. 163–185). doi:10.1007/978-94-017-9960-7_8

Socio-economic and Agricultural Vulnerability Across Districts of Karnataka

K.V. Raju, R.S. Deshpande and B. Satyasiba

Abstract Vulnerability assessments can play a vital role in the design of appropriate adaptation and mitigation policies targeted towards climate change and its impacts on ecosystems, and those who depend upon these resources for their livelihoods and well-being. Vulnerability is often reflected in the state of the economic system as well as the socio-economic features of the population living in that system. The current paper attempts to build a picture of the socio-economic context of vulnerability by focusing on indicators that measure both the state of development of the people as well as its capacity to progress further. The result of agricultural vulnerability index suggests indicators like cropping intensity, gross irrigated area and commercial crop area that are the major drivers in determining the vulnerability of districts of Karnataka. The socio-economic index depicts indicators like per capita income, population density and percentage of literacy rate that are the major drivers and contribute to the overall livelihood vulnerability of districts.

Keywords Climate change · Vulnerability index · Principal component analysis · Agricultural vulnerability · Socio-economic vulnerability · Karnataka · India

This paper was submitted to the International Conference on Climate Change and Social-Ecological—Economical Interface-Building: Modeling Approach to Exploring Potential Adaptation Strategies for Bioresource Conservation and Livelihood Development held on May 20–21, 2015, at Institute for Social and Economic Change, Bangalore, India.

K.V. Raju (✉)
ICRISAT Development Center, International Crop Research Institute for the Semi-Arid Tropics, Patancheru, Telangana, India
e-mail: kv.raju@cgiar.org

R.S. Deshpande
ICSSR, Bangalore, India
e-mail: rs.kalbandi@gmail.com

B. Satyasiba
CEENR, ISEC, Bangalore, India

S. Nautiyal et al. (eds.), *Climate Change Challenge (3C) and Social-Economic-Ecological Interface-Building*, Environmental Science,
DOI 10.1007/978-3-319-31014-5_11

1 Introduction

Vulnerability to natural hazards varies widely across communities, sectors and regions. The socio-economic vulnerability is determined by the internal structure of any social system that decides sensitivity of societies and communities to the incidence of hazards. The internal structure also helps to cope with damages from external shocks. It poses the important research question as to why there are different levels of vulnerability within a particular society, even in the context of similar hazards. The possible answer may be that the individuals and groups differ in terms of equality, entitlement capacity, institutions, political and cultural aspects that are responsible factors for the differential vulnerability. For example, marginalized communities are forced to live in susceptible regions that exposes them to floods/droughts and different diseases as compared to other people.

2 Vulnerability

A growing body of literature over the past two decades has identified climate change as the prime issue to global environmental degradation, and has analyzed the associated vulnerability and biodiversity loss (IPCC 2007). According to Fussel (2007), climate related vulnerability assessments are based on the characteristics of the vulnerable system spanning over physical, economic and social factors. The Intergovernmental Panel on Climate Change (IPCC), in its Second Assessment Report (IPCC 1996), defines vulnerability as "the extent to which climate change may damage or harm a system." It adds that vulnerability "depends not only on a system's sensitivity, but also on its ability to adapt to new climatic conditions"; and vulnerability depends on the level of economic development and institutions. Watson et al. (1996) argued that socio-economic systems "typically are more vulnerable in developing countries where economic and institutional circumstances are less favourable". In addition, social scientists tend to view vulnerability as representing the set of socio-economic factors that determine people's ability to cope with stress or change (Allen 2003) while climate scientists often view vulnerability in terms of the likelihood of occurrence and impacts of weather and climate related events (Nicholls et al. 1999).

IPCC defines vulnerability in terms of systems as "the degree to which a system is susceptible to, or unable to cope with, adverse effects of climate change, including climate variability and extremes" (IPCC 2007). Vulnerability is a function of the character, magnitude and rate of climate variation to which a system is exposed, its sensitivity and its adaptive capacity (IPCC 2007).

It is well understood that poor people in the poorest countries are the most vulnerable to the impacts of anthropogenic climate change (Stern et al. 2006). The poor are adversely impacted by climate change because they live in heavily impacted countries and locations within those countries; depend on natural resource-based livelihoods that are disproportionately affected by climate change; and have the weakest ability to adapt to the impacts. Small and marginal farmers are more vulnerable to both the current and future climate change impacts given their high dependence on agriculture, strong reliance on ecosystem and rapid population growth.

Vulnerability assessments can play a vital role in the design of appropriate adaptation and mitigation policies targeted towards climate change and its impacts on ecosystems, and those who depend upon these resources for their livelihoods and well-being. Every human community in the globe has a tendency to become adversely affected by the changes in climate, regardless of the communities' contribution to that change. This tendency is simply known as vulnerability of that particular community to climate change impacts. Agricultural and social vulnerability explicitly focuses on those agricultural, demographic and socio-economic factors that increase or attenuate the impacts of hazard events on local populations.

People who live in arid or semi-arid regions, in low-lying coastal areas, in water limited or flood-prone areas, or on small islands are particularly vulnerable to climate change (Watson et al. 1996). It is clear that climate change will, in many parts of the world, adversely affect socio-economic sectors, including water resources, agriculture, forestry, fisheries and human settlements, ecological systems and human health with developing countries being the most vulnerable (IPCC 2001). Developing countries have lesser capacity to adapt and are more vulnerable to climate change damages, just as they are to other stresses. This condition is most extreme among the poorest people (IPCC 2001).

There is an increasing need to develop indicators of vulnerability and of adaptive capacity both to determine the robustness of response strategies over time and to understand better the underlying processes (Adger et al. 2004). At the district level, vulnerability assessments contribute to setting development priorities and monitoring progress. Sectoral assessments provide details and targets for strategic development plans. In Karnataka, agricultural farmers and agricultural labourers form 56 % of the total workforce (Government of Karnataka 2005) and this is considered as one of the driving forces in determining the socio-economic vulnerabilities of communities in Karnataka. In the present context, a district-wise socio-economic and agricultural vulnerability profile of Karnataka was developed.

3 Objectives, Method and Data

The key objectives of this assessment are: (a) To assess vulnerability of agricultural sector across the districts of Karnataka; (b) To estimate the socio-economic vulnerability of the districts of Karnataka.

The data pertaining to various socio-economic and agricultural indicators were collected and compiled from different sources such as Census of India (2011) and Statistical Abstracts of Karnataka (Directorate of Economics and Statistics), 2008–09, 2009–10 and 2010–11. To understand the agricultural and socio-economic profile, the study analyses important indicators across the districts of Karnataka. This has been done by consultation with experts and based on previous studies (Table 1).

Vulnerability to climate change is a comprehensive process affected by a large number of indicators. However, it is not possible to consider all the available indicators, so only the most significant and representative indicators relevant to Karnataka state were selected in the development of vulnerability indices. Indicators considered in this study are

- ***Agricultural indicators***: Net sown area (3 years average), cropping intensity, area under commercial crops to the total cropped area (TCA), percentage irrigated area to TCA (3 years average), number of tractors/1000 ha area sown, total fallow land (3 years average) and agricultural credit cooperative societies/lakh population.
- ***Socio-economic indicators***: Population density, percentage population of scheduled caste (SC) and scheduled tribe (ST), literacy rate, percentage of marginal landholder (<1 ha), percentage of non-workers, livestock units per lakh population, per capita income (3 years average), cropping intensity and percentage irrigated area to TCA (3 years average).

4 Agricultural Vulnerability Index

4.1 Net Sown Area

Agricultural activities play a dominant role in shaping the livelihood across the districts of Karnataka. Net sown area is an important indicator of agricultural development in a district. The net sown area refers to the particular area sown once during an agriculture year. In the present study we have incorporated a 3-year average of the net sown area. Districts from northern Karnataka have more net sown area than the southern districts of Karnataka. The districts of Gulbarga and Gadag have the highest net sown area of 87.04 and 83.26 %, respectively. The districts of Uttara Kannada and Shimoga have the least net sown area of 11.04 and 26.21 %, respectively. Figure 1 presents district-wise percentage of net sown area to the total geographical area districts with high net sown area are above the median value and districts with low net sown area are below the median value.

4.2 Commercial Crops

Commercial crops are high value crops which are of crucial importance to the economy of a district. It was observed that in Karnataka state, over the years, cultivation of commercial crops has increased. The economic value for commercial

Table 1 Indicators used in earlier vulnerability studies

Study/Author(s)	Major indicators used	Indicators used in the present study
Vulnerability to agricultural drought in Western Orissa: A case study of representative blocks (Swain and Swain 2011)	Socio-economic indicators used: irrigation, crops, marginal farmers, land use pattern, literacy rate, population density, institutional factors, forest area, total geographic area, barren and other fallow land	Irrigation, marginal farmers, literacy rate, population density, fallow land
Analysis of vulnerability indices in various agro-climatic zones of Gujarat (Hiremath and Shiyani 2013)	Density of population, literacy rate, cropping intensity, irrigation, forest area, food crop, non-food crop, net sown areas, livestock population, main workers, cultivators, marginal workers, non-workers	Density of population, literacy rate, cropping intensity, irrigation, net sown areas, livestock population, non-workers
A simple human vulnerability index to climate change hazards for Pakistan (Khan and Salman 2012)	Density of population, literacy, sanitation, electricity, livestock	Density of population, literacy, livestock population
Environmental benefits and vulnerability reduction through Mahatma Gandhi National Rural Employment Guarantee Scheme (IISc 2013)	Groundwater depth, cropping intensity, irrigation intensity, net area irrigated, number of days of irrigation water availability, area under food grain production, crop yields, livestock population, soil organic carbon and soil erosion, migration, wage rates, percentage change in the number of days of employment, livestock population	Cropping intensity, irrigation, livestock population
Climate change impact on livelihood, vulnerability and coping mechanisms: a case study of West Arid Zone, Ethiopia (Senbeta 2009)	Land size, livestock number, literacy, sex, gender and age	Livestock number, literacy
Quantitative assessment of vulnerability to climate change (ICRISAT 2009) (http://www.icrisat.org/what-we-do/impi/training-cc/october-2-3-2009/vulnerability-analysis-manual.pdf, accessed in Aug 2013)	Cropping intensity, irrigation area, total food grains (tons) net sown area, literacy rate, density of population, livestock, population, total food crops area, total no food crops area, life expectancy	Cropping intensity, irrigation area, net sown area, literacy rate, density of population, livestock population

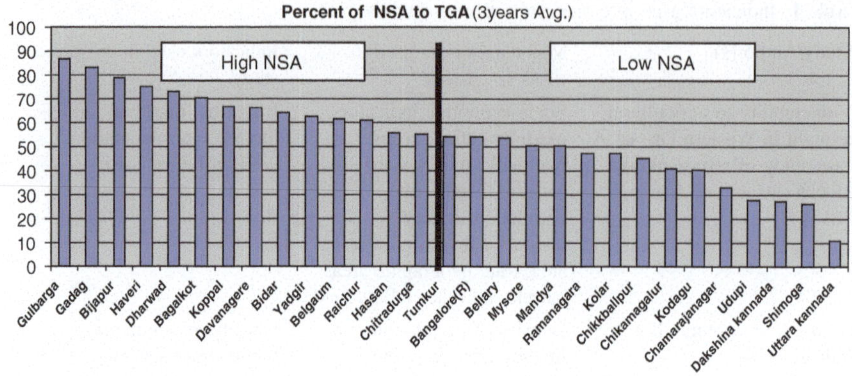

Fig. 1 Percentage net sown area of (*NSA*) to the total geographical area (3 years average) in districts of Karnataka (based on Directorate of Economics and Statistics, Ministry of Agriculture, 2008–09, 2009–2010)

crops has encouraged farmers to grow them. The major commercial crops are sugarcane and cotton among others. The northern districts have been observed to perform better in terms of commercial crops as compared to the southern regions of Karnataka (Fig. 2). This may be due to the fact that quality of land in central and northern regions of Karnataka is more suitable for cotton and sugarcane cultivation. The districts of Belgaum and Haveri ranked first and second in the state in terms of area under commercial crops to TCA, with 24.91 and 24.90 %, respectively. Likewise, Mysore, Dharwad and Bagalkot districts also have a considerable percentage of area under commercial crops. The districts of Dakshin Kannada and Bangalore Rural have the least percentage of area under commercial crops. Figure 2

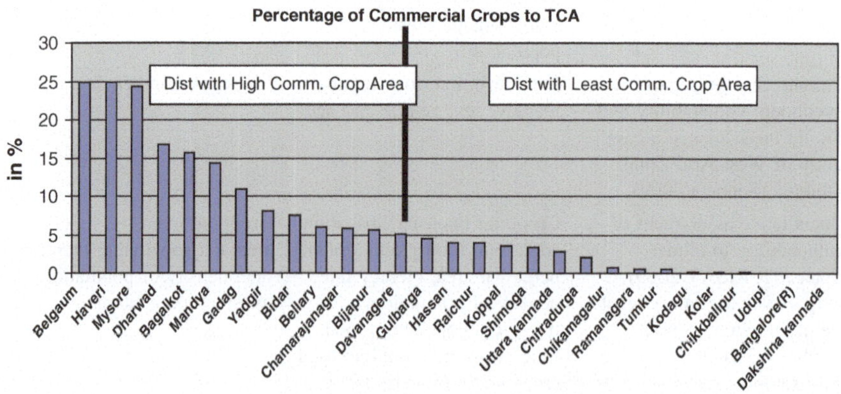

Fig. 2 Percentage area under commercial crops to the total cropped area (3 years average) in districts of Karnataka (based on Directorate of Economics and Statistics, Ministry of Agriculture (2008–09, 2009–10)

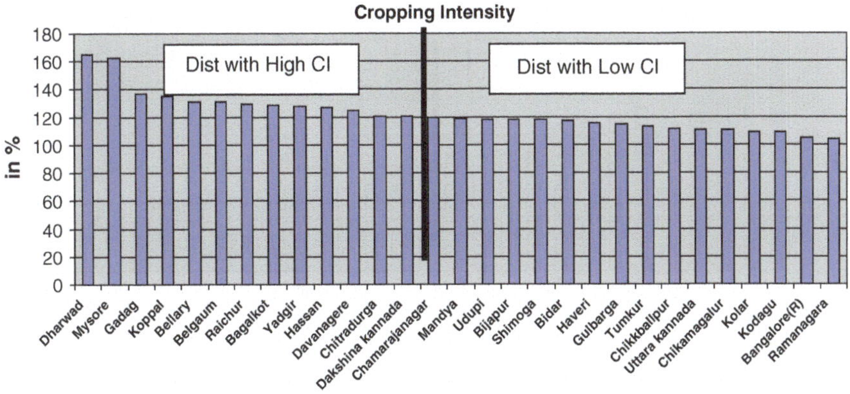

Fig. 3 Cropping intensity (*CI*) across the districts of Karnataka (based on Directorate of Economics and Statistics, Ministry of Agriculture, 2010–11)

presents the percentage of area under commercial crops to TCA across the districts of Karnataka (Fig. 3).

4.3 Cropping Intensity

Cropping intensity refers to cultivation of more than one crop in the same field during the same agricultural year. Higher the cropping intensity essentially means more number of crops cultivated in a year. Normally, districts with more irrigation water availability have higher cropping intensities. In addition, mechanization of farmlands has also had considerable effects on increasing cropping intensities. Dharwad district was observed to have the highest cropping intensity of 164.74 % followed by Mysore with 162.33 %. Districts like Bellary, Raichur and Koppal were also found to have good cropping intensities. The government has encouraged farmers to adopt water conservation methods and has focused on developing irrigation provisioning facilities in these regions thereby leading to increased cropping intensities in these districts.

4.4 Gross Irrigated Area

Irrigation water availability is essential for climate resilient agricultural production. In Karnataka, a large portion of area is under rainfed agriculture. Karnataka is one of the states with less irrigated area in the country, i.e. 32 % of gross irrigated area to TCA (Ministry of Agriculture, GoI, 2010–11). Among the different districts, Shimoga has the highest area of about 62 % of its TCA under irrigation and the

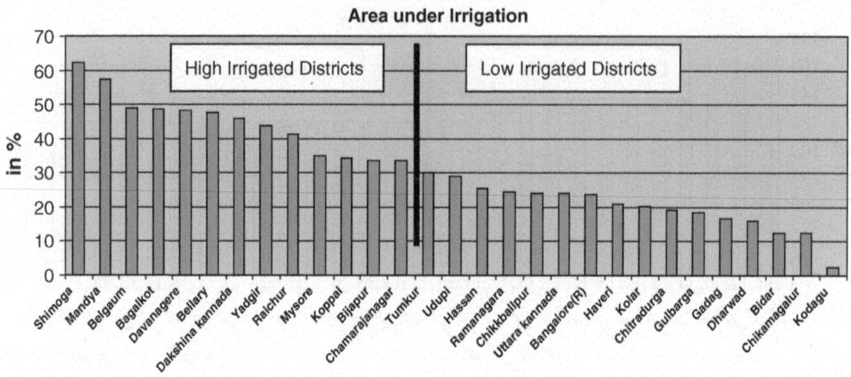

Fig. 4 Gross irrigated area as a percentage of total cropped area (3 years average) in districts of Karnataka (based on Directorate of Economics and Statistics, Ministry of Agriculture, 2008–09, 2009–10)

least in the state is in Kodagu district (2.34 %). In the northern region of Karnataka, Belgaum district has 48.9 % of TCA under irrigation and in southern Karnataka, the district of Mandya has 57.52 % area under irrigation, which is the second highest in the state. The districts of northern Karnataka namely Bagalkot, Bellary, Yadgiri, Raichur, Koppal and Bijapur have significant area under irrigation after Belgaum. Bidar district has the least area under irrigation in the northern region (12.77 %). The existing major reservoirs in the northern regions have the potential to bring additional area under irrigation. Figure 4 presents the percentage of irrigated area to TCA across the districts of Karnataka.

4.5 Fallow Land

Agricultural development is dependent on proper utilization of available land resources. Fallow land refers to the cultivable area which has not been cultivated for a period of time. The districts, where large areas have been left fallow, can be considered as districts where there is underutilization of land resources. Raichur has the highest percentage of fallow land to its total geographical area. It has about 26.46 % of its geographical area under fallow land, followed by Yadgiri with 17.04 % of area under fallow land. The districts of Uttara Kannada and Kodagu have only 1.89 and 2.08 % of area under fallow land, respectively, the least in the state. In northern Karnataka, the districts of Haveri and Gadag also have less area under fallow land, about 3.37 and 4.16 %, respectively. Figure 5 depicts the average percentage of area under fallow land over 3 years across the districts of Karnataka.

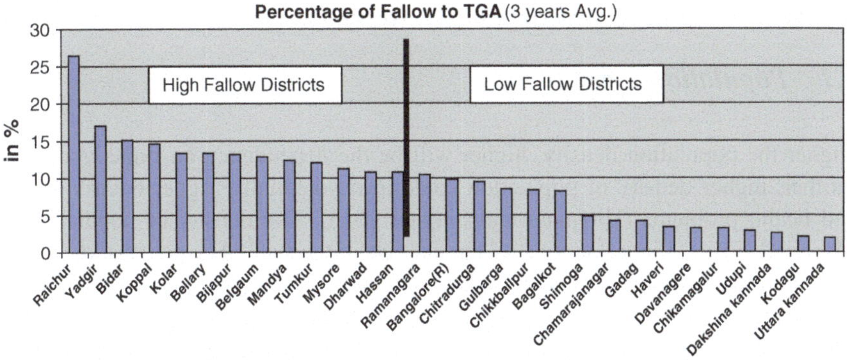

Fig. 5 Average percentage of fallow land to total geographical area (3 years average) across districts of Karnataka (based on Directorate of Economics and Statistics, Ministry of Agriculture, 2008–09, 2009–10)

4.6 *Agricultural Credit Cooperative Societies*

The role of institutions providing financial support to farmers is extremely important for sustainable agricultural production, as it ensures supply of agricultural credit and funds during the different stages of crop production and has the potential to deliver goods and services. Agricultural finance is a critical factor that impacts the development of agricultural activities in the state. The agricultural cooperative societies are considered, as the grass root financial institutions for the farming communities and provide credit to cultivars in time. At the state level, Belgaum district has the highest number of agricultural cooperative societies/lakh population, where it has 16 cooperatives. Likewise in southern Karnataka, Mandya district also has many cooperative societies per lakh population—about 13 agricultural cooperatives. The districts of Udupi and Dakshin Kannada have only 4 and 5 cooperatives per lakh population, respectively, which is the least in the state. Figure 6 represents the number of agricultural credit cooperative societies per lakh population across the districts of Karnataka.

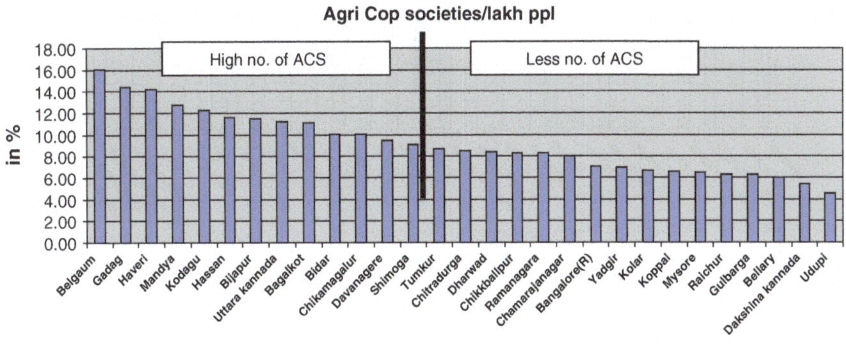

Fig. 6 Agricultural cooperative societies (ACS) (no. lakh population) in various districts of Karnataka (based on the Department of Co-operation cited in DES, 2010–11)

5 Socio-economic Vulnerability Index

5.1 Population Density

Higher the population density, higher will be the dependency on finite resources. Further, higher density of population could also potentially trigger environmental and health problems (Hiremath and Shiyani 2013). The density of population in Mysore district is highest among all the districts in the state (476 persons/km^2). The lowest population density in the state of about 135 persons/km^2 is in Kodagu district. In the northern districts, the main reason for migration is severe drought occurrence for consecutive years. Figure 7 gives the details of population density across the districts of Karnataka.

5.2 SC and ST Population

The social class of SC and ST are considered as the deprived section of the society (Karade 2008). The districts in southern Karnataka like Kolar, Chikkaballapura, Chamrajnagar and Chitradurg have the highest percentage of this category. Kolar has the highest of 30.32 % SC population in the state, followed by Chamrajnagar with 25.42 %. The population of SC is more in southern districts of Karnataka compared to the northern districts. In northern region of Karnataka, Raichur has 19.03 % of ST population, highest in the state, followed by Bellary district. The ST population in the southern Karnataka district of Chitradurg also constitutes a large percentage (18.23 %), while other districts like Chikkaballapura, Davengere and

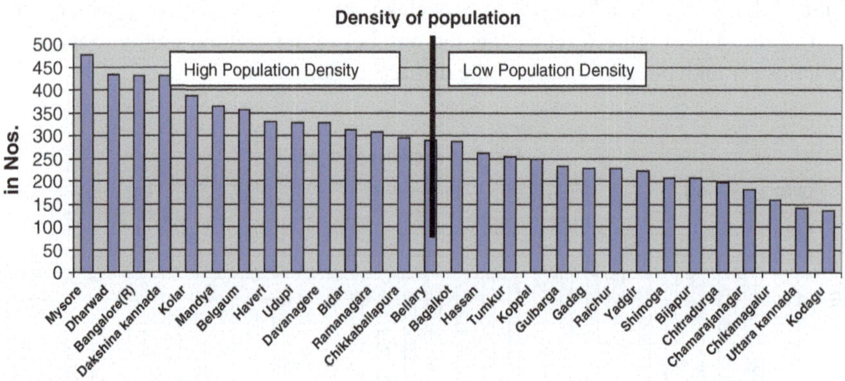

Fig. 7 Population density in different districts of Karnataka (based on Census of India, 2011) (*Note* Bangalore Urban is not included here as it is an outlier)

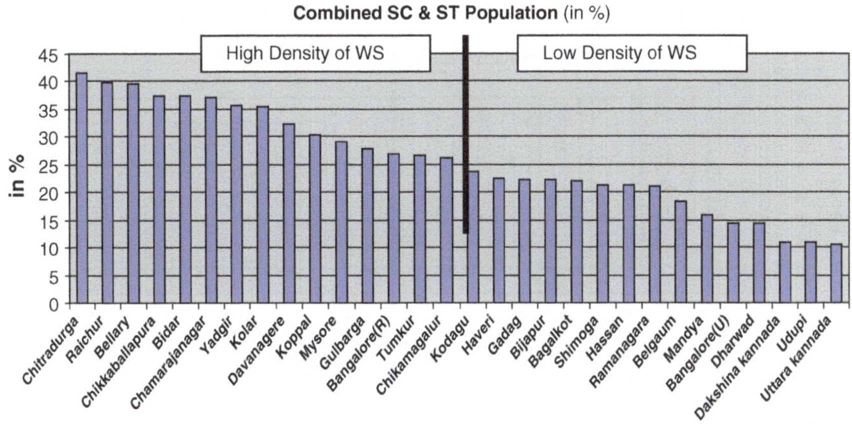

Fig. 8 Percentage of **total** scheduled caste (*SC*) and scheduled tribe (*ST*) population in the districts of Karnataka (based on Census of India, 2011)

Chamrajnagar and Kodagu too have high percentage of ST population. Total of SC and ST is highest in Chitradurg district while the percentage is lowest in Uttara Kannada (Fig. 8).

5.3 *Literacy Rate*

Higher literacy rates could enable communities to diversify their employment and income sources, enhancing standards of living and increasing their resilience towards any kind of shock and stress. This is due to the fact that higher the literacy rate, higher the adaptive capacity, higher the appropriation of opportunities and higher the awareness to face any pressure. The literacy rate in Karnataka state has considerably increased in recent years. The coastal district of Dakshin Kannada has the highest literacy of 88.57 % in the state. The other coastal districts such as Udupi and Uttara Kannada have recorded 86.24 and 84.06 %, respectively. The southern district of Bangalore Urban marks second with 87.67 %. Bidar district has 70.51 % and Yadgiri has 51.83 % literacy rate in the Hyderabad–Karnataka region. Yadgiri, Raichur and Bellary districts have lower literacy rates as compared to the rest of Karnataka. Poverty and lack of socio-economic development are the main reasons for poor literacy rate. Figure 9 shows the literacy rates across the districts of Karnataka.

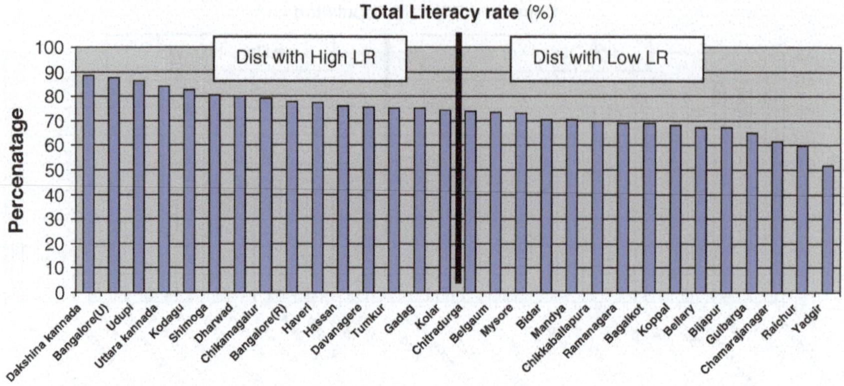

Fig. 9 Literacy rate (*LR*) across the districts of Karnataka (based on Census of India, 2011)

5.4 Marginal Landholders

The size of landholdings is an important indicator for overall agricultural and socio-economic development. In Karnataka, majority of the farmers belong to the marginal landholdings category; Udupi district has the highest percentage (79.8 %) of marginal landholders (Fig. 10). Mandya, Ramanagara, Dakshin Kannada and Bangalore Urban districts also have significant percentage of marginal landholders. The size of landholding is greater in the northern districts as compared to the southern districts of Karnataka.

5.5 Non-workers

Higher the percentage of non-workers, higher will be the dependency rate. Higher dependency rate suggests the district is more vulnerable. This is due to the fact that

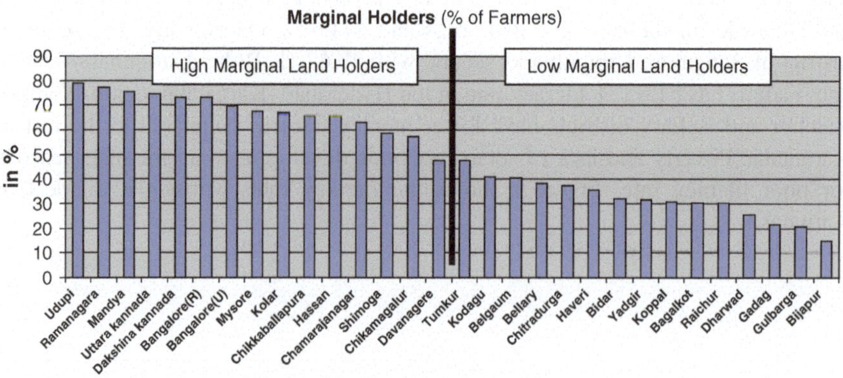

Fig. 10 Percentage of marginal landholders in different districts of Karnataka (based on the Directorate of Economics and Statistics, Agriculture census 2010–11)

the number of persons having income source is less. The district of Bidar has 58.75 % of non-workers, and is the highest in the state (Fig. 11). Other districts like Gulbarga, Uttara Kannada, Bijapur, Bagalkot, Mysore, Shimoga and Udupi also have a high percentage of non-workers. Figure 11 shows the percentage of non-workers across the districts of Karnataka.

5.6 *Livestock Units*

Livestock practices are considered as an important source of livelihood for rural communities. Agriculture and livestock are an integral part of the farmers' community. Livestock provides enormous opportunities to farmers to support their sustainable livelihood. The livestock units per lakh population is highest 58,431 in Yadgiri district and the lowest in Bangalore Urban district (1693) (Fig. 12). Districts like Belgaum, Tumkur and Chitradurg have higher numbers of livestock in absolute terms, however, the livestock units per lakh population is considered, Yadgiri ranks first. The southern districts of Karnataka are dominated by cross-breed cattle, since dairy development has taken place as a major source of economic activity in these districts. The central and northern Karnataka districts are dominated by sheep and goat. The districts of Dharwad, Bijapur, Bagalkot and Belgaum have more number of buffaloes.

5.7 *Per Capita Income*

One of the factors affecting the standard of living of people is per capita income. Higher the average per capita income, lesser is the levels of economic vulnerability.

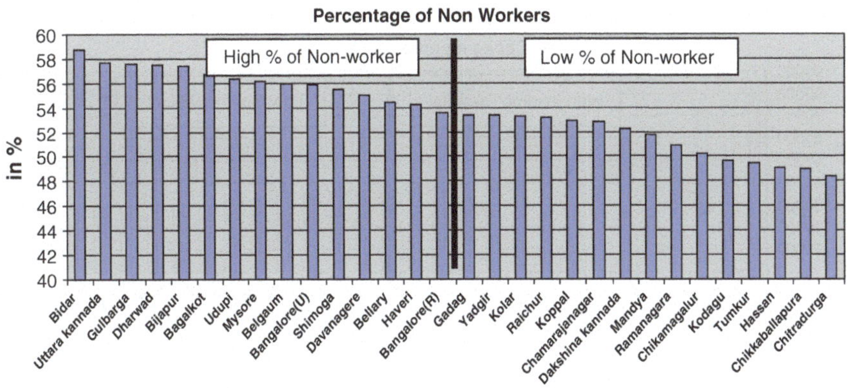

Fig. 11 Percentage of non-workers across the districts of Karnataka (based on the Census of India, 2011)

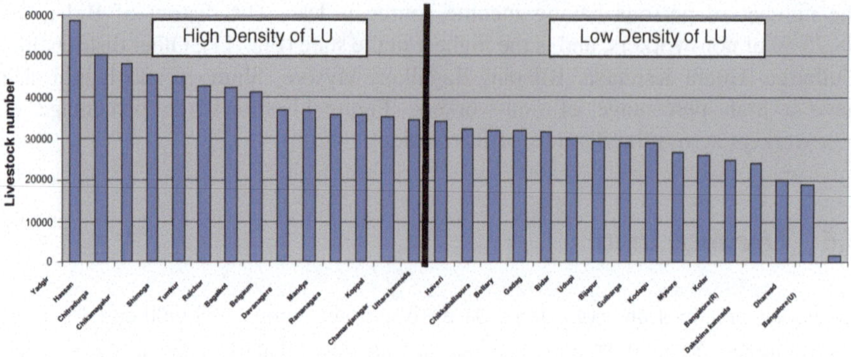

Fig. 12 Livestock units (*LU*) per lakh population in the districts of Karnataka (based on Department of Animal Husbandry and Veterinary Services, Livestock Census 2007) (*Note* 1 livestock unit (*LU*) = 1 cow = 1 buffalo = 5 sheep = 5 goats)

Bangalore Urban district has the highest per capita income of Rs. 139,033. Bangalore Rural and Kodagu have an average per capita income of Rs. 78,587 and Rs. 75,767, respectively. The Hyderabad–Karnataka region districts like Bidar, Gulbarga, Raichur, Yadgiri and Koppal have low per capita income in the state. Bidar district has the least (Rs. 26,905) per capita income in the state. In south Karnataka, the districts of Chamrajnagar and Mandya also have low per capita income. Figure 13 gives the details of per capita income (average of 3 years: 2008–09, 2009–10 and 2010–11) across the districts of Karnataka.

Table 2 presents all the districts of Karnataka, grouped into high and low incidence, indicating higher than the median value and lower than the median value, respectively, for all the indicators considered for assessment of livelihood and agricultural vulnerability.

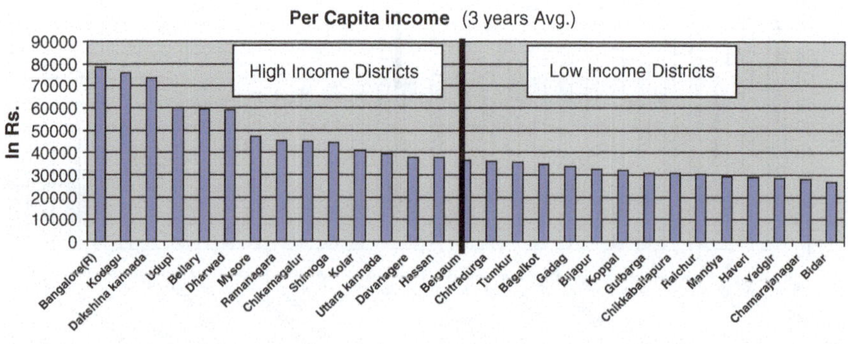

Fig. 13 Average per capita income across the districts of Karnataka (based on Directorate of Economics and Statistics published in Economy survey reports of Karnataka, 2010–11, 2011–12 and 2012–13) (*Note* Bangalore Urban is not included here as it is an outlier)

Table 2 Districts with indicators grouped into high and low incidence

No.	Indicators	High incidence (from median value)	Low incidence (from median value)
Socio-economic and livelihood			
1	Population density	Mysore, Dharwad, Bengaluru-R, Dakshin Kannada, Kolar, Mandya, Belgaum, Haveri, Udupi, Davengere, Bidar, Ramanagara, Chikkaballapura, Bellery, Bagalkot	Hassan, Tumkur, Koppal, Gulbarga, Gadag, Raichur, Yadgiri, Shimoga, Bijapur, Chitradurg, Chamrajnagar, Chikmagalur, Uttara Kannada, Kodagu
2	% SC and ST population	Chitradurg, Raichur, Bellary, Chikkaballapura, Bidar, Chamrajnagar, Yadgiri, Kolar, Davengere, Koppal, Mysore, Gulbarga, Bengaluru-R, Tumkur, Chikmagalur	Kodagu, Haveri, Gadag, Bijapur, Bagalkot, Shimoga, Hassan, Ramanagara, Belgaum, Mandya, Bengaluru-U, Dharwad, DK, Udupi and Uttara Kannada
3	% literacy rate	Dakshin Kannada, Bengaluru-U, Udupi. Uttara Kannada, Kodagu, Shimoga, Dharwad, Chikmagalur, Bengaluru-R. Haveri, Hassan, Davengere, Tumkur, Gadag, Kolar	Chitradurg, Belgaum, Mysore, Bidar, Mandya, Chikkaballapura, Ramanagara, Bagalkot, Koppal, Bellary, Bijapur, Gulbarga, Chamrajnagar, Raichur, Yadgiri
4	% of marginal land holder (<1 ha)	Udupi Ramanagara, Mandya. Uttara Kannada, Dakshin Kannada, Bengaluru-R, Bengaluru-U, Mysore, Kolar, Chikkaballapura, Hassan, Chamrajnagar, Shimoga Chikmagalur, Davengere	Tumkur, Kodagu, Belgaum, Bellary, Chitradurg, Haveri, Bidar, Yadgiri, Koppal, Bagalkot, Raichur, Dharwad, Gadag, Gulbarga, Bijapur
5	% of non-workers	Bidar, Uttara Kannada, Gulbarga, Dharwad, Bijapur, Bagalkot, Udupi Mysore, Belgaum. Bengaluru-U, Shimoga, Dhabngr, Bellery, Haveri Bengaluru-R	Gadag, Yadgiri, Kolar, Raichur, Koppal, Chamrajnagar, Dakshin Kannada, Mandya, Ramanagara, Chikmagalur, Kodagu, Tumkur, Hassan, Chikkaballapura, Chitradurg
6	Livestock units per lakh population (livestock unit = 1 cattle = 1 buffalo = 5 sheep = 5 goats)	Yadgiri, Hassan, Chitradurg, Chikmagalur, Shimoga, Tumkur, Raichur, Bagalkot, Belgaum, Dhbngr, Mandya, Ramanagara, Koppal, Chamrajnagar and Uttara Kannada	Haveri, Chikkaballapura, Bellary, Gadag, Bidar, Udupi, Bijapur, Gulbarga, Kodagu, Mysore, Kolar, Bengaluru-R, Dakshin Kannada, Dharwad, Bengaluru-U

(continued)

Table 2 (continued)

No.	Indicators	High incidence (from median value)	Low incidence (from median value)
7	Per capita income (3 years average)	Bengaluru-R, Kodagui, Dakshin Kannada, Udupi, Bellary, Dharwad, Mysore, Ramanagara, Chikmagalur, Shimoga, Kolar, Uttara Kannada, Davengere, Hassan,	Belgaum, Chitradurg, Tumkur, Bagalkot, Gadag, Bijapur, Koppal, Gulbarga, Chikkaballapura, Raichur, Mandya, Haveri, Yadgiri, Chamrajnagar, Bidar
8	Cropping intensity (CI)	Dharwad, Mysore, Gadag, Koppal, Bellary, Belgaum, Raichur, Bagalkot, Yadgiri, Hassan, Davengere, Chitradurg, DK, Chamrajnagar	Mandya, Udupi, Bijapur, Shimoga, Bidar, Haveri, Gulbarga, Tumkur, Chikkaballapura, Uttara Kannada, Chikmagalur, Kolar, Kodagu, Bengaluru-R, Ramanagara
9	% of irrigated area to total cropped area (3 years average)	Shimoga, Mandya, Belgaum, Bagalkot, Davengere, Bellary, Dakshin Kannada, Yadgiri, Raichur, Mysore, Koppal, Bijapur, Chamrajnagar and Tumkur	Udupi, Hassan, Ramanagara, Chikkaballapura, Uttara Kannada, Bengaluru-R, Haveri, Kolar, Chitradurg, Gulbarga, Gadag, Dharwad, Bidar, Chikmagalur, Kodagu
10	Total area under crops such as mango, grapes, pomegranate, citrus, papaya, cashew nut and others	Kolar, Dakshin Kannada, Udupi, Tumkur, Ramanagara, Chikkaballapura, Bijapur, Shimoga, Mysore, Bengaluru-R, Dharwad, Chamrajnagar, Belgaum, Koppal, Hassan	UK, Chitradurg, Bellary, Bagalkot, Kodagu, Dhvngr, Chikmagalur, Haveri, Mandya, Gul, Bengaluru-U, Bidar, Yadgiri, Gadag, Raichur
Agriculture			
1	Net sown area (3 years average)	Gulbarga, Gadag, Bijapur, Haveri, Dharwad, Gglkt, Koppal, Davengere, Bidar, Yadgiri, Belgaum, Raichur, Hassan Chitradurg, Tumkur	Bengaluru-R, Bellary, Mysore, Mandya, Ramanagara, Kolar, Chikkaballapura, Chikmagalur, Kodagu, Chamrajnagar, Udupi, Dakshin Kannada, Shimoga, Uttara Kannada
2	Cropping intensity	Dharwad, Mysore, Gadag, Koppal, Bellary, Belgaum, Raichur, Bagalkot, Yadgiri, Hassan, Dhvngr, Chitradurg, DK, Chamrajnagar	Mandya, Udupi, Bijapur, Shimoga. Bidar, Haveri, Gulbarga, Tumkur, Chikkaballapura, Uttara Kannada, Chikmagalur, Kolar, Kodagu, Bengaluru-R, Ramanagara

(continued)

Table 2 (continued)

No.	Indicators	High incidence (from median value)	Low incidence (from median value)
3	% area under commercial crops to TCA	Belgaum, Haveri, Mysore, Dharwad, Bagalkot, Mandya, Gadag, Yadgiri, Bidar, Bellary, Chamrajnagar, Bijapur, Dhngr, Gulbarga	Hassan, Raichur, Koppal, Shimoga, Uttara Kannada, Chitradurg, Chikmagalur, Ramanagara, Tumkur, Kodagu, Kolar, Chikkaballapura, Udupi, Bengaluru-R, Dakshin Kannada
4	% irrigated area to TCA (3 years average)	Shimoga, Mandya, Belgaum, Bagalkot, Dhvngr, Bellary, Dakshin Kannada, Yadgiri, Raichur, Mysore, Koppal, Bijapur, Chamrajnagar, and Tumkur	Udupi, Hassan, Ramanagara, Chikkaballapura, UK, Bengaluru-R, Haveri, Kolar, Chitradurg, Gulbarga, Gadag, Dharwad, Bidar, Chikmagalur, Kodagu
5	No. of tractors/1000 ha area sown	Kolar, Ramanagara, Dharwad, Dakshin Kannada, Shimoga, Chikkaballapura, Dhvngr, Udupi, Kodagu, Chikmagalur, Hassan, Belgaum, Uttara Kannada, Bengaluru-R, Tumkur	Bellary, Bagalkot, Mysore, Mandya, Haveri, Raichur, Chitradurg, Koppal, Gadag, Yadgiri, Bijapur, Gul, Chamrajnagar, Bidar
6	% of total fallow land to total geographical area (3 years average)	Raichur, Yadgiri, Bidar, Koppal, Kolar, Bellary, Bijapur, Belgaum, Mandya, Tumkur, Mysore, Dharwad, Hassan, Ramanagara	Bengaluru-R, Chitradurg, Gulbarga, Chikkaballapura, Bagalkot, Shimoga, Chamrajnagar, Gadag, Haveri, Dhvngr, Chikmagalur, Udupi, Dakshin Kannada, Kodagu and Uttara Kannada
7	No. of agricultural cooperative credit societies/lakh population	Belgaum, Gadag, Haveri, Mandya, Kodagu, Hassan, Bijapur, UK, Bagalkot, Bidar, Chikmagalur, Dhvngr, Shimoga, Tumkur	Chitradurg, Dharwad, Chikkaballapura, Ramanagara, Chamrajnagar, Bengaluru-R, Yadgiri, Kolar, Koppal, Mysore, Raichur, Gulbarga, Bellary, Dakshin Kannada, Udupi

6 Approach to Vulnerability Assessment

Vulnerability is often reflected in the state of the economic system as well as the socio-economic features of the population living in that system. The current section of the report attempts to build a picture of the socio-economic context of vulnerability by focusing on indicators that measure both the state of development of the people as well as its capacity to progress further. In addition, attempt has been made to construct a vulnerability index for each district of Karnataka and rank them in terms of their performance on the index. The index attempts to capture the comprehensive scale of vulnerability by considering some of the key indicators (that serve as proxies) for the assessment (Table 3).

There is consensus among researchers to address vulnerability issues at the regional level (Hiremath and Shiyani 2013). Therefore districts have been taken as a unit for developing vulnerability indices. In the next step we have selected important indicators for the vulnerability assessment. After the selection of indicators, data pertinent to the selected indicators were compiled (Table 3). In the next step, a Principal component analysis (PCA) was conducted to identify variability among the selected variables and finally vulnerability indices were developed. Figure 14 presents the details of the method adopted for assessment of vulnerability across the districts of Karnataka.

6.1 Principal Component Analysis (PCA)

A PCA was conducted to identify the variability among selected variables (indicators) for this study. PCA is a data reduction methodology that identifies smaller number of components that explains most of the variance observed in the larger data set. The goal is to arrive at a minimum number of components that will adequately account for the covariation among the larger number of analysis variables. PCA is a tool that converts a number of potentially correlated variables into a set of uncorrelated variables that capture the variability in the underlying data. It is a statistical method that thus transforms a given data set to a smaller number of uncorrelated variables called principal components (PCs). The first PC accounts for a large share of variability in the data, and each succeeding component accounts for as much of the remaining variability as possible. PCA approach provides several potential advantages in the aggregation of spatially explicit, potentially incommensurable variables. When the original variables are correlated then the higher order PCs will capture more of the total variability in the data than any individual original variable. Excluding the lower order PCs reduces the dimensionality (number of variables) of the data while minimizing the loss of information (Smith 2002). PCA thus helps reduce from a large number of individual indicators to a small number of composite, unitless indices (PCs) while reducing the trade-off between richness of information and communicability.

Table 3 Indicators used in this study and rationale

No.	Indicators	Rationality
Socio-economic and livelihoods		
1	Population density	Higher the density, more dependency on finite resources and lower the availability of resources
2	Percentage of SC and ST population	They are among the poor and vulnerable (both socially and economically)
3	Literacy rate	Higher the literacy rate, higher the adaptation capacity, higher the appropriation of opportunities and higher the awareness to face any shock and stress
4	Percentage of marginal land holders (<1 ha)	Higher the proportion of marginal farmers, lesser the production and income and thus higher the vulnerability
5	Percentage of non-workers	More unemployment, more dependency, lower will be the earning capacity and income as compared to expenditure and thus higher vulnerability
6	Livestock units per lakh population	Animal husbandry is an important source of livelihood for rural communities. Livestock provides alternate sources of income to farmers, thus supporting their sustainable livelihood
7	Per capita income (3 years average)	Higher per capita, higher standard of living
8	Cropping intensity	Higher cropping intensity means that a higher portion of the net area is being cropped more than once during one agricultural year. Higher cropping intensity, greater is the efficiency of land use
9	Percentage irrigated area to total cropped area (3 years average)	Irrigation protects crop production in the light of climate variations and extremes like drought
10	Total area under fruit crops	Alternative source of farm based income
Agriculture		
1	Net sown area (3 years average)	Represents the total area sown with crops
2	Cropping intensity	Higher cropping intensity means that a higher portion of the net area is being cropped more than once during one agricultural year. Higher cropping intensity, greater is the efficiency of land use
3	Percentage of area under commercial crops to TCA	Higher the area under commercial crops, lesser vulnerability (market linkage) High value cash crops represent one potential avenue of crop intensification
4	Percentage of irrigated area to TCA (3 years average)	Irrigation has the potential to provide higher yields than rainfed agriculture and reduce the insecurity of crop production

(continued)

Table 3 (continued)

No.	Indicators	Rationality
5	No. of tractors/1000 ha area sown	More tractors for agricultural activities (mechanization), increase the efficiency in production
6	% fallow land as proportion of total graphical area (3 years average)	Higher fallow land means more non utilization of resources
7	No. of agricultural cooperative credit societies/lakh population	It is a grass roots financial institution and provides credit to farmers in times of need, securing crop production

Above indicators have been selected based on consultation with experts, and studies reported (Hiremath and Shiyani 2013; Khan and Salman 2012; Swain and Swain 2011; ICRISAT 2009)

Vulnerability is the degree to which a system is susceptible to, or unable to cope with adverse effects of climate change, including climate variability and extremes. Vulnerability is a function of the character, magnitude and rate of climate change to which a system is exposed, along with its sensitivity and adaptive capacity.

Exposure is the nature and degree to which a system is exposed to significant climatic variations

Sensitivity is the degree to which a system can be affected, negatively or positively, by changes in climate. This includes change in mean climate and the frequency and magnitude of extremes. The effect may be direct (for example, a change in crop yield due to a change in temperature) or indirect (such as damage caused by increased frequency of coastal flooding due to sea-level rise).

Adaptive Capacity is a system's ability to adjust to climate change (including climate variability and extremes), to moderate potential damage, and to take advantage of opportunities or to cope with consequences.

Source: IPCC, Fourth Assessment Report, 2007

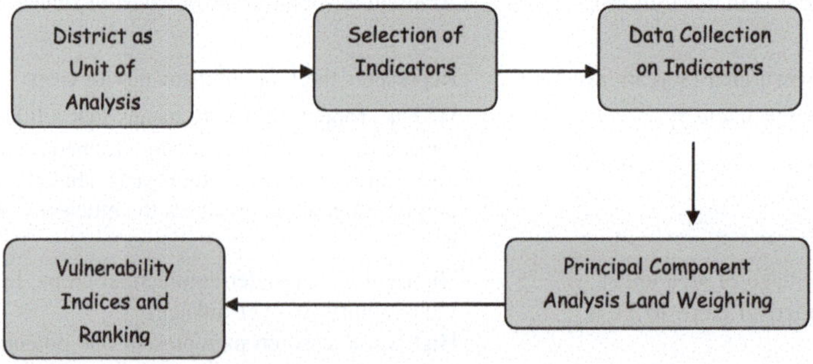

Fig. 14 Framework of assessment of vulnerability

PCA helped with generation of the weights, based on the assumption that there are common factors that explain the variance in the vulnerability. Varimax rotation was performed on the results of the PCA to maximize the variance accounted by the first component. Only components with Eigen values >1 were included in the analysis.

Table 4 Total variance explained by principal components for agricultural vulnerability

Component	Initial Eigen values			Rotation sums of squared loadings		
	Total	% variance	Cumulative %	Total	% variance	Cumulative %
1	2.192	31.309	31.309	1.815	25.924	25.924
2	1.566	22.369	53.678	1.659	23.699	49.623
3	1.292	18.454	72.133	1.576	22.510	72.133
4	0.836	11.946	84.079			
5	0.543	7.754	91.832			
6	0.385	5.504	97.337			
7	0.186	2.663	100.000			

Table 5 Total variance explained by principal components for socio-economic and livelihood vulnerability

Component	Initial Eigen values			Rotation sums of squared loadings		
	Total	% variance	Cumulative %	Total	% variance	Cumulative %
1	3.531	35.307	35.307	3.086	30.856	30.856
2	1.825	18.247	53.553	2.032	20.315	51.171
3	1.328	13.284	66.837	1.567	15.666	66.837
4	0.995	9.953	76.790			
5	0.795	7.949	84.739			
6	0.651	6.507	91.246			
7	0.355	3.545	94.792			
8	0.234	2.341	97.132			
9	0.163	1.629	98.762			
10	0.124	1.238	100.000			

Based on Statistical Packages for Social Sciences output, the findings of the study for the agriculture and socio-economic based classification on vulnerability indicators revealed three components, each with Eigen values greater than 1 (Tables 4 and 5). These two PCA results explain 72 and 67 % of the total variation in the two data sets.

There is subjectivity in assigning weights to indicators in vulnerability assessments. In order to overcome this problem, we employed the PCA technique through which we reduced the number of variables and also obtained weights (Eigen values) for the PCs. In the present study, weights are not therefore arbitrarily assigned but determined endogenously from the data matrix. The weights of the PCs are the corresponding Eigen values (Tables 6 and 7; Figs. 15 and 16).

Table 6 Agricultural vulnerability index (Principal component analysis) for Eigen value (*E*)

No.	District	Component 1 (*E*1) 2.192	Component 2 (*E*2) 1.566	Component 3 (*E*3) 1.292	Composite index	Rank
1	Kolar	−0.85511	−1.11939	−0.73926	−0.90742	1
2	Ramanagara	−0.87804	−1.27245	−0.15289	−0.81482	2
3	Chikkaballapura	−0.82441	−1.07869	−0.05907	−0.70746	3
4	Bangalore(R)	−1.16193	−0.16957	−0.49661	−0.68398	4
5	Chikmagalur	−1.3684	−0.58613	0.77913	−0.57639	5
6	Kodagu	−1.71026	−0.42877	1.3934	−0.51883	6
7	Shimoga	0.62477	−2.61467	0.15007	−0.50123	7
8	Tumkur	−0.62399	−0.1713	−0.47779	−0.44621	8
9	Chitradurg	−0.80222	0.36051	−0.21221	−0.29071	9
10	Chamrajnagar	−0.4255	−0.39351	0.06777	−0.28938	10
11	Raichur	0.60885	0.43638	−2.4042	−0.2155	11
12	Bellary	0.65088	−0.41453	−1.20227	−0.15362	12
13	Koppal	0.16297	0.71543	−1.23653	−0.02376	13
14	Gulbarga	−1.05421	1.79857	−0.46937	−0.01994	14
15	Hassan	−0.14221	−0.02496	0.29463	0.005911	15
16	Davengere	0.39099	−0.87883	0.54725	0.037198	16
17	Bidar	−0.73042	1.47056	−0.29887	0.06251	17
18	Yadgiri	0.5131	0.66057	−1.3445	0.083579	18
19	Bijapur	−0.17298	1.34544	−0.04336	0.331043	19
20	Mandya	1.18009	−0.61425	0.38476	0.420188	20
21	Bagalkot	1.08404	0.13255	0.59092	0.662823	21
22	Dharwad	1.29482	0.3963	0.11332	0.713913	22
23	Mysore	2.00556	−0.00639	−0.39	0.768772	23
24	Gadag	−0.03265	1.6015	1.63404	0.900507	24
25	Haveri	0.32733	0.91954	2.21027	0.992708	25
26	Belgaum	1.93892	−0.06391	1.36136	1.170081	26

Source Based on Directorate of Economics and Statistics, Ministry of Agriculture, 2008–09, 2009–10 and 2010–11

Note Bangalore (U) is not included in the construction of vulnerability index. This is due to the fact that Bangalore (U) is the capital city of Karnataka and since last one decade agricultural activities have been reduced rapidly. It was observed that in some districts of Western Ghats region, agricultural indicators are performing very poor and also affecting index values of other districts while constructing vulnerability index. Therefore, we consider Bangalore (U), Uttara Kannada, Dakshin Kannada and Udupi as outlier districts and excluded from agricultural vulnerability index analysis

Composite Index = $E1$ * fact1 + $E2$ * fact2 + $E3$ * fact3/$E1$ + $E2$ + $E3$

Weight 1 = $E1/E1$ + $E2$ + $E3$, Weight 2 = $E2/E1$ + $E2$ + $E3$, Weight 3 = $E3/E1$ + $E2$ + $E3$

Weight for Factor 1 = 0.434059, Weight for Factor 2 = 0.310099 and Weight for Factor 3 = 0.255842

Table 7 Socio-economic and livelihood vulnerability index (Principal component analysis for Eigen value (*E*)

No.	Districts	Component 1 (*E*1) 3.53	Component 2 (*E*2) 1.82	Component 3 (*E*3) 1.32	Composite index	Rank
1	Yadgiri	−1.4647	−1.2176	0.0553	−1.0952	1
2	Chitradurg	−0.7278	−0.7424	−1.6660	−0.9182	2
3	Raichur	−0.8808	−1.3046	0.0125	−0.8190	3
4	Chamrajnagar	−0.7335	−0.2774	−0.5385	−0.5702	4
5	Chikkaballapura	−0.5914	0.3343	−1.5872	−0.5364	5
6	Tumkur	−0.7033	0.6696	−0.9432	−0.3761	6
7	Chikkamagalur	−0.1782	0.0340	−1.4241	−0.3678	7
8	Koppal	−0.4457	−0.6864	0.2939	−0.3644	8
9	Hassan	−0.6769	0.5739	−0.6387	−0.3278	9
10	Bidar	0.2911	−1.5959	−0.0977	−0.3013	10
11	Gulbarga	0.1991	−1.4325	0.1460	−0.2569	11
12	Bellary	−0.1886	−0.7484	0.3989	−0.2247	12
13	Gadag	0.2242	−1.1114	0.2768	−0.1299	13
14	Ramanagara	−0.3164	1.0652	−1.0845	−0.0917	14
15	Davengere	−0.3537	−0.0738	0.5905	−0.0897	15
16	Bagalkot	−0.4631	−0.3315	1.3301	−0.0708	16
17	Haveri	0.1526	−0.4582	−0.1074	−0.0658	17
18	Bijapur	−0.0531	−0.5579	0.8446	−0.0125	18
19	Kolar	−0.2010	1.3257	−1.0409	0.0489	19
20	Kodagu	0.9383	−0.4084	−1.6853	0.0493	20
21	Mandya	−0.6866	1.0867	0.6148	0.0561	21
22	Belgaum	−0.2871	0.2015	1.4086	0.1831	22
23	Shimoga	−0.6208	1.1761	1.1372	0.2191	23
24	Mysore	0.1396	−0.0234	1.4892	0.3632	24
25	Bangalore (R)	0.7940	0.4613	−0.8811	0.3703	25
26	Uttara Kannada	0.3738	0.9583	0.5949	0.5773	26
27	Dharwad	1.1340	−0.7334	1.8957	0.7754	27
28	Udupi	0.5460	1.6941	0.6713	0.8843	28
29	Dakshin Kannada	0.5001	2.5343	0.5759	1.0705	29
30	Bangalore (U)	4.2800	−0.4116	−0.6419	2.0211	30

Source Based on Census of India 2011, Agricultural Census 2010–11, Directorate of Economics and Statistics, Ministry of Agriculture, 2008–09, 2009–10 and 2010–11 and Livestock Census 2007

Note Composite Index = $E1 * \text{fact1} + E2 * \text{fact2} + E3 * \text{fact3}/E1 + E2 + E3$

Weight 1 = $E1/E1 + E2 + E3$, Weight 2 = $E2/E1 + E2 + E3$, Weight 3 = $E3/E1 + E2 + E3$

Weight for Factor 1 = 0.528276, Weight for Factor 2 = 0.27304 and Weight for Factor 3 = 0.198683

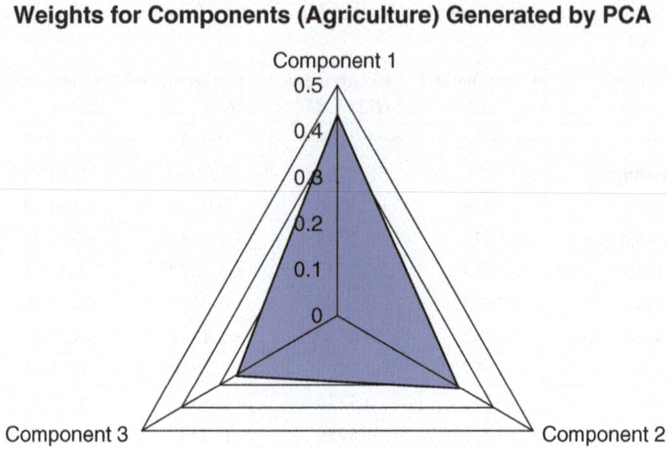

Fig. 15 Weights for agricultural component indicators generated by PCA

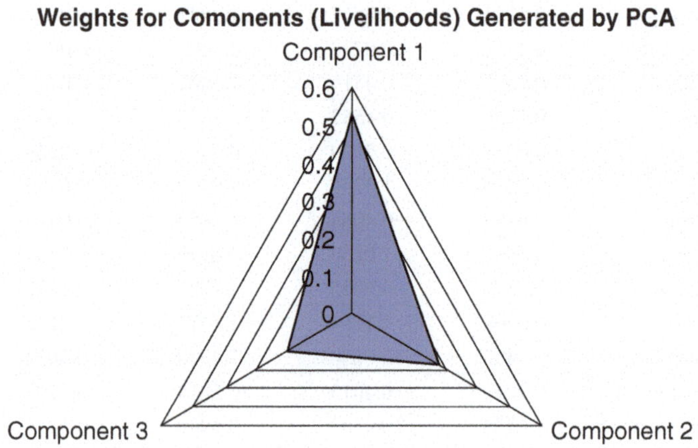

Fig. 16 Weights for socio-economic component Indicators generated by PCA

7 Agricultural Vulnerability Index for the Districts of Karnataka

In the present study, 7 indicators were considered for the development of agricultural vulnerability index (Table 3). Based on PCA, agricultural vulnerability index values for all the districts of Karnataka are given in Table 6. Rank 1 indicates maximum vulnerability and the vulnerability decreases with increasing rank. Figure 17 depicts the agricultural vulnerability of the districts of Karnataka. Areas in red are the most vulnerable districts and those in green are the least vulnerable

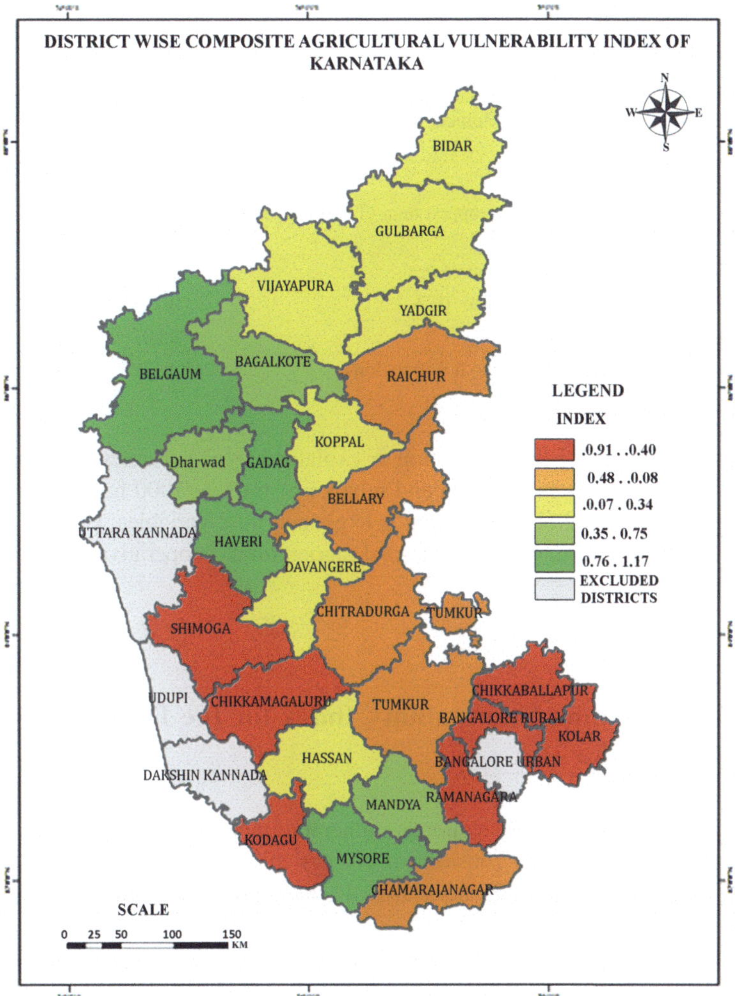

Fig. 17 Agricultural vulnerability index (*Note* Areas in *red* are the most agriculturally vulnerable districts and those in *green* are the least vulnerable districts of Karnataka. Bangalore (U) is not included)

districts of Karnataka. Table 8 gives the details of significance of variables that explain variation in each component.

- The rotated factor analysis generated 3 components which account for approximately 72 % of the total cumulative variance in agricultural vulnerability.
- In component 1, 26 % of variation is explained by 3 variables, namely, % of gross area irrigated, cropping intensity and % of commercial crops to TCA.

Table 8 Rotated component matrix

Agricultural indicators	Component		
	1	2	3
% of net sown area to total geographical area (TGA) (3 years average)	0.190	0.836	0.085
Cropping intensity	0.754	0.269	0.089
% of gross irrigated area to total cropped area (TCA) (3 years average)	0.685	0.369	0.220
% of fallow land to TGA (3 years average)	0.286	0.237	0.779
% of area under commercial crops to TCA	0.793	0.284	0.415
No. of tractors/1000 ha area sown	0.007	0.772	0.103
Agricultural cooperative societies/lakh population	0.172	0.140	0.850

- In component 2, 24 % of variation is explained by 2 variables, namely, % of NSA to total geographical area and number of tractors/1000 ha area sown.
- In component 3, 22 % of variation is explained by 2 variables, i.e. % of fallow land to total geographical area, and agricultural cooperative societies/lakh population.

8 Socio-economic Vulnerability Index for the Districts of Karnataka

For development of the socio-economic vulnerability index, 10 important indicators described in Table 3 were considered. As agriculture is a dominant livelihood activity, a few agricultural indicators have also been included in the development of this index.

In Table 7 rank 1 indicates maximum vulnerable district and vulnerability decreases with increasing rank. PCA shows that Yadgiri, Chitradurg, Raichur, Chamrajnagar and Chikkaballapura are the top five socio-economically vulnerable districts. Bangalore (U), Dakshin Kannada, Udupi, Dharwad and Uttara Kannada are the least socio-economically vulnerable districts of Karnataka. Figure 18 depicts the socio-economically vulnerable districts of Karnataka, with red and green coloured areas representing most and least vulnerable districts, respectively. Table 9 gives the details of significance of variables that explain variation in each component.

- The rotated factor analysis generated 3 components which account for approximately 67 % of the total cumulative variance in socio-economic and livelihood vulnerability.

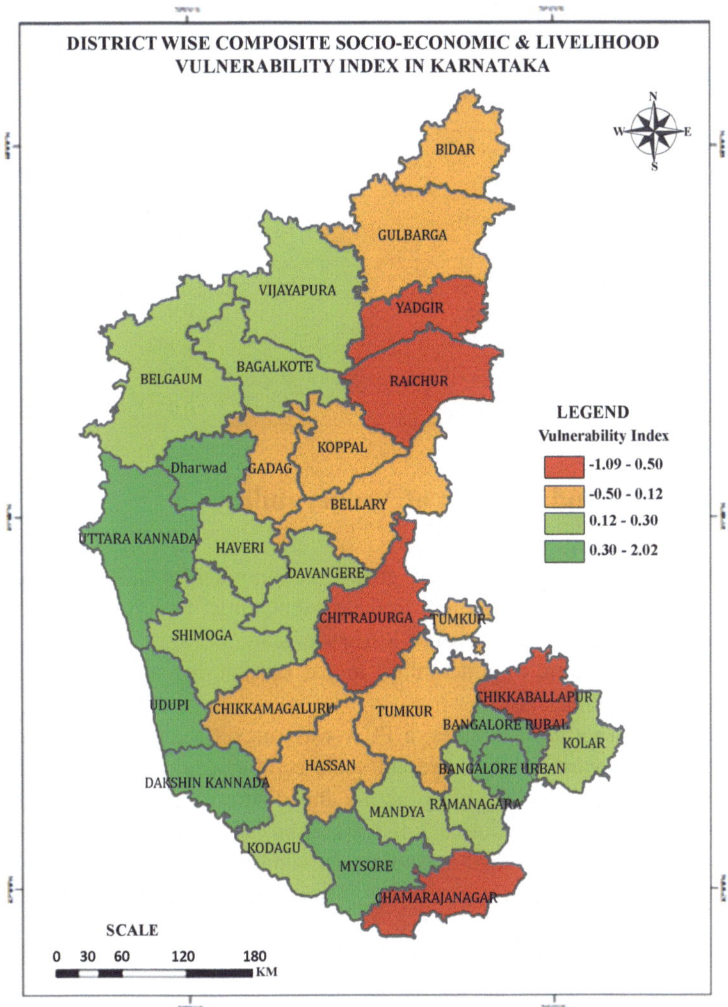

Fig. 18 Socio-economic vulnerability index. (*Note* Areas in *red* are the most socio-economically vulnerable districts and those in *green* are the least vulnerable districts of Karnataka)

- Factor 1 that accounts for the largest variance (about 31 %) includes population density, percentage of literacy rate, livestock unit/lakh population and per capita income.
- In component 2, 20 % variation is explained by 3 variables, namely, percentage of SC and ST population, percentage of marginal land holders and total area under fruit crops.
- In factor 3, 16 % of variation is explained by 3 variables, namely, percentage of non-workers, cropping intensity and percentage of irrigated area.

Table 9 Rotated component matrix (a)

Socio-economic indicator	Component		
	1	2	3
Density of population	0.825	0.040	0.076
% SC and ST population	0.467	0.529	0.404
Total literacy rate (%)	0.628	0.558	0.001
% of marginal land holders	0.172	0.802	0.229
% of non-workers	0.329	0.224	0.739
Livestock units per lakh population	0.850	0.105	0.107
(3 years average) per capita income	0.875	0.202	0.112
Cropping intensity	0.142	0.302	0.668
% gross irrigated area to total cropped area (3 years average)	0.382	0.272	0.563
Total fruit crops area	0.034	0.728	0.110

9 Discussion and Drivers of Vulnerability

Karnataka state is one of the fastest growing economies in India. Agriculture in Karnataka is predominantly rainfed. In the present study, two vulnerability indices were developed at the district level of Karnataka state, i.e. agricultural vulnerability index and socio-economic vulnerability index, considering all the 30 districts of Karnataka for the analysis.

In order to derive these indices, a PCA was run on a data set of 10 carefully selected indicator variables to represent socio-economic vulnerability and 7 indicators for agricultural vulnerability across the districts of Karnataka. The PCA generated three components for each index that broadly represented the underlying themes of agriculture and socio-economic vulnerability present in the larger data set. The findings suggest

- Agricultural vulnerability
  - Kolar, Ramanagara, Chikkaballapura and Bangalore (R) are the most agriculturally vulnerable districts of Karnataka.
  - Belgaum, Haveri and Gadag are the least agriculturally vulnerable districts.
- Socio-economic and livelihood vulnerability
  - Yadgiri, Chitradurg, Raichur, Chamrajnagar and Chikkaballapura districts are the most vulnerable among all districts of Karnataka.
  - Bangalore (U), Dakshin Kannada, Udupi, Dharwad and Uttara Kannada are the least vulnerable districts of Karnataka.

The result of agricultural vulnerability index suggests indicators like cropping intensity, gross area irrigated and commercial crop area are the major drivers in determining the vulnerability of districts. The livelihood vulnerability index analysis suggests Yadgir, Chitradurg, Raichur, Chamrajnagar and Chikkaballapura are the most vulnerable districts in Karnataka. The livelihood index depicts indicators like per capita income, population density, percentage of literacy rate and livestock units/lakh population, which are the major drivers and contribute to the overall livelihood vulnerability of districts.

References

Adger, W. N., Brooks, N., Bentham, G., Agnew, M., & Eriksen, S. (2004). New Indicators of Vulnerability and Adaptive Capacity. Technical Report 7, Tyndall Centre for Climate Change Research. Norwich UK.

Allen, K. (2003). Vulnerability reduction and the community-based approach. In Pelling (Ed.), *Natural disasters and development in a globalizing world* (pp. 170–184). London, UK: Routledge.

Fussel, H.-M. (2007). Vulnerability: A generally applicable conceptual framework for climate change research. *Global Environmental Change, 17*, 155–167.

Government of Karnataka. (2005). *Human Development Report, Karnataka*. Bangalore, India: Government of Karnataka.

Hiremath, D., & Shiyani, R. L. (2013). Analysis of vulnerability indices in various agro-climatic zones of Gujarat. *Indian Journal of Agricultural Economies, 68*(1), 122

ICRISAT. (2009). Quantitative assessment of vulnerability to climate change. (http://www.icrisat.org/what-we-do/impi/training-cc/october-2–32009/vulnerability-analysis-manual.pdf. Accessed Aug 2013).

IISc. (2013). *Environmental benefits and vulnerability reduction through Mahatma Gandhi National Rural Employment Guarantee Scheme*. Synthesis Report 2013. Bangalore, India: Indian Institute of Science.

IPCC. (1996). *Climate change 1995 second assessment report of the intergovernmental panel on climate change*. Geneva: Switzerland.

IPCC. (2001). *Climate change 2001: The scientific basis*. Contribution of Working Group I to the Third Assessment Report of the Intergovernmental Panel on Climate Change. Cambridge, UK: Cambridge University Press.

IPCC. (2007). *IPCC Fourth Assessment Report: Climate Change 2007 (AR4)*. Cambridge, UK: Cambridge University Press.

Karade, J. (2008). *Development of Scheduled Castes and Scheduled Tribes in India*. UK: Cambridge School of Publishing.

Khan, F. A., & Salman, A. (2012). A simple human vulnerability index to climate change hazards for Pakistan. *International Journal of Disaster Risk Science, 3*(3), 163–176.

Nicholls, R. J., Hoozemans, F. M. J., & Marchand, M. (1999). Increasing flood risk and wetland losses due to global sea-level rise: Regional and global analyses. *Global Environmental Change, 9*, S69–S87.

Senbeta, A. F. (2009). *Climate change impact on livelihood, vulnerability and coping mechanisms: A case study of West Arid Zone, Ethiopia*. MSc thesis submitted to Lund University Masters Program in Environmental Studies and Sustainability Science (LUMES), Lund, Sweden.

Smith, L. I. (2002). A tutorial on principal component analysis. In H. Anton (Ed.), *Elementary linear algebra 5e* (Vol. 2012). New York: Willey. ISBN, 0-471-852236.

Stern, N., Peters, S., Bakhshi, V., Bowen, A., Cameron, C., Catovsky, S., et al. (2006). Stern review: The economics of climate change. New York: OUP.

Swain, M., & Swain, M. (2011). Vulnerability to agricultural drought in Western Orissa: A case study of representative blocks. *Agricultural Economics Research Review, 24*, 47–56.

Watson, R. T., Zinyoera, M. C., & Moss, R. H. (1996). *The regional impact of climate change: An assessment of vulnerability*. A Special Report of IPCC Working Group II. Cambridge, UK: Cambridge University Press.

Knowledge, Perception and Socioeconomic Vulnerability of Urban and Peri-urban Households to Heat Waves in Pakistan

Khuda Bakhsh, Sara Rauf and Azhar Abbas

Abstract Heat waves cause a large number of morbidity and mortality cases each year in Pakistan. Although the institutions cannot overcome heat waves, its intensity in the forms of morbidity and mortality can be minimized by taking precautionary measures. We examine knowledge and perception of population regarding heat waves in Faisalabad. Results show that 53 % respondents in peri-urban areas are involved in open field jobs and such respondents are more vulnerable to heat waves in the absence of no precautionary measures. Further, long working hours increase misery of such individuals. Access to health services is poor, since visits made by health workers to households are very few per year. Unfortunately, 20–30 % individuals having knowledge on heat waves do not follow instructions to avoid adverse effects of heat waves. Strengthening the role of print and electronic media, social networks, and health institutions can save many lives in the heat prone areas in Pakistan.

1 Background

Heat waves are defined as a situation of hot temperature with increased humidity. Outcomes of such situation are in the form of distress and uncomfortable feelings to the people (Das and Smith 2012). As per the definition of the Australian Bureau of

K. Bakhsh (✉)
Department of Management Sciences, COMSATS Institute of Information Technology, Vehari, Pakistan
e-mail: kbakhsh@ciitvehari.edu.pk

S. Rauf · A. Abbas
Institute of Agricultural and Resource Economics, University of Agriculture, Faisalabad, Pakistan

A. Abbas
Leibniz Centre for Agricultural Landscape Research (ZALF), 15374 Müncheberg, Germany

S. Nautiyal et al. (eds.), *Climate Change Challenge (3C) and Social-Economic-Ecological Interface-Building*, Environmental Science,
DOI 10.1007/978-3-319-31014-5_12

Meteorology, heat waves situation is an abnormal high and low temperature for three or more consecutive days in any region.

Over the last few years, the world has suffered from heavy losses as a result of increased numbers, duration, and intensity of the weather and disastrous events associated with climate change which are projected to intensify in the future (Meehl and Tebaldi 2004). Like other countries, Pakistan also experiences different types of climate-related hazards such as floods, drought, earthquake, heat waves, etc. Heat waves cause several deaths during summer in Pakistan. The recent worst case happened in Karachi where dozens of deaths were reported in a few days in the summer 2015.

Many developed and developing nations have suffered from heat waves due to their destructive nature during the recent years. Western Europe experienced the deadliest heat waves causing almost 70,000 human deaths in 2003 (Robine et al. 2008). Heat waves resulted in more than 500 deaths in Australia (Akompab et al. 2013) and 15,000 deaths in Russia in 2009 (Dole et al. 2011). Heat waves occurrence is frequent in Asian countries as well, as they resulted 3000 deaths during the summer 2003 (Dos and Smith 2012) and 500 deaths in 2013 in India (TNN 2013).

Southern parts of Pakistan, particularly southern Punjab and Karachi have faced severe heat waves in recent years. Duration of heat waves also extended in these years since 1980, causing a severe damage to limbs and lives (Vidal and Sethna 2013). During the last decade, several deaths were reported in the national newspapers due to heat waves in Pakistan. In 2007, extreme heat waves took the lives of 232 people, in addition to several hospitalizations (Deutsche Presse Agentur 2007).

Impact of heat waves is enormous for vulnerable segments of the society. Heat waves make the situation worse for elderly individuals and children particularly (Sheridan 2007; Hansen et al. 2011). Those already having health problems are vulnerable to heat waves and chances of mortality and morbidity are too high among such people (Vandentorre et al. 2006). Kosatsky et al. (2009) argue that vulnerability to heat waves is too high among individuals suffering from heart and lung problems and therefore, mortality and morbidity rate increases for these people (Kovats and Hajat 2008). In Pakistani context, impoverished people such as laborer and farm workers are at high health risk due to heat waves. Such individuals have no and/or very few means to use precautionary measures while working under starching heat. Thus, admission in hospitals increases for heat stroke, hyperthermia, lung and heart problems. Furthermore, poor people are badly affected as they have little access to health facilities. Limited medical facilities in the underdeveloped districts increases the misery of poor people.

2 Knowledge and Perception About Heat Waves

The destructive health impacts of heat waves are mitigated by improving environmental conditions and infrastructure designs as well as controlling exposure to heat waves (Menne and Matthies 2009) and adopting protective measures.

However, level of knowledge among individuals helps in making appropriate arrangements for tackling climate-related risks, such as heat waves. The source of information and interactions with society contribute in building capacity of knowledge and perception of general public (Carvalho 2007). Role of awareness in building capacity of communities on heat waves is well evident in the literature (Sheridan 2007; Chowdhury et al. 2012). Sharples (2010) provides evidence on the importance of information through print and electronic media and their impact on perception to specific issue. It is argued that households with high income level are expected to have high perception level (Deressa et al. 2011). As a result, knowledge and perception building of households lead to adaptation, thereby reducing impacts on human health (Akompab et al. 2013; Liu et al. 2013).

In addition to perception level, socioeconomic factors affect adaptation to heat waves. These factors may include family size, access to information, credit, and extension services (Falco et al. 2011; Deressa et al. 2011); education, gender, and temperature (Deressa et al. 2011). Consequently, effective policy formulation addressing the issue of heat waves may be achieved through knowledge about these factors and hence would provide sound basis for effective planning for the mitigation of detrimental impacts (Sherwood and Huber 2010).

Understanding public attitudes and perceptions towards heat waves is important as they may help in boosting focus through public education and communication strategies for this issue under a changing climate. Furthermore, having an understanding of public attitudes may be important since any success to address potential consequences associated with heat waves may depend on public views about the phenomenon. Such information can help to design and formulate strategies to reduce the adverse impacts of heat waves. There is little evidence on perception of and adaptation to heat waves in Pakistan, although many studies are conducted in other countries. This research presents information on knowledge, perception, and adaptation to heat waves in a selected district of Pakistan.

3 Faisalabad Case Study

Faisalabad district in the province of Punjab was selected as a case study district, as historical data show a considerable population of this district affected by heat waves in the form of several deaths and senselessness in Faisalabad (Pak Tribune 2004). Faisalabad is located in the central Punjab, Pakistan. This is among the most populated cities of Pakistan. Its population as per 2013 estimates is 7085 thousands. The district has 289 union councils, out of which 166 are rural and 123 are urban which shows majority of rural population in the district. Out of 751 thousands households, 37 % have access to safe drinking water without bacteria, 83 % use sanitary means of excreta disposal, and 90 % have physical access to health facilities within half an hour distance. All this highlights poor environmental conditions prevailing in the district. Although, these conditions are somewhat better than those prevailing in many southern districts of the province, yet these statistics are not satisfactory. Considering educational facilities, there are only 1274 government primary schools with enrolment of 227,905 (134,081 male and 93,824 female). Student–teacher ratio is very poor as there is only 4703 teaching staff (2716 male and 1987 female). Consequently, literacy rate is only 51.9 % as per Population Census 1998. The situation is even worst among female with a literacy rate of 42 %. Being the third largest populated district, health facilities are not fully available to all households. There are only 26 hospitals with 3342 beds in Faisalabad district. In spite of low patient to hospital ratio, annually 7.6 million patients receive treatment.

3.1 Socioeconomic Characteristics

We conducted survey for the collection of information on knowledge about heat waves, their perception about impacts and adaptation from urban and peri-urban residents of Faisalabad. Socioeconomic profile of the respondents is presented in Table 1 to examine their possible vulnerability to heat waves. Employment status and nature of job determine vulnerability of the respondents because heat waves may have strong effects on those working in open environments. As given in Table 1, more than 40 % of the respondents with urban background were employed whereas, for peri-urban respondents, this percentage was 67.3. The low percentage of employment of the urban respondents may be due to the fact that most respondents are students and businessmen. On the other hand, higher employment percentage of the peri-urban respondents may be due to the fact that they may have started working early due to financial constraints.

We find five types of primary occupation reported by the respondents. They include student, businessman, job holder, at-home work, and agriculture. The 'at-home' category of respondents included the housewives, elderly, and retired persons. 'On-job' included those doing private and public jobs. As given in Table 1, majority (38 %) of the respondents of urban background work at home for their earnings. Respondents with employed status were the second largest (26.7 %) and students were the third largest respondent type in urban area. In the case of peri-urban areas, majority of the respondents were employed (33.7 %) whereas agriculture (23.8) was the second major occupation reported by the respondents. The student and businessman categories had minor proportions, i.e., 11.9 and 9.9 %, respectively. A higher employed percentage of the peri-urban respondents is due to the fact that most respondents were working on daily wages due to low education level.

Nature of the job of the employed respondents and their working hours are given in Table 1. Majority of the urban respondents had an office job (78.7 %) with 8.5 mean working hours. Most of the employed peri-urban respondents had a field job (52.9 %) with 8.36 mean working hours. The average time consumed by urban respondents to reach the work place was 19.33 min whereas for the peri-urban respondents, it was 24.37 min. The transportation mode used to reach the work

Table 1 Socioeconomic characteristics of the respondents

Characteristics	Units	Urban	Peri-urban
Employment status	No. (percentage)		
Employed		61 (40.7)	68 (67.3)
Unemployed		89 (59.3)	33 (32.7)
Primary occupation	No. (percentage)		
Student		32 (21.3)	12 (11.9)
Businessmen		21 (14.0)	10 (9.9)
On-job		40 (26.7)	34 (33.7)
At-home		57 (38.0)	21 (20.8)
Agriculture			24 (23.8)
Job nature	No. (percentage)		
Field		13 (21.3)	36 (52.9)
Office		48 (78.7)	32 (47.1)
Working hours	Hours (percentage)		
Field		9 (1.63)	8.36 (2.19)
Office		8.5 (2.41)	7.94 (1.41)
Commutation time to work	Minutes (SD)	19.33 (16.69)	24.37 (21.05)
Mode of transport	No. (percentage)		
By foot		14 (23.0)	24 (34.8)
Car		14 (23.0)	2(2.9)
Bike		19 (31.1)	22(31.9)
Public transport		11 (18.0)	14(20.3)
Cart		–	5(7.2)
Others		2 (2.9)	3(4.9)
Income group	No. (percentage)		
Low		11 (7.3)	24 (23.7)
Middle		38 (25.3)	44 (43.6)
High		101 (67.3)	33 (32.7)

place by majority of the urban respondents was motorbike (31.1 %) whereas, most of the peri-urban respondents traveled by foot (34.8 %) to reach their work place. Similarly, a relatively higher percentage (23 %) of the urban respondents also used cars as means of transport whereas this percentage was relatively low (2.9 %) in peri-urban areas. Other means of transportation was bicycle. Higher commuting time and poor mode of transportation indicate vulnerability of peri-urban respondents to heat waves compared to urban respondents.

Employing the criteria used by the Household Income and Expenditure Survey (2011), respondents were divided on the basis of income into various categories. Results indicated that majority of the urban respondents belonged to high income group (67.3 %), followed by middle income group (25.3 %) and low income group (7.3 %). In case of peri-urban respondents, the majority belonged to middle income

Table 2 Availability of health facilities for the respondents

Characteristics	Units	Urban	Peri-urban
Family members in health department	No. (SD)	0.32 (0.71)	0.75 (5.98)
Relatives in health department	No. (SD)	2.12 (3.90)	0.80 (1.22)
Distance to a health center	Minutes (SD)	11.00 (7.51)	18.73 (13.53)
Visits of health workers per year	No. (SD)	3.32 (2.66)	3.69 (4.15)

group (43.6 %). As a result, peri-urban respondents are relatively poor compared to urban respondents and thus they have less means to take precautionary measures against heat waves.

3.2 Access to Health Facilities

Information about access to health facilities for family member and relatives/friends is given in Table 2. The access to such facilities by family members and friends is a means of information exchange as well in the study area which in turn leads to increase in knowledge about the severity and impact of heat waves. Moreover, the distance from a nearest healthcare center and numbers of visits by health workers are important factors in creating awareness among the respondents. The results indicate that urban respondents have 2.12 relatives/friends who are working in health department whereas in case of peri-urban areas, this ratio is 0.80. The distance to a healthcare center in the urban area was less in terms of time required to reach (11 min) as compared to the peri-urban areas (18.73 min). This is basically due to the presence of better infrastructure within urban areas. On the other hand, average number of visits by health workers is almost same in both the areas.

3.3 Knowledge About Heat Waves and Its Sources

Respondents were prompted to share the mode of information about the forthcoming problems related to heat waves. These sources are shown in Fig. 1. In the urban areas, television and/or radio are found to be the main source (68 % responses) for getting information about heat waves. Family and friends are source of information for 27 % respondents; others (24 %), internet (21 %), and newspapers (16 %) respondents. The other sources include experience, intensity of heat, etc. Similarly, in the peri-urban areas, television/radio are the main source of information as it was cited by 58 % respondents, family/friends (24 %), others (21 %), newspapers (18 %), and internet was the least source of information reported by the respondents (7 %).

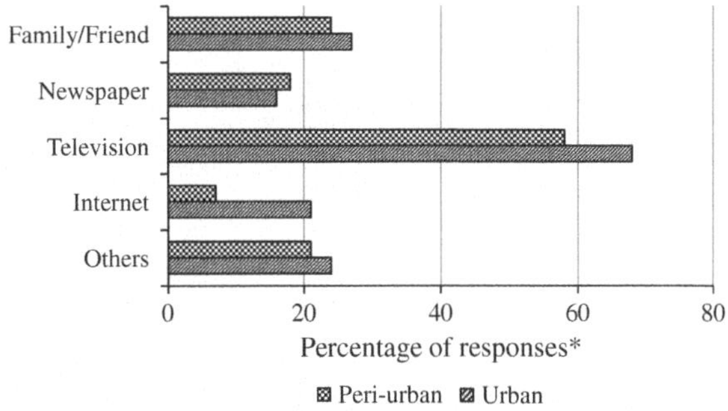

* Multiple responses are possible, therefore percentage may be more than 100 %

Fig. 1 Source of information about forthcoming heat waves. *Asterisk* Multiple responses are possible, therefore percentage may be more than 100 %

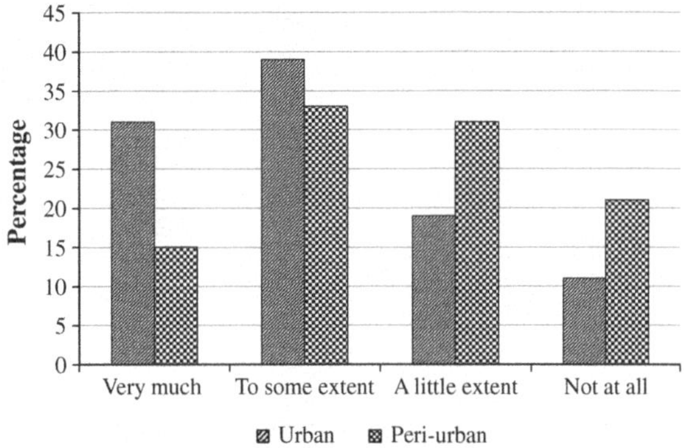

Fig. 2 Extent of news followed by the respondents regarding heat waves

In addition, respondents were asked about the extent to which they followed the news regarding heat waves from the sources they mentioned. In this regard, Fig. 2 demonstrates the responses of urban and peri-urban respondents. The extent of following news by urban respondents was found to be higher compared to the respondents from peri-urban areas. Only 11 % of the urban respondents did not follow any news about the heat waves at all. This percentage was 21 for the respondents from peri-urban areas. The higher extent of adopting information in the urban areas may be due to higher educational levels and more interaction with the relatives/friends in health department.

Figure 3 shows level of information about heat waves and their consequences among respondents of the study area. In urban area, nearly half (47 %) of the respondents reported to be fairly informed about heat waves and their consequences, 24 % utterly informed, 23 % slightly informed, and 6 % not informed at all. While in the peri-urban area, 35 % respondents reported that they were fairly informed, 30 % slightly informed, 11 % utterly informed, and 24 % not at all informed. The level of information about heat waves and their consequences among respondents from the urban areas was comparatively higher due to higher educational level and access to various sources of information.

Fig. 3 Level of information about heat waves and their consequences

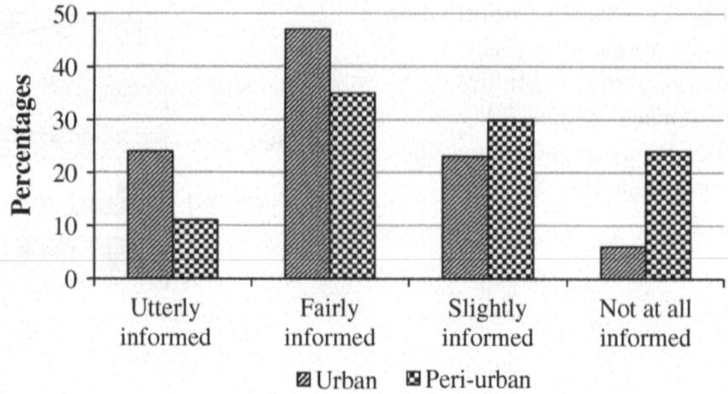

Fig. 4 Extent of personal worriedness about the effects of heat waves

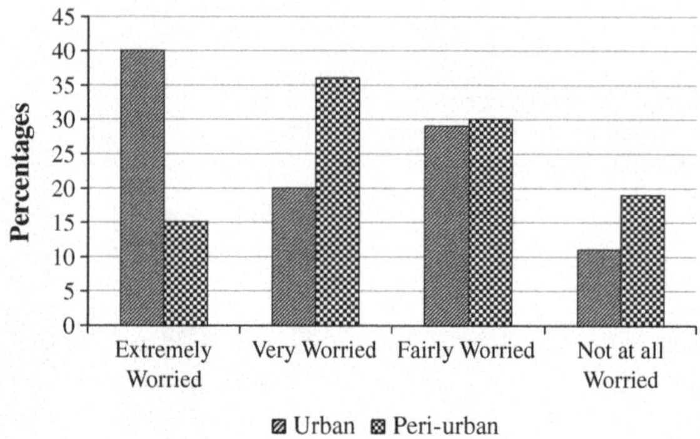

The concern level of respondents about effects of heat waves on themselves is given in Fig. 4. Respondents were asked how much worried were they about the effects of heat waves. Responses recorded show that majority of urban respondents (40 %) were extremely worried, followed by 29 % who were fairly worried whereas 20 % claimed to be very worried, and 11 % were not at all worried about the effects of heat waves. On the other hand, in the peri-urban background, 36 % respondents were reported as very worried, followed by 30 % fairly worried about the effects of heat waves whereas 15 % were extremely worried, and 19 % were not at all worried about the effects of heat waves. The difference between urban and peri-urban respondents regarding concern level about the effects of heat waves was found approximately the same except for one option (extremely worried).

In case of knowledge relating to heat waves, majority of the urban respondents (80 %) had good knowledge about the heat waves and its impacts on health whereas it was 71.3 % for peri-urban respondents, being slightly less than that of the urban respondents. In terms of individual statements used to quantify the knowledge level, more than two third of the urban respondents (i.e., 77 %) as well as peri-urban respondents (i.e., 82 %) reported that '*heat waves are caused by high temperature and low rainfall.*' Similarly, higher percentage of urban (93 %) and peri-urban (83 %) respondents expressed that '*extreme heat exposure is responsible for heat-related illnesses.*' Moreover, a higher proportion of urban (75 %) and peri-urban

Table 3 Knowledge about the effects of heat waves (Percentage)

Statements regarding knowledge about heat waves	Urban			Peri-urban		
	Yes	No	Don't know	Yes	No	Don't know
Heat waves are caused by high temperature and low rainfall	77	11	12	82	7	11
Extreme heat exposure is responsible for heat-related illnesses	93	5	2	83	7	10
Is excess sweating during a heat wave, a sign of heat stress?	84	9	7	77	7	16
Individuals suffering from heart problems are highly vulnerable to illness due to heat waves	75	17	8	73	7	20
The elderly and young ones are more vulnerable during a heat wave	67	29	4	67	12	21
Hepatitis is the disease linked with heat waves in the country	16	62	22	24	31	45

(73 %) respondents reported that '*individuals suffering from heart problems are highly vulnerable to illness due to heat waves.*' Majority of the urban respondents (62 %) disagreed with the statement that '*hepatitis is the disease linked with heat waves in the country*' whereas majority of the peri-urban respondents (45 %) did not have any information about this aspect. The detailed responses to the statements on knowledge about the effects of heat waves for the urban and peri-urban areas are given in Table 3.

3.4 Perception of Heat Waves Among Respondents

Based on knowledge of heat waves, people build perception and take precautionary measures thus leading to adaptation to heat waves. After discussing knowledge and its sources, we examine perception of the respondents about heat waves. Both urban and peri-urban respondents behave differently as they possess different knowledge levels. Thus, perception level among both types of the respondent is not similar as the information on perception of the respondents to heat waves as given in Table 4. It shows that more than 78 % of urban respondents have 'high perceived vulnerability' compared to the peri-urban respondents (63.4 %). This implies that the urban respondents give high weightage to heat waves and hence they may be adapting to heat waves rapidly. Considering the perceived severity of heat waves, around 55 % of peri-urban respondents perceive that heat wave events are bigger threat to their health compared with 50 % of the urban respondents. In contrast, we hypothesize that perceived benefit is usefulness of an adaptive measure in a person's opinion to avoid or mitigate the risk. We find that more than half of the urban respondents (55.3 %) are found as having 'high perceived benefit' whereas more

Table 4 Perception of the respondents and heat waves

Level of perception	Urban		Peri-urban	
	Low	High	Low	High
Perceived vulnerability	21.3 (32)	78.7 (118)	36.6 (37)	63.4 (64)
Perceived severity	49.3 (76)	50.7 (74)	44.6 (45)	55.4 (56)
Risk perception	56.7 (85)	43.3 (65)	66.3 (67)	33.7 (34)
Perceived benefits	44.7 (67)	55.3 (83)	52.5 (53)	47.5 (48)
Perceived barriers	42 (63)	58 (87)	61.4 (62)	38.6 (39)
Cues to action	36 (54)	64 (196)	25.7 (26)	74.3 (75)

Figures in parenthesis are frequencies

than half of the peri-urban respondents (52.5 %) reported having ‘low perceived benefit’ of adaptive measures to heat waves. Perceived barriers to efficient heat waves adaptation narrated by the respondents are owned hurdles an individual perceive while adopting a new behavior. Majority of the urban respondents are found assuming ‘high perceived barriers’ whereas ‘low perceived barriers’ is reported by 61 % peri-urban respondents. Cues to action are the situation, actions, people, things that move people to change behavior. Results show that ‘high cues to action’ are reported by 64 and 74.3 % urban and peri-urban respondents, respectively.

4 Conclusions

Heat waves cause many deaths in different parts of Pakistan in each year. Vulnerability to heat waves among the masses has increased overtime due to poor performance of related institutions and lack/less interest of the concerned authorities. Aged peoples, students, and farm workers are highly exposed to heat waves as the present study shows that 21 and 53 % respondents work in the field in urban and peri-urban areas for a period of above 8 h daily. This speaks of the exposure of such persons to starching heat waves. It is also found that people follow information and instructions given by the experts, although the extent of following information and instructions is very low as 30 % respondents are found not acting upon instructions. A 20 % of the respondents reporting no worries due to heat waves indicates dearth of lack of knowledge. Access to and provision of health services particularly during summer is vital in generating awareness among the masses. However, we find that health workers visit households 3–4 times per year. This highlights to widen and strengthen provision of health services. Further, health workers need to be rigorously trained especially relating to preventive measures for heat waves.

People are concerned over adverse effects of heat waves on human health, also indicating level of information and knowledge. Uncomfortable, mental tiredness and sick feelings are commonly reported consequences of heat waves as they highly affect work performance and productivity as well, in addition to extreme cases of morbidity and mortality. Considering perception, we find that the urban respondents assume that they are highly vulnerable to the effects of heat waves as compared to the peri-urban respondents. The peri-urban respondents perceive higher severity of heat waves effects than the urban respondents.

Under the wake of increasing risk of heat waves in Pakistan, electronic media, i.e., radio and television are the important sources for disseminating information and generating awareness among the vulnerable masses. Easy and low cost access to radio and television makes it possible to build capacity of the masses in relation to heat waves. Social and family networks are also important factors in building capacity of households on heat waves. The need is to devise policies to generate awareness and build knowledge of the key individuals in the social circle. Religious and local leaders have an important role in this direction and training and building capacity of such individuals would also help to achieve objectives of minimizing morbidity and mortality due to heat waves in the coming years.

References

Akompab, D. A., Bi, P., Williams, S., Grant, J., Walker, A. I., & Augoustinos, M. (2013). Awareness of and attitudes towards heat waves within the context of climate change among a cohort of residents in Adelaide, Australia. *International Journal of Environment Research and Public Health, 10*, 1–17.

Carvalho, A. (2007). Ideological cultures and media discourses on scientific knowledge: Re-reading news on climate change. *Public Understanding of Science, 16*, 223–243.

Chowdhury, P. D., Haque, C. E., & Driedger, S. M. (2012). Public versus expert knowledge and perception of climate change induced heat wave risk: A modified mental model approach. *Journal of Risk Research, 15*, 149–168.

Das, S., & Smith, S. C. (2012). *Awareness as an adaptation strategy for reducing mortality from heat waves: Evidence from a disaster risk management program in India*. SANDEE Working Paper, 72–12.

Deressa, T. T., Hassan, R. M., & Ringler, C. (2011). Perception of and adaptation to climate change by farmers in the Nile basin of Ethiopia. *The Journal of Agricultural Science, 149*, 23–31.

Deutsche Presse Agentur. (2007). Heat wave claims 40 more lives in Pakistan. Available online at http://reliefweb.int/report/pakistan/heatwave-claims-40-more-lives-pakistan

Dole, R., Hoerling, M., Perlwitz, J., Eischeid, J., Pegion, P., Zhang, T., et al. (2011). Was there a basis for anticipating the 2010 Russian heat wave? *Geophysical Research Letters, 38*, 1–5.

Falco, S. D., Veronesi, M., & Yesuf, M. (2011). Does adaptation to climate change provide food security? A micro perspective from Ethiopia. *American Journal of Agricultural Economics, 93*, 829–846.

Hansen, A., Bi, P., Nitschke, M., Pisaniello, D., Newbury, J., & Kitson, A. (2011). Perceptions of heat-susceptibility in older persons: Barriers to adaptation. *International Journal of Environmental Research and Public Health, 8*, 4714–4728.

Household Income and Expenditure Survey (HIES) 2011. Government of Pakistan. Available athttp://www.pbs.gov.pk/sites/default/files/pslm/publications/hies10_11/complete_report.pdf, Accessed August 2015.

Kosatsky, T., Dufresne, J., Richard, L., Renouf, A., Giannetti, N., Bourbeau, J., et al. (2009). Heat awareness and response among Montreal residents with chronic cardiac and pulmonary disease. *Canadian Journal of Public Health, 100*, 237–240.

Kovats, R. S., & Hajat, S. (2008). Heat stress and public health: A critical review. *Annual Review of Public Health, 29*, 1–15.

Liu, T., Xu, Y. J., Zhang, Y. H., Yan, Q. H., Song, X. L., Xie, H. Y., et al. (2013). Associations between risk perception, spontaneous adaptation behavior to heat waves and heatstroke in Guangdong province, China. *BMC Public Health, 13*, 1–14.

Meehl, G. A., & Tebaldi, C. (2004). More intense, more frequent, and longer lasting heat waves in the 21st century. *Science, 305*, 994–997.

Menne, B., & Matthies, F. (2009). Prevention and management of health hazards related to heat waves. *International Journal of Circumpolar Health, 68*, 8–22.

Pak Tribune. (2004). Death toll of heat wave victims reaches 158. Available online at http://www.paktribune.com/news/print.php?id=110790

Robine, J. M., Cheung, S. L. K., Le Roy, S., Oyen, H. V., Griffiths, C., Michel, J. P., et al. (2008). Death toll exceeded 70,000 in Europe during the summer of 2003. *ComptesRendusBiologies, 331*, 171–178.

Sharples, D. M. (2010). Communicating climate science: Evaluating the UK public's attitude to climate change. *Earth and Environment, 5*, 185–205.

Sheridan, S. C. (2007). A survey of public perception and response to heat warnings across four North American cities: An evaluation of municipal effectiveness. *International Journal of Biometeorology, 52*, 3–15.

Sherwood, S. C., & Huber, M. (2010). An adaptability limit to climate change due to heat stress. *Proceedings of National Academy of Sciences of United States of America, 107*, 9552–9555.

TNN. (2013). Cloud cover gives Hyderabad respite from heat wave. Available online at http://articles.timesofindia.indiatimes.com/2013-0528/hyderabad/39579407_1_heat-wave-celsius-cloud-cover

Vandentorren1, S., Bretin, P., Zeghnoun, A., Mandereau-Bruno, L., Croisier, A., Cochet, C., et al. (2006). August 2003 heat wave in France: Risk factors for death of elderly people living at home. *European Journal of Public Health, 16*, 583–591.

Vidal, J., & Sethna, R. (2013). Pakistan can expect worse heat waves to come, meteorologists warn. Available online at http://www.theguardian.com/environment/2013/jun/14/pakistan-heatwave-meteorologist

The Conflict Between Economy and Environment—An Environmentally Extended Social Accounting Matrix for India

Barun Deb Pal

Abstract This paper is an extension of existing ESAM for India for the year 2006–07 by incorporation of local air pollutants. The ESAM multiplier model described here analysed the conflict between economic growth and environmental emissions. The available emission data in India shows only the direct contribution of any activity on total emission inventory of India, whereas this multiplier model enables us to understand the direct- and indirect-induced effects of economic growth on GHG emission. This is true from the fact that the thermal electricity, cement, and iron and steel sectors are highly energy intensive, and hence their impact on emission is very high as compared to other sectors of the Indian economy. But it is also interesting to observe that the sectors which are low energy intensive have significant impact on overall emission inventory through the indirect-induced impact. On the other hand, the income-pollution multiplier presented in this study brings forth the issue that rural poverty alleviation will not impact much to the environment unless it is happening through rural to urban migration. Finally, these linkages would help policy makers to understand sector-specific policy requirement for environmental emission control.

Keywords Social accounting matrix · Air pollution · Environmental accounting · Environmental social accounting matrix · Environmental policy analysis · Multiplier analysis

1 Introduction

The developing countries face sustainability problems reflected by high levels of basic needs, un-satisfaction and growing environmental problems. India is one among them. In India, the ongoing population growth along with a move towards

B.D. Pal (✉)
Centre for Economic Studies and Policy, Institute for Social and Economic Change, Bangalore 560072, India
e-mail: palbarundeb@gmail.com

S. Nautiyal et al. (eds.), *Climate Change Challenge (3C) and Social-Economic-Ecological Interface-Building*, Environmental Science,
DOI 10.1007/978-3-319-31014-5_13

urbanisation and industrialisation has placed significant pressure on India's infrastructure and its natural resources. Deforestation, soil erosion, water pollution and land degradation are serious policy concerns in rural India. On the other hand, the rapid industrialisation and urbanisation in India's booming metropolises are straining the limits of municipal services and causing serious air pollution.

As most of these environmental problems can be directly attributed to the structure of production and consumption, it would seem urgent and necessary to find ways to make explicit those problems within an accounting framework. In order to do this, it is important to develop a consistent framework which will incorporate the economic and environmental indicators. It is in such a context that this paper should be interpreted. More concretely, this work is an attempt to respond to the need to include, explicitly and directly, the two sets of indicators into a system which accounts for their relations to the economic system and provides the basis for diagnoses and eventually for policy making.

1.1 Approaches Available to Integrate Economy and Environment

In the context of integrating economy and environment, the Systems of Environmental and Economic Accounting (SEEA) was initiated by United Nations in the year 1993 and the latest available database is for the year 2014. But this database does not provide any information about India. In India, various researchers have estimated economic values of environmental goods and services by following shadow price methods, hedonic price theory and contingent valuation methods (Murty et al. 2004; Markandya and Murty 2000; Murty and Gulati 2004; Roy et al. 2008). Apart from these Ministry of Environment and Forests (MoEF) of government of India, The Energy Data Directory (TEDDY) of The Energy Research Institute (TERI) publishes quite regularly the activity-specific environmental data for India for various years.

However, the above-mentioned initiatives taken by the Indian researchers and government consider only the direct relationship between economic activities and environment. Apart from the direct relationship, there can be indirect and induced impacts of economic activities on environment as the economic activities are interdependent to each other. This issue of indirect and induced impacts on environment is crucial while considering demand or input management options for environmental conservation. An economic activity may be insignificant contributor of pollutants from its own production systems but its input consumption pattern may result significant indirect implications on environmental pollution. Now this issue of indirect-induced impact can be captured if we can integrate environmental indicators in a social accounting matrix (SAM) framework.

1.2 Rational of Integrating Environmental Indicators in a Social Accounting Matrix

A Social Accounting Matrix (SAM) depicts the entire circular flow of income for an economy in a (square) matrix format. It shows production leading to the generation of incomes which, in turn, are allocated to institutional sectors. In addition, it shows the redistribution of income leading to disposable income of institutional sectors. These incomes are either spent on products or saved. Expenditures by institutions lead to production by domestic industries as well as supply from imports (Pradhan et al. 2006; Pal et al. 2012; Saluja and Yadav 2006). Hence, the advantage of using a SAM to incorporate both the economic and environmental indicators is that their interrelations can become more apparent and transparent. Moreover, the environmentally extended SAM (ESAM) follows same principle like a SAM, and hence it is suitable for multiplier analysis (Keuning 1992). The multiplier derived from an ESAM would give an understanding about the direct- and indirect-induced impact of a policy on economic growth as well as on environment. Furthermore, this ESAM can be applied as a balanced data source for the computable general equilibrium (CGE) model for environmental policy analysis (Xie 1996). Therefore, the extension of SAM can be considered as the logical step in the efforts to simultaneously account for the interrelationship between economic and environmental activity.

1.3 Challenges in Constructing ESAM for India

Economic analysis with the help of environmentally extended social accounting matrix is popular in Netherlands, Bolivia, Chile, China and UK (Keuning 1992; Alarcon et al. 1997; Gallardo and Mardones 2013; Xie 1996; Shmelev 2013). In case of India, the ESAM constructed for the year 2006–07 by Pal and pohit. (2014) is the first ESAM to describe the conflict between economic activities and GHG emissions and natural resource depletion. However, this ESAM does not take into account local pollutants (Carbon monoxide (CO), suspended particulates matters (SPM), nitrogen dioxide (NO_x), sulphur di oxide (SO_2), etc.) in its framework, whereas these local pollutants determine the domestic air quality and have severe implications on human health. Moreover, this ESAM has been constructed with data available from the MoEF for the year 2007 but this data is not available for every years. On the other hand, data for local pollutant is available for limited sectors and most of them are not directly matched with the sector of input–output table published by CSO (TERI 2009; www.mospi.nic.in). Therefore, scarcity of data poses challenge to construct an ESAM for India with detail account of local and global pollutants. Hence, our primary objective in this study is to describe in detail the method of constructing ESAM under a data constraints situation. Since the latest ESAM available for India is for the year 2006–07 (Pal and pohit. 2014), we are keen to extend this existing ESAM for India for the year 2006–07 by incorporating local air pollutants along with global pollutants.

Apart from that extension of existing ESAM for India for the year 2006–07, our next objective in this study is to provide an analytical application of ESAM in the context of policy analysis. Here we have focused on pollution income multiplier effect to understand the impact of change in socio-economic status of the households into the environmental pollution. Hence, this study makes a significant contribution from the accounting as well as analytical point of view for the policymakers as well as the academicians across the globe.

Following this introductory section the rest of the paper is organised as follows: Sect. 2 describes the methodology and data sources to construct ESAM for India. Section 3 deals with ESAM multiplier analysis and Sect. 4 brings forth some policy implications. Finally, the last section deals with some concluding remarks of this study.

2 Method of Extending ESAM

2.1 Estimation of Environmental Data

As described earlier our purpose in this study is to extend the existing ESAM for India for the year 2006–07. In this context the sect oral description and schematic structure of this ESAM are given in Appendix 1 and 2, respectively. However, the type of pollutants we have taken into account in our proposed extension of ESAM can be classified into two parts, viz. global pollutants which cause global warming or climate change, and local pollutants which cause damages in air quality of a particular region. Among all the global pollutants, carbon dioxide (CO_2), nitrus oxide (N_2O) and methane (CH_4) are crucial and they are together called green house gases (GHGs). On the other hand, the suspended particulate matter (SPM), nitrogen dioxide (NO_x), sulphur dioxide (SO_2) and carbon monoxide (CO) are major air pollutants causing damage to the regional air quality in India. As the existing ESAM has already taken into account GHG emissions our task here is to construct rows and column local air pollutants for our proposed ESAM for the year 2006–07.

Now this is a well-documented fact that the energy consumption is key for the air pollution in India (INCCA-MoEF 2010). Therefore, we have decided to estimate energy-based emission data for India for the year 2006–07. Now these estimations of energy-based emissions of our selected pollutants are based on default energy-specific emission factors published in IPCC guideline (IPCC 2006). According to Keuning (1992), the air pollutants are mainly originated from the domestic production, from the import and from consumption activities. But in case of India the data on substances through the import and consumption is not available. So we have taken only the production-based supply of these damaging substances in India. Again it has been observed that the substances generated from the production sectors are mainly due to energy consumption and also due to production process.[1] But the energy-based emission is most in the environment. Therefore, we have to estimate only the energy-based emission for the 36 production sectors of the Indian economy for the year 2006–07.

Now the energy-based emission coefficients for CO_2, N_2O, CH_4, CO, SO_2 and NO_x are available from the guideline of Inter government panel for climate change (IPCC) of the year 2006 and the same for SPM is obtained from the London Atmospheric Emissions Inventory (LAEI) report of the year 2003. On the other hand, the emission coefficients obtained from IPCC (2006) are different for different sectors but the emission coefficient obtained from LAEI (2003) is uniform for every sector.

On the other hand the emission coefficients as given in IPCC (2006) and LAEI (2003) provide the emission per unit of energy consumption and this energy consumption is expressed in terms of physical unit, i.e. in terajoule. So to estimate the sector-wise energy-based emission for India for the year 2006–07 the energy consumption data in physical unit is necessary. Though the SAM constructed for the year 2006–07 provides the sector-wise energy consumption data for the year 2006–07, it gives energy consumption only in money terms not in physical unit. Therefore, we have obtained the energy consumption data as available from the Central statistical Organisation (CSO) of India for the year 2006–07. But the data available from CSO gives the energy consumption in terms of tons and so we have to convert these data into terms of terajoule. Here to convert this energy consumption data from tons into terajoule, we have used the conversion factor given by IPCC (2006). Now the sector-wise energy consumption data obtained in this way is multiplied with the emission coefficients given in IPCC (2006) and the LAEI (2003) to obtain the sector-wise emission of the substances for India for the year 2006–07. Hence, in this way we have obtained the column of the air pollutants account for our ESAM of the year 2006–07.

After estimating the column of the substances account, our next task would be to construct the row of this air pollutants account for our ESAM of the year 2006–07. But to construct the row of this substances account we have to estimate the sector-specific absorption or abatement of each substance for the year 2006–07. It is also important to note here that the existing SAM does not take into account the abatement of pollutant as it focused on GHG emissions and data for abatement is not available in India for GHG abatements. However, in the following section we have made an attempt to estimate abatement for local pollutants which are taken into account for this ESAM.

2.2 *Estimation of Abatement or Absorption*

In India there is a pollution control board which controls the emission level of the production sectors. As a controlling measure this pollution control board imposes a

---

[1] IPCC (2006).

Table 1 Mapping between TERI sectors and ESAM sectors

TERI Sectors	ESAM sectors
Aluminium	ALU
Copper	OMN
Zinc	OMN
Electricity	NHY
Oil refinery	PET
Cement	CEM
Steel	IRS
Fertiliser	FER
Caustic soda	CHM
Distillery	CHM
Pulp and paper	PAP
Sugar	FBV
Sulphuric acid plant	CHM

standard level of pollution for every sector. Now according to the pollution emission act of India every sector has to follow the emission standard imposed by the pollution control board of India. But this pollution control board imposes emission standard for local pollutants only (i.e. CO, NO_x, SO_2, SPM, etc.) and there is no such standard available for green house gases (CO_2, CH_4 and N_2O) emissions in India. Therefore, we have tried to estimate the abatement supply for the pollutants like CO, NO_x, SO_2 and SPM only.

In India, The Energy and Resources Institute (TERI) gives percentage of pollution abatement for the non-transport sectors for every year.[2] These non-transport sectors are also major pollution sectors in India. But the sectors described by TERI are not directly matched with the sectors of our ESAM. So a map of concordance between TERI's sectors and the sectors of our ESAM is made and this is given in Table 1.

This table shows that there are only 10 sectors of our ESAM for which the abatement data is available. Now as the data on abatement supply for the other sectors is not available to us, we have assumed zero abatement supply for those sectors of our ESAM.

Again, in case of transport sector the data is available only for the road transport motorised (RTM) sector. In this case also we have assumed zero abatement supply for the other transport sector. Now the pollution abatement data for RTM sector is available from the research paper of Murty and Gulati. (2005). In this paper the authors have estimated pollution abatement per vehicle for the year 2002 and this estimation was done for two representative states of India, viz. Andhra Pradesh and Himachal Pradesh. Now it is observed from this paper that the pollution abatement per vehicle is quite similar between these two states. So we have assumed this state level data as the average pollution reduction per vehicle for India for the year 2006–07 and estimate the abatement supply for the road transport sector.

[2]TERI (2003–04), TERI Energy Data Directory and yearbook.

Therefore, in the above-mentioned way we have obtained the row and column of the damaging substances account of our ESAM for India for the year 2006–07. But this ESAM also includes depletable substances in the substances account. So we have to estimate the row and column of this depletable substances account of this ESAM and this estimation procedure is given below.

2.3 Estimation for Depletable Natural Resources

In this case we have considered crude oil and coal as the depletable natural resources. The production data in physical unit on crude oil and coal has been taken as measure of the quantities of depletion of these two types of resources and the data is available from CSO (2006).[3] Now the data obtained in this way can be interpreted as 'free' intermediate consumption (without direct cost) used in the production process of the crude oil and natural gas sector. So to construct the row of this depletable substances account we have put the data in the row of this depletable substances account corresponding to the column of crude oil and coal sectors.

On the other hand to construct the column of this depletable substances account, we have obtained the data on new discoveries of crude oil and coal reserve in India in the year 2006–07. In this case we have obtained this data on new discoveries from TERI (2005). Hence, in this way we have constructed the row and the column of the depletable substances account of our ESAM for India for the year 2006–07.

Thus we have obtained the row and column for each of the substances to construct our ESAM for India for the year 2006–07. Now apart from this substances account there is also another account, i.e. account for environmental themes which is also important for our ESAM for India for the year 2006–07. Below we have described the method of estimating environmental themes for India for the year 2006–07.

2.4 Estimation for Environmental Themes

In this case we have to estimate the inventory of the substances in the Indian economy for the year 2006–07. Now to estimate this inventory we have estimated the net generation of the substances in the Indian economy for the year 2006–07. The net generation in this case is obtained in the following way:

$$\text{Net Generation of Substances} = \text{Total Generation of each substances} - \text{Total abatement of each substances.}$$

Finally, on the basis of the method described throughout this study, we have obtained the complete extension of ESAM for India for the year 2006–07. This

[3]CSO (2006), statistical abstract of India.

complete extended ESAM is given in Appendix 3 of this paper. Now to understand the relation between the environmental pollution with the level of output, we have estimated the direct pollution coefficient directly from the ESAM data and this is given in Table 2. The direct pollution coefficient (shown in Table 2) is nothing but

Table 2 Direct pollution coefficient matrix 2003–04 (unit tons/lakh of output)

	CO_2 (t)	N_2O (t)	CH_4 (t)	SO_2 (t)	NO_x (t)	SPM (t)	CO (t)
PAD	0.0004	0.0000	0.0001	0.0029	0.0004	0.0003	0.0001
WHT	0.0004	0.0000	0.0002	0.0022	0.0003	0.0002	0.0001
CER	0.0003	0.0000	0.0001	0.0014	0.0002	0.0001	0.0000
CAS	0.0002	0.0000	0.0002	0.0018	0.0002	0.0002	0.0001
ANH	0.0047	0.0000	0.7150	0.0001	0.0000	0.0001	0.0002
FRS	0.0000	0.0000	0.0000	0.0007	0.0001	0.0001	0.0000
FSH	0.0000	0.0000	0.0000	0.0047	0.0006	0.0005	0.0001
COL	0.2611	0.0002	0.0537	0.0033	0.0019	0.0025	0.0001
OIL	0.0706	0.0000	0.0010	0.0008	0.0002	0.0001	0.0000
GAS	0.0895	0.0001	0.0065	0.0018	0.0008	0.0009	0.0001
FBV	0.1061	0.0001	0.0000	0.0022	0.0010	0.0010	0.0003
TEX	0.1785	0.0001	0.0000	0.0035	0.0017	0.0019	0.0006
WOD	0.6228	0.0003	0.0001	0.0056	0.0036	0.0048	0.0017
MIN	0.2713	0.0002	0.0001	0.0031	0.0008	0.0024	0.0009
PET	15.0714	0.0005	0.0030	0.0112	0.0061	0.0074	0.0003
CHM	0.6999	0.0003	0.0001	0.0069	0.0037	0.0045	0.0016
PAP	1.7649	0.0012	0.0004	0.0189	0.0124	0.0168	0.0060
FER	2.9970	0.0006	0.0003	0.0232	0.0100	0.0103	0.0033
CEM	7.9705	0.0054	0.0019	0.0803	0.0556	0.0769	0.0276
IRS	9.6929	0.0060	0.0020	0.0893	0.0616	0.0849	0.0304
ALU	4.6625	0.0028	0.0010	0.0420	0.0290	0.0401	0.0144
OMN	1.0465	0.0006	0.0002	0.0105	0.0067	0.0089	0.0032
MCH	0.2448	0.0001	0.0000	0.0029	0.0016	0.0019	0.0007
NHY	15.7510	0.0083	0.0028	0.1278	0.0860	0.1174	0.0420
HYD	0.0000	0.0000	0.0000	0.0000	0.0000	0.0000	0.0000
NUC	0.0000	0.0072	0.0025	0.1138	0.0754	0.1022	0.0365
BIO	0.0062	0.0000	0.0000	0.0000	0.0000	0.0000	0.0000
WAT	0.0317	0.0000	0.0000	0.0005	0.0001	0.0003	0.0001
CON	0.0030	0.0000	0.0000	0.0043	0.0011	0.0005	0.0001
RTN	0.0000	0.0000	0.0052	0.0300	0.0349	0.0038	0.0022
RLY	0.0623	0.0001	0.0000	0.0000	0.0000	0.0000	0.0000
AIR	0.0000	0.0000	0.0000	0.0000	0.0000	0.0000	0.0000
SEA	0.0002	0.0000	0.0000	0.0000	0.0000	0.0000	0.0000
HLM	0.0000	0.0000	0.0000	0.0007	0.0001	0.0001	0.0000
SER	0.0151	0.0000	0.0101	0.0007	0.0001	0.0002	0.0005

the amount of pollution generated for each level of output produced by the production activity. On the other hand this direct pollution coefficient also measures the direct effect of economic activities on the environmental pollution.

Now on the basis of Table 2 the direct pollution coefficients corresponding to each type of pollutants are higher for the sectors like NHY, CEM and IRS sector. Therefore, on the basis of the above discussion these three sectors have the large direct impact on the environmental pollution as compared to the other production sectors of the Indian economy. Again the data shows that the NHY sector has the highest direct pollution coefficient for each type of pollutants as compared to the other sectors. Hence, this NHY sector has the highest direct effect on environmental pollution in India. For example, the direct pollution coefficient of the NHY sector corresponding to CO_2 is 15.75; this implies the NHY sector generates 15.75 t of CO_2 to produce Rs. 1 lakh value of its output. So in this way we can understand the direct effect of the economic activities on the environmental pollution with the help of direct pollution coefficients.

Again if the output of the production sector increases then there are also indirect and induced effects on the environmental pollution. These indirect and induced effects cannot be captured with the help of direct pollution coefficients, but can be captured by the multiplier analysis. The multiplier analysis is given in the following section.

3 Multiplier Analysis for Environmental Impact Analysis

In this case we have first estimated the SAM multiplier matrix and then relate that matrix with the environmental data given in the ESAM.

Let A be the domestic expenditure coefficient matrix and Y be the matrix of sector-wise gross output. Also, suppose that the matrix X be a matrix which indicates the expenditure to the rests of the world. Therefore, the SAM can be written as follows:

$$Y = AY + X \tag{1}$$

$$\text{or} \quad Y = (I - A)^{-1}X \tag{2}$$

This shows that Y (i.e. Y_1, Y_2, Y_3 …) can be derived from X (i.e. X_1, X_2, X_3 …) through a generalised inverse $(I - A)^{-1}$. Now if we denote the $(I - A)^{-1}$ matrix as M, then Eq. (2) can be written as

$$Y = MX \tag{3}$$

where M is the SAM multiplier matrix, with a representative element m_{ij} as total (direct + indirect + induced) impact on account i due to change in exogenous injections in account j.

Now the SAM multiplier obtained in Eq. (3) does not take into account the environmental aspects. So this M matrix cannot say anything about the consequences on environment as a result of the exogenous injection into the economy.

But this can be easily remedied by making assumption about the link between gross output and pollution generation in each production activity. The standard practice in this regard is to use the direct pollution generation coefficients for each sector and this is given in Table 5. Now on the basis of Table 5 and the SAM multiplier, we have tried to estimate the pollution trade-off multiplier for India for the year 2003–04. In this case the method of estimating pollution trade-off multiplier is described below.

3.1 Pollution Trade-off Multiplier

These pollution trade-off multipliers measure the direct and indirect impact on pollution generation level due to exogenous change in sectoral output, household income, etc. In this study we have estimated two types of pollution trade-off multipliers: (1) output-pollution multiplier and (2) income-pollution multiplier.

The output-pollution multiplier is obtained by post multiplying the activity × activity block of M matrix with the direct pollution generation coefficients matrix. This can be written as

$$T_{k\times\text{activity}} = P_{k\times\text{activity}} \cdot M_{\text{activity}\times\text{activity}} \tag{9}$$

where

$T_{k\times\text{activity}}$	Output-pollution multiplier matrix
$P_{k\times\text{activity}}$	Direct pollution coefficient matrix
$M_{\text{activity}\times\text{activity}}$	Output multiplier matrix
k	Number of pollutants

On the other hand, the income-pollution multiplier is estimated as

$$I_{k\times\text{households}} = P_{k\times\text{activity}} \cdot M_{\text{activity}\times\text{households}} \tag{10}$$

where

$I_{k\times\text{households}}$	Income-pollution multiplier matrix
$M_{\text{activity}\times\text{household}}$	Income multiplier matrix

3.2 Analysis of Results Relating to Output-Pollution Multiplier

The output-pollution multiplier can be used to analyse the impact of sectoral growth on the pollution emission in India under the given technological situation of the year 2006–07. The output-pollution multiplier is shown in Table 3. Each cell entry of Table 3 shows the direct, indirect and induced effect on generation of the pollutants due to exogenous injection to the production sectors. Higher the value of this output-pollution multiplier implies larger the impact on pollution generation. Therefore, the analysis of this output-pollution multiplier would help us to find out the leading sectors of the economy which have highest impact on the environment.

According to the data presented in Table 3 we find that the values of output-pollution multiplier for thermal electricity (NHY), cement (CEM) and iron and steel (IRS) sectors are higher as compared to the other sectors of the Indian economy. For example, let us describe the total (direct, indirect and induced) impact of thermal electricity sector on the CO_2 generation. In case of thermal electricity sector the output-pollution multiplier corresponding to CO_2 generation is 57.06 (see Table 3). This implies that if the output of the thermal electricity sector is increased by Rs. 1 lakh then the total CO_2 generation in the environment will be increased by 57.06 t.

Again, if we look at the direct pollution coefficients matrix, given in Table 2, we can observe the direct impact of the production sectors on the environment. It has been observed from Table 2 that the CO_2 generation is increased by 24.53 t directly due to expansion of thermal electricity sector by Rs. 1 lakh. Now if we compare this direct effect of the thermal electricity sector with its total effect, we find that the extra 22.53 t of CO_2 generates indirectly in the environment if the thermal electricity sector's output expands by Rs. 1 lakh. Hence, this is the indirect and induced effect of thermal electricity sector on the environmental pollution.

On the other hand, the values of the direct pollution coefficients are very small for the agricultural sectors of the Indian economy (see Table 2). This shows that the direct effect of the agriculture sectors on the environment is negligible for each type of pollutants. But it can be observed from the output-pollution multiplier matrix that the total effect on NO_x generation is significant in agriculture sector as compared to the manufacturing sectors of the economy (see Table 3). This kind of result is obtained due to indirect effect of agriculture sectors on the environment. Therefore, it can be said that the sectors that have negligible direct impact may have significant indirect effect on the environment. Therefore, in the above way the ESAM multiplier model helps us to analyse the total impact on energy use as well as on environmental pollution.

This indirect effect on environmental pollution arises due to the backward and forward linkage of the production sectors of the economy. Growth in a sector with high backward linkage provides stimulus to other sectors by requiring more inputs, whereas growth in a sector with high forward linkage stimulates higher outputs in other sectors by providing more inputs to them. In case of Indian economy the thermal electricity sector is a key source of the electric supply sector. The share of thermal electricity sector in total electric supply is about 86 % in the year 2006–07 (Pal et al. 2015). So the other production sectors of the economy are strongly dependent on the thermal electricity sector for their electricity requirement, and hence the thermal electricity sector has a high forward linkage. As there is high forward linkage, the expansion of thermal electricity sector will lead to expand the

Table 3 Output-pollution multiplier of the year 2006–07 (tons/Rs. lakh of output)

	PAD	WHT	CER	CAS	ANH	FRS	FSH	COL	OIL
CO_2	27.8648	28.3430	26.0133	26.6439	26.0535	14.4298	26.1899	23.9649	5.9948
N_2O	0.0071	0.0072	0.0066	0.0068	0.0067	0.0037	0.0066	0.0064	0.0015
CH_4	0.3930	0.3503	0.3816	0.4028	1.0787	0.1957	0.3481	0.3515	0.0819
SO_2	0.1515	0.1533	0.1421	0.1451	0.1427	0.0793	0.1445	0.1314	0.0330
NO_x	0.1020	0.1032	0.0972	0.0989	0.0988	0.0546	0.0973	0.0906	0.0225
SPM	0.1045	0.1065	0.0987	0.1004	0.1000	0.0552	0.0988	0.0940	0.0228
CO	0.0379	0.0385	0.0360	0.0365	0.0368	0.0202	0.0359	0.0335	0.0083
	GAS	FBV	TEX	WOD	MIN	PET	CHM	PAP	FER
CO_2	20.8176	25.6492	26.2091	23.8174	8.2914	14.3754	24.3537	25.5293	29.6803
N_2O	0.0054	0.0066	0.0068	0.0063	0.0022	0.0037	0.0063	0.0070	0.0065
CH_4	0.2794	0.3801	0.3333	0.3024	0.1013	0.1708	0.2876	0.2690	0.2668
SO_2	0.1142	0.1404	0.1435	0.1304	0.0454	0.0795	0.1315	0.1399	0.1464
NO_x	0.0786	0.0964	0.0989	0.0903	0.0300	0.0532	0.0898	0.0964	0.0945
SPM	0.0806	0.0977	0.1005	0.0929	0.0326	0.0556	0.0929	0.1033	0.0973
CO	0.0291	0.0356	0.0366	0.0339	0.0118	0.0179	0.0337	0.0375	0.0346
	CEM	IRS	ALU	OMN	MCH	HYD	NHY	NUC	BIO
CO_2	41.5332	42.5254	18.5374	21.0683	23.0113	26.0375	57.0687	24.6752	26.1910
N_2O	0.0125	0.0132	0.0058	0.0058	0.0063	0.0068	0.0175	0.0136	0.0068
CH_4	0.3114	0.2848	0.1267	0.2203	0.2464	0.3483	0.3127	0.3262	0.3602
SO_2	0.2234	0.2308	0.1012	0.1153	0.1256	0.1427	0.3051	0.2489	0.1428
NO_x	0.1546	0.1595	0.0700	0.0792	0.0865	0.0986	0.2078	0.1684	0.0990
SPM	0.1810	0.1911	0.0837	0.0858	0.0933	0.1011	0.2529	0.1981	0.1002
CO	0.0654	0.0689	0.0302	0.0311	0.0339	0.0369	0.0907	0.0712	0.0366

(continued)

Table 3 (continued)

	WAT	CON	RTM	RNM	RLY	AIR	SEA	HLM	SER
	WAT	CON	RTM	RNM	RLY	AIR	SEA	HLM	SER
CO_2	27.4426	27.9460	27.5484	25.9567	28.5008	26.0605	26.5534	25.8009	26.0892
N_2O	0.0073	0.0074	0.0059	0.0065	0.0077	0.0067	0.0065	0.0067	0.0068
CH_4	0.3508	0.3150	0.3010	0.3469	0.3141	0.3378	0.3180	0.3433	0.3651
SO_2	0.1503	0.1535	0.1548	0.1426	0.1518	0.1403	0.1360	0.1412	0.1431
NO_x	0.1035	0.1046	0.1207	0.1007	0.1045	0.0974	0.0944	0.0972	0.0987
SPM	0.1075	0.1099	0.0920	0.0975	0.1115	0.0987	0.0957	0.0988	0.1007
CO	0.0392	0.0398	0.0338	0.0357	0.0404	0.0360	0.0347	0.0361	0.0372

Table 4 Income-pollution multiplier of the year 2006–07 (tons/Rs. lakh of income)

	RNASE	RAL	ROL	RASE	ROH	USE	USC	UCL	UOH
CO_2	126.87	130.79	129.27	128.60	128.12	127.54	125.77	130.57	130.80
N_2O	0.04	0.04	0.04	0.04	0.04	0.04	0.03	0.04	0.04
CH_4	1.87	1.95	1.93	1.91	1.90	1.89	1.85	1.94	1.95
SO_2	0.80	0.82	0.81	0.81	0.81	0.80	0.79	0.82	0.82
NO_x	0.56	0.58	0.57	0.57	0.57	0.56	0.56	0.58	0.58
SPM	0.53	0.54	0.54	0.54	0.53	0.53	0.53	0.54	0.54
CO	0.19	0.20	0.20	0.19	0.19	0.19	0.19	0.20	0.20

output of the other sectors of the economy. On the other hand the thermal electricity sector requires both energy and non-energy inputs for its production process and these are supplied by the other production sectors of the economy. If we look at the column of the thermal electricity sector in our ESAM, we can see that the most of the non-energy inputs of this sector are supplied from the energy-intensive sector. Again these energy-intensive sectors are also polluting sectors in the Indian economy (see Appendix 3). Now the output of these energy-intensive sectors would increase due to the backward linkage effect of the thermal electricity sector. Hence, in this way the total environmental pollution in the environment will increase if the thermal electricity sector expands. In the same way we can also analyse the total impact of each production sector to the generation of damaging substances.

In the above discussion we have described the output effect on the environmental pollution. Apart from this output effect there is also income effect on environmental pollution in India. By income effect we mean the impact on environmental pollution due to change in household income. Now this income effect on the environmental pollution can be worked out by estimating the income-pollution multiplier.

3.3 Analysis of Income-Pollution Multiplier

In this study we have estimated the income-pollution multiplier for the household classes described in our ESAM (see Appendix 1). The estimated income-pollution multiplier is given in Table 4.

The income-pollution multiplier shown in Table 4 gives the total effect on environmental pollution with the increase or decrease in income of the households. Table 4 shows that the total effects on environmental emission of the RAL household class are higher than the other household classes in the rural area. On the other hand the total effects on environmental pollution are higher for the UCL households' class than the other households' classes of the urban are. Now if we take an example of CO_2 emission we see that if income of the agricultural labour class (RAL) is increased or decreased by Rs. 1 lakh then the total CO_2 generation in the environment will be increased or decreased by 130.79 t (see Table 4) in India.

Table 5 Energy-income multiplier of the year 2006–07 (tons/Rs. lakh of income)

	Coal energy		Crude oil energy		Natural gas		Hydro energy		Nuclear energy	
	Direct Impact	Indirect and induced impacts	Direct impact	Indirect and induced impacts	Direct impact	Indirect and induced impacts	Direct impact	Indirect and induced impacts	Direct impact	Indirect and induced impacts
RNASE	0.0088	0.0576	0.0000	0.1924	0.0290	0.0182	0.0941	0.0300	0.0143	0.0046
RAL	**0.0133**	**0.0572**	**0.0000**	**0.2028**	**0.0438**	**0.0191**	**0.1420**	**0.0313**	**0.0216**	**0.0048**
ROL	0.0116	0.0568	0.0000	0.2029	0.0383	0.0186	0.1242	0.0306	0.0189	0.0047
RASE	0.0087	0.0570	0.0000	0.1987	0.0285	0.0183	0.0926	0.0302	0.0141	0.0046
ROH	0.0084	0.0567	0.0000	0.2021	0.0278	0.0181	0.0902	0.0300	0.0137	0.0046
USE	0.0093	0.0569	0.0000	0.1986	0.0306	0.0180	0.0993	0.0299	0.0151	0.0045
USC	0.0057	0.0568	0.0000	0.1989	0.0189	0.0175	0.0613	0.0291	0.0093	0.0044
UCL	**0.0143**	**0.0572**	**0.0000**	**0.2057**	**0.0471**	**0.0187**	**0.1529**	**0.0318**	**0.0232**	**0.0048**
UOH	0.0081	0.0557	0.0000	0.2082	0.0266	0.0180	0.0862	0.0300	0.0131	0.0046

Table 6 Average MPCE of households in rural and urban areas

HHs class	MPCE (Rs.)	HHs class	MPCE (Rs.)
RNASE	731	USE	1217
RAL	515	USC	1513
ROL	659	UCL	722
RASE	743	UOH	1788
ROH	1076		

Source NSSO 63rd round, Report No. 527

Similarly, if the income of the UCL class is increased or decreased by Rs. 1 lakh then total CO_2 emission in India will be increased by 130.57 t. Now the impacts on environmental pollution for the RAL and UCL household classes are higher due to their impact on energy use. In Table 5, we have shown the direct, indirect and induced impacts on energy for every household classes.

The above table describes the issue that impacts on primary energy consumption due to increase in income of the household income. As shown in the above table, the indirect and induced effects are higher than the direct effects. This is because of linkage effects of the households on energy sector and further the linkages of energy sectors with rest of the sectors in the economy. Quite obviously it is observed from Table 5 that the commercial primary energy use per unit of income is higher for urban households than the rural households. But the indirect-induced impacts on coal are higher for rural households than the urban households. It is again surprising to see that the direct energy coefficient for Indian household classes is lower for coal energy than other forms of energy but the indirect and induced effects for coal are higher than the other energy commodity irrespective of the households' class. Therefore, these above-mentioned results again reinforce the statement that Indian economy has coal-dominated energy basket which is in other way contributing to the environment in the form of air pollution.

Again if we look at the monthly per capita consumption expenditure (MPCE) of the household classes (see Table 6) we will find that the household classes RAL and UCL belong to the relatively poor household classes in terms of their average monthly per capita consumption expenditure (MPCE). Therefore, from Table 6 it is observed that the impact on environmental pollution for the relatively poor households' classes is higher as compared to the relatively richer household class.

4 Policy Analysis Using ESAM Multiplier: Some Illustrative Examples

On the basis of above discussion it will be more interesting to do the following impact analysis in India and these are as follows:

1. The impact of poverty alleviation programme on environmental pollution and
2. The impact of rural urban migration on environmental pollution.

Again in both of the above cases the people of one household class are moving to another household class. Therefore, by considering the above two impact analyses our main objective is to find out the impact on environmental pollution if the households move from on household class to another. Now the whole analyses are described below.

4.1 Poverty Alleviation Impact on Environment

The underlying assumptions behind this analysis are as follows:

1. The poverty alleviation measures are taken separately for rural and urban areas.
2. The poverty alleviation measure will increase the income of the households and they will move from poor household class to relatively rich household class.
3. There is no rural to urban migration. This implies the rural household will move within the rural area and the urban households will move within the urban area.
4. As our ESAM multiplier model static, the total population supply is assumed to be fixed at 2006–07 level in the rural and urban areas. Therefore, the movement of people from poor household class to rich household class will reduce the number of poor people in the economy. On the other hand the number of rich people will be increased in the economy.
5. The average per capita income for each household group is constant. So fall in population will reduce the total income of the poor household class and the rise in population will increase the total income of the rich household class.

On the basis of these assumptions, below we have presented an illustrative case by assuming a certain amount (say 1 lakh) of people are moving from poor household class to relatively rich household class in the rural and urban areas as a result of poverty alleviation programme. In this case, 1 lakh people are moving from RAL class to immediate rich class, i.e. ROL household class in the rural area, and the same amount of people are moving from UCL to immediately rich USE household class in the urban area. Now if this happens then on the basis of the above assumptions the total income (aggregate of all the individuals) of the RAL and UCL households' classes will fall and the income of the ROL and USE households' classes will rise. Again it is observed from the income-pollution multiplier matrix that the increase in income increases the total environmental pollution and the decrease in income decreases the total environmental pollution in the environment. Therefore, the environmental pollution in the one way will reduce and on the other way it will raise due to separate rural and urban poverty alleviation measures. But in this case we will only estimate the net impact on the environmental pollution and the estimated results are given in Table 7.

The data given in Table 7 reveals that the movement between the UCL and USE household class has the large impact on environmental pollution than the movement between the RAL and ROL class. For example, let us consider the net impact on CO_2 emission. The net increase in CO_2 emission will be 2.63 mt if the 1 lakh

Table 7 Impact of rural and urban poverty alleviation on the environment

	RAL to ROL	UCL to USE
CO_2 (t)	168,810	2,627,280
N_2O (t)	47	736
CH_4 (t)	2483	38,774
SO_2 (t)	1068	16,656
NO_x (t)	751	11,676
SPM (t)	712	11,047
CO (t)	259	4016

people move from UCL to USE household class. On the other hand if the same amount of people moves from RAL to ROL household class then the net increase in CO_2 emission will be 0.2 mt. Therefore, if the income of the 1 lakh poor people in the urban area is increased then their impact on the environment will be almost twice than that of rural areas. Hence, it can be inferred that the urban poverty alleviation is costly in terms of environmental pollution in India.

4.2 *Impact of Rural Urban Migration on Environment*

The analysis of impact of rural–urban migration on environmental is based on the basic concept of famous Harris–Todaro model of rural–urban migration (Harris and Todaro 1970). According to the Harris–Todaro model, the rural to urban migration occurs due to surplus labour in the rural agricultural sector and the wage rate differential between the rural and urban areas. Therefore, according to this model the people from agricultural sector will migrate to the urban sector and they will receive more income after migration.

Now on the basis of this Harris–Todaro model, here we have assumed two possible cases of rural to urban migration as illustrative analysis for this study. First, we have assumed that 1 lakh people of the RAL household class will migrate to UCL class and second we have assumed that the 1 lakh people of RAL household class will migrate to USC class. Therefore, we have to estimate the impact on the environmental pollution if the 1 lakh people move from RAL household class to UCL household class and if the same amount of people move from RAL to USC household class. Now to estimate this impact we have followed the same procedure as we have followed in Sect. 4.1. The results estimated in this way are given in Table 8.

It has been observed from Table 8 that if the 1 lakh people of RAL household class migrate to UCL class, the total CO_2 emission will be increased by 3.01 mt. On the other hand if the same people of RAL household class migrate to USC class then total CO_2 emission will be increased by 5.75 mt. Therefore, Table 8 shows

Table 8 Impact of rural to urban migration on environment

	RAL to UCL	RAL to USC	RAL to ROL
CO_2 (t)	3,014,420	5,752,300	168,810
N_2O (t)	831	1603	47
CH_4 (t)	44,792	83,935	2483
SO_2 (t)	18,952	36,329	1068
NO_x (t)	13,285	25,461	751
SPM (t)	12,493	24,073	712
CO (t)	4547	8756	259

that the net increase in emission of the pollutants is more in case of rural–urban migration than the rural–rural movement in India. This implies that the independent poverty alleviation measure in the rural area is more fruitful in the Indian economy than the poverty alleviation through urbanisation.

Therefore, from the above discussion we have seen that there is a positive relationship between the environmental pollution and the economic growth. But in what extend the environment will be affected depends on how the economy is growing. Again with the economic growth the technological condition is also changing over time in the Indian economy. But as this SAM multiplier model is a static model, there is no scope to analyse the impact on environment due to the technological change in the Indian economy with the help of single-year multiplier model. Now this will be possible by considering the structural change impact on the environment and this can be done with the help of the multiplier model in our research endeavour.

5 Concluding Remarks

In this work, an attempt has been made to analyse and quantify the conflict relation between the economic development of a region and the deterioration of the environment. Although the construction of an ESAM is not an easy task and many inadequate assumptions must be maintained, the matrix and the results are not flawless but provide a good tool for comparisons and indications for decision making. Several times in developing plans, the negative effects on the environment are ignored due to the lack of knowledge of the magnitude of the destruction or other reasons. Therefore, keeping in mind the negative effects which originate in certain activities and household consumption on the environment, it seems superfluous to stress on the importance of efforts to generate the relevant information on emission and depletion and putting it in a consistent framework.

As far as environmental accounts are concerned, one may include other types of environmental indicators, like waste generation and the depletion of natural resources other than gas and oil. Furthermore, it may involve the inclusion of stock data on natural resources and on accumulated pollutants. In addition, attempts might be made to disaggregate the production and commodity accounts in the money-metric part of the ESAM in order to show explicitly the production of goods and services associated with environmental protection and restoration. Also, attempts to disaggregate the tax account for institutions, in order to make visible environment-related tax policies, could be mentioned in this connection.

Finally, it seems only logic to recommend the construction of ESAM-based models and to explore the development of the required algorithms to solve them. This is only way one can assess the robustness of method, and together with the extended SAM, to develop CGE models which include the aspects of sustainability.

Appendix 1

Description of production sectors and households of ESAM 2006–07

A.	Production activities	
	Sector code	Description
1	PAD	Paddy rice
2	WHT	Wheat
3	CER	Cereals, grains, etc., other crops
4	CAS	Cash crops
5	ANH	Animal husbandry and production
6	FRS	Forestry
7	FSH	Fishing
8	COL	Coal
9	OIL	Oil
10	GAS	Gas
11	FBV	Food and beverages
12	TEX	Textile
13	WOD	Wood
14	MIN	Minerals n.e.c
15	PET	Petroleum and coal production
16	CHM	Chemical, rubber and plastic production
17	PAP	Paper and paper production
18	FER	Fertilisers and Pesticides

(continued)

(continued)

A.	Production activities	
	Sector code	Description
19	CEM	Cement
20	IRS	Iron and steel
21	ALU	Aluminium
22	OMN	Other manufacturing
23	MCH	machinery
24	HYD	Hydro
25	NHY	Non hydro
26	NUC	Nuclear
27	BIO	Biomass
28	WAT	Water
29	CON	Construction
30	RTM	Road transport motorised
31	RNM	Road transport non motorised
32	RLY	Rail transport
33	SEA	Air transport
34	AIR	Sea transport
35	HLM	Health and medical
36	SER	All other services
B.	Description of households	
	Household class	Description
1	RNASE	Non-agricultural self employed
2	RAL	Agricultural labour
3	ROL	Other labour
4	RASE	Agricultural self-employed
5	ROH	Other households
6	USE	Self-employed
7	USC	Salaried class
8	UCL	Casual labour
9	UOH	Other households

Appendix 2

Schematic structure of ESAM 2006–07

			Production	Factors of production	Institutions	Indirect taxes	Capital account	Rest of the world	Substances (GHG)	Depletion of natural resource		Environmental theme
										Renewal of energy resource	Renewal of land	
			1	2	3	4	5	6	7	8	9	10
Production		1	Intermediate consumption		Consumption of goods and services		Change in stocks and capital formation	Exports	Emission of pollutants from production			
Factors of production		2	Payment for factors					Net factor income from abroad		Discovery of energy capital	Renewal in land through conservation	
Institutions		3		Value added income	Transfer from other institutions	Total tax receive		Net current transfers	Emission of pollutants from consumption			
Indirect taxes		4	Taxes on intermediate		Taxes on purchase		Taxes on investment					
Capital account		5		Depreciation	Savings			Foreign savings				
Rest of the world		6	Imports									
Substances (GHG)		7	Absorption of substances in production		Absorption of substances in consumption							Accumulation of substances

(continued)

(continued)

			Production	Factors of production	Institutions	Indirect taxes	Capital account	Rest of the world	Substances (GHG)	Depletion of natural resource		Environmental theme
										Renewal of energy resource	Renewal of land	
			1	2	3	4	5	6	7	8	9	10
Depletable substances	Depletion of energy resources	8		Extraction of energy stock								Net reduction in natural stock
	Depletion of land	9		Depletion of land due to land use change					Emission from land use change			Net depletion in land
Environmental theme				GHG inventory								

Appendix 3

Environmental social accounting matrix of India 2006–07 (Rs. lakh)

	PAD	WHT	CER	CAS	ANH	FRS	FSH
PAD	3,604,537	56,673	297,765	333	45,486	6	513
WHT	63,062	2,703,447	377,231	4	13,612	0	6
CER	53,742	128,243	1,813,875	9	2,555,577	18	7
CAS	4048	14,766	47,603	798,319	1295	0	0
ANH	586,772	81,354	954,233	657,037	26,038	1	83
FRS	132	24	60	0	40	230	0
FSH	223	571	1182	0	0	0	246,158
COL	47	35	74	15	202	0	0
OIL	0	0	1	0	2590	0	0
GAS	0	0	0	19	257	0	0
FBV	45,025	6549	18,793	14	272,968	0	11,467
TEX	21,267	21,413	15,603	3338	2474	393	140,281
WOD	81	209	433	43	236	11	6893
MIN	1	3	9	1	503	0	0
PET	440,637	197,486	428,076	200,894	1292	3538	159,517
CHM	3040	3257	5886	2217	19,365	1043	9975
PAP	846	1023	1691	419	644	164	0
FER	1,551,704	1,351,880	1,533,062	951,908	194	44	219
CEM	0	0	0	0	33	1	0
IRS	0	1	7	1	1314	1	6856
ALU	1	3	8	1	662	19	362
OMN	13,173	8489	8675	4004	2950	3088	105,725
MCH	63,300	74,441	81,915	13,137	6681	633	9
NHY	303,163	292,496	154,790	47,095	919	63	43
HYD	56,174	54,198	28,682	8726	170	12	8
NUC	8532	8232	4356	1325	26	2	1
BIO	62,499	9197	108,434	69,230	13,710	776	9
WAT	62	53	71	33	16	28	0
CON	299,358	186,107	241,647	95,450	4927	3305	42
LTR	465,228	308,397	416,869	214,952	470,359	10,505	62,115
RLY	218,678	54,027	77,352	32,208	28,198	744	2773
AIR	33,234	21,421	15,620	7991	2153	46	2312
SEA	1017	1500	3189	362	32,522	25	57
HLM	0	0	0	0	0	0	0
SER	848,437	522,990	860,076	350,991	2,020,029	9655	56,472
Lab	3,689,555	2,645,224	13,081,859	4,985,874	7,939,400	155,015	1,870,083
Cap	1,115,629	799,304	3,842,846	1,408,910	7,823,192	166,197	1,535,925
Land	2,546,449	1,825,601	9,153,113	3,501,862			
RNASE							

(continued)

(continued)

	PAD	WHT	CER	CAS	ANH	FRS	FSH
RAL							
ROL							
RASE							
ROH							
USE							
USC							
UCL							
UOH							
PVT							
PUB							
GOV							
ITX	−1,139,709	−1,310,720	−1,108,597	−581,101	31,009	11,379	−202,612
CAC							
ROW	53	31,699	831,853	283,794	34,943	626,973	15,046
TOT	14,960,001	10,099,591	33,298,342	13,059,415	21,355,984	993,915	4,030,348
CO_2 (t)							
N_2O (t)							
CH_4 (t)							
SO_2 (t)	39,528	19,905	41,117	21,275	1113	621	17,178
NOx (t)	5138	2588	5347	2765	247	81	2231
SPM (t)	4268	2161	4462	2297	969	67	1840
CO (t)	1214	671	1380	652	4573	18	446
Oil (mt)							
Coal (mt)							

	COL	OIL	GAS	FBV	TEX	WOD	MIN
PAD	0	0	0	480,042	41	8	3
WHT	0	0	0	915,729	74	20	1
CER	0	0	1	3,051,005	7501	503	81
CAS	0	0	2	5,426,983	2,056,049	341	201
ANH	0	1	3	1,601,763	372,582	325	153
FRS	0	0	4	19,532	290	55,073	37
FSH	0	0	0	458,661	40	2	1
COL	11,356	2	957	29,140	27,400	5525	23,948
OIL	0	54,713	1376	194	12	300	58
GAS	0	0	5	3010	36,448	183	1484
FBV	0	0	2	4,300,382	16,609	498	218
TEX	88	3	19	106,573	5,501,608	1518	3789
WOD	29,899	3	2114	170,905	78,689	14,647	2562
MIN	0	16	70	4328	6850	1884	26,313
PET	42,221	96,751	11,260	406,413	407,654	7406	63,514
CHM	305,214	68,299	28,662	1,241,295	2,158,888	38,087	121,069
PAP	4436	3	325	442,279	160,128	19,331	1797

(continued)

(continued)

	COL	OIL	GAS	FBV	TEX	WOD	MIN
FER	0	0	3	61,026	1978	6754	80
CEM	0	857	53	76	370	140	218
IRS	2	1302	329	1671	17,261	8151	12,877
ALU	0	23	116	3941	11,634	3771	49,536
OMN	109,983	157,816	19,255	49,225	196,681	18,950	64,147
MCH	208,316	193,424	25,725	275,599	558,968	7848	33,712
NHY	105,965	28,621	15,172	218,487	616,474	13,510	61,600
HYD	19,635	5303	2811	40,484	114,228	2503	11,414
NUC	2982	806	427	6149	17,350	380	1734
BIO	0	1	13	94,233	1534	185,645	130
WAT	1859	0	133	6695	3641	46	404
CON	30,402	299,685	18,053	333,190	411,837	2372	58,265
LTR	145,944	54,946	17,441	1,665,994	2,120,591	44,524	45,926
RLY	8572	3718	1848	93,918	29,913	4081	14,546
AIR	697	1147	153	106,956	7532	3512	1299
SEA	869	543	114	19,230	37,301	242	278
HLM	0	0	0	0	0	0	0
SER	179,594	192,200	27,914	5,985,966	4,626,498	126,727	171,480
Lab	950,571	880,277	422,077	2,468,376	3,448,211	295,948	832,851
Cap	2,387,623	2,272,012	423,125	3,621,031	3,414,704	215,792	2,092,439
Land							
RNASE							
RAL							
ROL							
RASE							
ROH							
USE							
USC							
UCL							
UOH							
PVT							
PUB							
GOV							
ITX	90,587	81,514	13,229	768,166	483,529	20,759	61,959
CAC							
ROW	1,059,286	14,830,033	751,027	2,686,079	1,315,523	78,402	8,023,533
TOT	5,696,101	19,224,020	1,783,816	37,164,725	28,266,621	1,185,711	11,783,655
CO_2 (t)							
N_2O (t)							
CH_4 (t)							
SO_2 (t)	16,908	13,213	2922	73,677	87,848	5925	33,083
NO_x (t)	9960	3438	1338	32,665	42,473	3850	7955

(continued)

(continued)

	COL	OIL	GAS	FBV	TEX	WOD	MIN
SPM (t)	12,740	1428	1450	34,448	47,991	5172	25,965
CO (t)	673	258	93	11,276	16,132	1840	9142
Oil (mt)		34					
Coal (mt)	361						
	PET	CHM	PAP	FER	CEM	IRS	ALU
PAD	0	20,196	1773	340	0	19	1
WHT	0	36,804	137	735	0	30	1
CER	880	308,401	19,754	4427	44	378	183
CAS	1637	728,827	4493	8404	111	924	729
ANH	308	157,594	882	2331	59	1459	1161
FRS	217	16,076	50,715	69	19	643	154
FSH	0	19,972	172	356	0	17	1
COL	252,434	139,313	53,150	32,516	217,084	2,065,513	457,147
OIL	18,673,034	211,974	556	1	2	20,936	1085
GAS	886	327,777	2027	381,218	35,930	317,837	26,419
FBV	1015	479,422	11,745	10,967	166	2712	2049
TEX	4247	368,217	11,427	9356	9504	11,516	4827
WOD	9784	268,310	67,587	24,950	28,946	16,799	5632
MIN	2631	193,568	4602	162,315	368,646	651,084	477,031
PET	1,020,947	1,182,041	84,364	791,718	126,628	734,155	145,332
CHM	546,658	15,269,445	421,793	1,275,236	137,084	308,761	232,203
PAP	14,496	962,974	617,782	8541	37,859	20,696	9488
FER	3612	225,431	525	577,350	17	5835	198
CEM	3105	7641	70	55	1419	4361	1758
IRS	4515	215,983	13,309	3104	3193	3,560,402	176,752
ALU	10,981	225,889	7481	5077	1372	3,380,076	935,163
OMN	20,287	564,638	12,863	9520	171,932	1,036,459	217,809
MCH	41,103	484,615	15,348	23,692	7106	342,501	157,092
NHY	294,141	789,216	97,953	79,410	248,364	1,113,522	190,742
HYD	54,502	146,236	18,150	14,714	46,020	206,328	35,343
NUC	8278	22,212	2757	2235	6990	31,339	5368
BIO	781	59,890	172,836	313	176	2237	555
WAT	209	11,973	115	1729	23	633	187
CON	192,032	328,759	9861	49,682	9669	103,161	68,902
LTR	120,362	1,709,841	195,932	258,956	127,948	611,392	161,034
RLY	548,377	166,346	24,430	40,166	147,032	955,566	161,873
AIR	15,347	31,377	5648	13,136	13,577	51,495	11,825
SEA	3792	52,835	5044	1304	655	3886	1032
HLM	0	0	0	0	0	0	0
SER	986,547	4,137,078	325,955	592,151	383,060	3,200,680	651,886
Lab	306,368	3,575,263	259,182	334,750	251,382	2,243,925	786,206
Cap	4,606,321	7,319,453	423,864	931,988	624,803	3,903,004	350,669

(continued)

(continued)

	PET	CHM	PAP	FER	CEM	IRS	ALU
Land							
RNASE							
RAL							
ROL							
RASE							
ROH							
USE							
USC							
UCL							
UOH							
PVT							
PUB							
GOV							
ITX	1,554,145	2,787,084	233,237	301,979	111,024	1,029,909	356,922
CAC							
ROW	3,604,127	8,735,582	820,494	548,325	629,557	3,333,828	7,439,439
TOT	32,908,109	52,288,252	3,998,014	6,503,117	3,747,403	29,274,019	13,074,201
CO_2 (t)							
N_2O (t)							
CH_4 (t)							
SO_2 (t)	362,393	324,634	60,080	132,711	267,151	1,925,426	470,909
NO_x (t)	197,448	174,446	39,503	57,196	185,142	1,328,034	325,384
SPM (t)	241,628	211,684	53,345	58,824	255,831	1,831,409	449,057
CO (t)	10,718	72,958	19,008	19,057	91,757	656,487	161,003
Oil (mt)							
Coal (mt)							
	OMN	MCH	NHY	HYD	NUC	BIO	WAT
PAD	1578	281	1309	0	0	5115	49
WHT	3020	546	2266	0	0	3042	93
CER	16,627	3852	10,779	0	0	213,747	388
CAS	34,391	9627	16,922	0	0	2881	611
ANH	45,847	12,678	4697	0	0	6689	178
FRS	35,230	2603	704	0	0	1832	41
FSH	1639	343	1226	0	0	223	50
COL	574,505	114,056	1,692,816	0	0	164	101
OIL	91,313	4544	40,917	0	0	209	619
GAS	70,721	35,564	325,683	0	0	27	163
FBV	22,239	6140	12,255	0	0	24,024	543
TEX	180,233	196,041	7383	0	0	3233	112
WOD	171,272	227,330	2250	0	0	350	99
MIN	1,175,277	213,464	0	0	6146	54	10
PET	863,213	521,420	1,654,974	0	0	28,271	2190

(continued)

(continued)

	OMN	MCH	NHY	HYD	NUC	BIO	WAT
CHM	2,018,159	2,645,906	117,745	0	1711	10,884	6332
PAP	186,071	248,553	27,143	7055	497	3004	1524
FER	6348	6177	3234	0	0	6528	1241
CEM	72,661	7136	31	24	0	9	0
IRS	3,869,196	6,330,835	18,877	2020	0	147	2159
ALU	2,161,226	4,271,787	18,924	2551	0	212	57
OMN	7,001,479	4,244,945	361,252	12,119	5426	23,057	4979
MCH	2,824,181	12,649,358	797,475	40,206	12,174	5690	4336
NHY	874,539	603,835	4,222,060	832	61,371	1508	22,460
HYD	162,046	111,887	782,319	154	11,372	280	4162
NUC	24,613	16,994	118,826	23	1727	42	632
BIO	119,407	8788	2418	0	0	7717	140
WAT	20,829	1912	21,709	0	315	211	141,549
CON	858,778	1,039,096	353,407	21,809	5453	25,876	116,775
LTR	1,316,451	1,293,680	389,306	11,580	5827	118,111	6722
RLY	524,171	211,815	531,289	12,148	7897	8184	656
AIR	41,876	12,219	53,022	406	776	635	47
SEA	32,578	33,574	4699	853	81	2670	183
HLM	0	0	0	0	0	0	0
SER	5,679,499	7,705,456	2,107,244	139,633	32,654	240,078	89,571
Lab	5,060,576	3,681,199	320,002	1,087,917	25,610	1,877,973	326,430
Cap	7,187,010	4,935,561	2,711,260	1,466,763	241,362	1,859,180	363,570
Land							
RNASE							
RAL							
ROL							
RASE							
ROH							
USE							
USC							
UCL							
UOH							
PVT							
PUB							
GOV							
ITX	2,600,023	4,055,593	−1,769,435	−32,813	834	0	8260
CAC							
ROW	30,835,793	11,813,863	0	0	0	0	0
TOT	76,764,616	67,278,659	14,966,987	2,773,280	421,233	4,481,857	1,107,034
CO_2 (t)							
N_2O (t)							
CH_4 (t)							

(continued)

(continued)

	OMN	MCH	NHY	HYD	NUC	BIO	WAT
SO_2 (t)	669,952	172,722	0	312,098	42,176	0	527
NO_x (t)	426,809	94,401	0	210,171	27,954	0	130
SPM (t)	567,985	115,742	0	286,860	37,887	0	315
CO (t)	201,516	40,028	0	102,669	13,540	0	110
Oil (mt)							
Coal (mt)							
	CON	LTR	RLY	AIR	SEA	HLM	SER
PAD	113	18	0	0	111	3612	718,905
WHT	80	1300	0	0	117	4530	397,700
CER	1,061,416	953,328	0	0	712	13,580	2,341,984
CAS	800	0	0	2	0	0	146,536
ANH	601,652	0	0	0	0	10,338	1,285,048
FRS	83,419	0	9	0	0	0	2992
FSH	43	0	0	0	0	0	17,098
COL	1970	0	4164	0	0	0	20,790
OIL	23	0	0	11	0	0	13,295
GAS	325	0	0	0	0	0	6777
FBV	424	18,874	0	0	837	0	2,467,947
TEX	93,123	79,766	910	50	941	17,421	152,856
WOD	919,489	1068	97	0	0	0	23,867
MIN	3,729,164	0	0	0	0	0	27,148
PET	3,355,210	12,200,344	268,248	97,767	43,643	70,748	950,541
CHM	1,141,838	2,324,087	8258	135,793	183,592	2,391,228	1,138,535
PAP	36,405	153,476	4697	951	460	14,464	287,397
FER	11,379	230	6	0	0	0	14,203
CEM	3,969,131	0	0	0	0	0	1738
IRS	10,273,278	891	409	0	0	0	207,244
ALU	3853	326	0	0	0	0	122,373
OMN	8,386,495	2,113,947	1,250,215	91,698	103,333	133,745	2,538,691
MCH	2,366,624	933,154	41,763	5363	12,665	104,722	2,176,907
NHY	895,618	28,247	695,572	2400	5157	15,150	647,088
HYD	165,952	5234	128,885	445	956	2807	119,901
NUC	25,206	795	19,576	68	145	426	18,212
BIO	285,604	4041	29	0	3	58	20,010
WAT	177,244	15,485	320	985	18,334	900	109,167
CON	3,790,532	642,021	866,250	36,259	36,649	222,340	3,491,684
LTR	4,110,131	2,382,395	91,440	60,647	77,416	227,590	3,952,304
RLY	981,304	532,260	568,506	1416	1228	1859	114,131
AIR	92,231	142,044	3437	1177	512	1043	57,923
SEA	25,850	41,715	1673	375	201	43,820	103,162
HLM	0	0	100,676	0	0	0	173,246
SER	9,607,585	8,210,058	232,405	112,249	176,900	761,775	19,477,882

(continued)

(continued)

	CON	LTR	RLY	AIR	SEA	HLM	SER
Lab	22,499,481	9,807,816	2,462,740	263,300	578,908	4,692,207	71,922,015
Cap	9,461,837	7,531,709	1,774,196	197,957	415,166	2,693,040	95,799,645
Land							
RNASE							
RAL							
ROL							
RASE							
ROH							
USE							
USC							
UCL							
UOH							
PVT							
PUB							
GOV							
ITX	2,468,171	2,627,536	224,044	44,498	48,704	414,210	790,126
CAC							
ROW	0	478,851	0	0	0	0	7,888,498
TOT	90,622,999	51,231,016	8,748,525	1,053,410	1,706,690	11,841,614	219,745,567
CO_2 (t)							
N_2O (t)							
CH_4 (t)							
SO_2 (t)	348,557	1,143,620	0	0	0	7715	139,731
NO_x (t)	91,724	672,469	0	0	0	1002	20,258
SPM (t)	39,670	526,552	0	0	0	826	32,526
CO (t)	5436	88,456	0	0	0	200	97,570
Oil (mt)							
Coal (mt)							

	Lab	Cap	Land	RNASE	RAL	ROL	RASE
PAD				921,172	1,929,082	465,192	2,615,038
WHT				527,070	1,103,765	266,169	1,496,249
CER				1,827,586	3,590,841	947,921	4,944,834
CAS				299,626	627,464	151,311	850,582
ANH				1,203,885	1,541,555	566,288	4,038,074
FRS				56,650	106,912	28,156	168,129
FSH				270,564	566,605	136,635	768,082
COL				2587	4806	1392	7050
OIL				0	0	0	0
GAS				8530	15,845	4591	23,244
FBV				2,337,903	4,286,378	1,252,981	6,672,925
TEX				1,086,663	1,857,987	535,414	3,279,454
WOD				5587	5428	2970	18,109

(continued)

(continued)

	Lab	Cap	Land	RNASE	RAL	ROL	RASE
MIN				0	0	0	0
PET				345,403	571,748	238,310	1,112,680
CHM				419,787	578,193	220,158	1,431,805
PAP				27,994	36,969	13,736	96,565
FER				0	0	0	0
CEM				0	0	0	0
IRS				0	0	0	0
ALU				0	0	0	0
OMN				330,690	364,709	172,486	1,016,206
MCH				386,716	426,498	201,709	1,188,372
NHY				149,390	277,486	80,407	407,068
HYD				27,681	51,416	14,899	75,427
NUC				4204	7810	2263	11,457
BIO				363,625	686,243	180,725	1,079,176
WAT				9931	18,447	5345	27,062
CON				170,198	280,178	90,104	537,936
LTR				1,616,071	2,134,167	792,934	5,560,509
RLY				77,718	102,633	38,133	288,570
AIR				18,739	24,745	9194	−48,923
SEA				58,125	76,759	28,519	307,632
HLM				534,718	1,300,190	354,222	2,104,424
SER				4,613,267	6,644,682	2,597,189	15,922,245
Lab							
Cap							
Land							
RNASE	13,686,272	11,301,847					
RAL	30,655,386	99,289					
ROL	9,597,032	591,941					
RASE	23,604,200	28,570,049	17,027,026				
ROH	6,014,659	17,120,543					
USE	17,416,862	19,042,479					
USC	63,190,125	4,147,757					
UCL	9,330,416	1,346,862					
UOH	2,282,323	6,309,825					
PVT		32,937,007					
PUB		9,545,700					
GOV		7,439,300		355,411	0	0	4,142,314
ITX				802,814	1,317,718	426,876	2,548,941
CAC		44,737,987		10,554,871	5,666,113	2,167,887	18,770,185
ROW							
TOT	175,777,274	183,190,587	17,027,026	29,415,175	36,203,372	11,994,117	81,461,420
CO_2 (t)							

(continued)

(continued)

	Lab	Cap	Land	RNASE	RAL	ROL	RASE
N_2O (t)							
CH_4 (t)							
SO_2 (t)	6,754,716						
NO_x (t)	3,972,148						
SPM (t)	4,855,400						
CO (t)	1,628,881						
Oil (mt)							
Coal (mt)							
	ROH	USE	USC	UCL	UOH	PVT	PUB
PAD	686,698	882,091	971,107	295,510	156,272		
WHT	392,909	504,706	555,640	169,082	89,414		
CER	1,532,174	2,474,139	2,817,419	662,037	460,271		
CAS	223,359	286,913	315,868	96,119	50,830		
ANH	1,205,631	2,093,656	2,433,331	405,042	396,006		
FRS	46,044	38,522	42,409	12,905	6825		
FSH	201,695	259,085	285,231	86,796	45,899		
COL	2297	3982	4539	1797	815		
OIL	0	0	0	0	0		
GAS	7574	13,128	14,966	5924	2688		
FBV	2,092,675	3,174,404	3,775,280	892,075	647,357		
TEX	1,057,461	1,576,595	1,917,142	342,432	326,420		
WOD	5942	11,599	15,773	1513	2298		
MIN	0	0	0	0	0		
PET	650,630	723,282	1,928,089	211,808	209,213		
CHM	465,963	850,403	1,564,865	188,901	230,639		
PAP	26,651	56,321	93,756	11,664	15,059		
FER	0	0	0	0	0		
CEM	0	0	0	0	0		
IRS	0	0	0	0	0		
ALU	0	0	0	0	0		
OMN	356,694	699,474	968,900	100,834	150,272		
MCH	417,126	817,979	1,133,052	117,917	175,731		
NHY	132,646	229,909	262,088	103,738	47,068		
HYD	24,578	42,600	48,563	19,222	8721		
NUC	3733	6471	7376	2920	1325		
BIO	295,545	247,261	272,214	82,835	43,821		
WAT	8818	15,284	17,423	6896	3129		
CON	181,655	295,588	465,387	69,489	84,079		
LTR	1,538,523	3,251,313	5,412,380	673,354	869,322		
RLY	73,988	156,357	260,284	32,382	41,806		
AIR	17,839	37,698	62,755	7807	10,080		
SEA	55,336	116,939	194,666	24,218	31,267		

(continued)

(continued)

	ROH	USE	USC	UCL	UOH	PVT	PUB
HLM	1,185,892	930,716	1,834,783	322,525	810,765		
SER	6,145,036	10,965,155	21,129,082	2,307,761	3,926,428		
Lab							
Cap							
Land							
RNASE							
RAL							
ROL							
RASE							
ROH							
USE							
USC							
UCL							
UOH							
PVT							
PUB							
GOV	1,419,397	0	2,379,634	3,975,924	634,321	14,434,600	
ITX	867,247	1,406,262	2,238,702	330,440	408,564		
CAC	5,912,228	10,750,880	25,845,192	1,007,065	227,683	20,665,500	9,545,700
ROW							
TOT	27,233,983	42,918,713	79,267,896	12,568,935	10,114,388	35,100,100	9,545,700
CO_2 (t)							
N_2O (t)							
CH_4 (t)							
SO_2 (t)							
NO_x (t)							
SPM (t)							
CO (t)							
Oil (mt)							
Coal (mt)							

	GOV	ITX	CAC	ROW	TOT	CO_2 (t)	N_2O (t)
PAD	102,487		101,482	595,043	14,960,001	6198	726
WHT	53,035		−115,959	533,924	10,099,591	4540	3
CER	185,878		551,811	742,389	33,298,342	9769	5
CAS	0		709,321	141,521	13,059,415	2685	2
ANH	340,150		438,181	282,915	21,355,984	100,890	75
FRS	61		4938	212,218	993,915	33	0
FSH	0		4929	656,848	4,030,348	1	0
COL	6326		−80,482	16,566	5,696,101	1,487,370	966
OIL	0		−6482	112,740	19,224,020	1,356,507	1
GAS	27,381		11,623	75,561	1,783,816	159,694	101
FBV	398,033		1,400,206	2,500,625	37,164,725	3,941,759	2388

(continued)

(continued)

	GOV	ITX	CAC	ROW	TOT	CO_2 (t)	N_2O (t)
TEX	328,519		594,522	8,394,477	28,266,621	5,045,242	3472
WOD	107		−1,003,963	45,794	1,185,711	738,430	401
MIN	0		−130,444	4,862,982	11,783,655	3,197,345	1983
PET	421,402		−2,900,636	2,787,765	32,908,109	495,970,492	16,357
CHM	908,608		4,992,972	6,114,416	52,288,252	36,596,899	15,679
PAP	80,533		94,443	167,706	3,998,014	7,056,201	4701
FER	213		−58,531	240,268	6,503,117	19,489,690	4082
CEM	0		−330,620	7135	3,747,403	29,868,507	20,371
IRS	0		1,842,627	2,699,301	29,274,019	283,751,262	175,561
ALU	0		543,543	1,313,233	13,074,201	60,957,998	37,014
OMN	539,350		21,869,319	21,128,635	76,764,616	80,335,702	47,395
MCH	580,572		30,745,595	6,497,608	67,278,659	16,472,236	8606
NHY	529,604		0	0	14,966,987	235,745,397	0
HYD	98,132		0	0	2,773,280	0	123,609
NUC	14,905		0	0	421,233	0	3018
BIO	0		0	0	4,481,857	27,715	54
WAT	457,828		0	0	1,107,034	35,073	23
CON	738,347		73,456,402	0	90,622,999	271,601	210
LTR	844,476		1,440,554	3,824,555	51,231,016	0	0
RLY	151,856		455,432	958,135	8,748,525	545,386	893
AIR	20,100		33,413	106,138	1,053,410	0	0
SEA	45,962		197,583	112,451	1,706,690	273	0
HLM	2,189,457		0	0	11,841,614	0	0
SER	33,042,548		4,512,415	27,110,384	219,745,567	3,323,369	1553
Lab				−251,300	175,777,274		
Cap				−2,726,500	183,190,587		
Land					17,027,026		
RNASE	3,307,145			1,119,912	29,415,175		
RAL	4,070,341			1,378,356	36,203,372		
ROL	1,348,497			456,647	11,994,117		
RASE	9,158,699			3,101,446	81,461,420		
ROH	3,061,914			1,036,868	27,233,983		
USE	4,825,346			1,634,026	42,918,713		
USC	8,912,081			3,017,933	79,267,896		
UCL	1,413,124			478,532	12,568,935		
UOH	1,137,160			385,081	10,114,388		
PVT	2,163,093				35,100,100		
PUB					9,545,700		
GOV		35,574,693		4,669,800	75,025,393		
ITX	594,826		8,762,980	795,880	35,574,693		
CAC	−7,072,703			−641,415	148,137,174		
ROW					106,696,599		
TOT	75,025,393	35,574,693	148,137,174	106,696,599			

(continued)

(continued)

	GOV	ITX	CAC	ROW	TOT	CO_2 (t)	N_2O (t)
CO_2 (t)							
N_2O (t)							
CH_4 (t)							
SO_2 (t)							
NO_x (t)							
SPM (t)							
CO (t)							
Oil (mt)							
Coal (mt)							

References

Alarcon, J. V., Van Heemst, J., & de Jong, N. (1997). *The social accounting matrix extended with social and environmental indicators-an application to Bolivia.* Working Paper No. 256, Institute of Social Studies.

CSO (2006). Statistical Abstract of Indian Economy: Central Statistical Organization, India.

Gallardo, A., Mardones, C. (2013). *Environmentally extended social accounting matrix for chile.* Environment, Development and Sustainability, Springer *15*(4), 1099–1127 August.

Harris, J. R., and Todaro, M. P. (1970). Migration, unemployment and development: A two-sector analysis. *American Economic Review 60*, 126–142.

INCCA (2010). India: Greenhouse Gas Emissions 2007, Indian Network for Climate Change Assessment, Ministry of Environment and Forests (MoEF), Govt. of India.

IPCC (2006). *IPCC Guidelines for National Greenhouse Gas Inventories*, Intergovernmental Panel on Climate Change (IPCC), Energy, Vol. 2.

Keuning, S. J., (1992). *National Accounts and Environment: The case for a system's approach.* National Accounts Occasional Paper, Nr. NA-53, Netherlands Central Bureau of Statistics, Voorburg.

LAEI (2003). London Atmospheric Emissions Inventory database, http://data.london.gov.uk/dataset

Markandya, A., & Murty, M. N. (2000). Cost–benefit analysis of cleaning the Ganges: Some emerging environment and development issues. *Environment and Development Economics*, *9*(1), 61–81.

Murty, M. N., & Gulati, S. C. (2004). *A generalized method of hedonic prices: Measuring benefits from reduced urban air pollution.* Institute of Economic Growth Working Paper E/254/2005, New Delhi.

Murty, M. N., Gulati, S. C., & Banerjee, A. (2004): *Measuring benefits from reduced air pollution in the cities of Delhi and Kolkata in India using hedonic property prices model.* Institute of Economic Growth Working Paper 249, New Delhi.

Murty, M. N., Gulati, S. C. (2005) *A Generalized Method of Hedonic Prices: Measuring Benefits from Reduced Urban Air Pollution.* Institute of Economic Growth Working Paper E/254/2005, New Delhi.

National Sample Survey Organisation (NSSO). (2000–01). *Household consumption expenditure and employment.* Government of India.

Pal, B. D., Pohit, S., Roy, J. (2012). Social Accounting Matrix For India. *Economic Systems Research, 24*(1), 77–99.

Pal, B. D., Pohit, S. (2014). Environmentally Extended Social Accounting Matrix for Climate Change Policy Analysis in India. *Journal of Regional Development and Planning 3*(1), 61–76.

Pal, B. D., Ojha, V. P., Pohit, S., Roy, J. (2015). *GHG emission and Economic growth – a CGE modeling approach for India*, Springer book.

Prdhan, B. K., Saluja, M. R., & Sing, S. K. (2006). *A social accounting matrix for India, concepts, construction and applications*. New Delhi: Sage Publication.

Roy, J. et al. (2008). *Natural resource accounting for West Bengal: Sector air and water*. Final report Submitted to the Central Statistical Organization, GoI.

Saluja, M. R., & Yadav, B. (2006). *The social accounting matrix of India*. Government of India: Planning Commission.

Shmelev, S. E. (2013). Environmentally Extended Input–Output Analysis of the UK Economy: Key Sector Analysis, QEH Working Paper Series – QEHWPS183. Queen Elizabeth House: University of Oxford.

TERI (2003–04, 2005, 2009). The Energy Data Directory of The Energy Research Institute, Various issues.

Xie, J. (1996). *Environmental policy analysis: A general equilibrium approach*. England: Avebury publication.

Climate Change and Uncertainty in Agriculture: Does Crop Insurance Help in India?

Meenakshi Rajeev, Manojit Bhattacharjee and B.P. Vani

Abstract Climate change has impacted production risk in agriculture in many ways especially by increasing the volatility in production. Agriculture, which still in India is weather dependent, faces more frequent losses than before and variability in losses is also rising. As about 80 % of the farmers are small and marginal in India, they need certain protection from such uncertainties for their sustenance. Undoubtedly, crop insurance can provide the required safety net and enable farmers to take other measures that help to reduce impacts of climate change. In this back drop, this paper examines the state of crop insurance in India using both secondary data from National Sample Survey Organisation (NSSO) and conducting a field survey in the state of Karnataka. The analysis of data shows that only a meager 4 % of households are covered under crop insurance. While the lack of information is the major reason of not opting for insurance, another large percentage of the households are not interested due to the unattractive design of the schemes. Our regression analysis further highlights that both economically and socially advanced classes have better adoption record. These findings send strong signals to the policy makers to re-think about the mitigation strategies planned by them for the poor farmer households to handle the challenges of uncertainty in the face of climate change in India.

Keywords Risks · Uncertainty · Mitigation strategies · Crop insurance

JEL Classification G2 · Q 14 · 54

M. Rajeev (✉) · B.P. Vani
Centre for Economic Studies and Policy, Institute for Social and Economic Change, Dr. V.K.R.V. Rao Road, Nagarbhavi, Bangalore 560072, Karnataka, India
e-mail: meenakshi@isec.ac.in

M. Bhattacharjee
St. Joseph's College, Bangalore, India

S. Nautiyal et al. (eds.), *Climate Change Challenge (3C) and Social-Economic-Ecological Interface-Building*, Environmental Science,
DOI 10.1007/978-3-319-31014-5_14

1 Introduction

Agricultural sector in India is at a crossroads with different forces operating on it simultaneously. Primarily, it is the predominance of the rain fed agriculture in many States that inhibits the growth, followed by other vagaries of the weather. Consequent high instability in productivity is but a natural outcome of this situation which sets back the development clock in the agricultural sector. Also in many rain fed states like Karnataka, protective irrigation does not play any significant role. Due to the domination of the low-value low-density crops, the farmers' income is continuously depressed, and given the periodic increase in prices of inputs (specifically of the cash inputs), farmers' net income tends to shrink continuously, putting them under financial stress.

A farmer needs to make investment every season in working capital, which makes timely credit a necessary condition for the success of production activities. There are several problems associated with the rural credit system in developing nations in general and India in particular, which are discussed by the scholars(see Bardhan (1989), Barro (1976),Basu (1983, 1984, 1989), Bhaduri (1977), Ghatak (1975)). Even after making right investments, a farmer may not get due returns because of unforeseen reasons, most often beyond her/his control. In other words, farm income being uncertain, appropriate risk mitigation strategies are necessary for stabilizing the income of the farmers.

There are mainly three types of risks emanating from as many sources of uncertainties. These are:

1. Production risks,
2. Price risks and
3. Input risks.

While Production risk may arise owing to two major factors viz., weather risk and risk from pests and diseases, price-related risk occur due to sudden change of demand and instability in expectation formulation. As is well-known farm households mainly face the price risk because production decisions are made far in advance of the date when output is realized. Input risk occurs when there is either a shortage of inputs or when their prices vary (see also Ramaswami et al. 2003; Deshpande 2008).

In this context the state of Karnataka is an important state to study as it is one of the driest state with under developed irrigation system. Farmers in the state face drought quite frequently and commit suicide due to crop loss. While all three types of risks appear to be present in Karnataka agriculture, production risk arising from uncertain weather is more significant. One of the significant ways a farmer can hedge climatic risk is through crop insurance. However, crop insurance is seen to be not prevalent amongst them even though the Government has undertaken several initiatives to introduce crop insurance schemes. It is therefore important to understand what are the factors that determine adoption of insurance by the farmers.

In this backdrop the purpose of the current paper is to discuss the details of some of the risk-related issues based on a primary survey conducted in the state of Karnataka.

However, before moving on to the results from the survey, the paper also presents an analysis of unit record data from National Sample Survey Organisation (NSSO) 59th round data on Situation Assessment Survey of the Farmers. This exercise provides a state wise picture of crop insurance adoption in addition to examining the determinants of the same by the farmers using a probit regression analysis.

In this backdrop, the paper unfolds as follows: The next section briefly discusses the findings from the NSSO Situation Assessment of Farmers Survey to highlight risk and mitigation-related issues. Section 3.2 elucidates the field survey details, and in this context, discusses the sampling techniques and basic sample characteristics. The next three sections discuss different types of risks faced by the farmers, based on the field survey. Findings on mitigation strategies particularly that of insurance are taken up in the penultimate section. A concluding section follows at the end.

2 Risk and Mitigation: Findings from NSSO Survey

Union ministry of Agriculture wanted to have a comprehensive assessment of the situation of farmers in the country at the beginning of the millennium. The purpose was to understand various aspects concerning farmers, which include farmers' levels of living, income and productive assets they possessed, farming practices and preferences they had, availability of resources, their awareness on technical developments and access to modern technology in the field of agriculture, etc. To provide information on these aspects to the ministry of agriculture, NSSO, as a part of 59th round, conducted the Situation Assessment Survey of farmers (SAS), the period of survey being January–December 2003. Though the survey provides rich macro-level data, there is not much analysis of unit record household level data (see also Bhattacharjee and Rajeev 2010; Rao and Tripathi 2001; Gothaskar 1988) from this survey.

The survey was limited to only the rural areas of the country, and the respondents were members of farmer households. A farmer household is defined as one which has at least one member as farmer, possessing some land, and is engaged in agricultural activities on any part of the land during the preceding 365 days. In all 51,770 households were surveyed in the central sample. Only seven states participated in the state sample, and Karnataka is not among them (NSSO 2003).

In order to ensure profitable production especially in case of agriculture, risk management undoubtedly is critical.

Consequently, working group on risk management in Agriculture (Government of India (2007)) has dealt with this issue in detail. The group classifies the sources of risk into following components:

(a) Production risk
(b) Price or market risk

(c) Financial and credit risk
(d) Institutional risk
(e) Technology risk
(f) Personal risk.

As these factors not only affect the income of farmers but also the viability of agriculture, understanding the possible strategies and mechanisms to mitigate risk assumes importance. World Bank in its World Development Report (2001) classifies the risks management strategies into informal and formal strategies and the following Table prepared by the Planning Commission (2007) on similar basis provides a clear picture of the possible management strategies in agriculture (see Table 1).

Table 1 Risk management strategies in agriculture

		Informal mechanisms	Formal mechanism	
			Market based	Publicly provided
Ex-ante strategies	On farm	Avoiding exposure to risk		Agricultural extension
		Crop diversification and inter-cropping		Supply of quality seeds, inputs, etc.
		Plot diversification		Pest management systems
		Mixed farming		Infrastructures (roads, dams, irrigation systems)
		Diversification of income source		
		Buffer stock accumulation of crops or liquid assets		
		Adoption of advanced cropping techniques		
		(Fertilization, irrigation, resistant varieties)		
	Sharing risk with others	Crop sharing	Contract marketing	
		Sharing of agricultural equipment, irrigation sources etc.	Futures contracts	
		Informal risk pool	Insurance	
Ex-post strategies	Coping with shocks	Reduced consumption patterns	Credit	Social assistance (calamity relief, food for work etc.)
		Deferred/low key social and family functions		Rescheduling loans
		Sale of assets		Agricultural insurance
		Migration		Relaxations in grain procurement procedures
		Reallocation of labour		Supply of fodder
		Mutual aid		Cash transfer

Source Report of the Working Group on Risk Management, Planning Commission (2007-2012)

Understanding the scope of the above strategies as well as the extent to which these have been followed by our farmers, calls for an in-depth discussion with the farming community. However, the Situation Assessment Survey has collected some information on the awareness created about a few Government initiatives such as the minimum support price, crop insurance schemes and so on. Table 2 presents the details of awareness about these programmes at the State level. It is found that the general awareness level of the Indian farmers regarding risk mitigation measures such as crop insurance is quite low. It can be seen that at the All India level, only 29 % of the households are aware of the Minimum Support Price and identical is the awareness level of Karnataka farmers also. The States which have higher awareness level are Haryana, Punjab and Kerala, where the share of households aware of these risk mitigation measures is over 61 %. For certain crops, since the

Table 2 Awareness of the risk management schemes and WTO

States	Share of households aware of minimum support price	Share of households who have insured crop	Share of households who are not aware of crop insurance	Share of households who are aware of WTO
Andhra Pradesh	29.23	6.77	75.05	5.47
Assam	21.61	0.19	63.81	10.49
Bihar	18.88	0.85	40.89	8.41
Chhattisgarh	35.29	7.26	66.08	1.41
Gujarat	25.44	19.81	48.21	5.81
Haryana	64.14	0.12	40.96	11.46
Jharkhand	12.49	0.57	66.86	10.69
Jammu & Kashmir	26.99	0.22	25.00	7.15
Karnataka	29.22	7.90	53.79	7.02
Kerala	61.10	5.06	29.24	44.06
Maharashtra	27.68	10.74	62.93	5.84
Madhya Pradesh	29.41	2.24	59.86	2.75
Orissa	12.37	7.66	76.68	2.34
Punjab	62.49	1.25	19.64	23.38
Rajasthan	10.51	0.65	54.74	2.30
Tamil Nadu	48.40	2.65	55.91	12.11
Uttar Pradesh	32.91	1.23	55.65	4.56
Uttaranchal	23.03	0.07	54.06	13.03
West Bengal	30.35	1.06	64.53	11.88
All India	29.16	4.04	56.63	7.73

Source Author's analysis of NSSO data

price-related risks would originate from the international markets, information on import policy and World Trade Organisation (WTO) related measures plays an important role. However, the data shows that the share of households that are aware of WTO stipulations is as low as is only 8 % at All India level. Karnataka's score is one percent less than the All India figure, i.e. 7 %. *Kerala is the only state where 44 % of the farmer households are aware of it. Punjab stands second with 23 % of households being aware of WTO related norms.* The rest of the Indian States have negligible share of households being aware of WTO (see Table 2).

Moving on to the aspect of crop insurance, it is found that only 4 % of the household had insured their crop at All India level while in case of Karnataka the figure is slightly higher revealing 8 % of the households obtaining crop insurance. The main reason for not being insured is the lack of awareness about this programme. It is found that 57 % of households at All India level were not aware of insurance facility and Karnataka again roughly shows the same picture, i.e. 54 % of the households were not aware of it. Expectedly, Punjab on the other hand shows a different picture where, only 20 % of the households are not aware of the programme; but surprisingly, even though majority of the households had awareness, only 1.25 % of the households had insured their crop. What actually is the reason for such low insurance coverage of farmer households even when they are aware of the facility? Is it because the programme is not user friendly? If so the usefulness of the crop insurance programme and how viable it is for different sections of the farmers need to be studied with the aid of primary and secondary data as also intensive field visits to find answers to the above questions. The next section provides a regression analysis based on Situation Assessment Survey of Farmers data.

3 Regression Analysis

To identify the factors that determine adoption of crop insurance by the farmer households in the state of Karnataka, a probit regression has been carried out. The dependent variable is a dichotomous variable, assuming a value 1 if a farmer household has insured crop, zero value has been assigned otherwise. We have considered the following set of independent variables.

3.1 Selection of Explanatory Variables

Having crop insurance depends on both demand and supply side factors. Demand for crop insurance may depend on the need to have crop insurance and on awareness or information about crop insurance. Need or desire to have crop insurance may depend on the risk faced by a household. A household is expected to have higher demand for crop insurance if risk faced by the household is high. As already mentioned in Sect. 1, there are three types of risk that a farmer may face, namely, production risk, price risk and input risk. Production risk may depend on the region in which the farmer produces crop. This is because rainfall as well as soil type varies across regions. To see whether region or production risk plays an important role in determining crop insurance, we have considered a dummy variable, where we have assigned a value 1 to households belonging to inland North region of Karnataka, zero value has been assigned otherwise. Inland north region of Karnataka is a dry semi-arid region and it includes northern districts of Karnataka, such as Bagalkot, Bijapur, Gulbarga, Bidar, Dharwad, Haveri, Chitradurga, Davangere, Gadak, Raichur, Koppal, etc.

Apart from production risk, a household may face price risk and input risk. Since price risk and input risk varies across crops, we have included dummy variables to capture it. Generally, cereal crops are expected to have less fluctuation in price, so we have assigned a value 1 if the farmer produces mainly cereal crop, zero value has been assigned otherwise. However, a significant relation between type of crop and crop insurance does not necessarily imply inducement of price risk to avail insurance. The insurance terms and conditions may be better for some crops, which in turn may induce households in having insurance for that crop.

In addition to the above-mentioned variables, risk faced by a household is expected to be less if a household has diversified into non-farm activities. Households which have diversified into non-farm activities are captured in the explanatory part with the help of a dummy variable.

Awareness about crop insurance may depend on the education of the household. Educated household are expected to have better information. Therefore, we have assigned a value 1 to households having at least one member with secondary education, zero value is assigned otherwise. Age can also play an important role in creating awareness. Households with more aged people are expected to have more awareness. In the present analysis, mean age of the household is captured as an explanatory variable.

Information on insurance benefits in India may vary with caste and gender (Rajeev et al. 2011). Households with female head and households belonging to poorer caste may have poorer social networking with respect to insurance leading to less presence of information.

Supply side factors are expected to vary across regions, which is already captured. For example, in semi-arid region, supply of insurance may be less because higher risk of crop failure.

3.2 *Probit Regression*

In Probit model it is assumed that the zero or one value assumed by the observed dependent variable, depends an unobserved latent variable (Y_i^*). such that if Y_i^* exceeds a critical value (I_i), a household avails insurance and we observe a value 1 and if Y_i^* is less than this threshold value, a household is without crop insurance.

The unobserved variable in probit model is a function of a set of explanatory variables (X_i's). The relationship between the latent variable and the explanatory variables is expressed as

$$Y_i^* = X_i'\beta + \varepsilon i,\ \varepsilon i \sim N(0, \sigma^2)$$

Given that the model follows a normal distribution, the probabilities of having or not having crop insurance is given as

$$P_i = P(Y = 1|X) = P(Y_i^* \geq I_i) = P\{X_i'\beta + \varepsilon_i > 0\} = F(X_i'\beta)$$
$$P(Y = 0|X) = 1 - P_i = 1 - F(X_i'\beta)$$

The parameters in this model are estimated with maximum likelihood procedure. If the coefficient of the explanatory variable, β is positive, it implies that the probability of availing insurance increases with increase in the explanatory variable. On the other hand, a negative coefficient would imply that the probability of having insurance decreases with respect to increase in explanatory variable. The likelihood function of the model is given below

$$L(\beta) = \prod_{i=1}^{N} P(Y_i = 1|x_i; \beta)^{Y_i} P(Y_i = 1|x_i; \beta)^{1-Y_i}$$

Since the present analysis is done using cross-sectional observations, there are chances of heteroscedasticity problem. To solve it robust standard errors has been used for hypothesis testing purpose instead of ordinary standard error.

3.3 Results

The results of the regression analysis are presented in the Table 3. The table shows that both demand side factors and supply side factors play an important role in determining crop insurance. It is observed that probability of availing insurance is significantly high in inland north regions of Karnataka, which is a semi-arid region. One expects supply of insurance to be less here. However, a positive sign implies that demand side factors and not supply mainly determines crop insurance. Higher presence of crop insurance in North Karnataka as well mean that presence of production risk induces households to avail insurance.

The regression result shows that there are differences in probability of availing insurance across caste and class. General caste households have a higher probability of availing insurance. Similarly one also observes that land owned by the household to be positively related with probability of availing crop insurance. Households belonging to higher caste as well as higher class may be financially more aware. Second, it may also happen that for having better social networking in financial markets, they have better information.

Table 3 Regression results: determinant of crop insurance of farmers in Karnataka (situation assessment survey of farmers data 59th round NSS)

Number of obs = 1975					
LR chi2(8) = 88.24					
Prob > chi2 = 0.0000					
Pseudo R square = 0.0839					
Dependent variable crop insurance = 1, others = 0	dF/dx	Std. Err.	z	P > z	x-bar
Explanatory variables					
Inland North region of Karnataka = 1, others = 0	0.0518264*	0.012076	4.35	0	0.473924
Women headed household	−0.017815	0.015809	−1.01	0.311	0.109367
Secondary education = 1, others = 0	0.023979**	0.01199	2.06	0.04	0.396456
Land owned	0.0078168*	0.001767	4.53	0	1.67028
Non agriculture	0.0385444**	0.021268	2.1	0.036	0.106329
General caste = 1, others = 0	0.0307976*	0.012161	2.63	0.008	0.391392
cereal crop = 1, others = 0	0.0067823	0.013592	0.48	0.629	0.84557
Mean age of the household	−0.0006511	0.000604	−1.08	0.282	29.3647
Predicted probability at X-bar = 0.0610632					
obs. P 0.0749367					

Note dF/dx implies changes in probability arising from 1 unit change in explanatory variable.
* implies significance at 1% level and ** implies significant at 5% level.
The probability of having insurance at the mean value of explanatory variables 6.1 %

It is observed that presence of non-farm activity increases probability of availing crop insurance. One should note that non-farm activity may crop up under several situations (see Haggblade et al. 2007). First, it may arise if the farm sector does not provide adequate income and employment opportunity (push effect). In this case, non-farm activity may emerge even if there is less demand for non-farm goods in the region. Second, non-farm activity may emerge when more households earn surplus in the agricultural sector. Surplus earning in agricultural sector generates demand for non-farm goods. Moreover, the non-farm sector provides a vent to invest surplus earning and it allows the farm household to release family labour for non-farm activities.

In the present context, it is possible that, due to poorer income generation in the agricultural sector in Karnataka, non-farm activity does not pull investment, it is mainly carried out by households for whom farm income and employment is inadequate. The positive relation between presence of non-farm income generation and crop insurance thus implies that the household faces higher risk in the agricultural sector.

The above regression was carried out using *Situation Assessment Survey of Farmers* data, which is dated. More importantly in the decade of 2000, India has seen high growth and the policy makers were expecting a trickledown effect. Has it happened for agriculture in general and crop insurance in particular? Adoption of crop insurance and related aspects are investigated through a survey which is presented below.

4 Sampling Technique and Basic Sample Characteristics

As mentioned, in order to understand the kind of risk faced by the farmers as also the mitigation strategies followed, a survey has been carried out in three districts in the state of Karnataka. As in the absence of proper mitigation strategies farmers tend to borrow at times of distress, we have therefore purposively selected three districts viz., Mandya, Chamarajanagar and Haveri which have rather high level of indebtedness as per NSSO survey (59th round). Mandya and Chamarajanagar are among the top five highly indebted districts of South Karnnataka, while Chamarajanagar has the highest indebtedness amongst all districts in Karnataka. Among the districts of North Karnataka, Haveri has the highest level of indebtedness.

Further, these three districts have been selected also considering their varied performances levels in agriculture sector. Dr. D.M. Nanjundappa committee report on regional imbalances provides a detail assessment of the Talukas of Karnataka in terms of their agriculture productivity and infrastructure. The above report puts, Mandya district under the group of districts with good agriculture infrastructure as also performance. Using the same report, one can place Haveri as a middle performing district and Chamarajanagar as a low performing district. Using the agriculture infrastructure and performance indices constructed in Dr. Nanjundappa committee report, one backward and one better performing Talukas from each district were identified. Thus, as is clear, the purposive sampling technique is used here to arrive at a balanced view on farmers' situation in Karnataka.

Subsequently as mentioned, from each district, two Talukas were selected and from each chosen taluka 50 households were selected where, in order to select our samples in each Taluka, we took the assistance of Raita Samparka Kendras (RSK, farmers' assistance centre) for the sampling frame. For example, in Mandya district we selected two Taluks viz., Maddur (a better performing one) and Mallavalli (as a backward taluka). There are four RSKs in Maddur Taluka and from the purview of each RSK, we selected two villages. Further, from each village we have selected about six households at random using the list of households provided by the RSKs. Number of sample households selected from Maddur and Mallavalli Talukas is 50 each, making a total of 100 households. Similarly, we have selected two Talukas from Chamarajanagar districts viz., Kollegal (having 5 RSKs) and Elandur (2 RSKs). We have selected five households from each village of Kollegal and in case of Elandur, to arrive at a sample size of 50 respondent households, we selected four villages from each RSKs and six households from each village. Following the same procedure, we have selected 50 samples each from Kollegal (a backward taluka) and Yelandur (a better performing Taluka). Similarly, in Haveri district, we have selected two Talukas Viz., Haveri (backward), and Ranibennur (better performing) and selected 50 households each.

An alternative approach would have been to select equal number of households from each village. However that would make some Talukas much less representative in our sample and therefore we decided against this approach.

The basic characteristics of our sample households are as follows. In regard to the economic condition of households, it is observed that on an average 55 % of the households are below poverty line (as per ration card classification) and 45 % above poverty line, showing that the sample has a good mix of poor and not so poor farmers. Sample also captures caste background of the respondent households; in particular, we have 16 % of schedule caste (SC) population and about 80 % of other backward caste (OBC) population in the sample while Muslims constitute about 2 % of our sample farmers. Principal occupation of the respondents is farming, though they have other supplementary occupations such as working as agricultural labourers, small business ownership, etc. There are about 40 % marginal farmers, 25 % small farmers and 35 % medium farmers in our sample. Our respondent households also earn supplementary income through other occupations of the family members and these occupations include daily wage labourer both in agricultural and non- agricultural (27 % households income) activities, animal rearing (8 % of households), small business (30 % of households) and other such occupations.

A structured questionnaire was personally canvassed by us to understand the nature of indebtedness and the risk and mitigation strategies of the farmers. A farmer in Karnataka faces risks arising out of vagaries of weather as well as input and market risks, issues relating to which are discussed below.

5 Production Risk

5.1 Uncertain Weather

As is well known, Karnataka is largely a drought prone state with comparatively low amount of rainfall and climate change can further enhance this risk. Two-thirds of Karnataka's geographical area is arid or semi-arid where out of 27 districts, 18 districts are drought prone with annual normal rainfall of less than 750 mm. The normal annual rainfall in the state is 1139 mm received over 55 rainy days' (see Karnataka Crop Insurance Study, September 2003). Irrigation facilities in the state are yet to develop adequately to address the problems of the farmers. Our survey includes both irrigated and rain fed areas, and there is stark difference in farmers' conditions between these two situations. It is worth noting that in the recent past, the state had faced disaster due to flood as well. Though the state of Karnataka has an early warning system to understand the climatic aberrations, it would be necessary to translate this into contingency plans for remedial action at the very first signs of climatic distress. In order to do so effectively, it will be necessary to provide institutional training to the farmers as well as enhance the capability of the farmers to be receptive to early warnings.

There is also severe shortage of agriculture extension officers who are supposed to impart technical knowledge to the farmers as many farmers reported during our survey that they hardly ever see an extension officer in their village. These officers

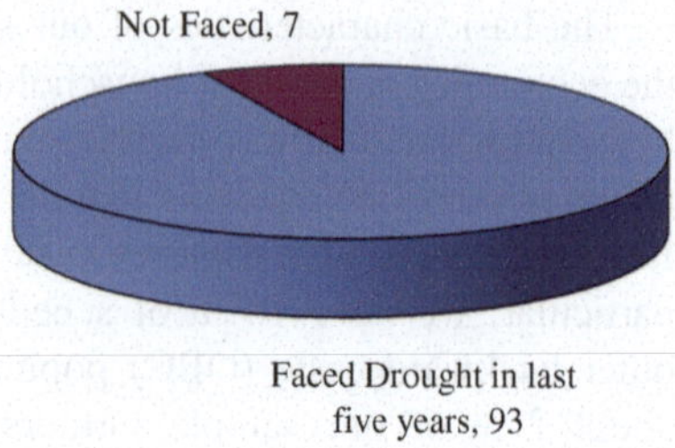

Fig. 1 Share of households faced substantial damage of crops due to drought or flood at least once in the last 5 years. *Source* Primary Survey

also should make the farmers aware of climate change and its impacts. Further, farmers also felt that extension officers themselves lacked the knowledge and that they should be trained in soil testing, identifying appropriate pesticides and other such necessary techniques.

During our survey, we have observed that weather-related disaster impacted almost all the farmers (Fig. 1). Only 7 % of households reported that though they faced adversities, their crops had not been destroyed substantially (see Fig. 1).

The average number of drought situation (which had severe impacts on crop) faced in preceding 5 years by farmer households was about 2; thus, one can observe that every alternate year one is faced with weather related adversities (Fig. 2). These situations may further aggravate with climate change. Given the intensity with which a farm household is subjected to whether related risk, it is essential to address the issue effectively. In this regard, it is worth mentioning that lack of irrigation facilities have severely affected farmers even in the parts of developed districts like Mandya or Chamarajanagar.

In case of severe drought, the Government declares aid to the farmers in that area. However, despite the noble intentions and efforts of the Government, such help seldom reaches the needy. In our sample, though almost all farmers faced weather related disasters, it is important to note that only a few (about 30 %) received Government help (see Fig. 3) and such lacuna in the implementation process needs to be seriously dealt with.

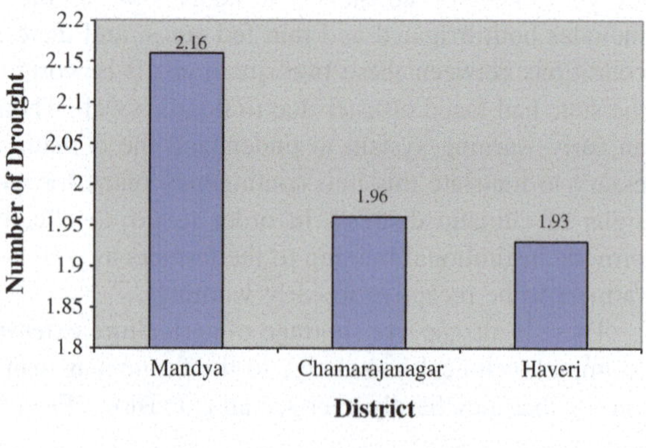

Fig. 2 Average number of droughts faced in the last five years. *Source* Field Survey

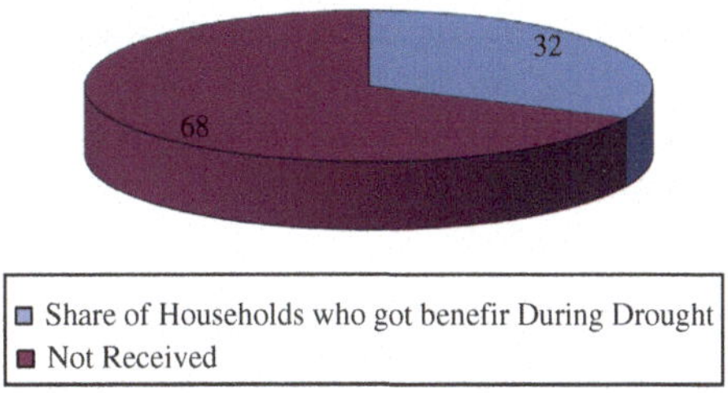

Fig. 3 Share of households who got benefited from Government during drought. *Source* Field Survey

Understandably, natural disasters drastically reduce a farmer's income. As far as mitigation strategies are concerned our survey shows that marginal and small farmers usually have little savings; so a large percentage of such farmers essentially need to borrow in order to meet their expenses (see also Rajeev et al. 2006). Economically better off friends and relatives often provide loans to the farmers in distress. Alternatively, farmers seek wage labour in agriculture or other occupations for their sustenance and of late, we have also observed that farmers with small land holdings are working under NREGA programme. A few farmers who have small business like petty shops fall back on the income earned from such non-farm activities. The survey also revealed that about 15 % of farmers had to sell assets for their sustenance. Figure 4 gives graphic details (bar chart) of the survival strategies adopted by farmers after visitations of drought.

It is clear that the strategies adopted by farmers to cope during weather related aberrations are rather weak, and a systematic approach to the problem is absolutely necessary to ameliorate farmers' distress. Developing non-farm activities is one such option, and the problems related to this are discussed in the sequel hereto. In

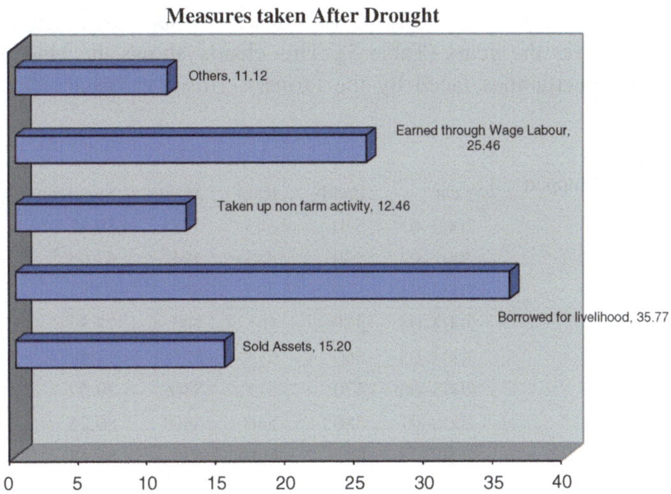

Fig. 4 Means of survival after drought (shown in terms of percentage of farmers). *Source* Field Survey

addition, irrigation can go a long way and we have seen the significant difference between the farmers having access to irrigation facilities and those who do not. Risk mitigation strategies like insurance coverage are essential; but at present insurance coverage is very meager, and this issue will be taken up for a detailed discussion in the sequel hereto.

6 Price-Related Uncertainties

As mentioned earlier, in addition to weather and disease related uncertainties, farmers also face price-related uncertainties, and the study covers this issue as well. The major crops covered in our study area are Paddy, Ragi, Maize, Sugarcane, and Cotton, and we have collected information on the minimum support price of these crops and presented in Table 4.

It is seen that the increase in minimum support price had been minimal till around 2008–09 and the initial jump in minimum support price was witnessed only during 2008–09. It was also revealed by our survey that though government declares minimum support price, procurement at the declared price is minimal. While many farmers wish to sell to the state agencies at the minimum support price, they are unable to do so due to lack of demand. Thus, they feel that declaring minimum support prices is not of much use to them as procurement would be far below the quantity available for sale.

The farmers often sell their produce at market prices mostly due to economic compulsions.

Therefore, it is necessary to highlight the trends of market prices of these crops for the regions of our interest and we have used both secondary data and primary information to shed light on this aspect. It is observed that market prices of crops have fluctuated over the years (Table 5). This clearly shows the kind of market situations and uncertainties faced by the farmers. However, data collected from

Table 4 Minimum support prices over the years

Year	Paddy	Ragi	Maize	Sugarcane	Cotton
2000–01	510	445	445	59.5	1625
2001–02	530	485	485	62.05	1675
2002–03	530	485	485	69.5	1675
2003–04	550	505	505	73.5	1725
2004–05	560	515	525	74.5	1760
2005–06	570	525	540	79.5	1760
2006–07	580	540	540	80.25	1770
2007–08	645	600	620	81.18	1800
2008–09	850	915	840	81.18	2500

Source Cost of Cultivation Survey

Table 5 Prices of different crops across the selected districts at different points of time

Year	Paddy		Ragi	Maize		Sugarcane	Cotton
	Mandya	Haveri	Mandya	Chamarajanagar	Haveri	Mandya	Haveri
2006–07	660	626	663	944	625	900	2188
2005–06	687	577	475	540	745	1100	1865
2004–05	637	697	401	659	688	–	2075
2003–04	557	706	444	500	492	–	2324
Mean	635	651	496	661	637	–	2113
Standard deviation	56.03	61.21	115.55	200.54	108.67	141.42	194.16

Note Prices for all crops are not available for all districts as these crops are not grown
Source Cost of Cultivation Survey

Table 6 Summary measures of prices from primary survey from different households: 2009–10

	Paddy	Ragi	Sugar	Maize	Mulbery	Cotton
Mean	877	883	1217	766	123	2514
Median	850	800	1100	770	120	2500
Mode	900	800	1100	800	120	2200

Source Field Survey

farmers in the recent year points to increase in market prices of all crops (compare Tables 5 and 6). This may corroborate the sharp increase in food price inflation during the last few years.

7 Risk Mitigation

7.1 Crop Insurance

One of the standard ways of mitigating risk in the face of climate change is through crop insurance. Unfortunately, however, use of crop insurance is not wide spread in Karnataka as we observe that among our sample households, only 17 % of the households are covered by insurance scheme while the rests have remained totally uncovered (Fig. 5) and consequently will be driven to a distress situation in the event of crop failure.

One of the major reasons for not taking crop insurance is lack of awareness among farmers and the absence of usable insurance scheme in the regions. Thus, we feel that formulation of appropriate crop insurance schemes as per need of specific regions is of significant importance. We also notice from our sample that around 40 % of the farmers are not interested in crop insurance (Fig. 6) as the area based approach of the insurance scheme and the resulting conditionalities do not appeal to them.

Indemnity under the "area approach" is offered as per the results of crop-cutting experiments for which each year a certain number of plots with the insured crops in

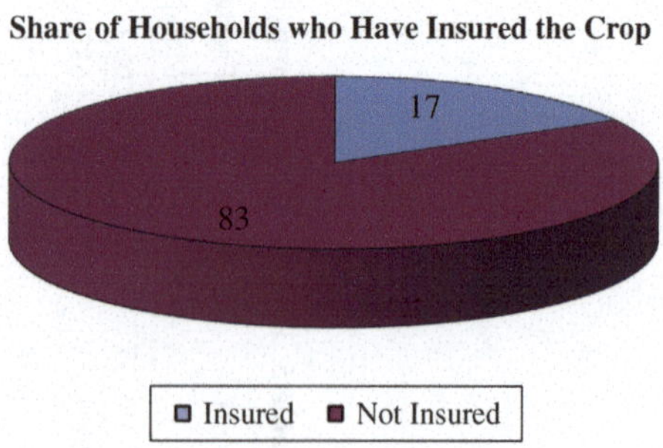

Fig. 5 Share of households with crop insurance. *Source* Field Survey

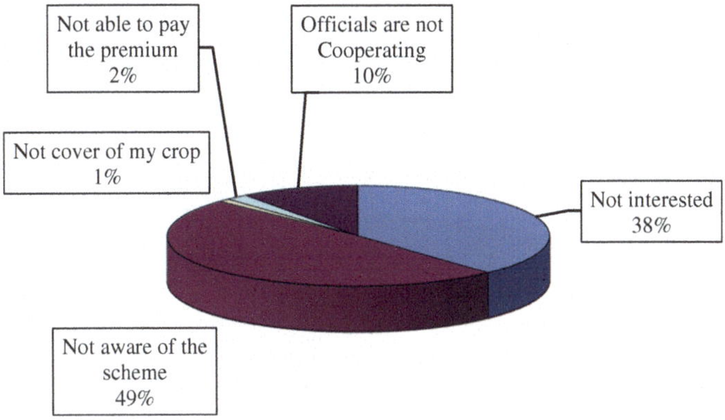

Fig. 6 Reasons for not taking up crop insurance. *Source* Field Survey

a certain "area" are taken as the indicators of an individual farmer's losses within that area. Insured farmers receive indemnity based upon the difference between the threshold yield and the yield of the crop-cutting experiments in their area. Crop yields naturally vary even over small areas and even localized natural calamities can cause distress. While such situations are not un-common farmers may not be compensated for their loss. Therefore, a farmer is wary about getting compensation, as he rightly feels that he/she may be paying insurance premium unnecessarily. If there is a large scale calamity, he would in most cases get government compensation without any insurance cover. Thus, a farmer considers an individual, his or her own land based insurance as the best option. However, such options have much higher level of premium implications which they fail to grasp. Given this scenario, it is necessary to take the farmers into confidence while formulating crop insurance policies. What can be the premium related implications of an individual based insurance should also be discussed; otherwise, the entire exercise is going to be futile.

8 Non Farm Activities

When a farmer faces a situation of depressed income, he can at least get over the distress condition, if he has any gainful nonfarm activities. Most of the farmers in our sample do not have non-farm activities. Therefore, the study has also sought information from the respondents about the possible non-farm activities they would like to take up.

An interesting fact revealed in our survey is that farmers seldom have the knowledge or imagination about diverse non-farm activities available (Fig. 7). Most farmers consider certain petty businesses like opening a small shop or buying capital goods such as a tractor or truck for renting is the only option. Thus, there is a

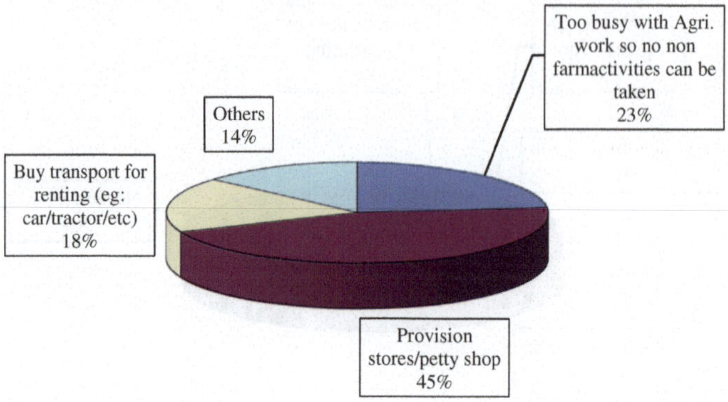

Fig. 7 Share of households showing interest in different types of nonfarm activities. *Source* Field Survey

need to provide training to the farmers as per the resource base of a region to develop meaningful nonfarm activities. This can go a long way in ameliorating their distress.

Currently, most farmers do not take up non-farm activities due to lack of information and resources. Also, it is rather difficult to get credit from formal sector for such activities without a proper project plan.

9 Conclusion

It is clear from the discussions above that the farmers of Karnataka in its entire agricultural regions face all the three major types of risks and this is a scenario that holds for the country as well. Availability of irrigation facilities, however is a critical element that determines the outcome of a weather related crisis; however, providence of this facility is rather poor in the state in spite of it being a dry region. Though risks are all pervasive mitigation strategies are rather weak for the agrarian community across the nation as can be seen from our analysis of NSSO data at the all India level and Karnataka experience. In particular, risk mitigation strategies in terms of crop insurance are highly under developed across the country in general and also in the state. In this context it is rather interesting to note that in our sample, around 40 % of the farmers are not interested in crop insurance.

This is because crop insurance is usually area based. Thus, even if a farmer's crop is destroyed his/her compensation would depend on whether the area in which the farmer is cultivating comes under the insurance coverage or not. Therefore, a farmer feels that he/she may be paying insurance premium unnecessarily. In this background, it is necessary to take the farmers into confidence while formulating the insurance policies. What can be the premium related implications of an individual based insurance should also be discussed; otherwise, the entire exercise would be futile. Development of non-farm activities is another major initiative that is necessary. Most farmers currently consider certain petty business like opening a small shop or buying capital goods such as tractor or truck for renting as the only option. Thus, there is a need to provide training to the farmers as per the resource base of a region to develop meaningful nonfarm activities. This can go a long way in ameliorating their distress during crop failure.

Acknowledgement We thank RBI and State Planning Board, Government of Karnataka for their support.

References

Bardhan, P. (Ed.). (1989). *The economic theory of Agrarian Institutions*. UK: Clarendon Press.

Barro, R. J. (1976). The loan market, collateral, and rates of interest. *Journal of Money, Credit and Banking, 8*(4), 439–456.

Basu, K. (1983). The emergence of isolation and interlinkage in rural markets. *Oxford Economic Papers, 35*(2), 262–280.

Basu, K. (1984). Implicit interest rates, usury and isolation in backward agriculture. *Cambridge Journal of Economics, 8*, 145–159.

Basu, K. (1989). Rural credit markets: The structure of interest rates, exploitation and efficiency. In P. Bardhan (Ed.), *The economic theory of Agrarian Institutions*. UK: Clarendon Press.

Basu, P. (2006). *Improving access to finance for India's rural poor*. Washington, D.C.: The World Bank.

Bhaduri, A. (1977). On the formation of usurious interest rates in backward agriculture. *Cambridge Journal of Economics, 1*(4), 341–352.

Bhattacharjee, M., & Rajeev, M. (2010). *Interest rate formation in informal credit markets in India: Does level of development matter?* Brooks World Poverty Institute Working Paper Series 12610. BWPI, The University of Manchester.

Chakravarti Committee Report. (1985). *Report of the committee to review the working of the monetary system*. Bombay: Reserve Bank of India.

Deshpande, R. S. (2008). *Farmers' Distress and Indebtedness: Miseries and Myths*. Banglore: Agriculture Development and Rural Transformation Centre, Institute for Social and Economic Change.

Ghatak, S. (1975). Rural interest rates in the Indian economy. *Journal of Development Studies, 11* (3), 191–201.

Gothaskar, S. P. (1988). On some estimates of rural indebtedness. *Reserve Bank of India Occasional Papers, 19*(4), 299–327.

Government of India. (2007). *Report of the Working Group on Risk Management*. New Delhi: Planning Commission (2007-2012).

Haggblade, S., Hazell, P. B. R., & Reardon, T. (2007). *Transforming the rural non farm economy: Opportunities and threats in the developing world*. New Delhi: Oxford University Press.

National Sample Survey Organisation. (2003). *Situation assessment survey of farmer households*. New Delhi: NSSO.

Patnaik, P. (2005). Financial liberalization and credit policy. In V. K. Ramchandran & M. Swaminathan (Eds.), *Financial liberalization and rural credit in India*. New Delhi: Tulika Books.

Rajeev, M., Ranade, R., & Deb, S. (2006). Why do poor farmers default less?: Case of Indian informal credit market a game theoretic exploration. Posted: November 17, 2006 http://mpra.ub.uni-muenchen.de/768/01/MPRA_paper_768.pdf

Rajeev, M., Vani, B. P. & Bhattacharjee, M. (2011). Credibility of equal access to credit: Does gender matter? *Economic and Political Weekly, XLVI.*

Ramaswami, B., Shamiika, R., & Chopra, S. D. (2003). Risk management in agriculture. Indian Statistical Institute, Planning Unit, Discussion Paper 03-08.

Rao, K. R., & Tripathi, A. K. (2001, May 12). Indebtedness of households: Changing characteristics. *Economic and Political Weekly*, 1617–1626.

World Development Report. (2011). *Conflict, Security, and Development*. Washington D.C.: The World Bank.

Climate Change and Natural Capital: Some Veiled Issues on Sustainable Livelihood in Agriculture Sector

Digambar Chand and Rajendra Gartia

Abstract Natural capital refers to the planet's stocks of water, land, air, and renewable and non-renewable resources (such as plant and animal species, forests, and minerals). It accounts for vital information regarding economic dependence and impact on natural world. Governments and businesses face the challenges of extracting the most from these scarce resources. Climate change, natural capital and economies have impact on each other. Economic activity drives climate change but both affect natural capital stock. We should calculate the economic cost of these services if we had to provide them ourselves. Many non-material benefits, such as spiritual or aesthetic enjoyment are also obtained from nature apart from food, fibre and fuel. Thus, it is necessary to understand the consequences of any climate change strategy on other dimensions of human well-being. Moreover, it also affects to livelihood. Agriculture is the main livelihood of Odisha. Every year it is affected by flood, drought, early/late arrival of monsoon, etc. As agriculture is highly dependent on climatic variations, sustainability of agriculture is a matter of question due to increased agricultural uncertainty. Therefore, food security poses a challenge to the world. Climate change is projected to have significant impacts on agricultural conditions, food supply and food security. The present research focuses on different practices for sustainable agriculture with respect to land, water, vegetation, farm animals and labour as a whole. A multi-stage random sampling technique was adopted for the collection of primary data. Test of significance of proportion has been adopted for drawing valid statistical inferences about the population parameters. It is found that there is no significant difference in proportion between the respondents over selection of right crop for the soil, maintaining optimal cropping intensity, rain water conservation, regular cleaning of channels, keeping birds, providing animal health check up. This paper concludes with the suggestions dealing

D. Chand (✉)
Department of Economics, Bamra Trust Fund College, Bamra, Odisha, India
e-mail: digambarchand@hotmail.com

R. Gartia
Department of Statistics, Sambalpur University, Jyoti Vihar, Burla, Odisha, India
e-mail: rajendra.gartia@gmail.com

S. Nautiyal et al. (eds.), *Climate Change Challenge (3C) and Social-Economic-Ecological Interface-Building*, Environmental Science,
DOI 10.1007/978-3-319-31014-5_15

with broad findings, inferences and broad outline of the strategy of fiscal policy, alternative choice of technology, regional subsidies, farmers compensation, public policy response in tariff structure as important factors for maintaining sustainable livelihood in agriculture.

Keywords Climate change · Natural capital · Farm management practices · Sustainable livelihood

1 Introduction

The economy of a country relies mostly on environment, which is not reflected by its national account. Thus, there is a distinction between 'Natural Capital' and 'Manufactured Capital'. Economics gives importance to 'Manufactured Capital' which is produced by four factors of production. There is no substitutability between manufactured capital and natural capital. Natural capital includes soil, water, air and all living things on which Human beings depend on, but the converse is not true. Poorly managed natural capital therefore becomes not only an ecological liability, but also a social and economic liability. Working against nature by overexploiting natural capital can be catastrophic not just in terms of biodiversity loss, but also catastrophic to humans, as ecosystem productivity and resilience decline over time and some regions become more prone to extreme events such as floods and droughts, famines and epidemics, etc. Natural capital accounts contain vital information regarding economic dependence and impact on natural world.

Land-based agriculture struggles to keep pace with the other sectors in India. Agricultural credit remains elusive, production risks continue untrammeled, as do market risks. In the resulting frustration and desperation, one of the most important coping mechanisms, our rural poor have is in turn to natural resources from forests. They do so every day, leaning on our forests for everything from food, fibre, fodder and fuel. More systemic changes in resource allocation need to be considered, such as targeted diversification of production systems and livelihoods to combat climate change. Climate variability and market risk, and with other policy domains are the related issues hover round sustainable development. Each issue has strong impact on sustainable development. Agriculture and allied activities are also major economic, social and cultural activity, and it provides a wide range of ecosystem services.

Importantly, agriculture in its many different forms and locations remains highly sensitive to climate variations, the dominant source of the overall inter-annual variability of production in many regions and a continuing source of disruption to ecosystem services. Here we use the term "adaptation" to include the actions of adjusting practices, processes and capital in response to the actuality or threat of climate change, as well as responses in the decision environment, such as changes in social and institutional structures or altered technical options that can affect the potential or capacity for these actions to be realized. There is an immense diversity of agricultural practices because of the range of climate and other environmental variables; cultural, institutional and economic factors; and their interactions. This means there is a correspondingly large array of possible adaptation options (Howden 2007).

1.1 Agriculture—Present Status

Odisha, bestowed with vast natural endowments is predominantly an agrarian state. Agriculture is the mainstay of state's economy and substance of life of its people. Agriculture and allied sectors contribute about 15.58 % of the gross state domestic product (GSDP) and continue to be the backbone of the state's rural economy. It provides employment, both direct and indirect, to about 60 % of the total workforce. Therefore, the economic advancement of the state depends to a large extent on development of the agriculture and allied sector. The rate of increasing production and growing need have, therefore, been in balance over the past four decades. The impressive gains in food production are a consequence of several factors. For one, some marginal land, formerly unused or underused for agriculture has been taken up for cultivation. There is very little of this resource available for future expansion and converting land such as that which supports tropical rain forests into farmland can have serious environmental consequences for the Earth as a whole. Furthermore, the soils that support this lush and varied growth are also not suitable for agriculture unless supplemented by major inputs of fertilizers (particularly phosphate fertilizers) and other amendments to modify their physical and chemical properties. The principal reason for the impressive gains in food production, however, has been the introduction of new varieties of grains and the associated management practices that accompany the new strains. A major component of this so-called 'green revolution' has been the requirement for a package of inputs beginning not only with the new seeds but also including an assured supply of water and adequate plant nutrition (usually supplied via inorganic chemical fertilizer and methods of pest control, again usually involving synthetic chemicals). Degradation of good agricultural land, conversion of forest and other unique habitats to agriculture and reliance on chemicals and other non-renewable inputs in the long term could have a negative impact on food security and on rural life in general. Even today, there is growing evidence that current practices are problematic in terms of long-term sustainability. The list agriculture-based environmental problems is a long one: loss of biodiversity and destruction of natural habitats, over-consumption of surface water and ground water, contamination of soil and water by organic biocides, nutrient-induced eutrophication of water bodies, microbial and nitrate contamination of drinking water supplies and release of excessive quantities of the green house gases such as carbon dioxide, methane and nitrous oxide associated with specific agricultural practices.

The economic element of agriculture relates to individuals as well as to the local community and the broader national society. In being a producer of food, each farm caries out agriculture, first of all, in order to be able to provide the physical needs of his/her own family, and paramount among these is the need for food. If the system does not make economic sense, if farming is not profitable, and if basic needs are not provided for the system is unsustainable. Society at large also requires that sustainable practices have a sound economic basis. Long-term subsidies to large groups of either producers or consumers distort the true picture of costs and can only be maintained by support from the profits of other sectors. Agricultural sustainability then demands that farmers continue to make a good living and that the population as a whole be supplied with an abundance of high quality food at reasonable cost.

Agriculture is a way of life. Many people are born into this way of life or choose it in preference to what others may describe as the advantages of life in the city. Sustaining rural life must then provide the basic services related to education, health, recreation, etc., that all humans have a right to expect. Each one of us is involved directly or indirectly in all aspects of agriculture. Agriculture is also a time- and labour-intensive occupation. When the resource base is deficient, the demands for long hours and heavy physical work can be especially severe. It is therefore an essential requirement for human well-being that requirements for being a successful farmer are not as harsh as to preclude opportunities for education, recreation and relaxation. This applies to all the members of the farm family as well as the rural community as a whole. Government policy has encouraged productivity at the expense of other factors, including a need to provide facilities and opportunities for all in the rural community. One consequence is the consolidation of land and resources in the hands of a small proportion of the total population, leaving many people with marginal holdings or even no land at all. Unless adequate support by way of accessible employment in other activities is available, an unsustainable population of dispossessed persons develops in the rural community. As is the case for sustainability in any setting, sustainable agriculture must consider and bring together sound practices in the environmental, economic and social spheres.

1.2 Sustainable Agriculture

The total geographical area of the state is 155.71 lakh ha. Out of this, the net sown area is about 56.91 lakh ha, which is nearly 36 % of the geographical area. The state has about 43.52 lakh operational holdings (Agricultural Census 2005–06) of which small and marginal farmers (having <2 ha land) own 86.16 % of the holdings. The emphasis is on increasing the productivity per unit land area and increasing cropping intensity. The cultivable area of the state is 61.80 lakh ha, of which 29.14 lakh ha are high land, 17.55 lakh ha medium land and 15.11 lakh ha low land. Irrigation facility in the kharif season is available to only 34 % of cultivated area. Although normal rainfall of the State is 1451.2 mm spread over 69 rainy days, there are seasonal fluctuations, both in forms of quantity and distribution. This greatly influences kharif crops. A number of developmental schemes are being implemented to assist farmers to adopt better and scientific methods of crop husbandry for enhancing production and productivity. Often it is observed during interactions and from feedback evaluation that the farmers are unable to tap the provisions under various developmental interventions due to their complete or partial ignorance of these and fail to optimally utilize the intended benefits.

Moreover, sustainable agriculture is the activity of growing food and fibre in productive and economically efficient manner, using practices that maintain or enhance the quality of the local and surrounding environment-—soil, water, air and all living things. It is also sustainable in supporting the health and quality of life of individual farmers, their families and the communities as a whole. Sustainable agriculture encompasses the elements of productivity, profitability, conservation, healthy safety and the environment. Sustainable agriculture is the need of the hour because of the urgency to develop farming techniques, which are sustainable from environments, production and socioeconomic point of view. It is more profitable in terms of money and soil conservation in the long run. Sustainable agriculture means not only the withdrawal of three things—synthetic chemicals, hybrid—genetically modified seeds and heavy agricultural implements but also application of multi-culture, intercropping, use of farmyard manure and rampants, mulching and integrated pests management. Recognizing that agriculture is a process of food and fibre production as well as a way of life, the categories are chosen so as to reflect its various dimensions.

- **Productivity**: For the needs of the farm family as well as to satisfy global food requirements, any sustainable agricultural system must be capable of producing high yields.
- **Stability**: It is necessary that the high level of productivity be maintained over an indefinite period of time. This requires that the quality of the resources on which production is based also be maintained and even enhanced.
- **Efficiency**: To be sustainable, all the resources required for agriculture—human, animal and material—should be used in a way that is not wasteful, but maximizes output per unit input. This is especially true of non-renewable resources.
- **Durability**: Any crop production process is from time to time subject to stresses of various types, such as those due to water or to pests. Sustainable systems are intrinsically resilient in the face of such stresses.
- **Compatibility**: Sustainable agriculture should fit in which the human, social and natural environments where it is located, maintaining and enhancing the health of these environments.
- **Equity**: Agriculture should promote a good quality of life among the various individuals involved in the farming activities and within families. This includes having consideration for the standard of living, health and education of all people in the community.

The overview and categories suggested here are meant provide a comprehensive assessment of agricultural sustainability (Vanloon 2005). Sustainable agriculture includes economically viable system that reduces use of the off- farm inputs such as chemical fertilizers and pesticides and relies more on on-farm resources. In order to be sustainable, agriculture needs to be (i) technologically feasible, (ii) economically viable, (iii) socially acceptable and (iv) environmentally sound. Moreover, sustainability in agriculture can be achieved broadly through—efficient management of natural resource base and integrated approaches to crop management. Agricultural production can only be sustained on a long-term basis if the resource base like land, water and forest on which it is based are not degraded (Mishra 2005).

1.3 Livelihood and Sustainable Agriculture—Indian Scenario

The diet of modern humans varied significantly depending on location and climate. The percentage of men with BMI <18.5 is marginally lower than women and stood at 34.2 % during 2005–06. According to World Health Organization (WHO), if 40 % or more of the population have a BMI <18.5 (which is quite close to Indian situation), it is regarded as a state of famine. If India is not in a state of famine, it is quite clearly in a state of chronic hunger, since only such hunger can lead to a situation where a third of the country's adults have a BMI under 18.5. In India, the estimates of poverty are based on consumption expenditure. There has been an increase in inequality in consumption expenditure in both rural and urban India, the disparity between states has declined. Based on the Tendulkar committee methodology after adjusting for inflation, the incidence of poverty for the year 2009–2010 was estimated to be 32 %. The share of food in total expenditure continued to fall throughout the three decades prior to 2004–2005 in both rural and urban India. The overall fall was from 73 to 55 % in rural areas and from 64.5 to 42.5 % in urban areas. In urban India not only has the share of cereals and pulses fallen, but there has been a steady fall in the share of other food groups as well; such as milk and milk products, edible oil, eggs, beverages and sugar. In rural India, however, the share of milk and milk products, eggs, fish, meat, fruits and nuts has increased one percentage point each, the share of vegetables has increased by 2.5 % points, and that of beverages, refreshments and processed foods has increased by two percentage points since 1972–1973. Apart from cereals, only the share of sugar and pulses (the latter largely during the last decade) has declined noticeably. However, the increase in share of non-cereals is not enough to compensate for the 22.6 % decline in cereal consumption (IHD Report 2011). Calorie consumption of the poorest quartile is significantly lower than the top quartile of the population. Further, calorie consumption of the bottom 50 % of the population has been consistently decreasing since 1987. On the other hand, cereal consumption is declining with no compensatory rise in non-cereal consumption and on the other, the calorie intake is much below the required level and has also been showing a declining trend for more than 20 years for the vast majority, thus reinforcing the magnitude and intensity of hunger in India (India Human Development Report 2011).

The share of agriculture in the total national income declined sharply from 50 % in 1950s to around 20 or 22 % in more recent years, i.e. 2007. It is a sign of development like developed nation like U.K., USA, etc. (i.e. around 5 %). But in India, agriculture continues to hold the key to higher GDP growth, employment expansion, reduction in poverty and the equitable distribution of income. Nearly 60 % of the population depend on agriculture for its livelihood. The non-agricultural sector has not been able to siphon off surplus labour from agricultural sector. This is a sharp contrast to the empirical experience of developed countries (Majumdar 2007).

In the absence of a continuous follow up to the green revolution and the dearth of a suitable technological breakthrough in Indian agriculture in the post green revolution era, there has been a continuous decline in total factor productivity in the Indian agriculture. Water and air pollution due to indiscriminate use of agro-chemicals such as inorganic fertilizers and pesticides, soil degradation, depletion of water resources, depletion of soil fertility and extinction of plant species are among the glaring problems that raise question whether the technology currently used is able to meet the challenge of current and future demand of production. The erosion of agricultural biodiversity threatens the long-term stability and sustainability of agriculture in many ways. It erodes the genetic base on which scientists depend for continuous improvement of crops. By opting high-yield varieties, farmers become increasingly dependent on the industry dominated market and the government. Low value of agriculture, low cost-benefit ratio and no clear-cut policy on agriculture prices affect agriculture (Roy 2012).

1.4 Livelihood and Sustainable Agriculture-Scenario in Odisha

Agriculture sector continues to be the backbone of the state economy in the sense that about 62 % population (as per 2011 census) still depend in varying degrees on this sector for their livelihoods. While the contribution of the agriculture sector to GSDP has come down to about 16 % during 2011–12 from a level of 53 % in 1980–81, the diversion of the workforce from farm sector to nonfarm sectors is very slow. The production of food grains in the state has increased from 51.04 lakh tonne in 1970–1971 to 102.10 lakh tonne in 2012–2013, but the increase in yield rate of food grains from 847 kg/ha in 1970–1971 to 1297 kg/ha in 2011–2012 is quite significant.

Over-dependence on paddy cultivation even in rainfed conditions is a limiting factor to agricultural growth because of the State's proneness to natural calamities particularly drought conditions. In terms of area, the share of paddy out of the net area sown has reduced from 64.36 % in 1950–1951 to 61.1 % in 2011–2012. The area under paddy cultivation is, however, still quite high. Other major constraints to adoption of modern agricultural practices in the state are low levels of capital formation and small sizes of operational holdings.

However, poverty estimates on the basis of 66th NSS round data and Tendulkar Committee Methodology for 2009–2010 are based on estimated poverty lines at Rs. 567.10 and Rs. 736.00 for rural and urban Odisha, respectively. The nutritional status of women in Odisha, as indicated in the NFHS-3 survey, shows that about 41.4 % of women have a body mass index BMI below 18.5, indicating a high prevalence of nutritional deficiency. Prevalence of severe malnutrition among children, mothers, old and indigent people is a matter of serious concern in the State. Sixty-two percent women suffer from anaemia against the national average of 55.3 % while 65 % children remain anaemic and suffer from chronic energy deficiency. There also appears to be a high correlation between infant mortality rate (IMR) and maternal mortality rate (MMR). Three factors explain a high level of IMR in Odisha: (i) poor availability of professional attendants at birth, (ii) high percentage of low birth weight babies and (iii) lack of professional pre-and post-natal care. Sixty-four percent infant deaths are attributed to neo-natal mortality. Premature deliveries result in 38.5 percent infant deaths. Pneumonia, respiratory infections in newborn babies, tetanus and diarrhoea result in 34.1 % infant deaths. Anaemia, which is caused due to malnutrition, suffered by both pregnant mothers and infants, explains 8.1 % infant deaths. Other causes account for another 19.3 % infant deaths. Health conditions depend on a number of factors including: (i) income and poverty levels, (ii) food security, food pricing and malnutrition, (iii) availability of professional medical attendants, paramedical professionals, quantity and quality of health infrastructure, (iv) socioeconomic development, literacy and health awareness and (v) physical and economic accessibility of private or public healthcare system. Rice production in the state has decreased from 6828 TMT in 2010–2011 to 5807 TMT during 2011–2012. The production of oilseeds has declined from 222 TMT in 2010–2011 to 166 TMT in 2011–2012. According to Livestock Census 2007, the livestock population of Odisha was 230.57 lakhs. Out of 230.57 lakhs livestock, 58.5 % were cattle and buffaloes, 38.8 % were small ruminants and 2.7 % were pigs. Besides, the state has also 206.00 lakhs poultry as compared to 648.83 million at an all India level.

1.5 Livelihood and Sustainable Agriculture in Western Orissa

The major pillars of human development are equity, empowerment, participation and sustainability. It is about creating an environment in which people develop their full potential and lead productive and creative lives, in accordance with their needs and interests. Reduction of inter-regional and inter-personal disparities is also a key concern that has considerably informed the human development debate. Good health is a livelihood asset that enables people to participate in work and socioeconomic development. Illness, on the other hand, causes misery and impoverishment. As per UNDP methodology, human development index (HDI) is an average of health index, education index and income index. There are significant social, regional and gender disparities in accessing public health in Odisha. Odisha published its first Human Development Report in 2004–2005. Wide variations have been observed in human development across districts in Odisha. Interior regions in general and tribal districts in particular have poor physical and economic access. These regions also bear the brunt of a resource crunch both in terms of a health budget deficit and neglected public health institutions. Low female literacy levels adversely impact reproductive child healthcare in tribal and other interior areas. Odisha is also a participant in implementing the GoI UNDP Project. However, about 22 % households have safe drinking water facilities within their premises. In many pockets, water quality is a serious issue. In some areas such as Nuapada, the content of fluoride in the ground water is higher than the recommended safe level. Frequent breakdown in tube wells and rural piped water supply units is another problem. Rural people in many parts still do not appreciate the value of safe drinking water. Sanitation habits of people in many districts are also poor. As per the 2011 Census estimates, only 22 % of total households in Orissa had access to basic sanitation facilities like toilets within the premises. Out of the remaining 78 % households, 1.4 % households use public latrines while the rest defecate in the open. As a result, the incidence of diarrhoea in interior Odisha is generally high and it is a major source of infant and other deaths. Several other pockets of southern and western Odisha are also socially and economically depressed. These regions are also frequently visited by natural calamities including severe drought sand floods. Persistence of heavy incidence of poverty in these regions is a cause of concern. The longer term problem related to malnutrition and poverty is referred to as chronic food insecurity, which is largely due to, continued lack of access to productive assets and employment. Agriculture plays an important role in tackling the problem of chronic food insecurity by providing livelihood to the poor.

The *transitory* or short term food insecurity is associated with instability in food production or food prices. Among other things agricultural households tend to cope with these uncertainties through crop diversification. However, for marginal land holders the scope for diversification is limited. Though the implementation of the special plan has brought several benefits to this region, as per the Tendulkar Committee methodology, rural poverty in this region came down by 21.0 percentage points from 73.4 % in 2004–2005 to 52.4 % in 2009–2010. Enrolment rate in primary schools in KBK districts has gone up to 103.81 % in 2010–2011 from 75.89 % in 1996–1997. Similarly, the enrolment rate in upper primary schools has risen from 56.39 % in 1996–1997 to 99.09 % in 2010–2011. The dropout rate in primary schools in KBK districts has been reduced to 2.3 in 2010–2011 and female literacy rate has increased to 45.49 % in 2011, while the overall literacy rate has increased to 57.17 % in 2011. Cotton is a major commercial crop in KBK districts in the khariff season.

The state's coastal areas are relatively better off, but the hinterland which also inhabits vast ST population in hilly and forest terrain, is quite backward. The majority of total ST population is concentrated in the western districts of Odisha.[1] Traditional agricultural system, indigenous knowledge were neglected, demeaned, discarded at last for establishment of capital-based agriculture. First green revolution has played an important role to achieve their aim. Now farmer is completely dependent on market for seeds, fertilizers, pesticides coupled with dependent on government and non-government institutions for begging loan. More importance on paddy and wheat destroyed our food basket with different varieties of food and self-sufficiency of food. There is acute shortage of Millets like *Desi Mandia, Bazra, Kudo, Koshla, Suan, Guruji, Kangu, Zuar, Gangei*. varieties of dal like *mung, biri,*

[1]India Human Development Report (2011), Oxford press, NewDelhi, p. 73.

Kandul, Harhar etc. varieties of oil like coconut, mustard, *Til, Methi,* etc., varieties of *Masala* and vegetables. Millets in our country linked directly with portentous food basket, food security and ecological protection. Crop failures occur due to Lack of irrigation facilities, scanty rainfall, natural calamities and crop diseases. Nowadays there is no other alternative provision for farmers in the occurrence of crop failure of paddy and wheat. The population of this region suffers from high morbidity on account of under nutrition as well as endemic, malaria and other life threatening diseases. The people living in this region lack in their awareness on various aspects of life including lack of access to various amenities and opportunities created and also to the natural as well as environmental resources around them. Ignorance and inadequate purchasing power to access the resources may have tremendous affect on their health and nutritional standard.

It is worth mentioning that despite the secular decline in its contribution to national income, the centrality of agricultural growth development cannot be de-emphasized. Agriculture continues to hold the key to higher GDP growth, employment expansion, reduction in poverty and equitable distribution of income. Despite the sustained growth of some 6 % years, the non-agricultural sector has not been able to siphon off surplus labour from the agricultural sector. This is in sharp contrast to the empirical experiences of developed countries. Such configuration of circumstances places agriculture and rural development in pre-eminent position in the development agenda of any Government. Further, since time immemorial people observed *Nua-Khai,*[2] *Cher-Chera,*[3] *Paus-Purnima.*[4] *Nua-Khai* is observed in the month of *Bhadraba Sukla Panchami*, which is the season of food insecurity. This festival as a ritual festival played a major role in promoting agriculture as a way of life. Like farmers who have already sown all their grains in agricultural fields, it is believed that even birds do not get food during this starvation period. Only '*Saria Dhan*'[5] a short period paddy variety is harvested during this time. The ancient origin of "*Nua Khai*" traced back at least to *vedic* times, when the *rishis* (Sages) had talked *panchayajna*, the five important activities in an annual calendar year of an agrarian society[6] (Jamiullah 2014).

---

[2]'Nua Khai' is such a food festival that nobody becomes underfed on this day. The festival "**Nua Khai**" is observed by the people in the month of September (Bhadraba in Odia Calender) where farmers even the animals and birds face difficulty to get their food. The people of Western Orissa give the "**Prasad**" prepared by new rice (Cultivated in Aatt) to their Adhistatri devi/Esta devi of the locality. They believe that God (Adhistatri devi/ Esta devi) will not deprive them to get food throughout the year.

[3]'Cher-Chera' deals with food distribution among 'haves' and 'have nots'.

[4]'Pausa-Purnima' is only food festival where no deity is worshipped. The concept 'Nara hi Narayan' is true in this sense. People only take varieties of dishes after harvest.

[5]'Saria Dhan'.

[6]These five activities have been specified as sitayajna (the tilling of the land) pravapana (the sowing of seeds) prabalambana yajna (the initial cutting of crops), Khala yajna (the harvesting of grains and prayayana). In view of this, 'Nuakhai' may be seen as having evolved out of the third activity, namely 'pralambana yajna'which involves cutting the first crop and reverently offering it to the mother goddess.

There is a difference between prices of agricultural and industrial commodities, by which the farmers are at loss. Increase in input costs and fraudulent practices in selling seeds, pesticides, black marketing of fertilizer and under rate of produces of farmers make the condition of farmers more deplorable. Erratic calculation of number of family members as a unit, recognition of agricultural labour as unskilled labour, compound interest on agricultural loan, improper weight and measure regulation, privatization of health and education and irrigation and above all, decreasing trend of investment on agriculture by the government lead agriculture for a total collapse if no rapid remedial measures are taken. Our contention is that this is equally important, if not more, to achieve food security in a cost effective manner. The government plays an active role to check food insecurity by certain welfare measures, which is deteriorating work culture because of the passive role played by people. The strategy should ensure that higher growth is accompanied by narrowing disparities in regional growth and of productivity differential among different farm size groups. Rising flow of credit to agriculture is normally associated with buoyancy in the farm sector. There is over-estimating credit flows to actual agricultural operations in recent years. Indirect finance to agriculture is a major proportion of the increase in agricultural credit by commercial banks to agricultural sector. Direct payment to input dealers, irrigation equipment suppliers and non-banking financial companies that lend to agriculture through farmers as loan are provided by commercial banks. Farmers are the facilitators. Nurturing of plants several times to increase crop productivity is important as food is made by plants. Fertile soil is important rather than hybrid seed. Diversity in farming is the single most important modern technology to achieve food security in a changing climate by giving importance to altering timing or location of cropping activities. Increase in agricultural production is constrained by limited land area that can be brought under cultivation. Further increase in intensity is, however, constrained by the extent to which irrigated area can be increased. The real intention of food security bill is to provide two square meals a day to the poor instead of nutritional food. Another source of leakage is diversion of food grain from fair price shops to the unregulated market because of the price difference between subsidized grain and grain sold through regular marketing channels, which make illegal profits. The paucity of adequate shops is the factor in tribal area. The problem of adulteration is added to the grave situation of targeted public distribution system. Movement restrictions have a direct bearing on household level food security. Increasing economic costs of handling food grains, record procurements in recent years and widening difference between the economic cost of food grains and the central issue price have led to rising food subsidies in recent years. With this backdrop of acceleration of food prices and sharpening of malnutrition, food insecurity and hunger among the poor household the urgent intervention of Government is necessary for sustainable agriculture.

Agricultural crops suffer because of natural calamities and consequential crop disease problems. Concerns of environmental pollution owing to unscrupulous use of plant protection chemicals has prompted scientists to advocate for Integrated Pest and Disease Management. Need based plant protection is one of the important pillars of IPM and IDM.

Since in India, agriculture—not merely a simple food producing technology but a complex and complete system in itself—evolved over several millennia in tune with the ecosystem and weaving people's life style and source of livelihood into one durable texture, farming with all its dimensions needed to be considered at the planning level (Dubey 2008).

2 Brief Review of Literature

The human economy rests within the ecosphere and cannot exist outside it, several concepts neglected or misunderstood by neoclassical economics become critically important. These ideas, beginning with the co-evolutionary paradigm, form the core of the ecological economics perspective. Ecological economics emphasizes the co-evolutionary development of human beings and the natural world. The natural world is ultimately irrelevant to human economic activity. Near-perfect substitutability implies that economic output can remain constant or even increase as natural resource inputs are reduced to zero, so long as labour and manufactured capital are increased enough to make up the difference (Prugh 1999).

Poverty is used in this study as an important indicator of individual vulnerability to climate extremes and to climate change, because poverty can be directly related to marginalization and lack of access to resources which are critical when faced with the risk of hazards and the resultant stress on livelihoods. Resource dependency is an element of individual vulnerability and is constituted by reliance on a narrow range of resources leading to social and economic stresses within livelihood systems. The role of credit in recovery from stress and disruption of livelihoods is particularly important where external assistance is not available for immediate injection of resources (Adger 1999).

The adaptation strategies farmers perceive as appropriate (as opposed to the strategies they actually carry out) include crop diversification; using different crop varieties; varying the planting and harvesting dates; increasing the use of irrigation; increasing the use of water and soil conservation techniques, shading and shelter; shortening the length of the growing season and diversifying from farming to non-farming activities (Hassan 2008).

Indigenous adapting to climate change strategies in the study area include, planting different varieties of crops, cultivating different crops, shortening growing season, changing the extend of land put into crop production, changing to irrigation the use of chemical fertilizer, improve in water maximization and mulching. Poverty and ignorance of various adaptation strategies are the major contributing factors to the impact felt by indigenous people. The knowledge and information gap concerning the effect of climate change, information dissemination, awareness programmes and training programmes calls for immediate action in order to relegate the impact of climate change in the study areas (Ishaya 2008)

3 Objectives of the Study

Agriculture as a livelihood is affected by climate change severely. Adaptation of different strategies and management practices of land, water, labour, vegetation and farm animals are important for sustainable development. At the juncture of leaving agriculture as a livelihood by the present generation, sustainability of agriculture has become a question. Therefore, the present research addresses on the study of the management practices of land, water, labour, vegetation and farm animals of the respondents of the sample villages and prescribe adaptation strategies.

4 Data and Methods

Sampling Techniques: A multi-stage random sampling technique was adopted for the collection of primary data. In the first stage a random sample of three district comprising of one tribal, one irrigated and one non-irrigated districts were selected out of ten districts of Western Odisha by the method of random sampling. Then one block from the corresponding district and one village from the corresponding block were selected by the method of simple random sampling. These three villages were selected to represent irrigated, non-irrigated and tribal villages from the corresponding blocks. In the next stage, the sample households were selected on stratified random basis to represent different land classes such as marginal, small, medium and large farmers' category. The distribution of sample households across the districts, blocks and the three villages are presented in Table 1.

To achieve the very objective of the research, the following hypotheses were developed and tested for drawing valid statistical inferences.

H_0: There exists no significant difference between the sample proportion and population proportions with regards to various attributes.
H_1: There exists significant difference between the sample proportion and population proportions with regards to various attributes.

Table 1 Sample village profile

Name of the districts	Name of the block	Name of the village	Total no. of house holds	Total no. of house hold surveyed
Sundargarh	Badagaon	Bhoi Pali	466	100
Sambalpur	Govindpur (Bamra)	Bamphei	432	100
Bargarh	Attabira	Sindurbahal	387	100
Total			1285	300

The test statistic is a Z test given by

$$Z = \frac{p - \hat{P}}{\sqrt{(\hat{P}\hat{Q}/n)}} \sim N(0, 1)$$

where p = sample proportion, $\hat{P}$ = population proportion = 0.5, $\hat{Q} = 1 - \hat{P} = 0.5$, n = sample size. The critical value of $Z_{0.05}$ = 1.96 (Two tailed test).

5 Findings and Discussion

The findings of the data analysis are presented in Table 2 as follows.

It is found from Table 2 that there is no significant difference in proportion between the respondents over selection of right crop for the soil, rain water conservation, regular cleaning of channels and providing health check up. There is significant difference in other statements and may be inferred as follows:

In case of **land management,** more importance should be given for the use of organic manure, traditional implements, crop rotation and soil bio-pesticides. Ploughing across the slope should be done for protection of soil erosion and water conservation. It is observed from the field study that farmers were selecting crops on the basis of type of the soil but did not give importance to soil health. Soil testing is important from time to time regularly and farmer's knowledge should be updated time to time.

In case of **water management,** construction of farm pond, regular cleaning of channels, lining of channels with concrete slabs should be given importance. Optimal tillage operation and irrigation to crops were also kept their place for better water holding capacity and secure production. Though, more investment, drip irrigation is required apart from diversion of excess water for productive purpose.

In case of **labour management,** skilled labour was required for operation modern tools and equipments in agriculture which needs training to farmers. Agricultural ITI should be established in different cluster of rural areas. Increasing literacy rate decreases the interest of family members from working in agricultural field due to less profit. Agricultural livelihood is going on declining. Thanks to UN for declaration of 2014 as the year of 'Family farm'. Family labourers, hired labourers, contract labourers for certain works are important for proper optimal engagement of labourers with proper supervision.

In case of **management of vegetation**, farm forestry and agro forestry should use trees as wind brakes. Trees should be planted on bounds as protection to some horticultural practices. They are used as natural fence for banana, papaya, etc. Apart from it, green leaves, recycling of crop wastes and bio-pesticides from the vegetations are also important for management of vegetation, which indirectly helps in water conservation.

Table 2 Test of significance of proportion

Sl. No.	Statements	*p*	*q*	|*Z*| Value	Remark (α = 0.05)
A	**Land management practices**				
1	Selection of right crop for the soil	0.53	0.47	1.15	Insignificant
2	Maintaining optimal cropping intensity	0.58	0.42	2.89	Significant
3	Use of organic manure	0.19	0.81	10.85	Significant
4	Use of traditional implements	0.40	0.60	3.35	Significant
5	Use of crop rotation	0.39	0.61	3.81	Significant
6	Use of soil bio-pesticides	0.17	0.83	11.32	Significant
7	Ploughing across the slope to prevent soil erosion	0.08	0.92	14.66	Significant
8	Ploughing across the slope for water conservation	0.04	0.96	15.93	Significant
B	**Water management practices**				
1	Rain water conservation	0.45	0.55	1.85	Insignificant
2	Optimal irrigation to crops	0.30	0.70	7.04	Significant
3	Optimal tillage operation to increase soil water holding capacity	0.31	0.69	6.70	Significant
4	Construction of farm pond	0.42	0.58	2.77	Significant
5	Regular cleaning of channels	0.54	0.46	1.50	Insignificant
6	Lining of channels with concrete slabs	0.26	0.74	8.31	Significant
7	Diverting excess water for productive purposes	0.18	0.82	11.20	Significant
8	Use of drip system	0.03	0.97	16.28	Significant
C	**Labour management practices**				
1	Use of family labourers	0.65	0.35	5.31	Significant
2	Working along with hired labourers for better turn out	0.62	0.38	4.04	Significant
3	Effective management (supervision)of the hired labourers	0.25	0.75	8.78	Significant
4	Going for operation-specific contract	0.59	0.41	3.23	Significant
5	Optimal engagement of labourers	0.65	0.35	5.08	Significant
6	Use of skilled labourers	0.25	0.75	8.54	Significant
D	**Management of vegetation**				
1	Going for farm forestry	0.15	0.85	12.01	Significant
2	Going for agro forestry	0.18	0.82	11.20	Significant
3	Use of green and green leaf manure	0.24	0.76	9.12	Significant
4	Growing trees as wind brakes	0.22	0.78	9.58	Significant
5	Growing and maintaining trees and plants as natural fence	0.43	0.57	2.54	Significant
6	Proper recycling crop wastes	0.34	0.66	5.43	Significant
7	Preparing bio-pesticides from the vegetation available in farm	0.05	0.95	15.47	Significant

(continued)

Table 2 (continued)

Sl. No.	Statements	p	q	$\|Z\|$ Value	Remark (α = 0.05)
E	**Management of farm animals**				
1	Keeping milch animals	0.07	0.93	14.90	Significant
2	Keeping work animals	0.35	0.65	5.20	Significant
3	Keeping birds	0.56	0.44	2.08	Significant
4	Providing balanced feed to the animals	0.26	0.74	8.31	Significant
5	Providing health check up	0.49	0.51	0.35	Insignificant
6	Providing healthy atmosphere in the animal sheds	0.39	0.61	3.70	Significant
7	Optimal utilization of animal power	0.31	0.69	6.58	Significant

Source Primary data 2012–13

In case of **management of farm animals**, providing balanced feed to farm animals, healthy atmosphere in animal sheds should be developed by farmers. Modern techniques in rearing farm animals should be developed. Awareness should be created among farmers by the government time to time. Keeping birds should be encouraged to meet the demand of eggs and meat. Optimal utilization of animal power should be used for work animals. It is observed from field survey that work animals are diminishing rapidly due to farm mechanization. Collection of urine and cow dung of farm animals was affected due to diminishing farm animals.

6 Conclusion and Recommendation

Sustainable agriculture entails an optimum production pattern of agricultural commodities moving upward the shifting growth path, not only without depleting the scare production resources but also enhancing and ensuring the conservation of these resources for continuous use for all time to come. The first priority should be to make handsome investment in agricultural research and development, especially, in the state-level research institutes and agricultural universities, to develop technologies that become instrumental in promoting desirable production pattern that economies on scarce resource use, restore and improve soil health, control water, soil and air pollution and enhance productivity and improve farm incomes.

Feeding the nation with food grains remained the main focus, and rightly so, yet the agricultural ecology was taken for granted and vote –bank-oriented electoral interests were nurtured, ignoring the much needed corrective and ameliorative measures. At best, the policy makers paid only lip service to the emerging disastrous condition. Soft paddling by the politician in power and politicized bureaucracy avoided rightful hard decisions. Now when the situation is visibly getting out of control, there is a lot of talk in the air about the diversification and sustainability of agricultural production pattern in the country. Unfocused input subsidies, in contrast to investment subsidies, distort production pattern and adversely affect the sustainability of agricultural production systems.

The following recommendation may be made for sustainable livelihood in agricultural sector:

(i) Soil testing and updating of knowledge of farmers should be given due importance.
(ii) Optimal tillage operation and irrigation along with diversion of excess water for productive purposes should be taken into consideration.
(iii) Agricultural ITI should be established in different cluster of rural areas for imparting training to farmers.
(iv) Green leaves, recycling of crop wastes, bio-pesticides from the vegetations are to be developed for management of vegetation apart from tree plantation.
(v) Compost pit, vermicompost pit should be developed with farm animal rearing. Cattle urine may be used as bio-pesticides in agricultural field.

References

Adger, W. (1999). Social vulnerability to climate change and extremes in Coastal Vietnam. *World Development, 27*(2), 249–269.

Dubey, V.K. (2008). *Sustaining traditional agricultural practices for food security*. Lucknow: International Book Distributing Company.

Hassan, R. (2008). Determinants of African farmers' strategies for adapting to climate change: Multinomial choice analysis. *AFJARE*, *2*(1).

Howden, S. (2007). http://www.pnas.org/content/104/50/19691.full.pdf. Retrieved May 17, 2015, from www.pnas.org, www.pnas.org

India Human Development Report. (2011). *India Human Development Report 2011: Towards social inclusion*. New Delhi: Oxford University Press.

Ishaya, S. (2008). Indigenous people's perception on climate change and adaptation strategies in Jema'a local government area of Kaduna State, Nigeria. *Journal of geography and regional planning, 1*(8), 138–143.

Jamiullah, M. (2014). Nuakhai-then and now. *Sociocrat*, pp. 8–9.

Majumdar, N. (2007). *Inclusive growth-development perspective in Indian economy*. New Delhi: Academic Foundation.

Mishra, B. (2005). Sustainable agriculture for food security. *Orissa Review*, pp. 34–36.

Prugh, T. (1999). *Natural capital and human economic survival*. USA: CRC Press LLC.

Roy, S. (2012). Role of Government towards food security in India: An assessment. In B. K. Chatterjee (Ed.), *Food security in India—A comprehensive plan for sustainable solution* (p. 310). New Delhi: Regal Publication.

Vanloon, G. (2005). *Agricultural sustainability—Strategies for assessment*. New Delhi: Sage Publication.

Economic Impacts of Climate Change in India's Cities

Kala Seetharam Sridhar

Abstract Given the significance of climate change, economic importance of cities, and lack of adequate research in India's context, in this paper, I deal with climate change impacts that are linked to market transactions and directly affect a city's value of output produced. I examine the impact of climate change which manifests in temperature differences and rainfall on city-level output in India. I estimate city output regressions for India's cities using several approaches. I find that overall, climate change thus far has impacted India's cities economically. I find climate change indicators such as rainfall positively impact nonagricultural output. This could be possibly due to the positive impact rainfall has on agricultural output, and the multiplier effects agriculture has on consumer confidence and nonagricultural output, in the context of developing cities. I find extreme temperature differences have a positive impact on nonagricultural output per capita, and this could be because when there are extreme temperatures, resources could move to nonagricultural uses. Coastal districts have higher levels of non-primary output per capita, when compared with their inland counterparts, which could be due to the fact that coastal cities are always ahead of others in terms of their economic growth due to their exposure to and proximity to international markets, when compared with inland districts. When we examine the impact of climate change on agricultural output, here are our observations in summary: when both climate change indicators—rainfall and extreme temperature differences are included, they do not have an impact on agricultural output; however, when rainfall is not controlled for, extreme temperatures have a positive impact on agricultural output, as they do with nonagricultural output, which probably refers to the volatility in agricultural output with extreme temperatures. Further, while coastal districts have a positive impact on nonagricultural output (which is plausible), they experience significantly decreased agricultural output, due to their salinity and other effects. The policy implications of the research and limitations of the data are summarized.

K.S. Sridhar (✉)
Centre for Research in Urban Affairs,
Institute for Social and Economic Change, Bangalore, India
e-mail: kala@isec.ac.in; kalaseetharam@gmail.com

S. Nautiyal et al. (eds.), *Climate Change Challenge (3C) and Social-Economic-Ecological Interface-Building*, Environmental Science,
DOI 10.1007/978-3-319-31014-5_16

Keywords Climate change—India · Cities · City output · District domestic product · Coastal cities · Inland cities

JEL Classification O18 · Q56 · R11 · R12

1 Introduction and Background

Ever since the Earth came into being there has been a climate system. The climate of a place is the average weather that it experiences over a period of time. While changes in the weather may occur suddenly and noticeably, changes in the climate take a long time to settle in and are therefore less obvious. The Intergovernmental Panel on Climate Change (IPCC) concluded in 2007 that warming of the climate system is now "unequivocal," based on observations of increases in global average air and ocean temperatures, widespread melting of snow and ice, and rising global average sea level (IPCC 2007). According to the National Oceanic and Atmospheric Administration's (NOAA) (2007) State of the Climate Report and the National Aeronautics and Space Administration's (NASA) (2007) Surface Temperature Analysis, the eight warmest years on record since 1850 have all occurred since 1998, with the warmest year being 2005.

According to the NOAA (2007) State of the Climate Report and the NASA (2007) Surface Temperature Analysis:

- Since the mid 1970s, the average surface temperature has warmed about 1 °F.
- The Earth's surface is currently warming at a rate of about 0.32 °F/decade or 3.2 °F/century.

Additionally (from IPCC 2007):

- The warming trend is seen in both daily maximum and minimum temperatures, with minimum temperatures increasing at a faster rate than maximum temperatures.
- Land areas have tended to warm faster than ocean areas and the winter months have warmed faster than summer months.

One of the expected impacts of climate change on South Asia is a general increase in both the mean minimum and mean maximum temperatures by 2–4 °C (see Sharma et al. 2006). A 10–15 % increase in monsoon precipitation in many regions, a simultaneous precipitation decline of 5–25 % in semi-arid and drought-prone central India and a sharp decline in winter rainfall in northern India is also projected (Ramesh and Yadava 2005).

2 Role of Cities in Climate Change

The role of cities in climate change is much debated. Cities are often blamed for contributing to the climate change that is hot enough to melt ice caps, change the precipitation mix to more rain, and possibly less snow, raise sea levels, and increase the frequency of severe storms. To the extent cities promote car use, urban sprawl is also often associated with climate change.

On the other hand, cities are viewed as playing a positive role in mitigating the negative impacts of climate change. The OECD is actively working with governments to highlight the role of cities to deliver cost-effective policy responses to climate change. Given cities are centers of innovation; they have the potential to advance clean energy systems, sustainable transportation, and waste management to reduce greenhouse gases. A number of projects at the OECD are advancing the understanding of the various roles that cities can play to respond to more efficiently and effectively to climate change.

Much of the risk is concentrated among low-income households, as described by Aromar Revi for India and Patricia Romero-Lankao for Latin America. The cities most at risk, as pointed by Sattherwaite (2007), are those which are

- least able to avoid the direct or indirect impacts because of their location—on the coast, by rivers or where cyclones or hurricanes are common;
- likely to be most affected by the impacts because of their physical form—poor quality buildings, and lack of water supply, and storm drainage infrastructure, which are likely to be inhabited by poorer residents;
- least able to cope with the impacts due to a lack of local government capacity and funding to rebuild or repair damage, restore services, and support households to rebuild homes and livelihoods.

The urban populations most at risk, as Sattherwaite (2007) points out, are:

- least able to avoid the direct or indirect impacts, because, for example, they live in poor quality homes and in areas with inadequate drainage systems, and are unable to move or change jobs if climate change threatens their livelihoods;
- most vulnerable to the impacts, such as infants and elderly people who are less able to cope with heat waves or unable to flee quickly when a disaster is imminent;
- least able to cope with illness, injury or premature death; or loss of income, livelihood or property.

Urban centers contain a large proportion of the people most at risk from the effects of climate change. Urban dwellers face the threat of damage to their livelihoods, property, quality of environment, and future prosperity: they suffer increasing risks from storms, flooding, landslides, drought, and overloading of water, drainage and energy supply systems. ***In this paper, I answer the question as to what the***

projected threats are to Indian cities from climate change, and how they might best adapt to such expected changes. Given half of the world's population started to live in cities in 2007, it is no exaggeration to say that the battle against climate change will be won or lost in cities. Since cities have high concentration of population density and economic activity, they are more vulnerable to climate change. India's cities are characterized by high density of population, housing stock, and poor infrastructure, which make them vulnerable to climate change. Given the most valued infrastructure is also located in cities than elsewhere, the economic and social costs of climate change will be much higher in cities. So climate change has impacts on the physical assets used within cities for economic production, on the costs of raw materials and inputs to economic production, the subsequent costs to businesses, and thus on output and competitiveness. For instance, at India's sub-national level, an annual Gross State Domestic Product (GSDP) compression of about 2 % has been estimated for Gujarat, of which drought makes up 57 %, cyclone and storm surge, 12 %, and inland flooding, 5 %, over a 100-year time horizon (Revi 2008). A recent report using data collected over 193 years shows that since the 1960s, there has been a dip in India's annual rainfall. For instance, in the plains of Kerala, peak rainfall declined from at least 3700 mm received in the 1920s and 1930s to around 2800 mm in the 1960s, a fall of 24 %. In the plains of Punjab, there is a similar fall, from 1100 mm in the early years of the twentieth century to around 687 mm now a decrease of 37.5 %. Part of this has been attributed to global warming (http://www.livemint.com/2008/08/17233753/India8217s-fluctuating-rain.html).

Hence, it is important to understand the economic impacts of climate change in India's cities. As the literature review will show, such an attempt has been made in a limited context. Given the significance of climate change, economic importance of cities and lack of adequate research in India's context, in this paper, I deal with climate change impacts that are linked to market transactions and directly affect a city's value of output produced. I examine the impact of climate change which manifests in temperature differences and rainfall on city-level output in India. I estimate the impacts of temperature differences and rainfall on Indian cities' output—both agricultural and non-primary.

I estimate city output regressions for India's cities using several approaches. Since only growing areas are usually designated as towns, to get rid of this selection bias, I estimate the output of districts that are larger geographical areas than cities, which is similar to the approach Beeson et al. (2001) take. I estimate the non-primary output of cities as dependent on rainfall, temperature differences and other socio-economic characteristics determining city-level output such as its size, human capital, infrastructure, economic base, and regulations affecting urban land use. Further, using a similar approach, I estimate the agricultural (primary) output of districts as dependent on climate change indicators, along with other factors.

3 Past Literature

There is an entire stream of the literature in this area, which devotes attention to the development of new models, methods, and strategies to estimate the impacts of climate change, given the uncertainties involved. Mendelsohn et al. (2000) develop a new climate-impact model—the Global Impact Model (GIM), which predicts that country specific results are likely to vary. Some papers (see Hallegatte et al. 2007) develop general heuristic tools to assess the economic impacts of climate change in urban areas. Deschênes and Greenstone (2007) estimate the economic impacts of climate change on US agricultural land by estimating the effect of random, year–to-year change in temperature and precipitation on agricultural profits. They find that the hedonic approach, which is the standard in the previous literature on the subject, to be unreliable because it produces results which are extremely sensitive to choice of sampling, weighting, and control variables.

Some modeling studies are applied to the Indian context. Using the example of Indian agriculture, O' Brien et al. (2004) present a methodology for investigating regional vulnerability to climate change in combination with other global stressors. A state-of-the-art regional climate modeling system, known as PRECIS (Providing Regional Climates for Impacts Studies) developed by the Hadley Centre for Climate Prediction and Research, is applied for India by Kumar et al. (2006) to develop high-resolution climate change scenarios. PRECIS simulations under scenarios of increasing greenhouse gas concentrations and sulphate aerosols indicate marked increase in both rainfall and temperature toward the end of the twenty-first century. This paper finds that warming is monotonously widespread over the country, but there are substantial spatial differences in the projected rainfall changes. They find that the west central part of India shows maximum expected increase in rainfall. They predict that extremes in maximum and minimum temperatures are also expected to increase into the future, but the night temperatures are increasing faster than the day temperatures. Another relevant finding from this study is that extreme precipitation shows substantial increases over a large area, and particularly over the west coast of India and west central India.

Although most models predict that higher temperatures will reduce grain yields as the cool wheat-growing areas get warmer, they do not examine the possibility that farmers will adapt to climate change, by making production decisions that are in their own best interests. A recent set of models which examines cross-sectional evidence from India and Brazil reported by Mendelsohn and Dinar (1999) find that even though the agricultural sector is sensitive to climate change, individual farmers do take local climatic conditions into account, and their ability to do so will help mitigate the impacts of global warming.

Certain studies arrive at monetized estimates of the impact of climate change, whereby impacts are expressed as functions of climate change and "vulnerability" (Toi (2002) is an example of such a study). Gbetibouo and Hassan (2005) employed a Ricardian model to measure the impact of climate change on South Africa's field crops and analyzed potential future impacts of further changes in the climate.

A regression of farm net revenue on climate, soil, and other socio-economic variables was conducted to capture farmer-adapted responses to climate variations. They found temperature rise positively affects net revenue, whereas the effect of reduction in rainfall is negative.

Hunt and Watkiss (2007) perform a review of the literature on studies of climate change on cities. It found an emerging literature on this subject, but that the studies are mostly qualitative in nature. The limited evidence reviewed in this study suggested that impacts are likely to be more important for developing country cities. The study also finds that most studies have focused on coastal cities and there is a lack of studies on inland cities. Moreover, in many cases, the studies only look at a single issue (or sector), most commonly sea-level rise.

Thus, while existing studies have examined the impacts of climate change in the context of India, this paper contributes to a more general debate regarding the economic impacts of climate change on India's cities, and motivates the formulation of suitable policies in India's cities to mitigate the impacts of climate change. It also takes into account the methodological caveats if we were to take the city as the unit of observation, and improves upon it, taking off from clues in past studies on the subject. I also use a unique data set of district domestic products (DDP), recently published by most Indian states.

4 Data, Model, and Methodology

In this paper, I assess the impact of climate change on cities using several approaches. In one approach, I run a simple regression of nonagricultural and agricultural output per capita as a function of several characteristics including indicators reflecting climate change. Since only growing areas are usually designated as towns, I estimate the population of districts that are larger geographical areas than cities. They are the equivalent of counties in the USA (see Beeson et al. 2001) for an investigation of this phenomenon in the context of the United States. Further, districts are designated even if they are growing, declining or stagnating.

Another possible approach to use in order to explain city growth is to use city output. This is because a city might economically grow, stagnate or decline even with increased population growth. In fact, Cohen (2004) points out that in sub-Saharan Africa as well as East Asia, change in the urbanization levels were very similar. However, the difference was that in sub-Saharan Africa it was unaccompanied by a corresponding increase in GDP per capita which stagnated, whereas in East Asia GDP per capita shot up. Taking this into account, I estimate city growth using plausible measures of city-level output. The basic model which I estimate may be stated as follows:

$$O_i = f(P_i, R_i, A_i MS_i, PS_i, H_i, C_i) \tag{1}$$

O_i is output, measured by the net DDP of district i. The vulnerability of cities to climate change is exacerbated by their size. The population of district i is used to represent its size (P_i). R_i and A_i refer to climate change characteristics such as rainfall and mean temperature differences during the year.[1] MS_i refers to the ratio of manufacturing to service employment in district i, PS_i refers to public services such as roads in the district. H_i ideally refers to human capital in the district. This is measured by the literacy rate. C_i refers to a dummy for coastal or inland districts, to assess the impact of coastal or riverine characteristics on output, primary or non-primary.

The economic base of the city (measured in terms of the proportion of manufacturing versus service employment) has impacts on city output. Specifically, the hypothesis is that due to India's booming services sector (which contributes nearly 53 % to the country's GDP (also see Paul and Sridhar (2015) who find that the service sector led the rapid surge in per capita incomes experienced by economically better performing states such as Tamil Nadu), higher service employment in a city's economic base might cause to increase city growth or output.

The quality of public services, through its impact on Tiebout kind of voting with the feet, is expected to impact city output positively. The measures chosen of public services is the average road (both surfaced and unsurfaced) length per 1000 population. This is representative of the quality of public services since roads are used by all, they provide access to jobs, markets for products, and better public services. Better quality of public services such as road length will unambiguously add to the city's economic output.

Human capital can be expected to impact city's economic output, because the accumulation of human capital creates a pool of skilled labor force and attracts firms and residents, leading to increases in city output. As mentioned earlier, this is measured by the literacy rate, for lack of reliable data on a better measure.

It is clear why indicators such as the difference between maximum and minimum temperatures and rainfall are used to represent climate change. According to existing studies, one of the expected impacts of climate change on South Asia is general increase in both the mean minimum and mean maximum temperatures by 2–4 °C (Sharma et al. 2006). A sharp decline in winter rainfall in northern India is also projected (Ramesh and Yadava 2005). Thus, gaps between maximum and minimum temperatures may be expected to impact output. Especially agriculture in India is rain dependent, so we would know if reductions in rainfall cause any reductions in agricultural output.

---

[1]The temperature differences are in average terms at the district-level as only the Census of India town directory contains information on annual maximum and minimum temperatures. Hence the temperature difference for all towns in a district was taken to be the average difference between the maximum and the minimum temperature for all towns in a district. Same was the case for rainfall since information on rainfall was reported only in the town directory. Hence rainfall for all towns in a district was aggregated for to arrive at rainfall at the district-level.

I focus separately on the effects of climate change on non-primary and primary output. Based on the past literature, the expectation is that climate change is most likely to have important impacts on cities in coastal or riverine locations, in resource-dependent regions, and in locations at risk at risk from extreme weather events, especially those undergoing rapid urbanization or those whose economies are closely linked with climate-sensitive resources (Hunt and Watkiss 2007).

Equation (1) was estimated at the district-level using a basic model. In this, we examined simply the district's net non-primary output (per capita for 2003–04) as being dependent on various exogenous factors indicated in Eq. (1). In order to do this, DDP estimates published by state directorates of Economics and Statistics in the various states were obtained, with the result that we obtained information regarding 500 districts in all states except a few which did not publish this.[2]

Using the district instead of the town as the unit of observation has a distinct advantage which is that only growing places are designated as towns, whereas even areas that are declining in terms of population are designated as districts, with the result that the selection bias that could arise with the choice of towns as the unit of observation, does not arise with respect to districts. The non-primary portion of the net DDP per capita at constant (1999–00 prices) was used as the measure of city-level output. While gross and net DDP estimates at current and constant prices were available for districts in all states except a few, the net DDP estimates for 2003–04 were chosen since for a few states such as Assam that was the most recent year for which the data were available. The net estimates were chosen to exclude any depreciation. The net DDP was computed in per capita terms to account for any scale effects in the case of large or highly urban or metropolitan districts such as Mumbai.

The exogenous variables summarized in Eq. (1) were constructed with utmost care. The literacy rate for 2001 for the districts was obtained from the Census of India. The remaining variables were constructed from the Census of India's town directories. The town directories contain data on maximum and minimum temperatures for all the 5000 towns in the country. The temperature difference was calculated between the maximum and minimum temperatures for every town and aggregated at the level of the district, the unit of observation here for various reasons discussed earlier; this was used as an exogenous factor influencing city economic growth, following the literature (for instance Haurin (1980) examines the influence of climate on population and migration). The difference between the maximum and minimum temperature for towns was aggregated (averaged) across districts containing them and in this way, district-specific estimates of these temperature differences were obtained for all districts.

[2]The states that did not publish DDP estimates are—Goa, Gujarat, Nagaland, Puducherry, and Tripura.

The rainfall at the town-level was aggregated at the district-level by aggregating rainfall data for all towns in the district. This approach suffers from the limitation that only urban districts are taken into account. We did not have information on predominantly rural districts which did not have even one town.

The ratio of employment in manufacturing to that in services indicates if the city's output is caused by one or the other sector. The ratio of manufacturing to services was computed for every state from the NSSO and was assumed to be the same for all districts in a state, given the 2001 town directories did not contain the manufacturing and service sector employment by town.

Information on coastal districts of India was obtained from the Census of India. All other districts were assumed as being inland districts (this was confirmed with the Census).

I use the Census of India's 2001 town directories for data on rainfall and temperature differences and other socio-economic characteristics for towns and districts, and newly published data by the directorates of Economics and Statistics in various Indian states for DDP. I obtained a listing of India's coastal districts from the Census of India, to enable me to do the estimations separately for coastal and inland districts.

Table 1 presents the descriptive summary of the data. Senapati district in Manipur has the lowest non-primary NDDP per capita, whereas Nalbari in Assam has the highest. Bokaro (Jharkhand) has the lowest primary (defined as agriculture, forestry and logging, and fishing) NDDP per capita, whereas North Cachar Hills in Assam has the highest.

Satara in Maharashtra has the maximum amount of rainfall. Aizawl in Mizoram had the highest literacy rate in 2001, with Dantewada in Chhattisgarh being the lowest on this count. As is clear, on average, a majority of districts are inland, with only 13 % of districts being coastal. Sivaganga in Tamil Nadu has the highest road length per 1000 population, consistent with what Paul and Sridhar (2015) find. Surprisingly, Mumbai has the least road length per 1000 of its population.

Table 1 Description of data

	N	Minimum	Maximum	Mean	Std. deviation
Non-primary NDDP per capita (in ₹)	500	4929.50	314,882.98	57,838.96	41,783.80
Primary NDDP per capita (in ₹)	487	660.83	22,359.49	5977.16	3491.69
Temperature differences (in °C)	451	6.00	46.20	23.86	10.90
Rainfall (in mm)	207	144.00	6063.00	1267.22	876.31
Literacy 2001 (%)	544	0.30	0.97	0.64	0.13
Coastal district (1 = yes; 0 = no)	546	0.00	1.00	0.13	0.34
Average road length per 1000 population	369	0.17	39.82	1.78	2.45
Ratio of manufacturing to services	546	0.05	0.70	0.43	0.15

5 Results from Estimations

As explained earlier, we performed several estimations to assess the impact of climate change and other characteristics on district output, nonagricultural to begin with, since that defines cities. Table 2 summarizes a couple of basic estimations specified by Eq. (1).

The findings from this estimation show statistically significant impacts of climate change indicators—temperature differences between the maximum and minimum, rainfall (in the specification in which it is included), human capital (the literacy rate), and the ratio of manufacturing to services on the city's nonagricultural output.

In terms of their relative magnitude, rainfall has the maximum impact on the city's non-primary output. Specifically, a 1 mm increase in the rainfall of a district increases per capita non-primary output by `16. This also implies that reduction in rainfall causes to decrease even nonagricultural output. Some possible reasons for this are that agriculture contributes to 17 % of India's GDP and provides employment to more than half of the population. Because of this, any impact of monsoons on agricultural growth influences prices, incomes, and GDP growth. Adequate rainfall, leading to higher agricultural and rural incomes keeps consumer confidence and spending strong. For instance, dealers of agricultural/farm equipment (such as John Deere) expect sales to pick up if rains are good.

In regions which are faced with extreme temperatures, the higher the temperature differences between the minimum and the maximum, the higher is the nonagricultural output, to the extent of `840 per capita. It is now well proven that extreme temperatures impact agricultural production negatively. When agriculture is

Table 2 Impact of climate change on district output per capita, all districts dependent variable: (constant) net non-primary district domestic product per capita

Parameter	With rainfall		Without rainfall	
	B (Std. error)	Sig.	B (Std. error)	Sig.
Literacy rate	**60,632.74 (31,555.94)***	**0.06**	24,243.18 (15,479.94)	0.12
Average temperature differences	**841.72 (481.56)***	**0.08**	**362.62 (177.07)***	**0.04**
Population	0.00 (0.00)	0.57	0.00 (0.00)	0.40
Ratio of manufacturing to service employment	**−37,169.94 (22,020.80)***	**0.09**	**−38,781.79 (15,760.89)****	**0.01**
Average road length per 1000 population	746.30 (831.64)	0.37	**2336.36 (763.20)*****	**0.00**
Rainfall	**15.51 (3.90)*****	**0.00**	–	–
Number of observations	140		339	
R^2	0.28		0.06	

Intercepts not reported

* Statistically significant at 10 % level of confidence

**Statistically significant at 5 % level of confidence

***Statistically significant at 1 % level of confidence

negatively impacted, there is a tendency to move resources to nonagricultural land uses; this may explain the positive impact of extreme temperatures on nonagricultural output.

Human capital (measured by literacy rate for 2001) has the expected positive impact on the district's output. A one percentage point increase in the literacy rate of the district increases its non-primary output per capita by `60,630. This is reasonable to expect. Human resources with specific skills are required to utilize natural resources to produce output, both agricultural and nonagricultural.

Finally, the ratio of manufacturing to services employment has a negative impact on nonagricultural output, demonstrating the silent services revolution that has taken over India's cities.

A specification reported in Table 2 excludes rainfall, since there was limited number of observations for this variable. Hence, removal of rainfall explains the larger sample size for the specification. This specification continues to exhibit the same robust signs for most variables as in the model with rainfall, except the human capital variable which becomes insignificant. The impact of average temperature differences on nonagricultural income becomes somewhat decreased, once rainfall is removed. The one variable which now has the expected impact on nonagricultural output is the road length per 1000 population which has a positive impact on nonagricultural output per capita. This is highly likely since roads represent access to markets, jobs, and public services; hence, they facilitate greatly the production of goods and services. Specifically, for every 1 km of extra (both pucca and katcha) road per 1000 population, there is increase in the non-primary output to the extent of nearly Rs. 2336 per capita.

A considerable number of studies discuss the impact of climate change in coastal or riverine locations). As discussed earlier, the impacts of climate change are usually more pronounced in coastal than they are in inland areas. In coastal districts/cities, the land available for productive economic activity (such as buildings used to house manufacturing or service firms) would be limited compared to that for inland districts. Hence, while coastal locations would lead to lower agricultural output, we nonetheless estimate the impact of coastal locations on nonagricultural output. For this reason, Table 3 summarizes the impact of climate change on district output per capita, by including a dummy for coastal districts of the country.

We estimate two specifications when we control for the effects of a district's coastal location, as before, one with rainfall included, and a second specification with rainfall excluded, to benefit from the larger number of observations that would result. In the first specification, with rainfall included, we find that the human capital measured by the literacy rate has a positive impact on nonagricultural output, which is highly plausible to expect, since human capital is the key factor which contributes to increases in output. As in the earlier case, average temperature differences (the difference between the mean maximum and mean minimum temperature) have a positive effect on increasing nonagricultural output, presumably

Table 3 Impact of climate change on district output per capita, all districts, controlling for coastal locations dependent variable: (constant) net non-primary district domestic product per capita

Parameter	With rainfall		Without rainfall	
	B (Std. error)	Sig.	B (Std. error)	Sig.
Literacy rate	**61,686.43 (32,774.39)****	**0.06**	12,482.82 (16,242.03)	0.43
Average temperature differences	**839.46 (483.69)***	**0.08**	**406.47 (177.18)****	**0.02**
Population	0.00 (0.00)	0.60	0.00 (0.00)	0.21
Ratio of manufacturing to service employment	−36,502.92 (22,736.89)	0.11	**−47,336.46 (16,147.05)*****	**0.00**
Average road length per 1000 population	740.13 (836.19)	0.38	**2326.44 (758.86)*****	**0.00**
Rainfall	**15.60 (3.99)*****	**0.00**	–	–
Dummy for coastal districts	−1020.58 (8158.98)	0.90	**15,217.22 (6923.05)****	**0.03**
Number of observations	140		339	
R^2	0.28		0.08	

Intercepts not reported
* Statistically significant at 10 % level of confidence
**Statistically significant at 5 % level of confidence
***Statistically significant at 1 % level of confidence

because as discussed earlier, when agriculture is negatively impacted, there is a tendency for output to move to nonagricultural sectors. The effect of rainfall on nonagricultural output continues to be robust, consistent with expectation that robust growth in agricultural output leads to strong consumer confidence and economic growth overall, by boosting demand for nonagricultural goods and services. In the specification in which rainfall is included, the dummy for coastal districts does not have any impact on nonagricultural output.

When rainfall is excluded from the model and when the dummy for coastal districts is included as well, temperature differences have a positive impact on nonagricultural output. This shows that districts which are characterized by extreme weather also have higher nonagricultural output per capita. Further, the ratio of manufacturing to services has a negative impact on nonagricultural output, which testifies to the fact that higher manufacturing leads to lower nonagricultural output per capita. Another finding which is noteworthy in this specification of Table 3 is that public services such as road length (per 1000 population) has a positive impact on nonagricultural output as we would expect, and robust as in the earlier case. Surprisingly, the dummy for coastal districts has a positive impact on non-primary output per capita (Table 3), which could be due to the fact that coastal cities are always ahead of others in terms of their economic growth due to their exposure to and proximity to markets, when compared with inland districts.

6 Impact of Climate Change on Agricultural Output

We note that agriculture in India continues to be highly sensitive to monsoon variability with 65 % of the cropped area being rainfed (Revi 2008). Since climate change is expected to increase the vulnerability of Indian agriculture, I estimate primary (agricultural) DDP as dependent on climate change characteristics such as temperature differences and rainfall, along with others, for all districts and separately by controlling for coastal locations.

Table 4 summarizes the estimates of the impact of climate change on *agricultural output* per capita in all districts. Here again, due to the availability of limited number of observations on rainfall, we estimated two specifications—one with and another without rainfall. Surprisingly, given their suggested impacts, climate change factors impact agricultural output. Average temperature differences have a positive impact on agricultural output, as they had on nonagricultural output. This could be a manifestation of the fact that extreme temperatures could cause higher variability in yields, depending on the type of the crop, which is consistent with other evidence. The negative coefficient on the coastal dummy (in both the specifications) strongly suggests that coastal locations have reduced agricultural output, than their counterparts in inland locations.

Even here, as in the case of the earlier regressions (of nonagricultural output), it is the literacy rate that has a positive and statistically significant robust impact on agricultural output. This is reasonable to expect since literacy rate is chosen as an

Table 4 Impact of climate change on agricultural output per capita, all districts dependent variable: (constant) net primary district domestic product per capita

	With rainfall		Without rainfall	
	Coefficients (Std. error)	Sig.	Coefficients (Std. error)	Sig.
Average temperature differences	−34.68 (27.29)	0.21	**115.56 (16.65)*****	**0.00**
Literacy rate	**4972.25 (1798.21)*****	**0.01**	**12757.62 (1494.50)*****	**0.00**
Average road length per 1,000 population	**−96.55 (45.12)****	**0.03**	**−140.18 (70.88)****	**0.05**
Population	0.00 (0.00)	0.12	0.00 (0.00)	0.15
Dummy for coastal districts	**−989.32 (447.69)****	**0.03**	**−1306.42 (657.71)****	**0.05**
Rainfall	0.18 (0.19)	0.35	–	–
Number of observations	133		332	
R^2	0.17		0.24	

Intercepts not reported
**Statistically significant at 5 % level of confidence
***Statistically significant at 1 % level of confidence

indicator of skills which contribute to better agricultural practices and increase agricultural output.

Road length has a negative impact (in both the specifications) on agricultural output, which is contrary to expectation. It is possible that given the poor state of roads, agricultural output is negatively impacted. However, there was not adequate data for us to determine the quality of roads. Coastal areas have less agricultural output per capita, when compared with inland districts, to the extent of nearly `1300 per capita (in the specification without rainfall), much less (a reduction of `989 per capita in coastal districts) when rainfall is included. This might be due to the salinity of the soil in coastal areas. It is also possible that warmer temperatures could increase the number of forest fires and outbreaks of pests (http://www.epa.gov/climatechange/impacts-adaptation/southeast.html) which could adversely impact agricultural production.

7 Summary and Policy Implications

We started off by noting that the existing literature does not estimate the economic impacts of climate change in India's cities, which contribute significantly to global warming due to their associated effects such as polluting transport, greater deforestation, lesser rainfall, or nature of industry. The literature has also not taken into account the effects of climate change in the especially vulnerable coastal or riverine locations. While a large number of recent studies in this area focus on modeling approaches, this paper makes an attempt to empirically quantify the impact climate change has on economic (both agricultural and nonagricultural) output in India's cities, using novel methodological techniques used in the past literature in the context of other countries such as the United States.

Making an attempt to examine these questions, I find that overall, climate change thus far has impacted India's cities economically. When we take into account nonagricultural output of districts (which are close approximations of cities) per capita, climate change indicators such as rainfall positively impact nonagricultural output. This could be possibly due to the positive impact rainfall has on agricultural output, and the multiplier effects agriculture has on consumer confidence and nonagricultural output, in the context of developing cities. Extreme temperature differences have a positive impact on nonagricultural output per capita, this could be because when there are extreme temperatures, resources move to nonagricultural uses. Coastal districts have higher levels of non-primary output per capita, to the extent of Rs. 15,200, when compared with their inland counterparts, which may certainly be attributed to the fact that coastal cities are always ahead of others in terms of their economic growth due to their exposure to and proximity to markets, when compared with inland districts. China's Special Economic Zones are designated based on this assumption.

When we examine the impact of climate change on agricultural output, here are our observations in summary: when both climate change indicators—rainfall and extreme temperature differences are included, they do not have an impact on agricultural output; however, when rainfall is not controlled for, extreme temperatures have a positive impact on agricultural output, as they do with nonagricultural output, which refers to the volatility in agricultural output with extreme temperatures. Further, while coastal districts have a positive impact on nonagricultural output (which is plausible), they experience significantly decreased agricultural output, due to their salinity and other effects.

8 Policy Implications

What are the policy implications of this research? The results emphasize that Indian agriculture continues to be dependent on conducive temperatures; if climate change increases the difference between maximum and minimum temperatures, then we certainly expect Indian agriculture to be more volatile. This could pose problems of food security. Coastal locations experience decreases in agricultural output per capita, due to well-established reasons—salinity, their rocky or sandy nature, depending on the type of coastline.

Frequently, as discussed at the beginning, cities exacerbate the impacts of extreme temperatures due to the use of vehicles and emissions thereof. Hence, Indian cities should encourage the use of nonmotorized traffic, which is environmentally friendly. Sridhar (2015) demonstrates several ways in which this can be done in Indian cities.

Coastal districts have higher nonagricultural output, which is understandable, given they are windows to the external world, and provide access to markets outside. Initiatives such as special economic zones in India are meant to capitalize on this advantage of coastal locations. Extreme temperatures increase nonagricultural output, which implies that when agriculture is adversely impacted, economic resources shift to nonagricultural activity. While we need more of both agricultural and nonagricultural output for greater overall welfare, the effect of extreme temperatures can be minimized. Vehicular emissions which lead to extreme temperatures by polluting the environment can be abated by encouraging the use of bicycles, pedestrianization, and mass transit systems. For this, mixed land use, not strict zoning, should be encouraged in the context of developing cities of India, as a paper by GIZ (2010) points out.

We find interestingly that non-agricultural output continues to be dependent on rainfall, due to the increased consumer confidence agriculture has on nonagricultural spending; hence, rainwater harvesting and methods of supplementing urban agriculture activities with water should be proactively pursued.

Further, coastal areas are characterized by salinity which restricts agricultural output in these areas, consistent with what other studies find. Hence, it is necessary to at least have an assessment of soil salinity to enable increasing agricultural output in such areas.

Several limitations of the data used in this paper should be mentioned. We found extreme temperatures do impact agricultural output positively, but this output could be volatile, which implies that if other years' data were to be used, the output may be negatively impacted (in this paper, agricultural and nonagricultural output for 2003–04 was used, and exogenous variables were used for 2001, for a lagged effect). It is to be noted that extreme temperatures in this paper are measured by the difference between average maximum and mean minimum temperatures at the level of the district, not at the level of the city, due to the selection problem which was discussed earlier.

It should also be noted that climate change might be a threat if it affects mean maximum and minimum temperatures eventually, many studies indeed confirm that this is the case (see Deschênes and Greenstone 2007). If there are other forms of climate change, such as the risk of flooding, unseasonal rainfall, and related events, which do not significantly impact temperatures, then the findings in this paper would not throw light on those aspects of climate change and their impacts.

The other limitation we are faced with is that several states did not report data on DDP for their districts for various reasons. Only when these data are complete, we would be able to say something definitive about the economic impacts of climate change.

Further, the data on rainfall and temperature differences are from the Census of India's town directory only for 2001. A better indicator to use may have been percentage changes in temperature over a long period of time to understand whether global warming has been really taking place. We expect that at least in the relatively more industrialized cities, global warming would be much more pronounced. Unfortunately the data on minimum temperatures required to make an assessment are not available free of cost from the Indian Meteorological Department. Additional funding would be required to obtain these data.

Finally, the estimations were conducted at the level of districts because there are econometric problems with the choice of cities as the unit of analysis (since only growing areas are designated as cities).Hence for purposes of the district-level data, the average rainfall and temperature differences are averaged for all cities within a district. Obviously future research in this area should build upon the limitations of this work.

References

Beeson, P., DeJong, D. N., & Troesken, W. (2001). Population growth in U.S. counties, 1840–1990. *Regional Science and Urban Economics, 31*, 669–699.

Cohen, B. (2004). Urban growth in developing countries: A review of current trends and a caution regarding existing forecasts. *World Development, 32*(1), 23–51.

Deschênes, O., & Greenstone, M. (2007). The economic impacts of climate change: Evidence from agricultural output and random fluctuations in weather. *The American Economic Review, 97*(1), 354–385.

Gbetibouo, G. A., & Hassan, R. M. (2005). Measuring the economic impact of climate change on major South African field crops: A Ricardian approach. *Global and Planetary Change, 47*(2–4), 143–152.
GIZ. (2010 October). Challenges of urban transport in developing countries—A summary. www.sutp.org
Hallegatte, S., Hourcade, J. C., & Ambrosi, P. (2007). Using climate analogues for assessing climate change economic impacts in urban areas. *Climatic change, 82*(1–2), 47–60.
Haurin, D. (1980). The regional distribution of population, migration and climate. *Quarterly Journal of Economics,* 294–308.
Hunt, A., & Watkiss, P. (2007). Literature review on climate change impacts on urban city centres: Initial findings. ENV/EPOC/GSP(2007)10. Paris, France: OECD.
IPCC. (2007). Summary for policymakers. In S. Solomon, D. Qin, M. Manning, Z. Chen, M. Marquis, K. B. Averyt, et al. (Eds.), *Climate change 2007: The physical science basis. Contribution of working group I to the fourth assessment report of the intergovernmental panel on climate change.* Cambridge, United Kingdom and New York, NY, USA: Cambridge University Press.
Kumar, K. R., Sahai, A. K., Kumar, K. K., Patwardhan, S. K., Mishra, P. K., Revadekar, J. V., et al. (2006). High-resolution climate change scenarios for India for the 21st century. *Current Science, 90*(3), 334–345. ISSN 0011-3891.
Mendelsohn, R., & Dinar, A. (1999). Climate change, agriculture, and developing countries: Does adaptation matter? *World Bank Research Observer, 14*(2), 277–293.
Mendelsohn, R., Morrison, W, Schlesinger, M. E., & Andronova, N. G. (2000). Country-specific market impacts of climate change. *Climatic Change, 45*(3–4), 553–569.
NASA. (2007). Earth's surface temperature analysis. Goddard Institute for Space Studies (GISS), New York City.
National Oceanic and Atmospheric Administration (NOAA). (2007). State of the Climate Report. Washington, DC: National Climatic Data Center, NOAA.
O'Brien, K., et al. (2004). Mapping vulnerability to multiple stressors: Climate change and globalization in India. *Global Environmental Change, 14*(4), 303–313.
Paul, S., & Sridhar, K. S. (2015). *The paradox of India's north-south divide: Lessons from the states and regions.* New Delhi: Sage.
Ramesh, R., & Yadava, M. G. (2005). Climate and water resources of India. *Current Science, 89* (5), 818–824 (Physical Research Laboratory, Ahmadabad).
Revi, A. (2008). Climate change risk: An adaptation and mitigation agenda for Indian cities. *Environment and Urbanization, 20*(1), 207–230.
Sattherwaite, D. (2007). *Climate change and urbanization: Effects and implications for urban governance.* UK: International Institute for Environment and Development.
Sharma, S., Bhattacharya, S., & Garg, A. (2006). Greenhouse gas emissions from India: A perspective. *Current Science, 90*(3), 326–333.
Sridhar, K. S. (2015, March 31). Walk (and pedal) the talk. *The Economic Times.* (http://www.economictimes.com)
Toi, R. S. J. (2002). Estimates of the damage costs of climate change, part II. Dynamic estimates. *Environmental and Resource Economics, 21*(2), 135–160.

Economic Impact of Air Pollution from Agricultural Residue Burning on Human Health

Surender Kumar and Parmod Kumar

Abstract This paper measures the value of health effects of air pollution for the Indian rural Punjab, where air pollution problem occurs from crop residue burning. Consumer choice model is used to get the monetary estimates of reduced air pollution level to the safe level. The paper uses data of 625 individuals collected from a household level survey conducted in three villages in Indian Punjab for 150 households. To obtain the monetary values, Tobit and Poisson models are used to estimate mitigation expenditure and workdays lost equations, respectively. Total annual welfare loss in terms of health damages due to air pollution caused by the burning of rice straw in rural Punjab amounts to INR 76 millions. If one also accounts for expenses on averting activities, productivity loss due to illness, monetary value of discomfort and utility and additional fertilizer, pesticides and irrigation, the losses would be much higher.

Keywords Air pollution · Residue burning · Mitigation expenditure · Workdays lost · Rural Punjab

Jel Classification Q510 · Q520 · Q530

1 Introduction

Epidemiological studies show that the contamination of air quality increases adverse health impacts (Ostro et al. 1995). Air pollution contributes to the respiratory diseases like eye irritation, bronchitis, emphysema, asthma, etc., which not

S. Kumar (✉)
Department of Economics, Delhi School of Economics, University of Delhi, Delhi, India
e-mail: skumar@econdse.org

P. Kumar
Institute for Social and Economic Change (ISEC), Nagarbhavi PO, Bangalore 560072, India
e-mail: pkumar@isec.ac.in

S. Nautiyal et al. (eds.), *Climate Change Challenge (3C) and Social-Economic-Ecological Interface-Building*, Environmental Science,
DOI 10.1007/978-3-319-31014-5_17

only increases individuals' diseases mitigation expense but also affects their productivity at work. Most of the studies valuing health impacts of air pollution remain confined to urban areas as air pollution is considered mainly the problem of urban areas in developing countries. Though health consequences from burning of agricultural residue are not fully understood, relative short exposure may be more of a nuisance rather than a real health hazard. Many of the components of agricultural smoke cause health problem under certain conditions (Long et al. 1998). This paper attempts to measure the value of health effects of air pollution for the rural Punjab, where air pollution problems happen from crop residue burning.

The rice and wheat system (RWS) is one of the widely practiced cropping systems in northern India. About 90–95 % of the rice area is used under intensive RWS in Punjab (Gadde et al. 2009). Widespread adoption of green revolution technologies and high-yielding variety of seeds increased both, crop as well as crop residue. In the past few decades intensive mechanization of agriculture has been occurring and combine[1] harvesting is one such input, particularly in the RWS. Note that in the RWS a short period of time is available between rice harvesting and wheat plantation, and any delay in planting adversely affects the wheat crop. This coupled with combine harvesting compels the farmers to burn the residue to get rid of stubble left out after the harvest. It is estimated that 22,289 Gg of rice straw surplus is produced in India each year out of which 13,915 Gg is estimated to be burnt in the field. The two states namely Punjab and Haryana alone contribute 48 % of the total and are subject to open-field burning (Gadde et al. 2009). Burning of straw emits emission of trace gases like CO_2, CH_4, CO, N_2O, NO_X, SO_2, and large amount of particulates which cause adverse impacts on human health. It is estimated that India annually emits 144,719 Mg of total particulate matter from open-field burning of rice straw (Gadde et al. 2009).

There are many studies in developed countries that estimate the value of adverse health effects of air pollution (Gerking and Linda 1986; Dockery et al. 1993; Schwartz 1993; Pope et al. 1995, etc.). Similar evidences are available from India and other developing countries (e.g., Cropper et al. 1997; Kumar and Rao 2001; Murty et al. 2003; Gupta 2008; Chestnut et al. 1997; Alberini and Krupnick 2000). These studies used either household health production model or damage function or cost of illness approaches to estimate the monetary value of health damage caused due to ambient air pollution.

Cropper et al. (1997) using dose-response model found that a 100-$\mu g/m^3$ increase in total suspended particulate matter (TSPM) leads to 2.3 % increase in trauma deaths in Delhi. Kumar and Rao (2001) estimated the household health production function using data of working individuals of the residential complex of Panipat Thermal Power Station in Haryana, India and find that individual

[1]The combine harvester, also called combine, is a machine that harvests grain crops. It combines into a single operation processes that previously required three separate operations, that is, reaping, binding, and threshing.

willingness to pay (WTP) varies between INR 12 and INR 53 per month for improving the air quality to WHO standards. Using a similar model, Murty et al. (2003) observed that a representative household gains about INR 2086 and INR 950 per annum due to reduced morbidity from reduction of air pollution to the safe level in Delhi and Kolkata, respectively. Similarly, Gupta (2008) estimates aggregate benefits of the magnitude of INR 225 million per year reducing air pollution to the safe level for the city of Kanpur, India. Note that these studies are restricted to measure the monetary value of reducing urban air pollution to the safe level since air pollution has been considered mainly the problem of urban areas. In the present study, we use a similar consumer choice model to get the monetary estimates of reduced air pollution level to the safe level for the rural Punjab.

We use data of 625 individuals collected from a household level survey conducted in three villages, namely Dhanauri, Ajnoda Kalan, and Simro of Patiala district of Punjab for 150 households. To get the monetary values we estimate two equations: one with mitigation expenditure and the other with workdays lost as dependent variables. Tobit and Poisson models are found to be suitable for estimating mitigation expenditure and workdays lost equations, respectively. We find that total annual welfare loss in terms of health damages due to air pollution caused by the burning of rice straw in rural Punjab amounts to INR 76 millions.

The paper is organized as follows: Sect. 2 presents the ambient air quality levels in the study villages during the period when harvesting of rice takes place. Section 3 gives details of the design of the household survey and analyzes the households' behavior. Section 4 describes the theoretical model and estimation strategy. The results are discussed in Sect. 5, while last section concludes the paper.

2 Ambient Air Quality Level in Study Area

In Punjab, it is common practice to openly burn agricultural residues in fields after harvesting crops by mechanical harvesters. Central and State Pollution Control Boards have been monitoring the ambient air quality for certain Indian cities for the last two decades. Monitoring of ambient air quality in rural areas is very sporadic and purpose specific. Given the severity of the problem, the Punjab Pollution Control Board (PPCB) conducted air pollution monitoring in three villages of Patiala districts, namely Dhanauri, Simro, and Ajnoda Kalan during November 1–3, 2006. Patiala is one of the agriculturally leading districts of Punjab which is rich in crops like rice and wheat. When monitoring was done in these designated villages, it was ensured that some burning of rice straw was happening in the fields of these villages. Monitoring stations for the sites were planned keeping in view the metrological conditions and environmental settings in terms of habited and non-habited areas and following parameters were monitored: metrological parameters (temperature, humidity, wind speed, and wind direction), particulate matters

Table 1 Descriptive statistics of emissions and metrological data

	PM_{10} ($\mu g/m^3$)	SO_2 ($\mu g/m^3$)	NO_X ($\mu g/m^3$)	Relative humidity (min) (%)	Wind speed (km/h)	Temperature (maximum) (°C)	Temperature difference (maximum minus minimum) (°C)
Mean	306.66	14.20	56.08	46.13	1.78	29.35	16.87
Standard deviation	16.60	2.96	12.32	0.97	0.40	0.60	1.05
Maximum	325.5	17.4	69	47.4	2.35	30.2	18.2
Minimum	284.75	10.25	39.25	45.1	1.425	28.8	15.6

Source Envirotech Instruments Private Limited (2006)

($PM_{2.5}$, PM_{10}, TSPM), gaseous pollutants (SO_2, NO_X, NH_3, CO, Ozone, THC, TC, and BTX), and heavy metals.[2]

Descriptive statistics of some of the important pollutants and metrological parameters is given in Table 1. The statistics shows that gaseous pollutants such as SO_2 and NO_X were within the safe limits put under National Ambient Air Quality Standards (NAAQS) and particulate matters either were measured in terms of SPM, PM_{10}, or $PM_{2.5}$ and they cross the limits set by the NAAQS.[3] In the study area all the particulates followed same pattern in whatever terms they are measured. The hourly peak values ranged between 300–350 and 24 h average concentration ranged between 200 and 300 $\mu g/m^3$. The ratio between peak and average values was found to be about 1.2, indicating almost uniform concentration over the monitoring period. The contribution of the burning to PM_{10} concentration appeared to be around 100–200 $\mu g/m^3$. In all the three monitoring sites, the difference in humidity and temperature levels was negligible and the wind speed was found to be in the range of 0–3.6 km/h. Low wind speed coupled with low wind direction fluctuation implies that the impact of polluting activities remains confined to its close vicinity.

3 Household Survey Design and Data

To measure the economic cost of pollution, we needed data on other socioeconomic health indicators in addition to pollutant emitted by paddy waste burning in the environment. The health indicators during and after the period of burning, measures adopted by people in the periphery to cope with the situation, and other socio, agricultural, income, and expenditure parameters were not collected by the PPCB

[2]For details, please see the report prepared by Envirotech Instruments Private Limited (2006) on 'Air Pollution Discharged from the Burning of Crop Residue in Agriculture Fields of Punjab' for Punjab State Pollution Control Board.

[3]CPCB has defined the NAAQS implying the safe level of pollutants for residential, rural, and other areas as follows: SO_2 60 and 80, NO_X 60 and 80 and PM_{10} 60 and 100 $\mu g/m^3$ as annual and 24 h averages, respectively.

survey. In order to further work on the economic cost, we resurveyed the same villages where PPCB conducted its exercise of measuring air pollutant before and after the burning. This exercise assumes that the findings of the PPCB exercise are still valid and no significant change has occurred neither in the incidences of burning nor in the pollutant emitted into the air by the exercise of burning in those areas.

Looking at some of the household and agricultural characteristics, the Nabha Tehsil under Patiala District has a total population of 251,326 with 75.33 % of it confined in the rural sector and 24.67 % in the urban sector. Out of operated area 26,395 acres of land is under wheat and 28,359 acres of land is under rice crop in the Nabha Tehsil. Data on the health status and socioeconomic variables of households for this exercise were collected through a household survey conducted for this study for the above-mentioned three villages in the month of May 2009 and is based on the recall memory. Selection of households in the respective villages was based on stratified random sampling.

The selection of villages was purposive as has been documented above. After selecting the villages, a list of all households including those who were cultivators, agricultural laborers and those who were working in the other formal or informal sectors like regular government or private services, self-business, and pension holders was worked out. Stratification was done for the cultivating households in terms of marginal farmers (≤2.5 acres), small farmers (2.51–5.0 acres), medium farmers (5.01–10.0 acres), and large farmers (above 10.0 acres). The farmers were selected on the basis of stratified random sampling method. From each village, approximately 10 farmers were selected for each category. Thus, total 40 farmers were selected from each village and 120 farmers were selected from all the three villages. In addition to cultivating households, a total number of 10 landless laborers were selected from every village. Thus total 30 numbers of agricultural laborers were selected from all the three villages. Therefore, the aggregate sample consists of a total number of 150 households surveyed for this exercise.

The questionnaire used for the household survey had twelve sections seeking detailed information on various aspects of inputs–outputs used by the farmers for agricultural practices, disposal of crop residues and socioeconomic characteristics. Sections 1 and 2 of the questionnaire provide information on individual household members' profile in terms of their age, education sex, occupation, marital status, etc. Section 3 deals with the information on end use of straw and health effects of the burning. It gives information on the current health status of individuals, symptoms of illnesses linked to air pollution exposure, averting and mitigation activities followed by all the members in a household during the designated months when the rice straw burning was happening. Subsections also provide information on whether a particular individual is suffering from any chronic diseases. There was question also on the general awareness of households about the illnesses that occur due to air pollution. Sections 4–7 seek information regarding agriculture productivity, input usage, stubble management, etc. Sections 8 and 9 provide information on medical expenditure and workdays lost during the designated period of rice straw burning. The former section provides information on the expenditure on formal medication such as fee paid to a doctor, expenses on the allopathic medicines, cost of hospitalization, etc., while the latter section provides information on expenditure incurred on informal medicines that Indian household generally takes without consulting any medical professional. Last two sections provide information on individual habits and households assets. Information on the habits includes whether an individual is habitual to smoking, alcohol drinking, or/and taking any other toxicants that affect health in general.

4 Methodology

4.1 *Theoretical Model*

Air quality affects the utility of individuals and an economic value exists. There are several ways to capture this economic value, viz., dose-response, revealed preferences, and contingent valuation methods. The dose-response method assumes a relationship between air quality and morbidity (and/or mortality). It puts a price tag on air quality without retrieving people's preferences for the good. But, such type of mechanical relationship of the dose-response function does not take into account consumer behavior. The revealed preference methods assume that the consumers are aware of the costs/benefits of air quality and are able to adjust their behavior to reveal their preferences. This necessitates the need to have estimates of willingness to pay (WTP) or willingness to accept (WTA) on the basis of a consumer choice model aimed at measuring the strength of association between health effects and contaminated air quality.

Suppose an individual maximizes his/her utility through expenditure on marketed goods and services, X.[4] The utility depends not only on X but also on the state of health, H, of an individual which is affected by the level of air quality (non-marketed good). It is further assumed that the contaminated air quality, P, is beyond the control of individuals, but individuals can at least partially reduce its effects through incurring defensive expenditure, D. The utility function is defined as

$$U(X, D; P) = H(D, P)U(X) \tag{1}$$

where $U_X > 0$, $U_{XX} < 0$, $H_D > 0$, $H_P < 0$, $H_{DD} < 0$, $H_{PP} < 0$.

The state of health affects individual's work performance and hence the wage income. Moreover, it is also possible that the contaminated air quality makes the individual so sick as to be completely incapacitated. During the time the individual is under this condition, he/she is absent from work and loses the wage income

[4]Harrington et al. (1989) take the individual utility as a function of expenditure on marketed goods and services, X and leisure time, L. Since in developing countries especially in rural areas people are living in the conditions of poverty; therefore, we assume that the individual utility is the function of marketed goods and services, X only.

completely. Therefore, 'sick time,' S, can also be assumed to be the function of defensive expenditure and contaminated air quality,

$$S = S(D, P) \tag{2}$$

where $S_D < 0$, $S_P > 0$, $S_{DD} > 0$, $S_{PP} > 0$

Equation (1) is the maximized subject to the following constraints:

The time constraint is

$$W + S = T \tag{3}$$

where W is the work time and T is the total time available. The income (resource) constraint is

$$I + wH(D, P)W \geq mS + D + X \tag{4}$$

where mS is the medical expenses which are assumed proportional to illness, S, I denotes nonwage income, and w is referred as wage rate.The Lagrangian of the problem is

$$\Pi = H(D, P)U(X) + \lambda[I + wH(D, P)(T - S) - D - X - mS] \tag{5}$$

The first-order optimization conditions are

$$\begin{aligned} \Pi_X &= H(D, P)U_X - \lambda = 0 \\ \Pi_D &= H_D U + \lambda H_D wW - \lambda HwS_D - \lambda - \lambda mS_D = 0 \end{aligned} \tag{6}$$

Using the envelope theorem, Harrington et al. (1989) obtain the WTP as

$$\mathrm{WTP} = -\left(\frac{H_D U}{\lambda} + H_D wW\right)\frac{H_P}{H_D} + (HwS_D + mS_D)\frac{S_P}{S_D} \tag{7}$$

and the marginal loss of social welfare (SW) associated with individual responses to deterioration in air quality, therefore, is

$$\begin{aligned} \frac{\partial \mathrm{SW}}{\partial P} = & -\frac{U}{\lambda}\frac{\mathrm{d}H}{\mathrm{d}P}(\text{Direct disutility of illness}) \\ & - w \times W\frac{\mathrm{d}H}{\mathrm{d}P}(\text{Lost work productivity}) \\ & - Hw \times W_P(\text{Value of lost time during illness}) \\ & + m\frac{\mathrm{d}S}{\mathrm{d}P}(\text{Medical expenses}) \\ & + D_P(\text{Defensive expenditure}) \end{aligned} \tag{8}$$

Equation (8) shows that the cost of illness caused by the contaminated air can be grouped into five categories. The term direct disutility is very subjective and it is very difficult to find its monetary value. The second term, the lost work productivity, measures the value of loss caused by the illness due to lower work productivity. This loss is caused when the sick person is present for work but is not able to work with his/her full productivity. The third term measures the loss in social welfare due to illness absence of individuals from work. The last two terms measure the expenses individual have to incur for defensive and mitigating activities due to contamination of air quality. In rural areas during survey we could not get figures on the defensive activities of individuals; therefore, we measure only two values: medical expenses and value of lost time during illness. Thus our measure of social loss due to contaminated air provides the lower bound of the value.

4.2 Estimation Strategy

To get the estimates of social welfare loss due to contaminated air in terms of health damages, we estimate the following two equations consisting of demand function for medical expenses (*mS*) and the workdays lost due to illness (*S*):

$$mS = \alpha_0 + \alpha_1 \mathrm{SPM} + \alpha_2 \mathrm{SO_2} + \alpha_3 \mathrm{SMOKING} + \alpha_4 \mathrm{DRINKING} + \alpha_5 \mathrm{PerCapitaAssets} + \alpha_6 \mathrm{SEX} + \alpha_7 \mathrm{AGE} + \alpha_8 \mathrm{EDUCATION} + \alpha_9 \mathrm{OCCUPATION} + \varepsilon_1 \tag{9}$$

and

$$S = \alpha_0 + \alpha_1 \mathrm{SPM} + \alpha_2 \mathrm{SMOKING} + \alpha_3 \mathrm{DRINKING} + \alpha_4 \mathrm{PerCapitaAssets} + \alpha_5 \mathrm{SEX} + \alpha_6 \mathrm{AGE} + \alpha_7 \mathrm{EDUCATION} + \alpha_8 \mathrm{OCCUPATION} + \varepsilon_2 \tag{10}$$

where

Medical Expenses (*mS*): Mitigating activities or medical expenses include expenses incurred as a result of air pollution-related diseases. These expenditures include costs of medicine (formal as well informal), doctor's fee, diagnostic tests, hospitalization, and travel to doctor's clinic during the two rice harvesting months.
Workdays Lost (*S*): S represent the number of workdays lost per person during the two rice harvesting months of October and November due to diseases/symptoms associated with air pollution.
Particulate matter (PM_{10}) and Sulfur Dioxide (SO_2): These are the averages of the ambient emission levels observed during the monitoring period measured in $\mu g/m^3$.

SMOKING: This is measured as dummy variable equal to 1 if the individual is having smoking habit, otherwise 0.
DRINKING: This is measured as dummy variable equal to 1 if the individual is having alcohol drinking habit, otherwise 0.
PerCapitaAssets: This is measured in Indian rupees (INR).
SEX: This is measured as dummy variable equal to 1 for male and 0 for female.
AGE: Age of the individual is measured in number of years.
EDUCATION: This is coded as follows: 1 = Illiterate; 2 = below primary; 3 = Primary; 4 = Middle; 5 = Secondary/Metric; 6 = Technical; 7 = Graduate; 8 = Post graduate and above.
OCCUPATION: This is measured as dummy variable equal to 1 if the individual is in the occupation of self-farming or agricultural laborer, 0 otherwise.

Note that the dependent variable in Eq. (9) is a censored variable, i.e., the dependent variable is zero for corresponding known values of independent variables for part of the sample. Therefore, we use Tobit model for estimating the demand for mitigating activities:

$$\begin{array}{ll} mS_i = \alpha + \beta x_i + u_i & \text{if RHS} > 0 \\ 0 & \text{otherwise} \end{array} \tag{11}$$

where mS_i refers to the probability of the ith individual incurring positive medical expenditure and x_i denotes a vector of individual characteristics, such as assets, age, sex, education, pollution parameter, etc.

In Eq. (10) the dependent variable is a count of the total number of workdays lost due to air pollution-related illness by an individual during the particular period; therefore, there are zeros for many observations. In this case, Poisson regression model is appropriate as it considers the predominance of zeros and the small values and the discrete nature of the dependent variable. The least square and other linear regression models do not take into account these features. The Poisson regression model can be stated as follows:

$$\text{prob}(Y_i = y_i/x_i) = \mu_i^{y_i} e^{-\mu_i}/y_i, \quad y_i = 0, 1, 2, \ldots \tag{12}$$

This equation is nonlinear in parameters; therefore, for estimation purpose by taking its natural log we convert it into an equation which is linear in parameters. Note that the Poisson regression model is restrictive in many ways. For example, the assumption that the conditional mean and variance of y_i, given x_i are equal is very strong and fails to account for over dispersion.[5] The data used in the estimation of Eqs. 9 and 10 and the background information are provided in Tables 2, 3, and 4.

[5]Similar estimation procedure is followed by Gupta (2008).

Table 2 Expenditure incurred due to problems faced during the crop stubble burning

Problem	Unit	Dhanauri	Ajnoda Kalan	Simro	Aggregate
Family members visited local doctor during October–November, 2008	Average number of members per household (hh)	2.93	2.12	2.82	2.63
Prescribed to any medicine during the 2 months of stubble burning (October–November, 2008)	Average number of members per hh	2.9	2.15	2.82	2.64
	Average number of days per hh	13.3	13.75	11.43	12.92
	Average amount spent per hh (INR)	280.33	335.77	504.76	360.26
Any member hospitalized during the 2 months of stubble burning (October–November, 2008)	Average number of members per hh	0.00	1.00	3.00	2.00
	Average number of days per hh	0.00	3.00	5.00	4.00
	Average amount spent per hh (INR)	0	300	1000	650

5 The Model Results

Tables 5 and 6 provide the results of parameter estimates of reduced form equations of mitigation expenditure and workdays lost. In the reduced form these equations are expressed as functions of a common set of socioeconomic variables and ambient air pollution expressed in terms of particulate matter (PM_{10}) and SO_2 levels.

The parameter estimates of mitigating expenditure equation are given in Table 5. We find there is a positive and statistically significant (at 10 % level) association between ambient PM_{10} level and the mitigating expenditure. This implies that individual have to spend higher amount of money to mitigate the adverse health effects when the particulate level is higher in the ambient environment. The relationship between mitigating expenditure and ambient SO_2 level is negative and statistically insignificant, as contrary to expectations. This might be happening as the ambient SO_2 level is within the NAAQS limits in the villages of Punjab.

As is expected, the coefficients of the variables such as smoking and drinking behavior of the individual are found to be positive and statistically significant. These personal habits coupled with the ambient air pollution make individual more

Table 3 Medical expenses incurred due to health problems

Village name	Medical expenses incurred during the last year (April 08–March 09)		Percentage of affected members observing severity of problem increasing at the time of crop stubble burning	Medical expenses incurred due to acute problem during the crop stubble burning (October–November 08)				Absence from work for each illness during October–November 2008	
	Chronic disease (per affected member) (INR)	Non chronic disease (per affected member) (INR)		Doctor/hospital/other charges (per affected member) (INR)	Medicine cost (per affected member) (INR)	Transportation/freight (per affected member) (INR)	Self medication (per affected member) (INR)	(No of days per affected member)	Money loss (per affected member) (INR)
Dhanauri	1667	767	100.00	71	162	77	80	3	300
Ajnoda Kalan	1750	1978	97.30	159	222	80	56	5	600
Simro	2000	478	95.45	116	200	71	92	10	1000
Total	1750	1145	97.75	119	196	77	82	5	610

Table 4 Variables used in the regression analysis

Variable	Mean	Standard deviation	Maximum	Minimum	Percent
Formal medical expenses (INR)	39.26	165.05	2700.00	0.00	
Informal medical expenses (INR)	19.46	66.62	450.00	0.00	
Workdays lost	0.06	0.72	15.00	0.00	
Age	31.35	18.50	90.00	1.00	
Education	3.14	1.77	8.00	1.00	
Per capita assets (INR)	64,469	78,377	539,467	250	
Male	54.41				
Occupation (farmers and agricultural laborers)					26.32
Smoking	2.12				
Drinking	5.88				
Toxicants	3.29				

Table 5 Tobit equation of total medical expenditure (left censured at 0)

Independent variable	Coefficient
PM_{10} (+)	0.046 (1.72)*
SO_2 (+)	−5.16 (−0.52)
SMOKING (+)	395.14 (2.62)***
DRINKING (+)	177.94 (1.71)*
Per capita assets (+)	0.0009 (2.65)***
SEX	−41.76 (−0.57)
AGE (+)	4.13 (1.74)*
EDUCATION (−)	−9.85 (−0.51)
OCCUPATION (+)	92.58 (1.16)
Constant	−678.69 (−2.73)***
Pseudo R^2	0.014
Log-likelihood	−1262.37
Wald Chi^2 (9)	35.74***
Uncensored observations: 141	Left censored observations: 484
Total observations	625

Notes Figures in parentheses are *t* values
*** Significance at 1 % level
* Significance at 10 % level

prone to asthmatic diseases and as a result they are required to spend more on mitigating activities. Similarly, we find there is positive and significant relationship between the age of individual and their mitigating expenses implying that the marginal effect of age on mitigating expenses is positive. We also observe that there

Table 6 Poisson equation of workdays lost

Independent variable	Coefficient
PM_{10} (+)	0.008 (5.59)***
SMOKING (+)	−14.66 (−0.01)
DRINKING (+)	−0.81 (−0.79)
Per capita assets (−)	−0.00,001 (−1.78)*
SEX	0.43 (1.07)
AGE	−0.011 (−0.97)
EDUCATION (−)	−0.71 (−5.07)***
OCCUPATION	0.32 (−0.67)
Constant	−5.02 (−3.98)***
Pseudo R^2	0.023
Log-likelihood	−170.93
Wald Chi^2 (8)	97.97
Total observations	625

Notes Figures in parentheses are t values
*** Significance at 1 % level
* Significance at 10 % level

is positive and statistically significant relationship between mitigating expenses and per capita assets. This might be happening because wealthier individuals do not hesitate to take mitigating activities if they are suspected to some diseases in comparison to people who have lesser assets.

Education raises awareness level of individuals with respect to environmental problems and related health damages and helps in taking informed preventing activity-related decisions. The coefficient of education is negative, as expected, though statistically insignificant, depicts that there happens to be a reduction in mitigation expenditure with the increase in education level. Similarly, the individuals who have to work in agriculture fields where burning of agricultural residue takes place are thought to be more prone to the adverse effects of pollution in comparison to their counterparts who are in other occupations such as salaried individuals. We use dummy variable equal to one for farmers and agricultural wage earners and zero for the individuals who are in other occupations. We find a positive association between occupation variable and medical expenditure.

Table 6 presents parameter estimates of the reduced form equation of workdays lost. As expected, the coefficient of PM_{10} variable is positive and statistically significant at 1 % level implying that the probability of losing workdays increases as the concentration of particulate matters in ambient environment increases. Education increases awareness level and helps in taking preventing action and as a result an individual is expected not to lose workday; therefore, we find that there is negative association between education level of individuals and workdays lost. Similarly, wealthier individuals could spend money on preventing activities and there is negative relationship between per capita assets and workdays lost.

Table 7 Welfare loss due to increased air pollution in Rural Punjab

	Representative individual (INR)	Rural Patiala district (millions INR)	Rural Punjab (millions INR)
Medical expenditure	2.17	2.35	36.52
Opportunity cost of workdays lost	2.35	2.54	39.57
Total welfare loss	4.52	4.89	76.09

5.1 *Welfare Loss*

The welfare loss in terms of health damage due to increase in the concentration of particulate matters from rice straw burning in the ambient environment can be estimated in terms of increase in the medical expenditure on mitigating activities and the opportunity cost of workdays lost and are presented in Table 7.

5.2 *Increase in Medical Expenditure*

To get the estimates of welfare loss in terms of increased medical expenditure we need to obtain the marginal effects. The marginal effects in the case of Tobit estimation could be computed by taking partial derivatives of mitigating expenditure equation with respect to PM_{10} and multiplying it by the probability of the dependent variable taking the nonzero values. If the ambient PM_{10} level is reduced from the level observed during the harvesting period of rice in rural Punjab to the safe level (i.e., a reduction of 207 $\mu g/m^3$ since the safe level defined under NAAQS is 100 $\mu g/m^3$ for the 24 h average), the estimated reduction in medical expenditure turns out to be INR 2.17 for the months of October and November for a representative person.

Total rural population projected for October 2008 based on Census 2001 is 1083 thousand and 16,839 thousand for the district of Patiala and the state of Punjab, respectively. Extrapolating this welfare loss for the entire rural population of Patiala and Punjab, it is estimated as INR 2.35 million and INR 36.52 million, respectively.

5.3 *Opportunity Cost of Increase in Workdays Lost*

To get the marginal effects of reduction in PM_{10} level on workdays lost, we differentiated partially the reduced form equation of workdays lost with respect to PM_{10}. The Poisson estimates show that one $\mu g/m^3$ increase in PM_{10} results in a marginal loss of 0.0,000,946 days for a representative individual in these two harvesting months. If the PM_{10} level is reduced from the current level to the safe levels during rice harvesting period, the estimated gain in workdays is 0.03. In monetary terms, the loss in terms of workdays lost for a representative individual is estimated to be INR 2.35 and for rural Patiala district and rural Punjab state it turns out to be INR 2.54 million and INR 39.57 million, respectively, assuming a wage rate of INR 120 per day.[6]

The total monetary loss (due to lost workdays and increased medical expenditures) caused in terms of health damages due to increase in ambient PM_{10} level beyond the safe level for the rural areas of Patiala district and Punjab state is estimated as INR 4.89 million and INR 76.09 million, respectively. These losses should be considered the lower bound of health damages caused by the increased air pollution level in rural Punjab. These estimates could be much higher if expenses on averting activities, productivity loss due to illness, monetary value of discomfort, and utility could also be considered. There is additional monetary cost of burning to the farmers in terms of additional fertilizer, pesticides, and irrigation as was shown by the survey results discussed in Sect. 4. One also has to add into the above cost the losses of soil nutrient, vegetation, bio-diversity, and accidents caused because of low visibility.

6 Conclusion

In this paper an attempt is made to estimate the monetary value of health damage caused by the smoke pollution emitted by the burning of rice and wheat stubble in the open fields in Punjab, India. We use data of 625 individuals collected from a household level survey conducted in three villages, namely Dhanauri, Ajnoda Kalan, and Simro of Patiala district of Punjab for 150 households. To get the monetary values we estimated two equations: one with mitigation expenditure and the other with workdays lost as dependent variables. Tobit and Poisson models are used for estimating mitigation expenditure and workdays lost equations, respectively.

Our household survey showed that paddy stubble burning leads to air pollution and several other problems. Irritation in eyes and congestion in the chest were the two major problems faced by the majority of the household members. Respiratory allergy, asthma, and bronchial problems were the other smoke-related diseases which affected household members in the selected villages. Almost 50 % of the selected households indicated that their health-related problems get aggravated during or shortly after harvest when crop stubble burning is in full swing during the months of October, November, and December. In the peak season, affected families had to consult doctor or use some home medicine to get relief from irritation/itching in eyes, breathing problem, and similar other smoke-related problems. On an

[6]A wage rate fixed for the state of Punjab under National Rural Employment Guarantee Act (NREGA).

average, the affected members suffered at least half a month from such problems and had to spend INR 300–500 per household on medicine. In addition, there were few examples where a family member had to be hospitalized for three to four days and additional expenditure was incurred. On an average, households spent around more than INR 1000 on the nonchronic respiratory diseases like coughing, difficulty in breathing, irregular heartbeat, itching in eyes, decreased lung function, etc., during the year 2008–09. However, out of this total expenditure, around 40–50 % was spent during the months of October and November during the time of crop stubble burning. There was an additional cost in terms of household members remaining absent from work due to illness.

We find that total annual welfare loss in terms of health damages due to air pollution caused by the burning of rice straw in rural Punjab amounts to INR 76 millions. These estimates could be much higher if expenses on averting activities, productivity loss due to illness, monetary value of discomfort, and utility could also be considered. There is additional monetary cost of burning to the farmers in terms of additional fertilizer, pesticides, and irrigation. One also needs to add the losses of soil nutrient, vegetation, bio-diversity, and accidents caused because of low visibility.

Acknowledgements This paper is part of a larger study, Policy instruments to address air pollution issues in agriculture—Implications for Happy Seeder technology adoption in India. The project was funded by the Australian Council of International Agricultural Research (ACIAR).

References

Alberini, A., & Krupnick, A. (2000). Cost of illness and willingness to pay estimates of the benefits of improved air quality: Evidence from Taiwan. *Land Economics, 76*, 37–53.

Chesnut, L. G., Ostro, B. D., & Vichit-Vadakan, N. (1997). Transferability of air pollution control health benefits estimates from the United States to developing countries: Evidence from the Bangkok study. *American Journal of Agricultural Economics, 79*, 1630–1635.

Cropper, M., Simon, N. B., Alberini, A., Seema, A., & Sharma, P. K. (1997). The health benefits of air pollution control in Delhi. *American Journal of Agricultural Economics, 79*(5), 1625–1629.

Dockery, D. W., Pope, C. A., Xu, X., Spengler, J. D., Ware, J. H., Fay, M. E., et al. (1993). An association between air pollution and mortality in six US cities. *New England Journal of Medicine, 329*(4), 1753–1808.

Gadde, B., Bonnet, S., Menke, C., & Garivait, S. (2009). Air pollutant emissions from rice straw open field burning in India Thailand and the Philippines. *Environmental Pollution, 157*, 1554–1558.

Gerking, S., & Linda, R. S. (1986). An economic analysis of air pollution and health: The case of St Louis. *Review of Economics and Statistics, 68*, 115–121.

Gupta, U. (2008). Valuation of urban air pollution: A case study of Kanpur city in India. *Environmental and Resource Economics, 41*, 315–326.

Harrington, W., Krupnick, A. J., & Spofford, W. O. (1989). The economic losses of waterborne disease outbreak. *Journal of Urban Economics, 25*, 116–137.

Kumar, S., & Rao, D. N. (2001). Valuing benefits of air pollution abatement using health production function: A case study of Panipat thermal power station India. *Environmental and Resource Economics, 20*, 91–102.

Long, W., Tate, R. B., Neuman, M., Manfreda, J., Becker, A. B., & Anthonisen, N. R. (1998). Respiratory symptoms in a susceptible population due to burning of agricultural residue. *Chest, 113*, 351–357.

Murty, M. N., Gulati, S. C., & Banerjee, A. (2003). *Health benefits from urban air pollution abatement in the Indian subcontinent*. Discussion Paper no. 62/2003, Institute of Economic Growth, Delhi, IEG website.

Ostro, B., Sanchez, J., Aranda, C., & Eskeland, G. S (1995). *Air pollution and mortality: Results from Santiago, Chile, Policy Research Department*. Working Paper 1453, World Bank, Washington, DC.

Pope III, C. A., Thun, M. J., Namboodiri, M. M., Dockery, D. W., Evans, J. S., Spieizer, F. E., et al. (1995). Particulate air pollution as a predictor of mortality in a perspective study of US adults. *American Journal of Respiratory and Critical Care Medicine, 151*(3), 669–674.

Schwartz, (1993). Particulate air pollution and chronic respiratory diseases. *Environmental Research, 62*, 7–13.

Environment Security and Mahatma Gandhi National Rural Employment Guarantee Scheme

Santosh Gawai and Vinod Sen

Abstract Mahatma Gandhi National Rural Employment Guarantee scheme (MGNREG Act 2005), has support as innovative and reconstructive help to the environment at all over rural India. MGNREGA scheme aim is enhancing livelihood security at national level through the providing 100 days employment to rural household in every year. In India, nearly 70 % population is working in agriculture and its allied activities. Present research paper is trying to look into the various aspects of the MGNREGA and its impacts on environment security. Through gram sabha, works have undertaken which are largely improving water resources, land improvement, soil conservation with sustainable protection. As we know that water is very essential to the environment, the scheme managed the irrigation system and improvement and renovation of traditional water sources, land development and drought. This scheme commenced the works to generate the ground water recharge, soil, biodiversity conservation, sustaining food products, halting land degradation and building resilience to current climate risk such as moisture, stress, delayed rainfall, drought and floods. MGNREGA has been hailed as a scheme that provides 'sustainable livelihoods' or 'green jobs' to crores of workers engaged in restoring rural environment. This is because the works created under MGNREGA are claimed to have massive potential to improve environmental indicators, for instance, by raising water levels, sequestering carbon dioxide, improving soil quality, etc. The scheme will give excellent outcomes after the better supporting of financial budget for scheme. The present research paper is trying to show the effect of MGNREGA works on environment and sustainable development, and employment generation to households and need of financial grant.

Keywords Environment security · Sustainable development · MGNREGA · Employment irrigation · Wage

S. Gawai (✉) · V. Sen
Centre for Studies and Research in Economics and Planning, School of Social Sciences, Central University of Gujarat, 382029, Gandhinagar, Gujarat, India
e-mail: santosh.from.cug.007@gmail.com

V. Sen
e-mail: senvinod79@gmail.com

S. Nautiyal et al. (eds.), *Climate Change Challenge (3C) and Social-Economic-Ecological Interface-Building*, Environmental Science,
DOI 10.1007/978-3-319-31014-5_18

1 Introduction

MGNREGA is recognized as an ecological act that creates sustainable development through regeneration of the natural resources. The National Rural Employment Guarantee Programme (NREGP) in India has completed over nine years of implementation. In October 2009, the programme was renamed as Mahatma Gandhi National Rural Employment Guarantee Programme (MGNREGP), on 140 birth anniversary of Mahatma Gandhi. The Act for rural India was passed by the government in 2005 and implemented initially in 200 backward districts in 2006. The Act guarantees 100 days of work to rural households who undertake unskilled manual work, with 33 % of all workdays reserved for women workers. The objectives of the programme as stated in the guidelines are to enhance livelihood security while producing durable assets, empowering women, reducing distress migration and promoting social equity. The programme has certainly created additional work—in the financial year 2008–2009, 45.1 million households were provided employment under this programme and 2163 million person days of work was created. Women's participation in this programme has been generally high on the average and increasing, though there are state level variations. This large participation by women is important in itself, and different in magnitude from previous public works programmes in India. The MGNREGA was motivated by the fact that India's recent economic growth had not reached large sections of the rural population and that this urban-based growth was deepening divisions in the economy and society. The 90s were termed as a period of 'job-less growth' as high GDP growth rates failed to generate adequate employment opportunities. In fact, during the mid to late 90s (1993–1994 to 1999–2000) when annual GDP growth rates increased and stood at over 7 % per annum, employment growth rate declined and was only a little over 3 % per annum. While the share of the agricultural sector in total employment is reducing, albeit at a low pace, over half of the entire labour force is still engaged in agriculture. But agriculture contributes to less than a quarter of GDP, and is characterized by low productivity and low earnings leading to significant working poverty in rural areas. In 2004–2005, the numbers of poor workers in rural India was estimated to be over 74 million (NCUES 2007). A large share of women workers, especially, works in the agriculture sector. Distress migration from rural to urban areas has been a major phenomenon. But it has failed to create growth and enhance productivity in all sectors as expected in a Lewis kind of model and instead contributed to large informal sectors in urban areas and urban working poverty. Inequality as measured by consumption/expenditure distribution, has risen during 1993–1994 to 2004–2005, and this has widened disparities between the urban and rural areas as well as within urban and rural. It is therefore important to understand and analyze if and to what extent participation in MGNREGP benefits women workers and reduces labour market inequalities between women and men (Dasgupta & Sudersan, 2011). Financial inclusion, with a focus on rural areas, has been one of the Reserve Bank of India's (RBI's) main priorities since 2005. The RBI's plan includes provision of banking services to habitations with a population of 2000 and above, by March 2012 (Strategy and Guidelines on Financial Inclusion, MoF, 21st October 2011) as per the 2001 census. In the wake of the RBI's efforts to ensure financial inclusion, the Ministry of Rural Development (MoRD) issued a circular on January 21, 2008 that mandated transfer of wage payments only to bank or post office accounts of Mahatma Gandhi National Rural Employment Guarantee Act 1 (MGNREGA) beneficiaries keeping in mind the objectives of transparency and financial inclusion. Schedule II, paragraph 31 of the Act was suitably amended on February 19, 2009 to state that 'the payment of wages shall be made through the individual or joint savings account of the workers in banks or post offices opened in accordance with the directions of the Central Government'. Wage payment through banks connects millions of people to banking, an inclusion to the financial system. This led to the omission of the provision of part payment in kind and payment on daily basis though not ruling out cash payments entirely. The introduction of bank payments could benefit women in several ways. The main idea behind these bank payments includes the perception that banks will increase the possibility of saving and a reduction in the possibility of malpractices by those who distribute wages in the villages. As per the Millennium development goals, (MDGs) Financial inclusion (FI) offers incremental and complementary solutions to tackle poverty, to promote inclusive development and to address the MDGs. Since then, nearly ten crores bank/post office accounts have been opened and around 80 % of MGNREGS payments have been made through this route. This has been hailed as the 'world's largest ever financial inclusion scheme'.

2 Review of Literature

Govt. of India (2012), report has shown the scheme active role to increase the environmental safety with less degradation of natural resources. Scheme has also given better employment to unemployed rural persons. Azam (2012), in this article shows the stronger evidence of increase in the public work, which is lead to reduce the environmental degradation with increasing females participation in the scheme works. Brief review on MGNREGA (2013), report has shown that most work undertaken under this scheme are environment friendly and try to protect the environment and better utilization of rural resources. Thadathil and Mohandas (2012), primary survey-based study has shown that the employment security and workers have advantage of employment at the time of their unemployment in the lean period. Scheme has the provision to word on demand especially when agriculture work is not taking place in summer time that will be helpful to agricultural labourers to get paid employment. Basu et al. (2009), Murugwel (2009), study focus on the implementation of the programme and how it has improved the agriculture productivity. Khera and Nayak (2009), work has deal with land development, land productivity, irrigation facilities and rural livelihood issues. This study has done with help of primary data which are collected from the two districts. On the basis of above review of the literature; it has been found that thousands of studies have been done on MGNREGA not only in India but also in abroad. Most of the studies are focusing on the evolution of the programme, women empowerment, women participation, and various aspect of right to work and other issues of MGNREGA. Only few studies have been done on MGNREGA and environment security issue. This research work provided an opportunity to explore the important issue of MGNREGA which is not touched by the previous studies.

3 Objectives of Study

1. To examine the active role of MGNREGA into the environment security in rural India.
2. To study the major impacts of the MGNREGA on the environment.
3. To find out the financial support to the scheme and its impact on beneficiary households.

4 Research Methodology

This study is based on secondary data which have been collected from various reliable sources such as reports on MGNREGA published by Ministry of Labour, central government documents, State government and report of National Sample Survey Organization (NSSO).

5 Agricultural Work and Provision Under MGNREGA Scheme

MGNREGA scheme has played an important role in the environment because scheme has initiated implementing various works under scheme. The scheme has mostly been helpful for agriculture all over India. The provisions under MGNREGA are given.

5.1 Agriculture Interrelated Work

Provision under scheme for rural areas i.e. '*Nalla*' repairing, is improving the Earthen structure; it also provides them the cement and plug for improving the farm facility and irrigation facility for handling Loose Bolder Structure. To manage the structure of graded bunding; as it is important to follow them. Within the certain selected area delivers compartment bunding for safety purpose. Under this section, provision of scheme is not directly affecting the agriculture but indirectly has impacts on farmers of rural areas like cleaning Vanrai Bandhara and also safety works done for vegetative bund in the scheme action in grass root level in Indian economy. In this way, scheme has given works to labour for the purpose of protection with management within 'Gramin Structure' and focused on another part i.e. Continuous Contour Trench only to understand the needs of rural peoples. MGNREGA has managed to bind rural development through the scheme. For the purpose of providing agriculture lands through government scheme under MGNREGA; it takes action in farm pond of saline land then this land will be capable for agriculture production. For taking care of public health, scheme is given facility striating of Nalla in the society and also contributes scheme in quick bund for livestock drinking water in villages (Biradar 2009).

5.2 Earthen Dam/Band/Irrigation Channel

Direct action of MGNREGA scheme in villages improves irrigation for farming through the loose soil dam as well as develops loose soil canals for the same purpose. Agriculture mostly depends on monsoon if the monsoon is not comes in the right time then storage tank helps for water uses in villages, also MGNREGA protects natural percolation tank which will be helpful for agriculture at the time of dry seasons. Through the scheme, the government has established a percolation channel to increase the agricultural production of villages, and also secure the village ponds in scheme with protection for underground bund. The scheme has spent lot of funds to renovate the canals which will support the village farms (Biradar 2009).

5.3 Public Forestry

MGNREGA is not only limited to agriculture but also has some advantages for public plantation with fodder; proper planning of non-fertile lands provides advantages to rural people and through additional planning for nurseries and roadside plantations it is also taking care of the environment (NREGA 2005).

5.4 Forestry

Other than the MGNREGA scheme, the government has been implementing various schemes for forest pond; it has resulted only in limited districts of India but the role has been played by MGNREGA all over India. Through MGNREGA, gram sabha managed the underground water to protect plantations as well as nursery at the village level. For minor works, common people can take benefits through MGNREGA scheme; for example, circular trenches around bamboo.

5.5 *Public Work*

One of main features of the MGNREGA scheme is to increase public work through provisions under scheme which will fulfil the implementations of MGNREGA, i.e. construction of village roads which will connect villages to district roads; it will be helpful for the development of rural infrastructure which will cover the internal roads in the village and join the national and state highways.

The scheme has primarily focused on water conservation, because we know India faces the problem of drinking water as well as the problem of irrigation. The government basically tries to reduce the problem of water as well as to protect the environment. In this regard, the government's spending of funds on water conservation has increased from 4.51 % in 2006–2007 to 37.71 % in 2012–2013. The MGNREGA scheme is wildly using its resources to water recharge and improve the irrigation facilities for agriculture due to which spending on these items are increasing every year (Table 1).

The benefit of irrigation facility has reached the backward sections of Indian society. The scheme is providing the connectivity to rural areas all over India. The expenditure on rural connectivity was highest 17.87 % in the year of 2011–2012 through this scheme. Financial provisions under the scheme are indirectly protecting the environment of the economy which leads to sustainable development. MGNREGA activity varies from one area to another due to the differences in resource endowments, which determine their potential to deliver environmental benefits and reduce vulnerability.

The MGNREGA work is basically planned to improve water sources, increase agricultural production, employment generation, increasing land quality, basic facility, reduction of environment problems, rural safety, reduction in pollution, etc. All the arrangements of the MGNREGA scheme directly or indirectly protect the rural environment in India, while implementation of the programme has been insured that the natural source security must be there through sustainable development.

In Table 2, data show the budget allocation to the MGNREGA programme and its share to wage payment to the beneficiaries of the schemes. The important factor is the budget of government for MGNREGA scheme all over India. The highest allocation of the fund to the MGREGA was in the year of 2010–2011 which was Rs. 40,100 crores, and provided employment to 5,561,812 lakh persons but the share for wages payment was least which was only 86 % of the total allocated fund of the years. Table 2 is also showing the wage share was the highest (75 %) in the year of 2012–2013.

Table 1 Financial provision under MGNREGA to various environmental activities

Year	2006–2007	2007–2008	2008–2009	2009–2010	2010–2011	2011–2012	2012–2013
Water conservation	4.51	8.73	12.79	23.43	24.26	44.09	37.71
Provision of irrigation facility to land owned by SC/ST/BPL and IAY beneficiaries	0.81	2.63	5.67	7.73	9.15	9.52	7.50
Rural connectivity	1.80	3.08	5.03	7.64	9.31	17.87	11.50
Land development	0.89	2.88	3.98	6.38	7.04	7.69	5.30
Any other activity	0.34	0.56	0.28	0.98	1.06	3.06	1.79

Source: (NSSO 2013)

Table 2 MGNREGA budget and employment generation

Years	Budget for MGNREGA in crores	100 days employments to HH (lakhs)	Exp. on wages in all India (% of wage exp. In total budget)
2006–07	11,300	1,261,286	5842.37 (66 %)
2007–08	12,000	3,602,405	10,738.47 (68 %)
2008–09	30,000	6,521,268	18,200.03 (67 %)
2009–10	39,100	7,083,663	25,579.32 (70 %)
2010–11	40,100	5,561,812	25,686.53 (68 %)
2011–12	40,000	3,896,589	24,860.91 (69 %)
2012–13	33,000	4,826,054	21,127.10 (75 %)

Source MGNREGA report, 2013

In the favour of more work days, the employment scheme needed lot of funds to provide green jobs. These work days in MGNREGA create huge potential for upgrading and creating sustainable rural infrastructure and eco-restoration through a wide range of activities related to water harvesting, rural connectivity, irrigation, flood control and protection works, rural sanitation apart from other works and aid workers in moving from wage employment to sustainable employment in the study area.

6 Conclusion

If the policy programme is designed and implemented on the basis of sustainable development, it creates inclusivity and environmentally sound way to reduce poverty as well as build shared prosperity among the present generation, in a further attempt to meet the needs of the future generations. The works undertaken through MGNREGA give priority to activities related to water harvesting, groundwater recharge, drought-proofing and flood protection. Its focus on eco-restoration and sustainable livelihoods will lead over time, to an increase in land productivity. This implies that MGNREGA is creating a huge potential for upgrading and creating sustainable rural infrastructure and eco-restoration through a wide range of works in the direction of environmental protection. But, the impact of MGNREGA has shown less reduction in environmental vulnerability. All assessed land development, water related, irrigation provisioning, rural sanitation, rural connectivity and direct and indirect increase in the development of village community. Infrastructure is one of the major causes for the development in rural areas. Construction of basic infrastructure facilities under MGNREGS, such as road connectivity can lead to the development of rural areas. This increases peoples' access to external environment, linkage to market access and the workers in moving from wage employment to sustainable employment.

By providing 100 days of work to a rural household, the Act provides an income supplement for poor households, especially during the slack agriculture season when demand for labour in agriculture is low. By increasing disposable income of poor households, the MGNREGA creates the conditions for a consumption-based growth path. It also promotes participation of the local people and the poor in choice of assets to be created. The programme can act as an automatic macro-economic stabilizer.

The lack of appropriate knowledge among practitioners is a major hurdle preventing MGNREGA from achieving its potential. Ponds, check dams, tanks, soak pits and other similar water conservation and harvesting structures are works, which are meant to lead to the green transformation of rural India. However, the study found that the engineering cadres upon whom all responsibility for technical soundness of MGNREGA works rests, are too few in number to be able to ensure that works being constructed will serve their purpose. Panchayat level functionaries like the mates, rozgar sevaks and secretaries, have little knowledge about watershed development, but it is they who take crucial decisions such as the location, direction, dimensions and design of works. As a result, the rate of success of projects related to water conservation and harvesting is poor. In this context, the role of cluster facilitation teams (CFTs) and barefoot engineers, who understand basic hydrogeology would be extremely useful. Integrating these practices into planning and designing MGNREGA works would go a long way in making it a truly 'green scheme'.

References

Azam, M. (2012). The Impact of Indian Job Guarantee Scheme on Labour Market Outcomes: Evidence from a Natural Experiment, Discussion paper No. 6548, word bank and IZA

Basu, A., Chau, N., & Kanbur, R. (2009). A theory of Employment Guarantee: Contestability, credibility and distributional concerns. *Journal of Public Economics*, *93*, 482–497.

Khera, R., & Nayak, N. (2009). Women workers and perception of the National Rural Employment Guarantee Act. *Economic and Political Weekly*, *44*(43) October 24.

Muthamizh Vendan Murugavel, D. (2009). Driver of Rural Economy. *Kurukshetra*, *58*(2), 15–18.

NREGA (2005). National Rural Employment Guarantee Act. Delhi: Govt. of India.

Biradar, R. S., & A. S. (2009). A Study report on Appraisal of NREGA Programme in Thane And Akola Districts of Maharashtra. Central Institute of Fisheries Education, Deemed University, (ICAR), 1–101.

Thadathil, M. S., & Mohandas, V. (2012). Impact of MGNREGS on labour supply to agricultural sector of Wayanad district in Kerala. *Agricultural Economics Research Review*, *25*(1), January-June 2012, 151–155.

Part III
Ecosystems and Production System Analyses

Analysis of Policies in Sustaining Sandalwood Resources in India

M. Srinivasa Rao, G. Ravikumar, P.R. Triveni, V. Soundara Rajan and Sunil Nautiyal

Abstract Sandalwood (*Santalum album* Linn.) is an important tree species in peninsular India because of its high economic value and is the best endemic tree in the world. Past regulations and policies have been identified as the main cause for the decrease in Sandalwood population, particularly in southern parts of India. Stringent regulations and policies which excluded farming communities from growing sandalwood resulted in scarcity of Sandalwood which in turn increased demand-supply gap. Human intervention, however, has decreased sandalwood adaptation capability for its sustainability. Due to high value of sandalwood and its oil and rising demand in domestic and international markets, prices have sky-rocketed. Smuggling of Sandalwood has created socio-economic and law and order problems in all Sandalwood producing states. Karnataka Forest (Amendment) Act, 2001 allows cultivation of sandalwood trees on private lands. This comes as a major policy change as far as sandalwood cultivation is concerned. Liberalization of policies of different States in southern parts of India encourages commercial plantation, but much remains in encouraging the corporate sector to embark on plantation of this economically important pride species. The severe shortage of sandalwood, hitting user industries like perfume, soaps and medicine, has encouraged the policy makers to make pragmatic changes in the policies and rules and make it more economical and to sustain this valued resource in India. The governments of Karnataka and Tamil Nadu amended the sandalwood laws in 2001 and 2002, respectively, and made the grower an owner of the wood. This amendment encouraged the farmers to take up cultivation on a commercial scale. This paper focuses on the policies in sustaining Sandalwood resources in India.

M. Srinivasa Rao (✉) · G. Ravikumar · P.R. Triveni · V. Soundara Rajan
Institute of Wood Science and Technology, Malleswaram, Bangalore 560003, India
e-mail: srinivasarm92@gmail.com

S. Nautiyal
Centre for Ecological Economics and Natural Resources (CEENR), Institute for Social and Economic Change (ISEC), Bangalore 560072, India

S. Nautiyal et al. (eds.), *Climate Change Challenge (3C) and Social-Economic-Ecological Interface-Building*, Environmental Science,
DOI 10.1007/978-3-319-31014-5_19

1 Introduction

Santalum album, the Indian Sandalwood, of the genus *Santalum*, is a semi-parasitic woody plant and produces highly aromatic oil, price of which is skyrocketing in the world market of perfumery. Traditionally, sandalwood oil has unique stability, and provides perfume with its best fixative property. The heartwood, sandalwood's fragrant timber, has long been used traditionally in important religious ceremonies not only in India but in many parts of the world. Smaller amounts of sandalwood are also used for decorative carving, craft and smaller furniture pieces. Sandalwood oil has antipyretic, antiseptic, antiscabietic, and diuretic properties. It is also effective in the treatment of bronchitis, cystitis, dysuria and diseases of the urinary tract. The oil has an important place in the indigenous system of medicine. It is considered a cure against migraine. Markets worldwide are currently facing a serious shortage of this increasingly expensive, high-quality Indian sandalwood (*S. album*)—largely as a result of unsustainable harvesting of the species over a long period of time in its native environments of India. Indian sandalwood tree is becoming endangered, and in an attempt to curb its possible extinction increased interest by both forest departments and private growers towards expansion of plantations has added to the resource building of this valuable tree (Venkatesha Gowda 2011).

2 Conservation Status of Sandalwood Species

There are more than 56 species and varieties of 'Santalum' mentioned in literature, based on morphological characters. However, currently only sixteen prominent species of Santalum family and their geographical occurrences have been identified in the world. They can be broadly grouped as East Indian sandalwood, Australian sandalwood, Hawaiian sandalwood and sandalwood of the Pacific Islands. Their geographical occurrences in the world are as follows:

Sl. No.	Species and varieties	Geographical occurrence
1.	*S. album* L	India, Indonesia, Srilanka
2.	*S. austrocaledonicum* Viell	New Caledonia, Vanuatu
3	*S. boninense* (Nakai) Tuyama	Bonin Islands
4	*S. lanceolatum* R.Br.	Australia
5	*S. macgregorii* F. Muell	Papua New Guinea
6	*S. Obtusifolium* R. Br.	Australia
7	*S. yasi.* Seem	Fiji, Tonga
8	*S. freycinetianum* Gaudich	Hawaii Islands (O'ahu.Moloka'l)
9	*S. haleakale.* Hillebr	Hawaiian Islands (Maul)
10	*S. ellipticum* Gaudich	Hawaiian Island

(continued)

(continued)

Sl. No.	Species and varieties	Geographical occurrence
11	*S. paniculatum* Hook & AM	Hawaii Islands
12	*S. fernaandezianum* F. Phil.	Jaun Fernandez Island
13	*S. insulare* Bertero	Society Islands (Tahiti)
14	*S. acuminatum* (R.Br.) A.D.C.	Australia
15	*S. murrayanum* (T. Mitch) C.A. Gardner	Australia
16	*S. spicatum* (R.Br.) A.D.C.	Australia

Many of these are under the threat of extinction due to over exploitation. East Indian sandalwood, *S. album* L., is the queen among the species yielding supreme quality essential oil much required by domestic and international perfume and allied industries (Ananthapadmanabha 2011).

Sandalwood plants are widely distributed in Southern parts of peninsular India mainly in the States like, Karnataka, Tamil Nadu, Andhra Pradesh and Kerala. The value of sandalwood was realized as early as 1792 when Tipu Sultan the then ruler of Mysore declared it as a "Royal Tree". Due to this historical reason sandalwood has ever since remained under government control. It is defined as a "forest produce", although it does not have any special provisions under "Indian Forest Act".

Sandalwood resource in India especially wild population is currently threatened mainly because of illicit felling, forest fire and grazing and to a certain extent spike disease coupled with heavy domestic and international demand with inadequate uniform regulation in the Southern states especially in Tamil Nadu, Karnataka and Kerala. Smuggling of Sandalwood has created socio-economic and law and order problems in all Sandalwood producing states.

India realised the value of sandalwood trade and took positive steps to protect the natural population of sandalwood. Measures to sustain this valued bio resources, took prominence and flaws in policy which endangered the species were amended. Government of Karnataka promulgated an amendment to Karnataka Forest Act in 2001 to encourage private domestication of sandalwood as a means to conserve and enhance the status of this resource. The amendment gave landowners full legal right to trees on their land, made them eligible to receive full value on extraction and made amendments on the sale of sandalwood through forest department and govt. departments. Tamil Nadu followed path with the Tamil Nadu Forest (Amendment) Act in 2002, the landowners were given the right to trees (Dhanya et al. 2010). Subsequently, Kerala Forest (Amendment) Act in 2010 regulated cutting and possession of sandalwood.

A need was felt to take sandalwood beyond the burden of forests and farmers through people's participation. Karnataka Soaps and Detergents Limited, Bangalore, which is a State owned Company in India with nearly a 100 year legacy in the production of natural sandalwood oil and sandalwood oil-based products is working towards promoting sustainable development of sandalwood for meeting the needs of society, present and future. Accordingly, the company integrated the practice of Sandalwood tree-based farming by joint cultivation on a Share and Prosper basis for reaping the rich benefits, both individually and for the community, by making efforts in converting unproductive waste lands into productive lands. In this endeavour the company is trying to interconnect the needs of the nature.

In India, the National Medicinal Plant Board & National Horticultural Mission are assisting sandalwood cultivation and inter-planting with other medicinal plants through their State Agencies. There has been a healthy response from private land owners for commercial cultivation and harvesting of commercially planted sandalwood particularly in Karnataka. Furthermore, many farmers and entrepreneurs are setting up plantations of Indian Sandalwood in the states of Gujarat, Andhra Pradesh, Madhya Pradesh, Maharashtra, Rajasthan and Assam on a large commercial scale of some 5000 ha.

The legal hassles created by sandalwood trade regulation results from the fact that there is no comprehensive regulation applicable to the entire country. Hence, clandestine trade is flourishing and sandalwood population declining to the extent that it is on the verge of extinction. Sandalwood is included in the Red List of International Union for Conservation of Nature (IUCN) as vulnerable and sandalwood growing states of India have stated that it faces extinction.

At this point, The Supreme Court of India, in the case of T.N. Godavarman Thirumalpad Versus Union of India in its writ petition (C) No. 202 of 1995 gave a milestone judgement. It directed the Central Government to formulate a policy for conservation of sandalwood including provision for financial reserves for conservation and scientific research for sustainable use of its biological diversity. Central Government would also formulate rules and regulations under Sects. 3 and 5 of Environmental Protection Act for effective monitoring, control and regulation of sandalwood industries and factories to ensure that no imported sandalwood is sold under the banner of Indian Sandalwood. Even products manufactured from or of import of sandalwood would be labelled adequately. The State governments were directed to shut down all unlicensed sandalwood oil factories. Accordingly, legislation similar to the Endangered Species Act, enacted in the United States is proposed. The Endangered Species Act, enacted in the United States protects both endangered species defined as those “in danger of extinction throughout all or a significant portion of their range” and “threatened species”, those likely to become endangered “within a foreseeable time”. The term species includes species and sub-species of fish, wildlife and plants even though the species as a whole may not be endangered.

India thus remains competitive in its attempt to grow more sandalwood both in situ and ex situ by inter plantation methods to bring back its lost glory. It is hoped that this initiative will meet both domestic and international market demands.

Sandalwood is recognized worldwide as one of the most valuable economically important commercial tree species. Due to over exploitation from the wild, human interferences and lack of sufficient plantation establishment, sandalwood resource declined worldwide.

The links between biodiversity conservation and sustainable development are now universally acknowledged and recognized in the Millennium Development Goals and the conclusions of the World Summit on Sustainable Development. Sustainable trade in natural resources contributes to local and national economies creating incentives to conserve species and their natural habitats. A sustainable trade in wild species from the natural forests can contribute significantly to rural incomes, and the effect upon local economies in developing countries can be substantial.

Realizing the global demand for sandalwood oil Government of India, Australia, Hawaii, New Caledonia and Sri Lanka have taken action against illegal harvesting and formulated rigid policies for conservation of sandalwood resource (Xiaojin et al. 2011). Strengthening Government initiatives, different organizations such as International Sandalwood Foundation of United States, Iliahi Foundation of Sandalwood of Hawaii promote research and take actions in conservation of sandalwood species in the world.

3 Felling and Transit Regulations for Tree Species Grown on Non-Forests/Private Lands in India

India has made impressive economic growth in recent times reflected in rise in income of people, which is leading to increase in consumption of wood and wood products like furniture, construction timber, paper and pulp, etc. This growing demand has resulted in large gap between demand and supply of forest products, which is met partly through agro-forestry and remaining by import of timber and allied products. Globally, there is increasing focus on sustainable forest management and some countries like USA, EU are developing legally binding processes for trade of certified timber from sustainably managed forests, which will have significant impact on international trade of forest products. This uniform policy and regulation will always help in conserving our natural heritage in the country by preserving the reminiscent natural forests with the vast variety of remarkable biological diversity and genetic resources of the country (Jain et al. 1998). An extract of felling and transit regulation for various tree species grown on forests/non-forest lands in different states are briefed in Annexure 1.

4 Production and Utilization of Sandalwood

Ecologically sandalwood has adapted to various agro-climatic and soil conditions for in situ regeneration. Heartwood yield again depends upon locality and age of tree. Karnataka and Tamilnadu in southern India account for nearly 80–90 % of sandalwood production in India. There has been a significant rather drastic reduction in production of sandalwood during the past years. Production of heartwood during 1930s through 1950s was around 4000 tonnes a year, which decreased to a meagre 500 tonnes of wood a year or even less. It is estimated that the Global demand for Indian sandalwood is around 5000–6000 tonnes/annum and for oil it is around 100–120 tonnes/annum (Dhanya et al. 2010).

S. album is also found in Srilanka and South East Asia (Timor, Indonesia, Malaysia, Cambodia, Vietnam, Burma, Thailnad and China), the Pacific (Papua New Guinea, Fiji, Vanatu, New Celodonia and Hawaii) and to a certain extent in areas northwest of Western Australia (Kanunurra) Expansion of Indian sandalwood plantation resource has been occurring in northern Australia over the past few years and is expected to make a significant market contribution to the otherwise declining resource in its natural distribution (Stemmermann 1990). In fact, 90 % of the world's *S. album* output is produced in India. Indonesia dominates the balance of world production and export.

5 Trade and Quality Issues of Sandalwood Oil

Sandalwood marks the beginning of trading in India. India was over the past decades producing large quantities of sandalwood oil, meeting over 80 % of world demand, but due to over exploitation and various other reasons the quantity of production has come down drastically. Different sandalwood species are indigenous to India, Indonesia, Sri Lanka, Bangladesh (*S. album*), and Australia (*Santalum spicatum* and *Santalum lanceolatum*), as well as to several Pacific Islands such as Hawaii (*Santalum ellipticum*), Fiji and Tonga (*Santalum yasi*), Papua New Guinea (*Santalum macgregorii*), Vanuatu and New Caledonia (*Santalum austrocaledonicum*) and French Polynesia (*Santalum insulare*) (Butaud 2004). The essential oil produced by sandalwood is obtained through steam distillation of chipped heartwood. The oil is a pale yellow to yellow viscous liquid that is insoluble in water. Sandalwood oil is used in perfumery, cosmetics and aromatherapy, pharmaceutical and food industries. Sandalwood is also used for woodcarving and as an ingredient for incense. Although Sandalwood stocks in India continue to decline, there is a possibility that recent policy changes in regulations especially in southern parts of India will provide incentives for large-scale development of private plantation and may boost production levels of *S. album* from India in the mid to long-term future (Ananthapadmanabha 2012).

Further, several countries namely China, Australia, Thailand, Costarica, Cambodia and Srilanka are also venturing into *S. album* plantation because of its fragrance characteristics and economic feasibility coupled with huge demand for natural products worldwide (Subasinghe et al. 2013).

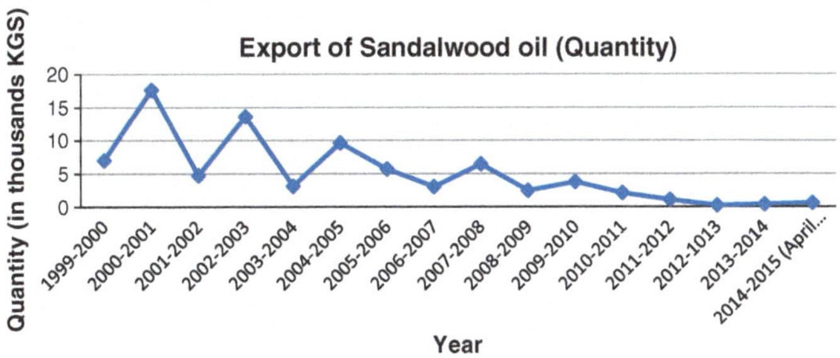

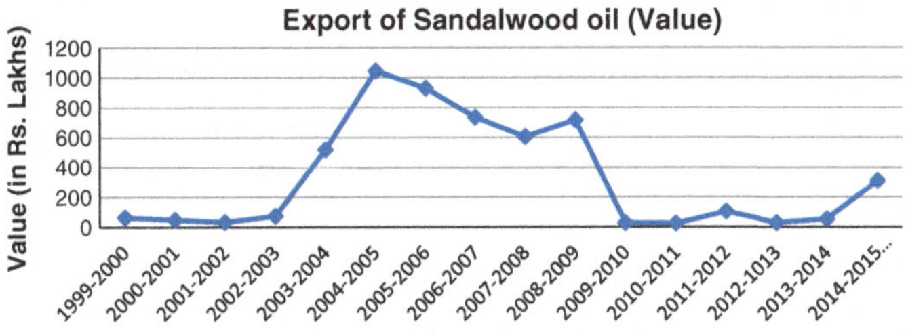

The price of Indian sandalwood and its oil has risen significantly in the past decade. 90 % of World's *S. album* production is in India. The gap between sandalwood oil supply and demand has widened over time (Soundararajan et al. 2015).

6 Variations of Sandalwood Oil Contents

The values of the oils are ranked in the world market of sandalwood. *S. album* tops the list as rest of the species produce low or no value in the oils or the other oil producing sandalwood species such as *Santalum haleakalae* have *become* very rare.

The most important concern among the sandalwood growers is about quantity and quality of oil that can be obtained from a harvested tree when compared to the oil obtained from wild sources of Indian origin. The valuable fragrant oil of the sandalwood tree is concentrated in its heartwood (Jain et al. 2003). The market value of a given volume of heartwood will depend primarily on the concentration and quality of its oil that is, the quality of sandalwood in the marketplace is typically determined by the relative proportions of α- and β-santalol as recognized in the international standard for Indian sandalwood (*S. album*) oil (ISO 3518:2002).

Heartwood oil concentration and quality in sandalwood vary between species and

Species	Origin	Oil content (%)	Santalol content
S. album	India, Indonesia	6–7	α-santalol: 41–55 % β-santalol: 16–24 %
S. yasi	Fiji	5	α-santalol: 37–39 % β-santalol: 26–28 %
S. austrocaledonicum	Vanuatu, New Caledonia	3–5	α-santalol: 48–49 % β-santalol: 20–22 %
S. spicatum	Western Australia	2	α-santalol: 15–25 % β-santalol: 5–20 %

Source Australian Agribusiness Group (2006)

can also be influenced by genetic, environmental and agro-climatic factors (Doran et al. 2005).

7 Comparison of Sandalwood Oil Content

S. album or Indian Sandalwood has the highest oil content of Sandalwood species (6–7 % oil content) compared to other commercially exploited species elsewhere in the World including *S. spicatum* (2 %) and *S. lanceolatum* (1 %). This explains the reason Indian Sandalwood is the most highly demanded of the Sandalwood species.

Several non-*Santalum* species (e.g. *Osyris tenuifolia, O. laceolata* from East Africa and *Amyris balsamifera* L. from West Indies) are also used as sources of "Sandalwood" type. The less fragrant wood and the distilled oil are different when compared to true sandalwood (Brand et al. 2004).

ISO (International Organization for Standardization) is a worldwide federation of national standards bodies (ISO member bodies). This International Standard specifies certain characteristics of the oil of sandalwood (*S. album* L.), in order to facilitate assessment of its quality. The work of preparing International Standards is normally carried out through ISO technical committees. Each member body interested in a subject for whom a technical committee has been established has the right to be represented on that committee. International organizations, governmental and non-governmental bodies, in liaison with ISO, also take part in the work. The main task of technical committees is to prepare International Standards. Attention is drawn to the possibility that some of the elements of this International Standard may be the subject of patent rights. ISO shall not be held responsible for identifying any or all such patent rights. On these lines, ISO 3518 was prepared by Technical Committee ISO/TC 54 for Essential *oils*. This second edition cancels and replaces the first edition (ISO 3518:1979) and also ISO 7610:1985, which have been technically revised. Similarly, to safeguard the quality of oil, The European Commission database with information on cosmetic substances and ingredients, lists several types of sandalwood and their derivates under their INCI names.

INCI name: *S. album* wood oil—masking and perfuming properties
INCI name: *S. austrocaledonicum* wood oil—masking, perfuming and flavouring properties.

ISO 3518:2002 [oil of sandalwood (*S. album* L.)] and ISO 22769:2009 [Australian Sandalwood *Santalum spicatum* (R.Br.) A.DC.], are the latest edition of International Standards which specifies certain characteristics of the oil of sandalwood with a view to facilitate the assessment of its quality. Quality standards prescribed by the ISO for *S. album* and *S. spicatum* are as mentioned below.

8 Quality Standards

(a) **East Indian Sandalwood oil (ISO 3518:2002; NF T 75-339)**

Aspect: limpid liquid, somewhat viscous
Colour: almost clear to yellow
Odour: heavy, sweet and very lasting odour
Physical characteristics:
Density: 0.968–0.983 @ 20 °C
Refractive index @ 20 °C: 1.503–1.508
Optical rotation @ 20 °C: −15° to −21°
Solubility in 70 % ethanol (v/v) @ 20 °C: less than 5 volumes of 70 % ethanol for 1 volume of oil
Chemical characteristics:
Ester number: max 10
Total alcohol content, calculated as santalol: 90 % minimum

(b) **Australian Sandalwood oil (NF T 75-248)**

Aspect: limpid liquid, somewhat viscous
Colour: colourless to light yellow
Odour: strong lasting characteristic odour of the wood
Physical characteristics:
Density: 0.968–0.978 @ 20 °C
Refractive index @ 20 °C: 1.504–1.510
Optical rotation @ 20 °C: −8° to −3°

Solubility in 70 % ethanol (v/v) @ 20 °C: less than 5 volumes of 70 % ethanol for 1 volume of oil
Chemical characteristics:
Acid number: maximum 5
Ester number: minimum 4.5; maximum 10
Ester number after acetylation: minimum 199

Total alcohol content, calculated as santalol not stated, but should be 90 % minimum.

9 Quality Specifications of Sandalwood Oil

The quality of sandalwood oil depends on the level of santalol content in the oil. Indian sandalwood is of the highest quality in the World market. Several Pacific sandalwoods are similar in quality, especially sandalwoods from Fiji (*S. yasi*) and New Caledonia and Vanuatu (*S. austrocaledonicum*).

In general, older sandalwood trees have more heartwood and produce higher quality of oil since santalol contents are higher. High-quality sandalwood oil comes from trees that are at least 30 years old. Sandalwood trees normally start producing heartwood at around 10–15 years. Growth of heartwood depends mainly on genetic factors and also on soil type, rainfall, exposure to sun light and different agro-climatic conditions. Trees that grow in shallow soil or in soil with a high level of stone inclusions, with distinct annual dry period and exposure to full sun develop heartwood more rapidly and could be harvested after at least 15–20 years. For trees that grow in deep, fertile soil, with high and evenly distributed rainfall throughout the year and a shaded canopy, harvesting should be delayed until the trees are at least 30–40 years old (Nagaveni and Vijayalakshmi 2003). Given its high price and low availability, there is a high risk of adulteration. Buyers are very wary about buying sandalwood oil. European buyers expect a good, reliable level of quality; most of them expect suppliers to follow the HACCP principles for food processing. The supply and demand of sandalwood, and the price of substitute products in the market, will be important factors in influencing market conditions for sandalwood products in the future.

The following requirements are also often referred to as mandatory in sustained resource management of sandalwood.

- Good Manufacturing Practices (GMP) is not obligatory for cosmetic ingredient producers, but compliance can provide a competitive advantage.
- Good Agricultural and Collection Practices (GACP), even though legally binding for medicinal plants, is a common practice for farming and wild collection of cosmetic ingredients, prior to processing. But unfortunately, GACP is not formulated for *S. album*, the prime species of India.

- Quality standards such as ISO are referred to in buyers' product specifications, to assure the quality of cosmetic ingredients. ISO22716 is a good reference. Even though it is intended for finished cosmetic products, it also includes guidelines for ingredients that manufacturers need to refer to.
- International Fragrance Association (IFRA) Standards form the basis of the globally accepted and recognized risk management system for the safe use of fragrance ingredients.

10 Results and Discussion

10.1 Policy, Regulation Incentives and Support Mechanism

- The policy/legislation of Government of Karnataka to abolish control over sandalwood cultivation has paved the way for community and private entrepreneurs to cultivate sandalwood which is in high demand. Karnataka Forest (Amendment) Bill, 2001 allows cultivation of sandalwood trees in private lands. This comes as a major policy change as far as sandalwood cultivation is concerned.
- A uniform policy for the entire Country on conservation strategies and transit policies of Sandalwood may help in improving overall status of the precious wealth of India (Nageswara Rao 2012).

10.2 Production

- Raising large-scale plantations in natural Sandalwood bearing areas will add to the resource building of this valuable tree species.
- Australian sandalwood industry is all set to dominate World supply of sandal oil. Highest priorities for *S. album* (East Indian Sandal wood) research has been evincing globally in the field of conservation, understanding plant physiology and genetics to maximize oil production and minimize maturation period, processing, markets mainly focusing to the industrial needs (Balachandran and Kichenamourthy 2007).

10.3 Process

- Certification system and product standardization are critical elements of supplier–buyer chains, particularly in high-value European sandalwood markets.

- Development of certification programs and product standards should therefore be a priority for Indian sandalwood industry and could be championed and coordinated by prime research institutions like Institute of Wood Science and Technology, ICFRE. MoEF and the Govt. of India.

10.4 Marketing

- The high price of Indian sandalwood oil in both domestic and international market has always been a reason behind frequent adulteration of this valuable material with other cheaper essential oils and resinous substances.
- European buyers expect a good, reliable level of quality; most of them expect suppliers to follow the HACCP principles for food processing and also they prefer suppliers who can demonstrate good standards in sustainability. This involves social and environmental responsibility as well as sustainable sourcing practices. Buyers may not require compliance with certification standards to prove sustainability. However, they expect the suppliers who demonstrate that the supplier incorporate sustainability component into their operations. Hence, good performance assessment in terms of sustainability with regard to sourcing, social and environmental responsibility should be incorporated and quality should be maintained and demonstrated to the buyers in the international market for its higher value.
- More research is needed in developing methods to identify genuine sandalwood oils from the counterfeit products, as the present market is flooded with such low quality artificial oils deliberately manufactured to capture the sandalwood market.
- In addition, more research should be conducted on the oil quality of hybrid sandalwood varieties although some hybrids such as *S. album* × *S. yasi* show promising improvements in vigour and oil contents.

Annexure 1

Acts/Rules for Felling and Transit Regulation of Sandalwood/Various Tree Species Grown on Forests/Non-forest Lands of Different States (Arun Kumar Bansal 2012; Subhash Chandra 2014)

Sl. No	State	Rules and other
1	Karnataka	• Karnataka Preservation of Tree Act 1976 and Karnataka Preservation of Tree Rules, 1977 • Under this important act and rules it allows for Establishment of tree authority and appointment of Tree Officer and Restriction of felling of trees and also Provision for compulsory planting trees in place of tree felled • Exemption for 11 species from transit rules—Eucalyptus, Casuarina, Subabul, Rubber, Coconut, Arecanut, orange, Erythrina, Glyricidia, Sesbania, Silver oak are also specified • Karnataka Forest (Amendment) Bill, 2001 allows cultivation of sandalwood trees in private lands. This comes as a major policy change as far as sandalwood cultivation is concerned. This particular policy/legislation amendment of Government of Karnataka to abolish control over sandalwood cultivation has paved the way for community and private entrepreneurs to cultivate sandalwood which is in high demand • This Karnataka Forest (Amendment) Act 2001 clearly states that 'every occupant or the holder of land shall be legally entitled to the sandalwood tree in his land' • The Government of Karnataka authorized Karnataka Soaps and Detergents Limited (KSDL) and Karnataka State Handicrafts Development Corporation (KSHDC) to buy sandalwood directly from landowners though applications still have to be filed and routed through Forest Department
2	Tamil Nadu	• Tamil Nadu Preservation of Private Forest Act, 1949 • The rules not to be applied for the cutting of sandal wood trees or thinning of private plantation to teak, Casuarina and eucalyptus manifested for the silvicultural improvement of the crop and the cutting and removal of fuel small timber reeds and green manure leaf for bonafide agricultural or domestic purposes • The cutting of sandalwood trees in a forest and their transportation outside are to be governed by the Madras Forest Act, 1882 • In granting permission for the cutting of trees by the selection method in case of private forest, the minimum girth height has been specified for the trees like Casuarina, Eucalyptus, Wattle, Silver Oak, Pines, etc. • Trees shall be permitted to cut only in one coupe in each block during a specified year • Tamil Nadu Sandalwood Transit Rules 1970 allows the sandalwood purchased from neighbouring states like Mysore, Kerala and Andhra Pradesh

(continued)

(continued)

Sl. No	State	Rules and other
		• Under this Rule exemption of small quantities of sandalwood (up to 5 Kgs) carried by a bonafide traveller for their own use and also allows the landlord to cut and carry the sandalwood grown on patta lands for his own bonafide personal use with proper certificate from the village munsif • The Tamil Nadu Sandalwood Trees on Patta Land Rules 2008 allows cultivation and the ownership of the patta land owner to produce the relevant documents and sell the same to the forest department and get 80 % of the sale price • This is the latest notification by Government of Tamil nadu encouraging the private land owner to take up cultivation of sandalwood in their private land
3	Kerala	• The Kerala Private Forests (Vesting and Assignment) Act 1971 • The Kerala Restriction on cutting and destruction of valuable Trees Rules 1974. Wherein permission for cutting or/and destruction of valuable trees shall be obtained from the authorized authority of the department • The Kerala Preservation of Trees Act 1986 is applicable to private forest • The Kerala promotion of Tree Growth in Non-Forest Area (Amendment) Act 2007 • Under this amended act every owner of non-forest land in a non-notified area shall have Right to cut and Transport any tree except sandal wood tree standing on his land. No tree standing in any area of non-forest land specified in the notification shall be cut
4	Puducherry/Pondicherry	• Pondicherry Timber Transit Rules 1983 • Issue of permit for the transportation of timber and registration is to DFO • Teak, Rosewood, Sandalwood and Red Sanders are protected wood and such species cannot be kept in possession or transported by any individual/farm without special permit
5	Andhra Pradesh	• The Andhra Pradesh Sandalwood Possession Rules 1969 • The Andhra Pradesh Sandalwood and Red sanders wood Transit Rules 1969 • Under these rules no person shall have in his possession any quantity of sandalwood in excess of ten Kgs, except under a licence granted by the District Forest Officer • The Andhra Pradesh Forest Produce Transit Rules, 1970 is applicable

(continued)

(continued)

Sl. No	State	Rules and other
		• AP Private Forest Rules in Agency area, 1977. The Andhra Pradesh Preservation of Private Forest Rules, 1978 • The Andhra Pradesh (Protection of Trees and Timber in Public Premises) Rules, 1989 • WALTA, 2000 provides for water land and tree • TP required for felled timber • In tribal area no felling can be done without permission of the collector • In Govt. land nobody can cut tree without permission of forest department • Exempted species are categorized district wise • The Andhra Pradesh Forest Produce Transit Rules, 1970 is not applicable to Red sanders and sandalwood, Orange, Tati, Casuarina, Guava, Seemathoma, Sapota, Coconut, Cashew, Eucalyptus, Neelagiri Jamaoil, Subabul, Seema Chinta, Ber, Rain tree nidraganneru, Juamun, Nallatumma, Mango, Panasa jack fruit, Bamboo, Myrobalam, Wood apple • During 2006 Principal Chief Conservator of Forests, Andhra Pradesh issued circular in addition to the guidelines and supplementary to the statutory Acts and Rules for the procedure to be grant of transit permits in patta lands for red sander wood • Under this, regulation and possession of Red sanders and Sandalwood should be dealt under Andhra Pradesh Sandalwood and Red sander wood transit Rules 1969 and Andhra Pradesh Red sanders wood Possession Rules 1989. These Rules provide detailed provisions for verification of the material, issue and checking of transit permit, registration of property marks and fixing of marks on the trees and felled logs etc.
6	Maharashtra	• Felling of trees in private lands is regulated by following three Acts: • Maharashtra Felling of Tree (Act 1964) • Maharashtra (Urban Areas) Preservation of Trees Act, 1975 • The Maharashtra Land Revenue Code 1966 • Permission for transportation of forest produce, so obtained after receipt of felling permission from the Tree Officer, is generally given by the concerned Dy. Conservator of Forests under the provisions of the Indian Forest Act 1927 and the rules made there under namely the Bombay Forest Rules, 1942 • Presently 16 species are included in the Schedule which are: 1. Hirda (*Terminalia chebula*) 2. Teak (*Tectona grandis*)

(continued)

(continued)

Sl. No	State	Rules and other
		3. Mahuwa (*Madhuca latifolia*) 4. Tamarind (*Tamarindus indica*) 5. Mango (*Mangifera indica*) 6. Jack (*Artocarpus integrifolia*) 7. Khair (*Acacia catechu*) 8. **Sandal** (***Santalum album***) 9. Bija (*Ptercarpus marsupium*) 10. Haldu (*Adina cardifolia*) 11. Tiwas (*Ougeinia dailbergoidies*) 12. Ain (*Terminalia tomentosa*) 13. Kinjal or Kindal (*Terminalia paniculata*) 14. Anjan (*Hardwickia binata*) 15. Jambhul (*Syzigium cumini*) 16. Mangrove Within the district of Sindhudurg only 17. Shisam (*Dalbergia latifolia*) 18. Shivan (*Gmelina arboria*) 19. Nana (*Lagerstroemia lanceolata*) 20. Behala (*Terminalia bellerica*) 21. Kazra (*Strychnous nuxvomica*) 22. Bhedus (*Euginia zeylanica*) 23. Pandhra ain (*Terminalia arjuna*) 24. Kajoo (*Anacardium occidentale*) • Exempted species under timber transit rules: babhul, Subabhul, Prosopis, Eucalyptus, Ashok, Moringa, Phoenix, Chiku, Bhendi, Acacia and Poplar • The Maharashtra (Urban Area) Preservation of Trees Act, 1975 is for regulating felling of any tree in Urban Area. Felling permission is granted by the "Tree Officer" appointed by the Tree Authority. This being the local area, provisions of Bombay Forest Rules, 1942 regarding transportation of forest produce are not applicable, hence no transit pass is required for local limits • The Maharashtra Land Revenue Code, 1966 regulated the felling in the non-forest areas other than Urban Area and for the species which are not included in the Schedule attached with above mentioned Act of 1964
7	Gujarat	• Permission to cut trees regulated by Saurashtra Felling of Trees (Infliction of Punishment) Act 1951. Rules framed under the Act in 1961 • Permission required from cutting trees by Revenue Office as empowered by the Act not below the rank of Tehsildar. Revenue Officer can inflict fine for violation • Permission is required for felling of Teak, black wood, Sandalwood, Khair, Mahuda, Timru, Simla, Sadad, Kanaj, Kanaji, Seven, Bio, Rohan, Ebony, Kadao, Kalam, Baldervo, Harde, Dhavada, Mango, Palmyra palm, Date tree and Jamun • The state has relaxed felling of some species Neem, Kasia, Kanji, Khijdo, Mango and Amla under the Act

(continued)

(continued)

Sl. No	State	Rules and other
8	Haryana	• Punjab land preservation Act, 1900 is applicable in Aravalli and Shivalik • No transit rules for agro-forestry spp. In remaining area • Tree felling is banned without permission from Forest Department • Species exempted eucalyptus and poplar, Ailanthus, *Melia azadirach* and *M. composata*, bamboo and mulberry Saw Milling Rules as per direction of Hou'ble Supreme Court applicable since 2006 • No TP required for Eucalyptus, Poplar and Mango etc. • Forest Corporation revises rates every six months for minimum support price. Market rates are better for farmers • No registration of tree plantations of farmers is required
9	Madhya Pradesh	• Vide Gazette Notification no. F.30-40-95-X.3 date 13.12.2000 the Madhya Pradesh Transit (Forest Produce) Rules 2000 • The transit pass will be issued by the Panchayat on the recommendations of the Panchayat Level Committee for the species Babool, siris, Neem, Ber, Palas, Jawan, Rennjha, Bamboo (except in the districts of Khandwa, Betul, Hoshangabad, Harda, Chhindwara, Seoni, Balaghat, Jabalpur, Katni, Mandla, Dindori, Shahdol) • Transit pass for the species other than those mentioned above is to be issued by Forest Officer on recommendations of Panchayat level Committee • Transport of privately owned timber under the Lok Vaniki Mission requires the procedures mentioned above • Registration in the office of DFO is required for the forest produce to be transported • Gram Panchayat will issue the transit pass for transporting the forest produce within district and the adjoining districts for the other destinations the transit pass is required to be issued by a Forest officer • The tree species exempted from transit pass for transport are Neelgiri, Casuarina, Subabul, Poplar, Israili Babul, Vilayati Babul • No transit pass shall be required for the removal of any forest produce for bonafide domestic consumption by any person and such forest produce which is exempted by the State Government from the operation of these Rules • No transit pass is required for removal of mineral from forest for which transit pass is not compulsory under these Rules

(continued)

(continued)

Sl. No	State	Rules and other
10	Odisha/Orissa	• Orissa Timber and other Transit Rules 1980 • Village Forest Rules, 1985 • Orissa Timber and other Forest Produce Transit Amendment Rules, 2006 • Exempted species under Orissa Timber transit Rules, 1980 Notification No. 2013 dated 8.2.99 • Permit to be taken from the committee for felling of trees and transit • Management of village forest by drawing of management plans • Transit permission shall be issued free of cost by DFO for all forest produce in transit land, rail or water • No transit permits for transport of minor forest produce within district except lac, tassar, myrabolans, gums and resin, root or Patalagaruda. Sal seed, tamarind and hill broom • Species to which provisions regarding farm forestry and forestry farming for the rural poor plantation are not applicable are kurum, panas, kasi sissoo, gamhar, amba, champa sal, teak asan • Species exempted *Bambusa nutan* (Sundarkanai), *Bambusa vulgaris, eucalyptus Hybrid* (*Nilgiri*), *Acacia auriculiformis, Cassia siamea, Casuarina equisetifolia,* Silver oak • No transit permit for timber grown outside India
11	Punjab	• Punjab Land Preservation Act, 1990 applicable • Agro-forestry/farm forestry crops have no requirement for transit pass like poplar, Eucalyptus and *Melia composita* • Transit Rules applicable in 5 Districts covering the area closed under Punjab Land Preservation Act (PLPA) 1900 in Ropar, Mohali, Hoshiarpur, Pathankot and Nawashahar (Shaheed Bhagat Singh Nagar). In closed are which cover private forests closed under Section—4 and 5 of PLPA, management is done by Forest Department and 100 % revenue is given to Farmers. Marking Fee is charged by the Dept. plantation is also done by Forest Department and normally Khair, Amla and bamboo is planted. Transit Pass is required in such cases as felling permit. Initially 10 year cycle has now reduced to 5 year cycle to increase people's participation and management plans is prepared • In 2000, Tree Apportionment Rules, 2000 were framed a Govt. Notification sharing revenue with farmers for raising roadside trees on Govt. land covering road side trees like NH and State Highways and Link Road

(continued)

(continued)

Sl. No	State	Rules and other
12	Uttar Pradesh	• Uttar Pradesh timber and other forest produce transit rules, 1976 • Total 20 species are exempted from transit pass like eucalyptus, poplar in various district state • District wise species are exempted from transit where reserve forest or etc. • Farmers are encouraged to plant exempted species • Private forests are also governed by these rules. The department can assist in preparation of management plans of private forests • Forest department facilitates individual farmers in growing private agro-forestry/farm forestry by taking up plantation activities at their cost. Promotion of kisan nurseries in western UP
13	West Bengal	• WB Private Forest Act, 1948 is in force which regulates permission for felling and transit of trees grown on private lands • WB Forest Produce Transit Rules, 1959 • WB Trees (Protection and Conservation in Non-Forest Areas) Act, 2006 • The West Bengal Trees (Protection and Conservation in Non-Forest Areas) Act 2006 • Provision for Tree Card for individual farmers • Provision for penalty and confiscation of materials felled and transported in violation of such rules • No tree exempted • No tree can be felled in non forest area except with the procedure laid out for obtaining permission for felling of trees with obligation to plant trees in lieu of trees felled • Compulsory tree plantation in certain areas including high rise multi unit buildings for residential, commercial, industrial or institutional use • No owner of a private forest in a notified area or other person shall fell or remove tree from such forests until the working plan in respect of such forest has been approved • Permission mandatory for 11 species: Khair, Semal, Sissoo, Tendu, Gama, Mahua, Champ, Sal, Mahogani, Teak and Mangroves. Remaining exempted • Permission for felling of a tree outside forest areas is required from competent authority with obligation to plant trees as prescribed. Compulsory plantation of trees in respect of development works • Appropriate forms prescribed for obtaining permission for felling of trees

References

Anon. (2006). Australian Agribusiness Group, Market Overview–The Australian Sandalwood Industry. Independent Assessment–November 2006.

Ananthapadmanabha, H. S. (2011). Sustainable supply of sandalwood for industry. In *Proceedings of the Art and Joy of Wood conference, Bangalore, India*, October 19–22, 2011.

Ananthapadmanabha, H. S. (2012). Indian sandalwood market trend production. In *Proceedings of International Sandalwood Symposium, Honolulu, Hawaii*, October 21–24, 2012. Arun Kumar Bansal, (2012), Report of the Committee constituted to study the regulatory regime regarding felling and transit regulations for tree species grown on private land 2012, Ministry of Environment, Forest and Climate Change, Government of India.

Arun Kumar Bansal. (2012). *Report of the Committee constituted to study the regulatory regime regarding felling and transit regulations for tree species grown on private land.* Ministry of Environment, Forest and Climate Change, Government of India.

Balachandran, N., & Kichenamourthy, S. (2007). Profile of natural stands of *Santalum album* L. in the Pondicherry region. *India. Sandalwood Research Newsletter, 22*, 4–9.

Brand, J. E., Jones, P., & Donovan, O. (2004). Current growth rates and predicted yields of sandalwood (*Santalum spicatum*) grown in plantation in south-western Australia. *Sandalwood Research Newsletter, 19*, 4–7.

Butaud, J. F. (2004). *Santalum insulare* (Bertero ex A. DC.): Distribution and ecology. *Sandalwood Research Newsletter, 19*, 1–4.

Dhanya, B., Viswanath, S., & Purushothman, S. (2010). Sandal (*Santalum album* L.) conservation in southern India: A review of policies and their impacts *Journal of Tropical Agriculture, 48*(1–2), 1–10.

Doran, J. C., Thomson, L., Brophy, J. J., Goldsack, B., Bulai, P., Faka'osi, T., & Mokoia, T. (2005). Variation in heartwood oil composition of young sandalwood trees in the south Pacific (*Santalum yasi, S. Album* and F1 hybrids in Fiji, and *S. yasi* in Tonga and Niue). *Sandalwood Research Newsletter, 20*, 3–7.

Jain, S. H., Angandi, V. G., Shankaranarayana, K. H., Rajeevalochan, A. N., Theagarajan, K. S., & Ragaswami, C. R. (1998). Identification of provenances of sandal in India for genetic conservation. In A. M. Radomiljac, H. S. Ananthapadmanabha, R. M. Welbourn & K. Salyanarayana Rao (Eds.), *Sandal and its products* (pp. 117–120). Canberra: Australian Centre for International Agricultural Research.

Jain, S. H., Angandi, V. G., Shankaranarayana, K. H., & Ravikumar, G. (2003). Relationship between girth and percentage of oil in sandal provenances. *Sandalwood Research Newsletter, 18*, 4–5.

Nagaveni, H. C., & Vijayalakshmi, G. (2003). Growth performance of sandal (*Santalum album* L.) with different host species. *Sandalwood Research Newsletter, 18*, 1–4.

Nageswara-Rao, M. (2012). Mapping sandal genetic resources in India: Threats and conservation strategies. In *Proceedings of International Sandalwood Symposium, Honolulu, Hawaii*, October 21–24, 2012.

Soundararajan, V., Ravi Kumar, G., Murugesan, K. (2015), Trade Scenario of Sandalwood and its valued oil. *International Journal of Novel Research in Marketing Management and Economics, 2*(3), 52–59, September–December 2015.

Stemmermann, L. (1990). Distribution and status of sandalwood in Hawaii. In *Proceedings of the Symposium on Sandalwood in the Pacific, Honolulu, Hawaii*, April 9–11, 1990.

Subasinghe, U., Gamage, M., & Hettiarachchi, D. S. (2013). Essential oil content and composition of Indian sandalwood (*Santalum album*) in Sri Lanka. *Journal of Forestry Research, 24*(1), 127–130.

Subhash Chandra. (2014). *Report on the regulatory regime regarding felling and transit regulations for tree species grown on non forests/private lands*. Ministry of Environment, Forest and Climate Change, Government of India.

Venkatesha Gowda, V. S. (2011). Global emerging trends on sustainable production of natural sandalwood. In *Proceedings of the Art and Joy of Wood Conference, Bangalore, India*, October 19–22, 2011.

Xiaojin, L., Daping, X., Zengjiang, Y., Ningnan, Z., & Lijun, Y. (2011). Preliminary analysis of growth and oil composition from a 6-year-old sandal (*Santalum album* L.) plantation in Gaoyao, Guangdong, south China. *Sandalwood Research Newsletter, 26*, 1–5.

Monitoring Condition Factor of the Dominant Fin Fishes in the Estuaries of Lower Gangetic Delta in the Backdrop of Climate Change

Ankita Mitra, Sufia Zaman, Prosenjit Pramanick and Shampa Mitra

Abstract The current global fisheries production is ~160 million tonnes. The quantum is gradually rising as a result of increases in aquaculture production. A number of climate-related threats to capture both the fisheries and aquaculture are identified, but we have high uncertainty in predictions of the future fisheries production because of uncertainty over the future global aquatic net primary production and the transfer of this production through the food chain to human consumption. Recent changes in the distribution and productivity of a number of fish species can be ascribed with high confidence to regional climate variability, such as the El Niño–Southern Oscillation. The future production may increase in some high-latitude regions because of warming and decreased ice cover, but the dynamics in low-latitude regions are governed by different processes, and production may decline as a result of reduced vertical mixing of the water column and, hence, reduced recycling of nutrients. There are strong interactions between the effects of fishing and the effects of climate because fishing reduces the age, size and geographic diversity of populations and the biodiversity of marine ecosystems, making both more sensitive to additional stresses such as climate change-induced salinity oscillation. The estuaries in the lower Gangetic delta at the apex of Bay of Bengal are noted for contrasting spatial variation of salinity. The Hooghly estuary in the western sector is relatively low saline compared to the Matla estuary in the central sector on account of receiving the fresh water discharge from the Farakka barrage. The central sector is hyper saline due to siltation of the Bidyadhari River since the late fifteenth century. The present study evaluates the condition factor of fourteen commercially important fin fish species (that constitute ~75 % of the fishes in the catch basket) collected from the Hooghly and Matla estuaries in the western and central sectors of the lower Gangetic delta, respectively. Relatively

A. Mitra (✉)
Department of Ecology and Environmental Sciences,
School of Life Sciences, Puducherry, India
e-mail: mitrankita92@gmail.com

S. Zaman · P. Pramanick · S. Mitra
Department of Oceanography, Techno India University,
Salt Lake Campus, Kolkata 700091, India

S. Nautiyal et al. (eds.), *Climate Change Challenge (3C) and Social-Economic-Ecological Interface-Building*, Environmental Science,
DOI 10.1007/978-3-319-31014-5_20

higher values of condition factor of all the species collected from the Hooghly estuary (compared to those collected from the Matla estuary) confirm the adverse effect of hyper salinity on the growth and condition factor of the species.

Keywords Salinity · Condition index · Fin fish · Climate change

1 Introduction

The lower Gangetic delta at apex of Bay of Bengal sustains a wide spectrum of fin fishes. The litter and detritus contributed by mangroves provide nutrition to the fish community by triggering the growth of plankton. The fish fauna of the present study area may be classified into residents and transients (migrants). The species whose individuals of different sizes are present during all the months of the year in any zone of the estuary are referred to as resident species. The important resident finfish species are *Mugil parsia, M. tade, Polynemus paradiseus, Polydactylus indicus, Otolithoides biauritus, Lates calcarifer, Hilsa toli, Arius jella, Harpodon nehereus, Setipinna taty, Ilisha elongata, Setipinna phasa, Coilia ramcarati, Pama pama* and *Sillaginopsis panijus*. The transient or migratory fishes enter and stay in the Bay of Bengal associated estuaries for a short period. Depending on their migratory pattern and direction, the migrants are divided into three categories (Jhingran 1982; Mitra 2013): (1) Marine forms that migrate upstream and spawn in freshwater areas of the estuary such as *Tenualosa ilisha, P. paradiseus, S. panijus* and *P. pama*, (2) Freshwater species, which spawn in saline area of the estuary such as *Pangasius pangasius*, and (3) Marine species, that spawn in less saline water of the estuary such as *A. jella*, *Osteogeneious militaris* and *Polydactylus indicus*.

Two important estuaries in the lower Gangetic delta are the Hooghly and the Matla estuaries. These two estuaries exhibit pronounced variation with respect to salinity. The Hooghly estuary in the western sector is relatively low saline compared to the Matla estuary in the central sector. This contrasting salinity may be attributed to discharge of fresh water through Farakka barrage in the Hooghly estuary and complete closure of fresh water discharge in the Matla estuary due to siltation of the Bidyadhari River since the late fifteenth century. The difference in salinity has profound influence on the condition factor of the fin fishes, which has been attempted to study in the present programme.

Our main aim is to determine the impact of salinity variation on the condition factor of the selected fin fish species as the price of the fish mostly depends on their growth (in terms of length and weight).

2 Materials and Methods

The entire network of the present study consists of the collection of 100 individuals of the selected species (*T. ilisha, Pama pama, Pampus* spp., *Ilisha elongate, L. calcarifer, P. pangasius, Liza parsia, Liza tade, Tenualosa toli, P. paradiseus, O. biauritus, Tachysurus jella, Sciaena biauritus, Eleutheronema tetradactylum*) from the two major estuaries in the study area namely Hooghly (in the western sector) and Matla (in the central sector). The sampling stations selected for the present study are Diamond Harbour (in the Hooghly estuary) and Canning (in the Matla estuary). Individual length and weight of the individuals of the selected species were measured to evaluate the condition factor (Chow and Sandifer 1991) as per the following expression:

$$K = \frac{\bar{w}}{(\overline{\mathrm{TL}})^3} \times 10^3$$

where K is the condition factor, $\bar{w}$ is the average weight (g) and $\overline{\mathrm{TL}}$ is the average total length (cm).

The secondary data of surface water salinity were obtained from the data bank cited by Ray Choudhury et al. (2014), Chakraborty et al. (2013), Mitra (2013), Sengupta et al. (2013), Mitra and Zaman (2014) and Mitra and Zaman (2015).

3 Results

It is evident that condition factor is relatively higher for the fin fish species collected from the Hooghly estuary compared to those collected from the Matla estuarine water (Table 1). This may be attributed to variation in aquatic salinity as evidenced from the secondary data bank. The Hooghly estuarine water is relatively hyposaline as compared to the water of Matla estuary in the central sector of the study area (Figs. 1 and 2).

Table 1 Condition factors of the selected fin fish species in the western and central sectors of the study area

Sl. No.	Commercially important fin fish	Western sector (Hooghly estuary)	Central sector (Matla estuary)
1	 *Tenualosa ilisha* (Family: Clupeidae)	0.942	0
2	 *Pama pama* (Family: Sciaenidae)	1.009	0.595
3	 *Pampus* spp. (Family: Stromateidae)	1.018	0
4	 *Ilisha elongata* (Family: Pristigasteridae)	0.895	0
5	 *Lates calcarifer* (Family: Centropomidae)	1.619	0.492
6	 *Pangasius pangasius* (Family: Pangasiidae)	0.884	0.578

(continued)

Table 1 (continued)

Sl. No.	Commercially important fin fish	Western sector (Hooghly estuary)	Central sector (Matla estuary)
7	 *Liza parsia* (Family: Mugilidae)	0.986	0.713
8	 *Liza tade* (Family: Mugilidae)	0.778	0.709
9	 *Tenualosa toli* (Family: Clupeidae)	0.992	0.679
10	 *Polynemus paradiseus* (Family: Polynemidae)	0.899	0.697
11	 *Otolithoides biauritus* (Family: Sciaenidae)	1.009	0.918
12	 *Tachysurus jella* (Family: Ariidae)	1.129	0.743

(continued)

Table 1 (continued)

Sl. No.	Commercially important fin fish	Western sector (Hooghly estuary)	Central sector (Matla estuary)
13	 *Sciaena biauritus* (Family: Sciaenidae)	1.099	0.992
14	 *Eleutheronema tetradactylum* (Family: Polynemidae)	0.994	0.788

Note The value 0 (zero) represents the nonavailability of the species in the estuarine water

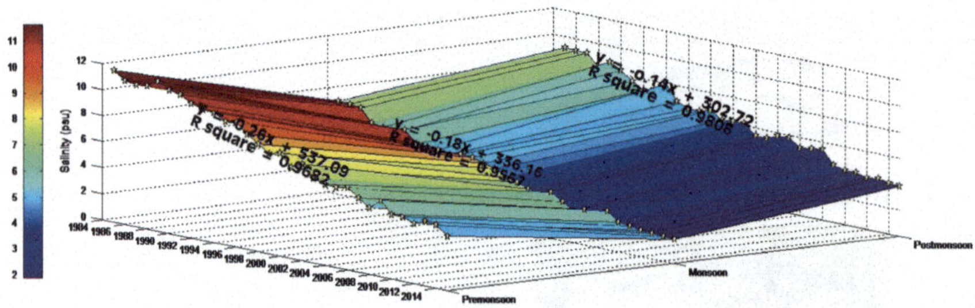

Fig. 1 Decreasing trend of surface water salinity in western sector of the study area over a period of 31 years

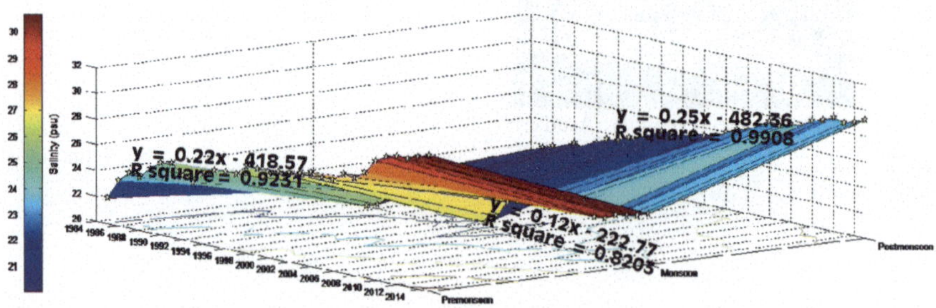

Fig. 2 Increasing trend of surface water salinity in central sector of the study area over a period of 31 years

4 Discussion

Climate change has both direct and indirect impacts on fish stocks which are exploited commercially preferably for the livelihood. Direct effects act on physiology and behaviour of fishes and alter their growth, reproduction, mortality and distribution. Indirect effects encompass events like alteration of aquatic productivity, biotic community structure and composition of the marine and estuarine ecosystems on which fishes depend for food and survival. Changes in primary and secondary production will obviously have a major effect on fisheries production, but it is not possible in the current state of knowledge to make accurate quantitative predictions of changes in global marine primary production solely due to climate change-induced salinity alteration (Mitra 2013).

In this paper, we used condition factor as proxy to assess the impact of salinity on the fish community of the lower Gangetic delta complex. Condition factor is an indication of the well being of an organism and is based on the hypothesis that heavier fish of a given length are in a better condition (Bagenal and Tesch 1978; Abowei and George 2009). It has been used as an index of growth and feeding intensity (Fagade 1979; Abowei et al. 2009), decreases with increase in length (Bakare 1970; Fagade 1979; Abowei 2009) and also influences the reproductive cycle in fish.

Our first order analysis from the data sets of two fish landing stations (Diamond Harbour in the western sector and Canning in the central sector of lower Gangetic delta region) clearly reflect a pronounced variation between the two sectors. Significantly higher values of condition index in fin fish were observed from the fish catch of Diamond Harbour, where the aquatic phase is congenial in terms of salinity (Mitra 2013). This congenial salinity may be attributed to Farakka barrage discharge situated in the upstream region of Ganga–Bhagirathi–Hooghly river system. 10-year surveys (1999–2008) on water discharge from Farakka dam revealed an average discharge of $(3.7 \pm 1.15) \times 10^3$ m^3 s^{-1}. Higher discharge values were observed during the monsoon with an average of $(3.81 \pm 1.23) \times 10^3$ m^3 s^{-1}, and the maximum of the order 4524 m^3 s^{-1} during freshet (September). Considerably lower discharge values were recorded during pre-monsoon with an average of $(1.18 \pm 0.08) \times 10^3$ m^3 s^{-1}, and the minimum of the order 846 m^3 s^{-1} during May. During post-monsoon discharge, values were moderate with an average of $(1.98 \pm 0.97) \times 10^3$ m^3 s^{-1} as recorded by earlier workers (Mitra 2013). This hyposaline condition supports migration of *T. ilisha* for breeding in the upstream region of Gangetic delta and also acts as the congenial nursery ground of several species of commercially important fin fish that has been reflected through relatively high condition factor values (Table 1).

In the central sector, the ingression of seawater and resultant salinity increase has completely reversed the picture with relatively low condition index values of economically important fin fish species (Table 1). The unavailability of fresh water in this sector of lower Gangetic delta due siltation and blockage of Bidyadhari River since the late fifteenth century (Mitra 2013; Mitra and Zaman 2014, 2015) may be the primary cause of (i) reproductive failure of economically important fin fish species in the hypersaline environment (ii) change in migratory route for breeding purpose (iii) loss of primary food supply (mainly plankton) due to adverse impact of salinity tolerance for that organism (plankton) and (iv) direct mortality due to extreme saline condition (Mitra 2013). All these reasons directly or indirectly lower the condition factor of fin fish species by amplifying the environmental stress.

A long-term study of some fifty years (encompassing all the major landing stations) is, however, needed to pinpoint the impact of salinity fluctuation on the condition factor of fin fish species in the lower Gangetic delta at the apex of Bay of Bengal.

References

Abowei, J. F. N. (2009). The abundance, condition factor and length–weight relationship of *Cynoglossus senegalensis* (Kaup, 1858) from Nkoro River Niger Delta, Nigeria. *Advance Journal of Food Science and Technology, 1*(1), 56–61.

Abowei, J. F. N., Davies, O. A., & Eli, E. E. (2009). Study of the length–weight relationship and condition factor of five fish species from Nkoro river, Niger Delta, Nigeria. *Current Research Journal of Biological Sciences, 1*(3), 94–98.

Abowei, J. F. N., & George, A. D. I. (2009). A study of the length–weight relationship and condition factor of *Callinectes amicola* (De Rochebrune, 1883) from Okpoka Creek, Niger Delta, Nigeria. *International Journal of Animals and Veterinary Advances, 1*(2), 66–72.

Bagenal, T. B., & Tesch, A. T. (1978). *Conditions and growth patterns in freshwater habitats* (pp. 75–89). Oxford: Blackwell Scientific Publications.

Bakare, O. (1970). Bottom deposits as food of inland freshwater fish. In S. A. Visser & K. Lake (Eds.), *A Nigerian man made lake. Kainji Lake studies, Vol. 1, ecology*. Ibadan, Nigeria: NISER.

Chakraborty, S., Zaman, S., Pramanick, P., Raha, A. K., Mukhopadhyay, N., Chakravartty, D., et al. (2013). Acidifications of Sundarbans mangrove estuarine system. *Dicovery Nature, 6*(14), 14–20.

Chaudhuri, T. R., Fazli, P., Zaman, S., Pramanick, P., Bose, R., & Mitra, A. (2014). Impact of acidification on heavy metals in Hooghly Estuary. *Journal of Harmonized Research in Applied Sciences, 2*(2), 91–97.

Chow, S., & Sandifer, P. A. (1991). Differences in growth, morphometric traits and male sexual maturity among Pacific white shrimp, *Penaeus vannamei*, from different commercial hatcheries. *Aquaculture, 92*, 165–178.

Fagade, S. O. (1979). Observations on the biology of two species of *Tilapia* from Lagos Lagoon, Nigeria. *Bulleetin of d'Institut Français d'Afrique Noire, 41*, 627–653.

Jhingran, V. G. (1982). *Fish and fisheries of India*. Delhi: Hindusthan Publishing Corporation (India).

Mitra, A. (2013). *Sensitivity of mangrove ecosystem to changing climate* (p. 323). Berlin: Springer. doi:10.1007/978-81-322-1509-7

Mitra, A., & Zaman, S. (2014). *Carbon sequestration by Coastal Floral Community*. India: The Energy and Resources Institute (TERI) TERI Press. Copyright The Energy and Resources Institute. ISBN 978-81-7993-551-4.

Mitra, A., & Zaman, S. (2015). *Blue carbon reservoir of the blue planet*. Berlin: Springer. ISBN 978-81-322-2106-7.

Sengupta, K., Chowdhury, M. R., Bhattacharyya, S. B., Raha, A. K., Zaman, S., & Mitra, A. (2013). Spatial variation of stored carbon in *Avicennia alba* of Indian Sundarbans. *Discovery Nature, 3*(8), 19–24.

Is There a Tomorrow?: The Story of Survival of Sunderbans Against Climate Change

Haimanti Pakrashi

Abstract Occurrences of earthquakes, floods, unseasonal rainfall, heat flashes, and ravaging forest fires are increasing in different parts of the world. The climate is changing, sea levels are rising all around the world; thus endangering the coastal communities. Their lives are increasingly becoming unsustainable. Are these phenomenons naturally occurring or Anthropogenic? These are some of the questions which scientists globally are looking answers for. New researches show that the global temperature will rise by 4 °C by the end of this century. The consequences for such an occurrence would be huge in the lives of all the inhabitants of this planet. Our lives might change forever. Scientists and scholars are trying to come up with solutions to deal with the incoming crisis, but unfortunately for the coastal communities life has already started changing. They are already feeling the wrath of climate change. In this paper, I am going to discuss the present socio-economic conditions of the island dwellers of Sunderbans. Sunderbans are a cluster of mangrove forested islands in the mouth of the Ganga Brahmaputra delta. The fragile ecosystem and the endangered animals of this region have received substantial academic attention but my paper talks about the daily struggle of the million plus poverty-stricken people who live on these islands. The threats of climate change like the rising sea level, extreme weather events like hurricanes, cyclones and floods, loss of livelihoods and finally people being forced to leave their coastal homelands for higher and greener pastures is already happening in these islands for quite some years now. This paper discusses about the problems faced by the people of Sunderbans engaged in farming and fishing. The embankment protecting their land is poorly maintained and gets damaged during the frequent storms and high tides, which eats up their land on a regular basis. Already a few of the islands have disappeared altogether. Mangroves are the only natural protection these islands have; thus afforestation measures are required on a large scale. Out of desperation, these people enter the reserved forest area kept for tigers and other animals and get eaten away. Is this a way to live? Is there a way forward for these island dwellers

H. Pakrashi (✉)
Centre for Regional Studies, University of Hyderabad, Hyderabad, India
e-mail: haimanti.pakrashi@gmail.com

S. Nautiyal et al. (eds.), *Climate Change Challenge (3C) and Social-Economic-Ecological Interface-Building*, Environmental Science,
DOI 10.1007/978-3-319-31014-5_21

other than leaving the only homes they have known and join the fast growing community of climate migrants?

Keywords Climate change · Sea level rise · Anthropogenic · Sunderbans

1 Introduction

Earth is the only planet which supports life amongst the other planets and it is only the human race who are aware of this fact. But that does not mean the millions of other living organisms of the earth, who are unaware of this fact, have any less right on the natural capital of this planet. The Earth has gone through several revolutionary environmental changes in the past. But, for the past 10,000 years it has remained quite stable and this period is called the Holocene period by the Geologists. Since the beginning of Industrial Revolution in the eighteenth century, the activities of humans have been driving the environmental changes, so much so that the scientists call it the Anthropocene period, that is, the Age of the Humans (Rockstrom et al. 2009). We are cutting down trees, acidifying oceans, increasing Green House Gases (GHGs) in the atmosphere to extraordinary levels, destroying natural habitats of other species, draining the water bodies, extracting minerals embedded in the earth's surface; and all this for our own development. As a result of these insensitivities, a large number of animals have become extinct or are under threat. So much so that scientists are of the opinion that, Humanity is causing the sixth great extinction wave of planet Earth; the earlier ones were caused by natural dynamics (Sachs 2010).

Unfortunately, it is not the habitats of the animals alone that we have destroyed. The future of the human race is also under duress because we have pushed the Planetary boundaries too far (Sachs 2010). Planetary boundaries are defined as the "safe operating space for humanity with respect to the Earth system and are associated with the planet's physical subsystems or processes" (Rockstrom et al. 2009, p. 1). Professor Rockstrom and his colleagues have identified nine Earth system processes individually and also defined a particular threshold for each of them. These nine processes are climate change, ocean acidification, stratospheric ozone depletion, global fresh water use, change in land use, chemical pollution, interference with the Nitrogen and Phosphorus cycles and atmospheric aerosol loading (Rockstrom et al. 2009). Regrettably, the threshold levels of climate change, rate of biodiversity loss and nitrogen cycles have already been crossed and for the other processes the boundaries are not very far from the danger zone.

Jeffry Sachs in his talk on Sustainable Development keeps on enforcing that in order to achieve sustainable development, economic development is a prerequisite, keeping in mind environmental sustainability, inclusive of social development and for all this, good governance in a must (Sachs 2010). He goes further to say it is vital to take environmental dangers seriously or else it can upend all the progress: destroy crops, flood cities and outbreak of droughts (Sachs 2010).

The International Conferences on climate change like the ones held in Geneva (1996), Kyoto (1997), Bonn (2001), Montreal (2005), Copenhagen (2009), Durban (2011) Doha (2012) and others discussed the "2 degree centigrade" rise is global temperature. But according to the latest reports, there is going to be a "4 degree centigrade" temperature rise. (Carrington 2013; Kemsley 2013; PTI 2014). This indeed further enhances the urgency for sustainable development and calls for immediate and drastic measures to reduce global carbon footprint and check the transfer of forested land into farm land or pastures. Forests absorb huge amounts of carbon dioxide and act as carbon sinks along with oceans. However, when large-scale deforestation is carried out, trapped CO_2 gets released in the atmosphere and adds to global warming (Friend 2013). Also, deforestation in such a mammoth scale is endangering biodiversity. The world oceans have also become far more acidic than ever before, threatening the survival of millions of marine flora and fauna (Sachs 2010). It has also been observed by Scholars that all over the world the water table have declined sharply in the past few decades. The polar ice caps have started melting alarmingly making the sea levels go up with the additional flow of water (Graham 1999). But a global fresh water crunch is being predicted in the coming decades. In fact, World Wars are expected in the future over access to water (Goldenberg 2014; Judge 2013). Rainfall patterns are also changing dramatically all over the planet (Pinto 2014; Ghoshal 2014) and widespread droughts are also being predicted in the near future (Sachs 2010). That these predictions and assessments are correct are being established by the fact that some of these disturbances can already be seen in different parts of the world (Pallavi 2014; Misra 2014; Manocha 2014).

Thus, the occurrences of unforeseeable events are increasing in number, the future of us and our planet is indeed in grave danger. Having established global warming and the subsequent sea level rise, scientists have started to consider the impacts in each sector and regions in a 4 degree temperature rise scenario (Warren 2011).

This paper is an attempt to show the effects of climate change and loss of biodiversity on this ecologically sensitive region, for this is the prelude to what perhaps all the coastal communities would be facing subsequently, if immediate steps are not taken otherwise.

1.1 Introducing the Study Area

The Sunderban group of islands hosts one of the largest mangrove forests in the world (Rawat and Wikramanayake 2015). These islands are situated at the mouth of the biggest delta in the world; the Ganga Brahmaputra delta formed by the deposition of silt brought down by these mighty rivers and their numerous distributaries. These islands are cris-crossed by numerous distributaries, creeks and canals making these islands difficult to access from the mainland. Being estuarine islands, they face the onslaught of tides from both rivers and the sea, which means water which inundates these low lying lands are both sweet and salty by nature. The silt deposited by the river is often washed away by the sea during high tides. Again, the eroded silt gets deposited at some other place. This process of land formation on one side and erosion on the other has been going on simultaneously for thousands of years (Hazra et al. 2002). In the months of March to September, the high water levels during tidal bores are the highest in the region (Danda 2007). The Sunderbans provide a unique ecosystem for countless varieties of mangroves because of its saline soil. It is also home to different animals and is also rich with wide varieties of fishes. Politically, these islands are shared by the Indian state of West Bengal and Bangladesh. The larger chunk lies in Bangladesh and the International boundary between India and Bangladesh is marked by the river Raimangal, which flows along these islands. It is a very porous border and there is a continuous flow of people from the other side of the border. In this paper, I have only discussed the Indian Sunderbans. In total, there are 102 islands in the Indian Sunderbans of which 54 are inhabited. The uninhabited ones fall under the reserved forest area and has been recognized as a UNESCO World Heritage Site.

Though these islands support a huge population, civic amenities are found to be abysmally missing here. The inhabitants are primarily engaged in farming or fishing activities. But most are subsistent farmers and fishermen. There are several ponds on these islands which are used for both domestic and agricultural use. Usually, the richer families own these water bodies. There is still no grid electricity in most of these villages. Transferring power across the numerous rivers is a tricky business. People thus use either solar powered lights or kerosene lamps. Some of the block headquarters have brick or concrete roads but the interior villages only have mud roads. Most of the houses here are mud houses and are of single storey. Primary schools are found in plenty but the dropout rates are very high. High schools or Colleges are very few and far. The medical set up found here is very poor, and thus even in emergencies they have to travel to the hospitals in the district headquarters. And many a times, the patient does not survive the long uncomfortable journey. Transport mainly suffices of overloaded, wooden motor or oar driven boats. Vans and cars ply on some of islands which have been connected by bridges with the mainland and have all weather roads.

The islands are administratively part of the North and South 24 Pargana districts in the State of West Bengal. In fact, the North 24 Pargana district is the second most densely populated district in India (India Census 2011).

1.1.1 Data Source

The data for this paper has been taken from my dissertation on the "*Problems of development and planning in the Sunderbans*" under Prof. Swapna Banerjee-Guha as part of my Masters in Development Studies from Tata Institute of Social Sciences, Mumbai (2010–2012). I collected data from five villages, namely, Beguakhali, Rangabelia, Pakhiralaya, Gosaba and Masjidbari of the South 24 Parganas on various socio-economic indicators. I spoke to various experts, government officials and read the works of both the parties and came to the conclusion that participatory development, afforestation of mangroves, regular maintenance of the embankments and improving the infrastructure of these islands might provide the residents comparatively a safer and more sustainable life on these islands.

1.1.2 The Objective of the Study

1. To look into problems of livelihood and the factors associated with the same.
2. To analyze the phenomenon of outmigration in relation to the vulnerability of the region both economic and physical environment.
   It was found that, increasingly, the youth are migrating from the islands in search of better opportunities in the cities. Climate has played a significant role in making life difficult for the people living on these islands.
3. To look into the prevalent development pattern in this unique region and inquire into the measures adopted by the local government bodies.

1.1.3 Methodology

Both quantitative and qualitative methods were used in this study. Primary data was collected using random sampling method by the author through questionnaire survey, covering information on different dimensions like embankment, practiced livelihood, climate change, etc. While secondary data included a variety of literature and reports by experts on Sunderbans.

1.1.4 Limitations Experienced by the Researcher

Availability and access to government data was very poor. However, the researcher tried to supplement them with primary and other secondary data. The paucity of time was also a major hindrance. Access to these islands is still difficult. In the pre monsoon and monsoon months, these islands become almost inaccessible from the mainland. Thus the data collection could only be done in the winter months. However, this researcher hopes to carry further research in this area in the future and get more insights on this region.

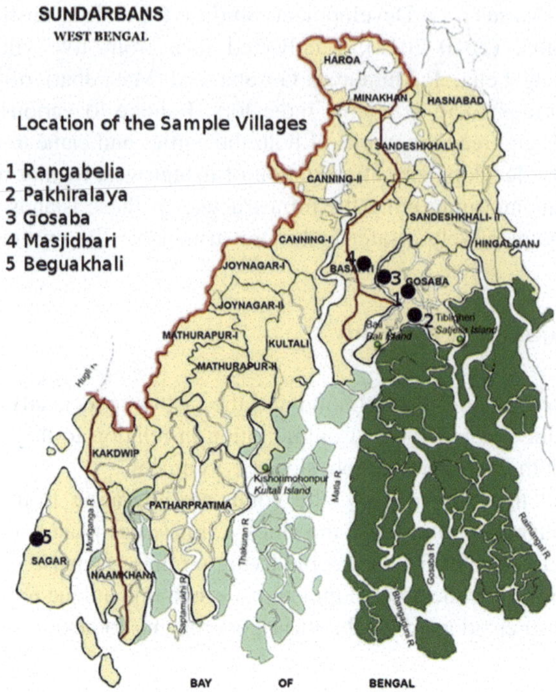

Location of the sample villages (Pakrashi 2012, p. 40)

1.2 The Sample Villages

The village of Beguakhali is located in the south west corner of Sagar Island. This island is the largest and the western most Sunderban island. It is subject to constant erosion and accretion. As the river Ganges meets the Bay of Bengal at Sagar, it is an important pilgrimage centre for the Hindus.

The villages of Rangabelia, Pakhiralaya and Gosaba of the Gosaba block and Masjidbari village in the Basanti block lie in the eastern part of the Sunderbans. These blocks are separated from each other by numerous water channels and rivers like Matla and Bidyadhari. The villages face the onslaught of cyclones and very high tides perennially.

Before going into the practiced livelihoods of Sunderbans, let us have a look at the land use pattern there, in order to understand the practiced livelihoods better.

Table 1 shows the land use pattern in Sunderbans.

Table 1 The land use pattern in the Sunderbans

Type of land	South 24 Parganas (in ha)	North 24 Parganas (in ha)	Total area of the Sunderbans (in ha)
Total cultivable land	2,433,320	61,514	304,834
Land for grazing and vegetables	1969	1858	3827
Unused cultivable land	2674	4154	6828
Forest land	82,300.04	32	82,332.04
Land with a second crop	76,203	25,270	101,473

Source State Gazette Bureau of statistics; Bairi (2004), Sundorboner Krishi: somoshya o unanyan, p. 138

1.2.1 The Livelihood Patterns in these Villages

In the nineteenth century, people from northern Bengal, Bihar and Orissa used to migrate to these islands twice a year to cultivate paddy on the land they owned there which had been distributed to them by the British (Danda 2007). However, with time, they started settling down on these islands. The Farukshiyar's Firman of 1757 granted the British the Rights to collect taxes from Bengal, Bihar and Orissa. They cleared the forests in these areas and converted them into farm lands. At that time none gave any thought to the negative impacts of deforestation. As a result in the next three hundred years, the two horned Rhinos, the Indian Cheetah, the Golden Eagle and the pink headed duck, species indigenous to the Sunderbans disappeared (Mukherjee et al. 1984, cited by Danda 2007).

1.2.2 Practice of Agriculture

Despite the geophysical adversity and the fact that 50 % of the population is landless, the primary source of livelihood in the Sunderbans is agriculture and their principal crop is paddy (Danda 2007). In West Bengal, farmers with land holding below 2.5 acres are called marginal farmers and the ones with land holding between 2.5 and 5 acres are called small farmers. The Sunderbans farmers are even below these levels. Thus, a farmer in the Sunderbans with a holding between 1 and 2 Bighas (0.13 and 0.27 ha) is considered a marginal farmer while one with land holding of 2–5 bighas (0.67 ha) is considered as a small farmer (Danda 2007). Now, let us take a look at the average land holding and the land to man ratio in the sample villages (Table 2).

Back in the nineteenth century itself, the expanding agricultural population soon found that practicing agriculture was in contradiction with the topography of the islands (Danda 2007). It was only possible to have a single harvest due to scarcity of freshwater in pre- and post-monsoon seasons. The British built embankments along the edges of the islands to stop the saline water from inundating the farm lands. They also built some irrigational canals; sluice gates were constructed in between these canals to maintain the water flow during the tidal bores (locally called Kotal) that come every fortnight. A few decades back more irrigation facilities were provided by

Table 2 The average land holding and land to man ration in sample villages

Village	Sample size	Landed peasants	Land owned (bighas) 1 bigha = 0.13 ha	Average family size	Land:Man ratio (in bighas)
Beguakhali	41	27	2	5	2:5
Gosaba	28	12	2.5–3	5	2.75:5
Masjidbari	13	5	2	4	1:2
Pakhiralaya	33	27	3–5	4	1
Rangabelia	43	22	2.5–3	5	2.75:5

Source Compiled from the primary data collected by Pakrashi (2012, p. 62)

the government with assistance from International Fund for Agricultural Development (IFAD) (www.ifad.org) to increase land under second harvest (Kanjilal 2000). Currently, only 20 % of the total area of the Sunderbans can produce two harvests. Efforts are on, to increase this proportion as this has proved to be quite profitable to the farmers (Bairi 2004). This has also made cultivation of certain winter crops possible. However, Danda (2007) points out that lack of maintenance has turned most of these canals dry. Since these are considered as government property, the locals do not have any authority to reconstruct them (Banerjee 1998). The villagers do not have the resource to re excavate the canals by themselves. They wait for the government to do the maintenance work, knowing it would be a long wait.

With passage of time, new technologies have come up. Some well to do farmers have bought diesel or kerosene driven generators which enables them to take water from the ponds or extract groundwater for a second harvest. But again, not every farmer in Sunderbans can afford a generator or pay for so much fuel. There are not many rules followed in regards to groundwater extraction and thus the farmers extract whatever amount they want, resulting in further depletion of groundwater aquifers in this ecologically fragile region. At times, the owner of a pond might loan it to a farmer without a pond, for a season. In return the owner is paid some rent or in kind by the borrower. At times, in order to make more profits farmers tend to grow the same crop every season which reduces the soil fertility. Hence, farmers are encouraged to grow pulses, as they provide nitrogen to the soil. In the Sunderbans, the soil salinity decreases during and after a good monsoon season and only starts increasing from December. During summer and winter times it becomes clayey and hard and hence is difficult to cultivate (Saha 2004). Historically the farmers used the indigenous Matla and Hamilton paddy variety which were more salt resistant. The indigenous variety of paddy which are used now, include Pankaj, Shalibon or Bipasha. But hybrid varieties are used by the majority for higher yield. But the hybrid ones require more water and fertilizer and are also more expensive. Other crops include a few vegetables, some fruits and betel leaves. However, their saline soil, expensive fertilizers, water scarcity and a skewed land to man ratio hinders these farmers from earning huge profits. And thus most of the farmers in Sunderbans are subsistent farmers.

Cattle farming of late has become less profitable in the Sunderbans. This is because of unavailability of fodder. Those who own a few animals, mainly use the products like milk and eggs for family consumption. Still some well to do households own a few hens, cows and sheep. There was only one poultry farmer in Beguakhali, Sagar who was a respondent is amongst the 160 families interviewed.

1.2.3 Fishing as a Livelihood

The interconnected and innumerable tidal canals and creeks that form an intricate network of waterways in Sunderbans are teeming with juvenile population of fish species (Banerjee 1998). Fishing is a major livelihood in the region, along with cultivation and crabs/prawn seed collection. Parts of the rivers lying about 2069 km^2 inside the Reserve Forest are considered ideal for riverine fishing depending on traditional methods. Boats locally known as naukas (country boats) are generally used by the fishermen in the region; at times trawlers (large motor boats) are also used (Patel et al. 2009). Fishermen of the South 24 Parganas have more access to the sea unlike their counterparts in the North 24 Parganas. Fishing and prawn seed collections are carried out by both men and women in inter tidal waters (Patel et al. 2009). Fishermen in most of these remote islands do not have storage facilities and it causes much hindrance to them. The fishing areas in this region include, (i) Inland fisheries having freshwater species like Rohu, Katla, Mrigal, Kalbose etc. (ii) Estuarine fisheries with brackish water fish culture locally called Bhasa bada, nona bheri fisheries (Banerjee 1998). The islands in Eastern Sunderban include small fishermen with country boats and they usually fish in the rivers or the creeks. Some of them get permission to fish in the Reserved Forest water bodies. Though fishing in the reserved forest waters is more profitable, it is also far more risky. There are Tigers on land or Crocodiles in the water. There have been countless incidents where fishermen have either been attacked by tigers, bears and crocodiles. It is estimated that currently there are over 1000 widows whose husbands have been killed by tigers in the remote islands of Sunderbans (www.tigerwidows.org).

Baby prawns are generally sold to agents who nurture them in brackish water fisheries and subsequently sell the catch to make huge profit. The villagers earn a pittance as compared to these agents. The harmful effects of prawn fishing have been discussed later in the paper.

1.2.4 Other Practiced Livelihoods

Apart from faming and fishing the people of Sunderbans are also engaged in collecting forest resources, such as wood and honey. However, these are rather risky occupations as people are frequently attacked by bears and tigers. Also, since 1980s restrictions have been imposed by the Government with regards to collection of forest resources. They have started driving diesel-driven cycle vans, which carries both passengers and goods. The earnings are meagre and vary on a daily basis. A large number of peasants have taken up this occupation in addition to farming to make two ends meet. Actually for the landless families, this is the only source of income.

Another significant occupation is zari work, done by the women mainly in several of the villages. They are employed by agents from Kolkata, who delegate them the work and purchase the entire output from them. In absence of electricity, the women work in candle light and subsequently damage their eyesight. The income is meagre, much less than the market rate; yet, since they can do this work from home, it suits them. This is certainly a good source of income for those women whose husbands have gone out to work outside Sunderbans as construction labourers, thus making them the only earning members at home.

The art of boat making has been prevalent in the Sunderbans since the days of the Mughals. Villages well known for this work are Basanti, Jharkhali, Mollakhali, Rangabelia, Canning, Gosaba, etc. Boats are usually made from the barks of trees like Babla, Khirish, Goran, etc. The handicraft sector in the Sunderbans is rather small. Beautiful and colourful mats, rags, hats, baskets, shades and other decorative items are made by the locals using shrubs and leaves or barks of Dates and Palm trees. Mathurapur and Mandirbajar two small towns in South 24 Pargana are well known for this art. Goods made from thermocol, iron, bamboo and clay are also produced in a few villages and towns. Sweets made from jaggery and dates from Joynagar and Canning (S 24 Pargana) are quite famous. Handicrafts could be a better alternative for these people, but it needs to be developed. There is neither any capital for starting a business, nor there exists any transport infrastructure to carry the finished goods to the market or sell them locally. The region lacks adequate banking facility, lack of credit and cooperative societies, according to Banerjee (1998) and these factors constrain the growth of these alternative livelihood options in the region.

Thus, we can say that the two major livelihood options for the people of Sunderbans are heavily dependent on nature, fraught with risks and from the looks of it not very sustainable. The rest, while not risky are in desperate need of some capital investments and development of infrastructure.

1.3 *Effect of Climate and Anthropogenic Changes on Sunderbans*

Till the year 1770, the Sunderbans of India and Bangladesh together was spread over an area of 36,000 km^2. Currently, it covers an area of 25,000 km^2, which means almost 11,000 km^2 has been lost due to erosion. According to Hazra, (Future Imperfect 2007) by 2020 the region would lose 20 % of its land. The Indian part consists of 9630 km^2 and the rest lies in Bangladesh. Out of the 9630, 4264 km^2 of wetland constitutes reserve forests, which in turn comprises of 2195 km^2 of mangrove forest and 2069 km^2 of tidal river. This means that an area of around 5336 km^2 has been cleared of forest and used for human settlements in 19 blocks

(13 in South 24 Parganas and 6 in North 24 Parganas). This reduction of forest cover has been caused by both natural factors and human intervention (Hazra et al. 2003). In recent years, certain changes have been noticed in the climate of this region. Scholars are of the opinion that the span of the summer season has increased while the span of winter has decreased. Rainfall has increased by one and half times from what it was 15 years ago. The monsoon season are now getting delayed by approximately 15–20 days (Das 2006). Surface air temperature databases of the Sunderbans and adjacent parts of the Bay of Bengal have been analyzed by the School of Oceanography, Jadavpur University. They have noticed an increasing trend in the yearly rise in temperature. Their finding corroborates with the existing global warming phenomena. Temperature increase is found to be at the rate 0.019 % per year (Hazra et al. 2002).

The Sunderbans is presently under the threat of severe coastal erosion due to relative sea level rise (Hazra 2002). The analysis of 50 years of data for Mean Sea Level (PSMSL) from three of the four data stations in the Hugli estuary shows sea level increase between +0.76 and +5.22 mm/year at different locations in the Sunderbans (Nandy and Bandypadhyay 2011). The tide gauge data of Sagar Island observatory for the period 2002–2009 indicated a rise in the Relative Mean Sea Level (RMSL) at the rate of 12 mm/year during the decade (Hazra 2010). Considering the record of past 25 years, the rate of relative sea level rise comes close to 8 mm/year, which of course is significantly higher than the rate of 3.14 mm/year observed during the previous decade (Hazra 2002). Besides global warming and subsequent thermal expansion of water, subsidence of the Bengal Delta at 2–4 mm/year due to compaction of silt and other local causes may be responsible for the exceptionally high rate of relative sea level rise in the Indian Sundarbans. The mean tide level in the Sunderbans seems to vary in close correlation with sea surface temperature (SST) in the Bay of Bengal (Delta Vision 2011, p. 27). Current projections estimate that the temperature in the Sunderbans will rise by 1 °C by 2050 (Hazra et al. 2002). Between 2002 and 2009, the SST showed a rising trend at the rate of 0.0453 °C/year and reached the highest level in 2009 which confirms with a earlier study done by Singh (Singh 2002, cited in Delta Vision 2011, p. 29), that had predicted a decadal increase of about 0.4–0.5 °C. Increase in SST has a bearing on chemical composition of sea water in terms of increased acidification and decreased dissolved oxygen levels (Hazra 2010, cited in Delta Vision 2011). Considering the present relative sea level rise at 3.14 mm per year, it is estimated that by the year 2050, the compound sea level elevation will become close to 1 m and this estimated rise of sea level will pose serious problems during the pre- and post-monsoon phase when most of the cyclonic storms occur (Hazra et al. 2002, p. 10). The threat to Sunderbans in a 2 degree rise in temperature was very big, now with the 4° rise, it going to be a Herculean task to save these islands.

There seems to be a direct correlation between the amount of erosion and submergence with the rate of relative rise and fall of sea level in different island segments as found by the School of Oceanography. The impact of such erosion–accretion processes and sea level rise induced changes is predicted to be visible along the coastal zones of the island system along the tidal creeks and mangrove swamps. In Bangladesh this rate is more than double due to higher rate of deltaic subsidence. All over the world, the sea level is threatening a large number of coastal communities and turning them into "climate refugees". In the Sunderbans, thousands of people are being displaced from their original habitat from islands of Lohachara, Baisnabra, Khasimara Char, Baghjara, Bedford and Ghoramara. These displaced people are becoming part of the vulnerable community of "climate migrants". The inhabitants of Sunderbans are losing their land on a daily basis. The island of Sagar itself has registered a net loss of 30 km^2 area over the past 30 years and around 30,000 people have been rendered homeless there (Hazra et al. 2002, p. 22). According to some estimates, at least one lakh people will have to be evacuated from the 12 threatened inner estuary islands of the Sunderbans in the next decade if the present rate of submergence continues (Nandy 2010).

The impact of such erosion–accretion is expected to affect the ecological community of the Sunderbans who depends upon the resource of mangrove forests for its livelihood due to the rapid degradation of mangrove forest (Hazra et al. 2002).

A scrutiny of the "land use land cover change" data in the decadal scale discusses the changes due to natural causes like erosion, accretion, submergence due to sea level rise and corresponding geomorphic changes which together account for 66 % of the land use land cover change in the study area. Anthropogenic causes include, increasing population pressure, urbanization (settlement area has increased from 1226 to 1666 km^2 while agricultural land has reduced from 2149 to 1691 km^2 during the period 2001–08 (Danda et al. 2011) and growing human need leading to unwise utilization of natural resources are identified as factors behind the conversion of agricultural land, mangrove or mudflats into aquaculture farms and reclamation. Declining fish catch is due to the over extraction of near shore biological resources at unsustainable levels (Fisheries Department, 2008, cited in Danda et al. 2011).

From an ecological perspective, the Sunderbans may be classified into three divisions

1. The area that has resisted human encroachment and is still under forest cover and thus immune from the effects of global pollution and green house effects.
2. The area that has partially been altered by human intervention and is used for forest resource collection by the locals and also as a tourist spot.
3. The area completely transformed by human intervention and which does not retain much of its original characteristics (Kanjilal 2000, p. 26).

Based on the opinion of the scholars (Kanjilal 2000), it can be said that the area which has been transformed for human habitation is the most endangered one. In case of the other two, precautions can still be taken in order to stop them from entering the vulnerable zone.

One study has estimated that there has been a 5 % loss of forest cover in the 20 years between 1989 and 2009. The same study has found that often degraded forest is being replaced by saline blanks further reducing forest cover. The creation of saline blanks is largely attributed to sporadic inundation of the upper reaches of the island during high tide, storm surges and subsequent drying (Danda et al. 2011).

The change in salinity regime over the last 50–100 years resulted in the extinction of various animal populations and endangered some more (Sanyal 1983, cited in Hazra et al. 2003).

1.3.1 Increase of Occurrences of Cyclones

Cyclones are infrequent in Bay of Bengal from January to March. But the period between April and May face several varieties of cyclones which often strike Andhra-Orissa-West Bengal-Bangladesh coasts. Pre- and post-monsoon storms are more severe than the storms of the monsoon season. In between 2007 and 2009, the northern part of Bay of Bengal has witnessed four cyclones, namely, Sidr, Nargis, Bijli and Aila. All these cyclones brought strong wind, heavy rainfall and flooding, resulting in severe coastal erosion, destroying the embankment in the Sunderbans. Scientists are predicting such grave events with more frequency and destructive nature (Sachs 2010). The latest super cyclone to strike the Sunderbans was Aila, which had struck on May 26th, 2009 at a maximum speed of 110 km/h. The farm lands were inundated with saline water, animals died, diseases broke out, freshwater sources were ruined, homes destroyed and embankments were broken. The people of Sunderbans are still reeling under its after effects. Such cyclones are only serving as an early warning to us about what the future might have in store.

1.4 Mitigation

The Sunderban islands and its rich biodiversity including the million plus people are under threat. The survival of these islands is not only important for its inhabitants but for a far more wider audience. Without these islands acting as buffer, the entire lower portion of eastern India and Bangladesh will get submerged since it is already a subsiding delta.

Mitigation processes unlike Adaptation processes, can be carried out on a global scale and include reduction of carbon emissions and other drivers of climate change.

The major mitigation processes required in Sunderbans are

1.4.1 Afforestation of Mangroves

The importance of mangroves is immense for the survival of these islands. They prevent the soil from getting washed away because of the dual action of the rivers and the Sea. They are natural safeguards for these islands and the mainland against the onslaught of frequent cyclones. They also provide shelter to several marine lives, which would otherwise get washed away. Cutting these forests also releases a huge amount of carbon stored in them. During my field work in Sunderbans, a lot of my respondents kept of highlighting the importance of the mangroves as their protectors and thus the urgent need to save them. But poverty and the struggle to live today prevents them from saving the world for tomorrow. Thus, afforestation of mangroves in and around these islands need to be done immediately and on a huge scale.

1.4.2 Changes Sought for in the Practiced Livelihoods

Agriculture is the primary occupation of the people of Sunderbans. The saline soil, scarce water and the need to feed their entire family makes them buy hybrid seeds and use lots of chemical fertilizers to get higher yield. The run off from these fertilizers pollute the soil, the estuary and the local ponds. Sachs talks about the growth of algae in these nitrogen rich water bodies and finally their conversion into dead zones caused by eutrophication. The farmers know about the harm they are inflicting but again they said that they do not have a choice.

Fishing is the second most important occupation of the people of Sunderbans. For domestic use they fish in the nearby creeks and rivers. For market sale they go out to the sea in trawlers. These trawlers stay out in the sea for long periods of time and come back with huge catches. Only rich fishermen can afford such expeditions. However, globally concerns have been raised about over fishing causing extinction of several marine species.

Collection of baby shrimps is also a big and risky business prevailing here which is also harming the habitat. But it is the only source of livelihood for thousands of landless. Though there have been numerous occasions where women who are primarily involved in this occupation, standing in knee deep water has been attacked by crocodiles yet they continue to do it. Countries like Thailand, Taiwan, the Philippines, Malaysia and Indonesia are also facing similar issues. In these countries, protests are common against the environmentally harmful prawn fishing industry. Small patches of mangroves in these countries are also under threat of destruction because of unscientific capture methods. To mitigate the adverse environmental impact some of these countries have developed "induced breeding techniques" which is not being used in the Sundarbans. The quest for earning more money in a shorter time frame is causing faster destruction of the ecological balance of the delicate ecosystem of the Sundarbans (Bandopadhyay 2000, p. 4).

Simply stopping all these harmful practices for the good of the environment might seem a good mitigation process but there is a dark economic reality behind the scene. These steps could lead to the loss of livelihood for thousands. The farmers and fishermen need to be educated further about the impacts of their activities on the ecosystem. Alternate livelihood options need to be explored and provided to these inhabitants for the greater good of the environment.

1.4.3 Fossil Fuel Use

The per capita carbon footprint of these people is quite low. They use wood or coal fired stoves for cooking. Since they do not have grid electricity, the better of households have installed solar powered lights. In fact, a lot of them have been using solar lights for almost three decades now. However, a lot of them cannot afford solar powered systems, they thus use kerosene lamps.

Some of the pumps used for extracting water are diesel or kerosene driven. But then again like I have already mentioned not everyone can afford them. Most of their lands are ploughed by hand or animals; it is only a handful who have diesel-driven ploughs (Energy Survey 2012).

The diesel-driven auto rickshaws are the only mode of transport on the islands apart from bicycles. Some NGOs tried to introduce environment friendly electric vehicles (EVs) and do away with these rickshaws. The problem which arose were that the mud or brick roads were proving to be difficult to navigate by these EVs and also their prices were too high. The places where there are only mud roads, even the cycle-vans cannot ply, the only option is then to walk. Thus, we can say, carbon emission from fossil fuel usage is not very high in these islands anyway.

There is no industrial water discharge occurring from these islands in the neighbouring rivers or the sea for there are no heavy industries.

These islands are surrounded with water on all sides. Since that water is saline, neither can they drink it nor can they use it for irrigational purposes. Thus, they depend on rainfall and groundwater. However, with the underground aquifers dipping at a very first rate, these people should be made aware of that fact and taught to use water economically. For most of the villages the drinking water comes from the mainland via pipes, while for the rest they use groundwater.

1.5 Adaptation

Global Adaptation strategies include switching over to flood resistant paddy, moving to higher grounds specially in low lying settlements, starting to use cleaner energy; moving away from fossil fuels, building smart cities and use of smarter and sustainable form of transport. Technological development is required in order to reduce poverty and improve life on earth yet if "Bussiness as Usual" goes on with fossil fuel use going up almost on a daily basis, adding on to the GHG emissions further, draining of global fresh water reserves drastically then it would be impossible to stay within the planetary boundaries (Sachs 2010).

The people of Sunderbans have started adapting to the changing climatic conditions in various ways.

They are shifting the farming time in anticipation of shifting of the monsoon season. However, this will lead to lowering of market demand and price. Many are diversifying into different weather resistant crops. Construction and renovation of ponds and canals for rain water harvesting and use in winter cultivation is being carried out. Construction of mud barrages around the island to protect it from incursion of saline water has also been taken up. However, lack of finance and absence of a proper institutional mechanism also act as major deterrents to its continuation. Reforestation of mangroves on the mud barrages is also being carried out to make it durable. Search is on for alternative livelihood options especially by those who are involved in baby prawn/shrimp fishing, timber smuggling, etc. Effective capacity building efforts through scientific and organizational intervention are also being taken up to support the indigenous adaptation activities (Sengupta and Rao 2003, pp. 10–11).

Again to meet the changing climatic conditions Nandy (2010) feels that some specific measures should be taken. Some of them are: use of salt tolerant paddy, introduction of salinity shock resistant fishes in the ponds and construction of disaster risk reduction shelters. In addition, a proper Climate Change plan and policies need to be worked out (Nandy 2010). In a nut shell, a lot of thinking and research is required to come up with adaptation strategies for these people.

1.5.1 Construction and Maintenance of Sustainable Embankments

In the nineteenth century, land was reclaimed at the limit of the low water level in Sunderbans by building embankments (Danda 2007). The islands were encroached upon, before the natural delta process had been completed. Nature builds new islands by silting up interlocking creeks. Gradually, the islands were interconnected by filling up the intervening channels and the land getting raised permanently above the high water level. But because of the embankment, the silt that would have been otherwise deposited on the islands got deposited in the creeks, raising their levels instead. Over the time the creek beds rose higher than the low lying reclaimed areas, turning those into vast stretches of permanent marshes (Mukherjee 1969, cited in Danda 2007). The embankment was supposed to act as natural barriers against the ingress of sea water. Unfortunately the embankments also extinguished any possibility of these tracts ever maturing naturally into lands habitable by humans, (Bhattacharya 1998) and this makes the situation worse. It needs to be mentioned that the heights of these islands are not more than 10–12 ft. from the sea level, yet during high tide, the water level goes up by 17–18 ft. The total length of the embankment in the Sunderbans is 3500 km. The entire stretch is made with mud and thus it is very weak. The soil was not suitable for sustaining the structures, maintenance was inadequate on a regular basis, (Danda 2007 and at that time the alignments of the embankments were also defective (Kanjilal 2000). Since the abolition of the Zamindari system in 1951 the embankments became the responsibility of the Drainage wing of the Irrigation Department, Government of West Bengal (Chakraborty 2005, cited in Danda 2007). The embankment not only acts as the lifeline of the people but also serve as the main road connecting different islands and providing shelter during the frequent natural calamities. Thus the villages of the region grew around the embankments making them the central nuclei of the settlements. However, the embankments are not very well maintained. The people blame the government for the neglect while the authorities cite difficulties raised by the people. Employing the locals in the maintenance work can be a good way to soothe both the parties and it is being done in some of the villages. However, the residents of Rangabelia, Pakhiralaya and Gosaba have complained, that the government does not employ them in the construction and maintenance of the embankment. Having lived by the embankment, they feel they are much more knowledgeable than the city bred engineers and contractors. In Beguakhali, however, locals themselves were involved in the maintenance work. It is also a means to earn some money. During the survey it was also found, that the embankment in Sagar is better maintained than the ones in Rangabelia, Pakhiralaya and Gosaba. The government alone cannot be blamed for the poor state of the embankment the villagers are also to be blamed. Often, bricks from the embankment are taken away by them to build homes. As the houses of the poor villagers are annually destroyed by storms or by the river, out of desperation, they take bricks from the embankment. Branches from mangroves are also cut to make new houses.

When asked whether they would give up their land for construction of embankment, they said they would, but as long as they are compensated for their loss. The families living beside the embankment have already lost a great amount of land to the river and are still losing. Thus, only if they are relocated and properly compensated with land, they can give up their existing land for building embankment. But then actually these villages hardly have any virgin land left and it is almost impossible to compensate them with new land elsewhere. This has become a major issue of conflict between the villagers and the authorities. Thus, policies are needed urgently, which would help the villages to rebuild their houses in the aftermath of cyclones so that they do not harm the embankment which is extremely important for the survival of these island dwellers and their livelihoods.

In the 1960s, the State government planned to get Dutch engineers and build dams and dykes in the Sunderbans in the Dutch style. However, structurally the Sunderbans is quite different from Netherlands. The deltas in the Netherlands are outward facing, while the ones in the Sunderbans are inward facing. The rate of sedimentation and inflow of saline water are much higher in the Sunderbans as compared to in the Netherlands. The waves go as high as 3.5 m in the Netherlands, whereas in the Sunderbans in high tides, waves go as high as 5 m. Hence the proposal was shelved (Dutta 2010).

Most of the houses here are made of mud and are of single storey. There are some brick, two storeyed houses, but their numbers are not very big. The reason behind this is that, due to the frequent flooding and land erosion, the people lose their land and homes on a fairly regular basis and thus they do not spend a lot of money on building houses. At times, households with a number of members working outside and who own land in the centre of the village take the risk of building a sturdy two storeyed houses or else they make do with their mud houses with thatched roofs.

1.5.2 Outmigration

I mentioned already that the farmers in Sunderbans are subsistence farmers. The fishermen risk their lives for getting a good catch. People become landless because of high tides. They loose their homes every time there is a devastating cyclone. Yes, climate change is definitely making their lives worse, but it would seem that their lives were never comfortable. Economic distress was always there. Thus for almost three decades now, male members from Sunderbans have been migrating out seasonally to big metro cities in search for greener pastures. Most of them find work in cities as unskilled construction labourers or artisans in textile or jewelery factories. They do go back to help during harvests. In the recent past this tendency to go out has increased. I found several instances where the entire family had out migrated after losing their land or house to a cyclone. Young women are also going out in large numbers. Some of them find genuine jobs and work as maids or care giver in cities while others end up in the hands of traffickers. During the course of interaction with the people of Sunderbans over a course of two years I found the elderlies, women with young children and school going children are mainly the ones left behind in these islands. The men keep coming and going. Though, some of the households benefit from the extra money, again at times, these men and women come back so sick working under inhuman conditions that their savings dry off paying their medical bills. But these people are not alone. All over the world climate refugees are increasing in numbers.

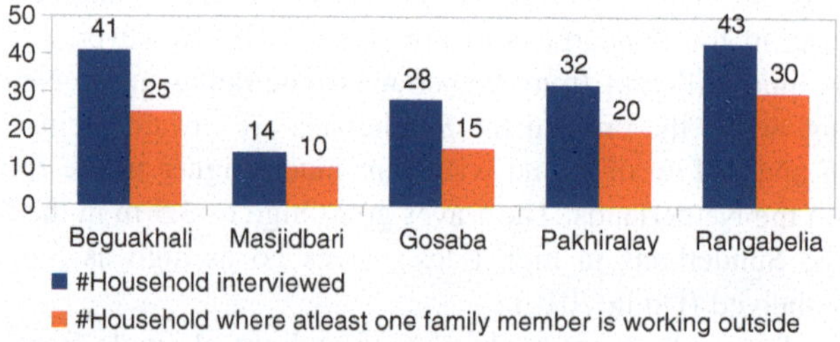

The Outmigration flow from the sample villages. *Source* Compiled from the primary data collected by Pakrashi (2012, p. 49)

1.6 Conclusion

The threat to Sunderbans is very real and urgent. More and more islands are drowning with rising sea level and the protective mangrove forest is disappearing slowly. The people and animals inhabiting these islands are losing their homes at an alarming rate. The rising sea level is inundating these low lying islands more frequently than ever before. This is resulting in increased salinity levels in the soil and subsequent reduction of its fertility. The farmers, in order to get higher yields in these adverse conditions are increasing their dependence on hybrid seeds and chemical fertilizers, which by itself is not a sustainable process.

Human adaptation to climate change induced impacts in one sector affects other sectors both in the same region and outside (Warren 2011). The fact is that the run off from these chemical fertilizers are polluting the surrounding water bodies and subsequently increasing the acid levels of the Sea. Increased rates of acidification of Oceans, creation of dead zones and over fishing together are leading to extinction of several species of fishes and harming the marine ecosystem all over the world. This in turn would hurt most the coastal communities who depend on the Sea for their livelihoods. This problem has already started affecting the Sunderban fishermen. "Human adaptation responses to sea-level rise range from managed retreat, building of dykes and construction of flood barriers" (cited in Warren 2011, p. 227), But with the sea levels going up considerably, it would be challenging to make the embankments' height match the height of the rising Sea. Also building such barriers harm the ecosystem for it hinders the growth of mangroves. Warren writes that in a 4 degree world, water scarcity would go up, soil fertility would reduce, crops would fail and this all would force people to migrate out from their homelands in search of greener pastures. Unfortunately all these climate change and anthropogenic causes induced effects are already showing in Sunderbans and is forcing the inhabitants to move out. Afforestation with indigenous trees, reduction of fossil fuel usage globally and stopping deforestation are the only ways to save ecologically fragile regions like the Sunderbans.

The Delta Vision Report published by World Wild Life Fund in 2011 suggested that all the inhabitants of the Sunderban islands should move out in waves and settle elsewhere. They argue that since the island building process was incomplete when the settlements began, the islands are thus unable to cope with the rising sea level and the current rate of erosion. But social scientists counter argue, for these unskilled or not formally trained folks of Sunderbans, such a drastic step would be impossible to manage. Yes, they do outmigrate but they come back during harvest or when they fall sick. But if these lands are completely taken away from them, they would simply become homeless in big cities.

Thus my question, Is there a tomorrow for Sunderbans? And even if the dwellers migrate out, will the islands survive?

References

Bairi, A. (2004). *Sunderbaner krishi: Samashya o unnayan sambhabana (Cultivation is Sunderbans: Problems and paths of development, Srikhanda Sunderban*. Kolkata: Deep Prakashan.

Bandopadhyay, D., (2000, November). Can the Sunderbans be saved? EPW (Vol. 35, No. 45).

Banerjee, A. (1998). *Environment, population and human settlements of Sunderbans Delta*. New Delhi: Concept Publishing Company.

Bhattacharya, S (1998): Sunderbans: Dying a slow death. The Hindu survey of the environment, 1998 (cited by Danda, 97, Chap. 4).

Carrington, D. (2013). Planet likely to warm by 4C by 2100, scientists warn. The Guardian, 31 December 2013. http://www.theguardian.com/environment/2013/dec/31/planet-will-warm-4c-2100-climate. Accessed on March 2, 2015.

Census of India. (2011). http://www.census2011.co.in/district.php. Accessed on December 25, 2015.

Danda, A. (2007). *Surviving in the Sunderbans: threats and responses* (PhD dissertation, University of Twete, The Netherlands).

Danda, A. A., Sriskanthan, G., Ghosh, A., Bandyopadhyay, J., & Hazra, S. (2011). *Indian Sundarbans Delta: A vision*. New Delhi: World Wide Fund for Nature-India.

Das, G. (2006). *Sunderbans: Environment and ecosystem*. Kolkata: Levant Book.

Dutta, S. (2010). *Bandher Badhan Bhange kishe, (How does the Embankment break). Prekhapot Sunderban*. Kolkata: Ajkal Publishers.

Energy Survey. (2012). Survey conducted by the author for an organization working on environment. The organization was trying to build environmental custodianship amongst the people. They wanted to know the energy usage of the residents of Brajaballavpur Gram Panchayat. The survey was conducted in the months of November–December with the help of questionnaires. We had surveyed around 500 households in that GP.

Friend, A. (2013). 4 degree temperature rise will end vegetation 'carbon sink'. 17 December 2013. http://www.cam.ac.uk/research/news/4-degree-temperature-rise-will-end-vegetation-carbon-sink#sthash.QBdR7xEK.dpuf. Accessed on March 3, 2015.

Future Imperfect. (2007). A climate witness video by WWF. http://www.youtube.com/watch?v=eLaaIBIBp10. Accessed on January 2012.

Ghoshal, D. (2014). India's vital monsoon rains are changing—and not for the better. Quartz, India. 10 August 2014. http://timesofindia.indiatimes.com/city/mumbai/Untimely-rain-sign-of-climate-change/articleshow/45509188.cms. Accessed on December 15, 2014.

Goldenberg, S. (2014). Why global water shortages pose threat of terror and war. The Guardian. 9 February 2014. http://www.theguardian.com/environment/2014/feb/09/global-water-shortages-threat-terror-war. Accessed on February 10, 2014.

Graham, S. (1999). Ice over the poles. Earth Observatory. http://earthobservatory.nasa.gov/Features/PolarIce/polar_ice.php. Accessed on March 7, 2015.

Hazra, S., Ghosh, T., DasGupta, R., & Sen, G. (2002). *Sea level and associated changes in the Sunderbans*. School of Oceanographic Studies, Jadavpur University: Kolkata.

Hazra, S. et al. (2003). A preparatory assessment of vulnerability of the ecologically sensitive Sunderban island system: West Bengal in the perspective of climate change.

Hazra, S. (2010). Temporal change detection (2001–2008) of the Sundarban. Unpublished report, WWF-India.

Judge, C. (2013). The coming water wars. US News. 19 February 2013. http://www.usnews.com/opinion/blogs/clark-judge/2013/02/19/the-next-big-wars-will-be-fought-over-water. Accessed on February 10, 2014.

Kanjilal, T. (2000). Who killed the Sunderbans, Tagore Society for Rural Development (TSRD), Kolkata. TSRD is one of the biggest and oldest NGOs working in Sunderbans. They have training centres for women, sanitary marts and several other projects running in different islands.

Kemsley, T. (2013). Global average temperatures set to rise by 4 degrees celsius by 2100. 31 December 2013. http://www.natureworldnews.com/articles/5487/20131231/global-average-temperatures-set-to-rise-by-4-degrees-celsius-by-2100.htm. Accessed on March 2, 2015.

Manocha, V. (2014). Centre warns of possible drought in western India. Down to Earth. 2 July 2014. http://www.downtoearth.org.in/content/centre-warns-possible-drought-western-india. Accessed on December 15, 2014.

Misra, A. (2014). Central America reeling under severe draught. Down to Earth. 11 August 2011. http://www.downtoearth.org.in/content/central-america-reeling-under-severe-drought. Accessed on December 15, 2014.

Nandy, S. (2010). Climate adaptation strategies—Study of great river Ganges in Sunderbans. In *13th International River Symposium*, Perth, Australia.

Nandy, S., & Bandopadhyay, S. (2011). Trend of sea level change in the Hugli estuary, India. *Indian Journal of Geo-Marine Sciences, 40*(6), 802–812.

Pakrashi, H. (2012). Problems of development and planning in the Sunderbans. *Masters dissertation for Masters in Development Studies.* Mumbai: Tata Institute of Social Sciences.

Pallavi, A. (2014). Drinking water shortage to hit Maharashtra as drought worsens. Down to Earth. 28 August 2014. http://www.downtoearth.org.in/content/drinking-water-shortage-hit-maharashtra-drought-worsens. Accessed on December 15, 2014.

Patel, V., & Rajagopalan, R. (2009). *Fishing communities issues in the Sunderbans Tiger reserve.* Chennai: International Community in support of Fish workers.

Pinto, R. (2014). Untimely rain sign of climate change. The Times of India. 14 December 2014. http://timesofindia.indiatimes.com/city/mumbai/Untimely-rain-sign-of-climate-change/articleshow/45509188.cms. Accessed on December 15, 2014.

Press Trust of India. (2014). Global temperatures to rise 4 degrees by 2100. 1 January 2014. http://timesofindia.indiatimes.com/home/environment/global-warming/Global-temperatures-to-rise-4-degrees-by-2100/articleshow/28232719.cms. Accessed on March 2, 2015.

Rawat G. S., & Wikramanayake, E. D. (2015). Mangroves: Southern Asia: Bangladesh and India, WWF. https://www.worldwildlife.org/ecoregions/im1406. Accessed on February 12, 2015.

Rockström, J., Steffen, W., Noone, K., Persson, Å., Chapin, F. S., Lambin, E. F., et al. (2009, September 24). A safe operating space for humanity. *Nature, 461.*

Sachs, J. (2010). Sustainable development. It is an online free course on Coursera. https://www.coursera.org/learn/sustainabledevelopment1/outline. In order to see the videos and get the readings on Sustainable Development delivered by Prof. Sachs, one needs to enroll for this course after going to Coursera. Click on this link https://www.coursera.org to go to their homepage.

Saha, D. (2004). Sunderbaner chirayat krishir udesshe bhinnamukhi fashal chakra, (The pattern of cultivation in the Sunderbans), Srikhanda Sunderbans. Kolkata: Deep Prakashan.

Sanyal, P. (1983). Mangrove Tiger Land: The Sundarbans of India. Tiger Paper–X (Vols. 1–4).

Sengupta, S., & Rao, P., (2003). *Adverse impacts of climate change: Perception and responses of local community; Sunderban Delta.* WWF report for UNFCCC workshop.

Tiger Widows. (2011). http://www.tigerwidows.org. They organize widows and form self help groups and also help in educating the women and children. They try to help these widows start their life afresh.

Warren, R. (2011). The role of interactions in a world implementing adaptation and mitigation solutions to climate change. *Philosophical Transactions of Royal Society A, 369*, 217–241.

The Population Decline of Indian Sandalwood and People's Role in Conservation—An Analysis

A.N. Arunkumar, Geeta Joshi, M.S. Rao, T.S. Rathore and V. Ramakantha

Abstract Indian Sandalwood (*Santalum album* L.) is one of the most important trees in Indian forestry. Time immemorial, Sandalwood has been an integral part of the culture and heritage of India. Description of Sandalwood appears in *Subhashitas*—collections of verses of wise sayings, code of conduct and ethics. The various uses of Sandalwood and oil have been mentioned extensively in the Indian literature and the aroma of wood and oil has been valued by all the major religions of the world. The fragrant heartwood and the essential oil obtained from it is a significant component in most of the auspicious occasion and a vital cog during cremation as the aroma that emanates from the burnt wood is considered to be the carrier of soul to its destination. The wood is used for carving and handicrafts while the oil has extensive uses in pharmaceutical industry and is an essential ingredient in all high valued perfumes world over. The Southern part of Karnataka and Northern part of Tamil Nadu is considered to be the natural distribution area of Sandalwood. Therefore, Karnataka State is also popularly known as *Gandhada Nadu* (Sandalwood state). For having been ordained with Royal tree status—a tag indicating the exclusive proprietorship of Government, not only hampered the Sandalwood tree cultivation, but also indirectly resulted in over exploitation and subsequently being categorised as vulnerable in IUCN Red List. The average production of Sandalwood in Karnataka between the years 1960–1961 and 1964–1965 was 2287.8 tons which drastically reduced to 366.74 tons by the end of the last five years of the twentieth century (1995–1996 to 1999–2000). The average production further fell to a mere 61.57 tons in the first decade of this century. The situation in Tamil Nadu is also very similar and the natural population has substantially reduced. It has been reported that commercially utilisable Sandalwood trees of girth more than 30 cm to a large extent is absent in its natural area of distribution in both the states. Considering this critical situation, in the early part of this century, the two Governments have relaxed their policies with reference to Sandalwood cultivation. This has encouraged farmers and entrepreneurs

A.N. Arunkumar (✉) · G. Joshi · M.S. Rao · T.S. Rathore · V. Ramakantha
Institute of Wood Science and Technology, Malleswaram, Bangalore, India
e-mail: anarunkumar@gmail.com

S. Nautiyal et al. (eds.), *Climate Change Challenge (3C) and Social-Economic-Ecological Interface-Building*, Environmental Science,
DOI 10.1007/978-3-319-31014-5_22

to cultivate Sandalwood. We discuss the future of Sandalwood and the role of people participation especially in conservation and utilisation of Indian Sandalwood.

Keywords Indian Sandalwood · Conservation · Production · Utilisation

1 Introduction

Indian Sandalwood (*Santalum album* L.) has been an integral part of Indian culture and is one of the prized woods in Indian Forestry. Globally, Sandalwood is known for its valuable heartwood (the inner dead wood part of the tree) and the oil that is distilled from the heartwood has a huge demand in the international market. In fact, the fragrance of the oil and the wood is so profound that it is venerated by, Hinduism, Buddhism and Islams, which are the three major religions of the world. According to various Sanskrit manuscripts, Sandalwood oil is considered to be one of the oldest known perfume materials (more than 4000 years). Burdock and Carabin (2008) mentions that way back Egyptians had been importing this wood and used it for medicine purpose, embalming and ritual burning to venerate the gods. The Sandalwood paste obtained by rubbing wood on the stone with a little water is used in spiritual and ritualistic practices. The wood is also used in death rituals, as it is considered that the soul is carried back to its eternal abode with the scent of Sandalwood.

Sandalwood tree is one of those few tree species which has been prudently utilised in *Subhashitas*—which are wise sayings that describe code of conduct and ethics (Arun Kumar et al. 2012). The stanzas which are composed in Sanskrit decipher various aspects of life as it guides and protects a person from wavering thoughts and actions. To describe the saying 'excessive familiarity breeds contempt and repeated visit brings disrespect'—it is mentioned that a tribal woman living in Malaya Mountain which has Sandalwood in abundance, uses Sandalwood as a fuel wood. The same Sandalwood is highly valued in the outside world as it is not easily available. To stress that good things are not common and has to be valued, it is described that "not every mountain has precious stones, not every elephant has pearls, Sandalwood is not found in every forest, and similarly saintly persons are not found everywhere". To understand the value of being a good or saintly person and useful to others, it has been written that "one should see how Sandalwood tree is selfless, the roots of the Sandalwood tree provide shelter to snakes, on the crown birds take rest while monkeys play on its branches and flowers are used by bees". The value of Sandalwood can be summarised by quoting what our own Nobel Laureate and great poet Gurudev Rabindranath Tagore described Sandalwood tree as—"The sandal tree as if to prove, how sweet to conquer hate, love, perfumes the axe that lays it low". All these indicate the value and divinity this important tree possess and its influence in our day-to-day life. Being a prized product of Indian forests, Sandalwood is commercially known as Indian Sandalwood and its oil Indian Sandalwood oil.

The wood is an excellent material for carving and the oil is being extensively used in perfumery and pharmaceutical industries. Sapwood of Sandalwood is white or yellow in colour and not scented, but is used for wood turning, particularly for toy making and is extensively now used in making *agarbattis* (incense sticks).

Sandalwood has inseparable history with Karnataka (the Princely State of Mysore, post independence period became Mysore State after reorganisation of states in 1956 and was subsequently renamed as Karnataka in 1973). Much before the reorganisation of this Princely State of Mysore, B.M. Srikantaiah, popularly known as *Kannada Kanva* for being the custodian of Kannada literature and the champion of the Kannada literary movement composed a poem on the Princely State of Mysore in which he describes that Princely State of Mysore as *Gandhada Nadu* (shrine of Sandalwood). The abundance of Sandalwood availability also corroborated this description and it became such a household name that the Kannada language film industry located in Bengaluru has a sobriquet as Sandalwood, which is a parlance to Bollywood for Hindi language industry located in Mumbai and Hollywood for English language film industry located in Los Angeles.

2 Research on Sandalwood

It is a small to medium evergreen partial root parasite tree; and has been extensively overexploited for its commercial value because of which International Conservation for Nature has categorised it as vulnerable. Perhaps one of those few trees in the realm of Indian trees in which research work has been carried out for more than 140 years considering the first report of being a partial root parasite was published by Scot (1871). It was in 1899 that Sandalwood spike disease was reported (Mc Carthy 1899). The spread of this was so dreadful and rampant that between 1903 and 1916, more than a million trees were removed/killed in Mysore State (7,00,000) and the erstwhile Coorg State and the estimated possible loss was to an extent of three and half million rupees during that time (Rajan 1992). This may also be one of the reasons prompting the Raja Rishi Nalwadi Krisharaja Wodeyar (1884–1940), Maharaja of Mysore to seriously look into the issue of Sandalwood spike disease and Forest Research Laboratory was established in 1938. With its reorganisation in 1956 as Regional Research Centre of Forest Research Institute and Colleges, an exclusive division named after Sandalwood Spike was started. In 1977, Sandal Research Centre (SRC) was set up to carry out research on wide-ranging aspects of silviculture, genetics and management of Indian Sandalwood which heralded a new beginning in Sandalwood research. As an initiative to strengthen the Sandalwood improvement research work, the first countrywide survey of Sandalwood was carried out during 1978–1980. It was reported that Sandalwood was distributed in an area of 9034 km^2, and ~ 90 % of Sandalwood being found in Southern part of Karnataka and Northern part of Tamil Nadu (Rai 1990) apart from the small population in Kerala and Andhra Pradesh, but spread of Sandalwood had been reported to be far and wide. Further research work demonstrated that Sandalwood exhibits considerable variability for different morphological traits. Variability was documented for the bark colour, texture, thickness and rust coloured bark being associated with fast growth (Kulkarni and Srimathi 1982) Seeds showed polymorphic characters in size, shape, germination and vigour. (Nagaveni and Anathapadmanabha 1986; Veerendra and Sarma 1990; Ramalakshmi and Rangaswamy 1998; Annapurna et al. 2005) and six types of leaves—ovate (common), lanceolate (12 %), elliptic, linear, small and big which were being discernible at juvenile stage (Kulkarni and Srimathi 1982; Bagchi and Veerendra 1985). Apart from the variability for the morphological traits, various studies using molecular markers have revealed that there is considerable genetic diversity in Sandalwood population (Angadi et al. 2003; Shashidhara et al. 2003; Suma and Balasundaran 2003, 2004; Nageshwar Rao et al. 2007; Jones 2008; Azeez et al. 2009).

3 Studies on Heartwood and Oil Content

Considering the importance and value of Sandalwood, two traits that play significant role are heartwood and oil content. Studies have been carried out by various authors and initially most of them being on the basis of field observations/opinions and very few of them have been supported with scientific data. Cameron (1894) suggested that Sandalwood tree attains commercial maturity at 27–30 years. Rama Rao (1904) found that there was variation in heartwood content between and within different girth classes of trees. Singh (1915) in his observation on 40 trees was of the view that elevation, age or locality has any definite relation with the oil quality. Even in the Fifth Silviculture Conference held at Dehradun in 1941, Mitchell stressed that it would take a long time to determine factors ideal for formation of heartwood and increasing the oil content and merits investigation. Bhatnagar (1965) reported that Sandalwood trees reach full maturity at an age of 50–80 years or more and a tree may reach physiological maturity without forming any heartwood. Even in the First All India Sandal Seminar held in Bangalore in 1977 and the subsequent Second All India Sandal Seminar that was convened highlighted the fact that selection of genotypes having higher heartwood content and research on heartwood formation must be given a priority. Srimathi and Kulkarni (1980) carried out an interesting study to understand the variation for heartwood content among different trees (n = 50) having similar diameter of 11 cm in the Forest Research Laboratory Campus. It was found that the coefficient of variation value was as high as 41.7 % for heartwood content. They opined that there is no fixed year when the heartwood starts forming and mentioned that it can start as early as 5–6 years in some trees and as late as over 15 years. Similarly, Jayappa et al. (1981) collected Sandalwood samples from different locations and found variability in Sandalwood oil for α and β Santalol content. Even though most of the studies revealed substantial insights about heartwood and oil, the major inherent flaw in most of these studies were that the samples were collected from the trees in which age was not known. The effect of age on the complex phenomenon of heartwood and oil formation was quite significant and could not be overlooked. In a study carried out by Brand et al. (2006) on fourteen-year-old Sandalwood trees ($n = 20$) revealed interesting observations. It was found that heartwood formation had not occurred in four trees at 30 cm height (20 %) and five trees at 100 cm height (25 %) above the ground. Similarly, a study carried out by Arun Kumar (2005) on 20-year-old Sandalwood trees grown in clonal germplasm bank ($n = 111$) not only showed considerable variation for heartwood and oil, but also found that in 14 % trees, heartwood had not been formed. Interestingly, it was also found that there was a strong positive relationship between tree diameter and heartwood diameter, while heartwood and oil content did not show significant relationship indicating that heartwood formation and oil accumulation are two independent processes (Arun Kumar et al. 2011). Considering all these studies, some of the grey areas in heartwood research are—heartwood and its formation is yet to be properly understood, interaction of Genetics and environment in heartwood formation is still not known and understanding biosynthetic pathway of oil formation is still not clear.

4 Sandalwood Production and Current Scenario

Sandalwood production in major Sandalwood growing states of Karnataka and Tamil Nadu has significantly dwindled over the period of time. In case of Karnataka, the average production of Sandalwood between the years 1960–1961 to 1964–1965 was 2287.8 tons which drastically reduced to 366.74 tons by the end of the last five years of the twentieth century (1995–1996 to 1999–2000). It further fell to a mere 61.57 tons in the first decade of this century. As is evident from the Fig. 1, during the period from 1995–1996 to 2013–2014, the quantity of Sandalwood produced and seized in Karnataka has already reached an alarming stage. In 1997–1998 the production was 316 metric tones and the seizure was 80.07 metric tones. The highest production of 605 metric tons was in 1999–2000. However, in the 2013–2014, the production was only 0.95 metric ton indicating the present precarious condition of Sandalwood in the state of Karnataka. It has been reported in 1997 that commercially utilisable Sandalwood trees of girth more than 30 cm to a large extent is absent in its natural area of distribution in Karnataka (Swaminath et al. 1998). The situation in Tamil Nadu is also very similar and the natural population has substantially reduced. It is pertinent to mention here that Sandalwood tree was considered as the government property in the states of Karnataka and Tamil Nadu. Realising the serious situation, Governments of Karnataka and Tamil Nadu have relaxed their policies which to a considerable extent has started encouraging growing of Sandalwood in private lands. With the decline in Sandalwood availability, the price has skyrocketed. The current cost of heartwood (June, 2015) that is being sold for general public at Cauvery Handicraft Emporium, the Karnataka State Government outlet is ~Rs. 1800 for 100 g of wood and the Sandalwood oil is being priced at Rs. 1800 for 5 g which amounts to ~US$ 5673/kg of oil. The Government of Karnataka recently

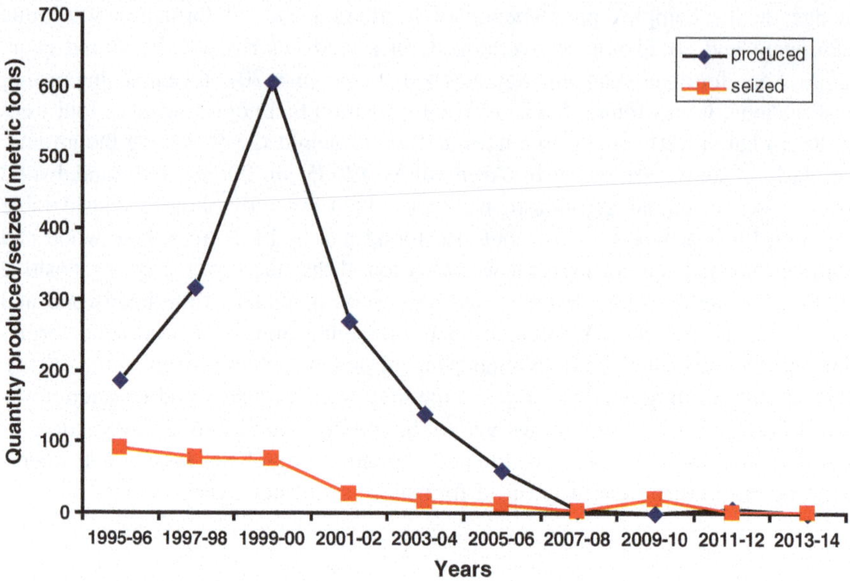

Fig. 1 Quantity of Sandalwood produced and seized in Karnataka from 1995–1996 to 2013–14. *Source* Annual Reports, Karnataka Forest Department

(21/10/2014) has fixed the price of sorted Sandalwood before being passed for sale, the maximum and minimum price varies from Rs. 8338/kg (China Budh and Ghotla) to Rs. 3056/kg (Hatri Chilta), respectively (Table 1).

As per the first all India Sandalwood survey, Sandalwood is present in many parts of India. Apart from peninsular Indian states, it is found in Orissa, Madhya Pradesh, Uttar Pradesh, Himachal Pradesh, Uttarakhand, Rajasthan, Gujarat and in Assam. From the legal perspective, some of the states in India that specifically mention about Sandalwood harvest are as follows (MoEF 2012):

Tamil Nadu: In case of Sandalwood grown by the Government, the harvesting/cutting of Sandalwood trees in a forest and subsequent transportation outside are governed by the Madras Forest Act, 1882. To facilitate Sandalwood cultivation, land owners have been allowed to own and grow Sandalwood trees on their land, with the restriction that the disposal should be through forest department.

Kerala: According to 'Promotion of Tree growth in Non- forest Areas Act, 2005', in the non forest area, the land owner is permitted to cut and transport any tree, other than Sandalwood tree growing on his land.

Maharashtra: Tree Felling in private lands is regulated by following three Acts: (i) Maharashtra Felling of Tree (Act 1964), (ii) Maharashtra (Urban Areas) Preservation of Trees Act, 1975; (iii) The Maharashtra Land Revenue Code, 1966. Permission for tree felling and transportation of forest produce is generally given by

Table 1 Classification of Sandalwood sorted before being passed for sale (as per the Karnataka Forest Manual Rule No. 95) and the fixed price by Government of Karnataka per kilogram of wood (as on 21/10/2014)

Sl. no.	Class	Description	Price per kilogram*
1	Vilayat Budh (Class I billets)	Sound billet weighing not less than 9 kg and not exceeding 112 pieces per ton	7869.00
2	China Budh (Class II billets)	Slightly inferior billet weighing less than 4.50 kg and not exceeding 224 pieces per ton	8338.00
3	Panjam (Class III billets)	Billets having small knots, cracks and hollows weighing not less than 2.2 kg and not exceeding 448 pieces per ton	7558.00
4	Ghotla (Billets of short length)	Includes short and sound pieces. There are no limits of weights and numbers per ton	8338.00
5	Ghatbadla	Billets with knots, cracks, small hollows, weighing not less than 4.5 kg and not exceeding 250 pieces per ton	7572.00
6	Bagardad	Consists of solid pieces without limit as regards dimensions, weight or number	7285.00
7	Roots (Class I)	Pieces weighing not less than 6.75 kg and not exceeding 150 pieces per ton	5334.00
8	Roots (Class II)	Consists of pieces weighing not less than 2.25 kg and not exceeding 448 pieces per ton	5389.00
9	Roots (Class III)	Consists of small and side roots below 2.25 kg in weight	5527.00
10	Jajpokal or Badla (Class I)	Consists of hollow pieces weighing not less than 3.10 kg and not exceeding 320 pieces per ton	6711.00
11	Jajpokal (Class II)	Hollow pieces weighing not less than 1.3 kg per ton	6375.00
12	Ainbagar	Consists of solid, cracked and hollow pieces weighing not less than 450 gm	7155.00
13	China Sali or Large Chilta	Consists of pieces and chips of heartwood weighing not less than 2.25 gm	7155.00
14	Ain Chilta	Consists of small pieces of heartwood	4756.00
15	Hatri Chilta	Consists of heartwood, and chips obtained by planing billets with Hatri or Randha (Plane)	3056.00
16	Milva Chilta	Consists of pieces and chips having fair proportions of heartwood and sapwood	2801.00
17	Basola Bukni	Consists of small heartwood and sapwood chips	2249.00
18	Saw dust	Sawn powder obtained while sawing the Sandalwood	976.60

* Includes FDT 12 %; VAT 14.5 %; IT 2.5 %; SC 10 % on IT; E.cess 3 %. The sapwood and bark costs 97.00 rupees and 33.00 rupees per kilogram. *Source* Karnataka Forest Department

the concerned Deputy Conservator of Forests under the provisions of the Indian Forest Act 1927 and the rules made there under namely the Bombay Forest Rules, 1942. Sandalwood is scheduled and covered under the Transit Regulation.

Table 2 A few pertinent Sandalwood provisions in the states of Karnataka, Kerala and Tamil Nadu

<table>
<tr><th>Sandalwood provisions</th><th>Karnataka State</th><th>Kerala State</th><th>Tamil Nadu State</th></tr>
<tr><td>Sandalwood status</td><td>Forest produce</td><td>Forest produce</td><td>Royalty</td></tr>
<tr><td>Ownership of Sandalwood cultivated in farm lands or private lands</td><td>Landowner</td><td>Not specified</td><td>Landowner</td></tr>
<tr><td>Permission for harvesting the Sandalwood tree</td><td>Only matured trees after obtaining permission from Forest Department</td><td>After obtaining requisite permission from the Forest Department</td><td>Harvested only by the Forest Department</td></tr>
<tr><td rowspan="3">Selling of Sandalwood</td><td>State Forest Department</td><td rowspan="3">State Forest Department</td><td rowspan="3">State Forest Department</td></tr>
<tr><td>Karnataka Soaps and Detergents Limited</td></tr>
<tr><td>Karnataka State Handicrafts Development Corporation</td></tr>
</table>

Source Dhanya et al. 2010

Andhra Pradesh: The Andhra Pradesh Forest Produce Transit Rules, 1970 is not applicable to two important species Sandalwood (*S. album*) and Red sanders (*Pterocarpus santalinus*).

Gujarat: In Gujarat, permission from the Forest Department is required to fell the Sandalwood tree grown in farm land.

Puducherry/Pondicherry: In the Union Territory, Sandalwood is a protected wood and special permit is needed for possession or transportation by any individual/farm.

A few pertinent Sandalwood provisions relevant to Sandalwood tree especially in the three Southern states namely, Karnataka, Kerala and Tamil Nadu are provided in Table 2. However, rules regarding Sandalwood is not mentioned in many states and also efforts to have uniform rules is essential not only for conservation but from cultivation and better ultilisation perspective. The policy makers have to seriously consider this; otherwise it may have cascading effect not only on Sandalwood cultivation, but also on survival of Sandalwood as a species.

5 Role of People in Conserving Sandalwood

Sandalwood is not an exacting species, except for proper drainage and very low temperature; cultivating Sandalwood in different parts of the country has been taken up with lot of enthusiasm. Accepting the harsh fact that availability of Sandalwood in its natural habitat is ruled out, it is imperative that cultivating Sandalwood far and

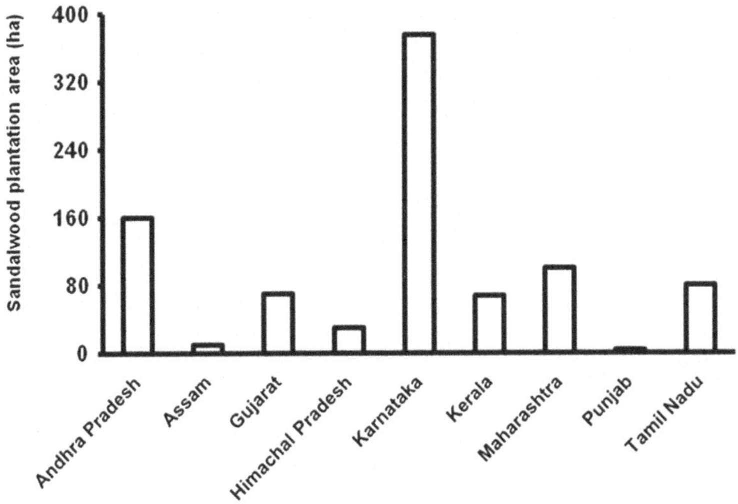

Fig. 2 Cultivation of Sandalwood in different states of India

wide is one of the important steps not only from conservation perspective but also for sustainable harvest and utilisation. The debate about the quality of the heartwood and oil from those trees grown outside its natural habitat is one issue which may often be questioned, but some of the preliminary results reveal that heartwood formation and oil accumulation has been observed in places such as Rajasthan and Uttarakhand where Sandalwood has been introduced. Sandalwood cultivation has been taken up in different states and the area of cultivation is increasing too. It is conservatively estimated that a total of ~900 ha of Sandalwood plantation has been raised in different states of India (Fig. 2) indicating the importance and commercial prospects.

What remains an important fact is, conservation of Sandalwood *per se* in its natural habitat seems to be a difficult option at this point of time. Also, research activities related to Sandalwood on the basis of natural population may not be possible in the near future. If conservation, utilisation and tree improvement research has to go hand in hand, the viable option left is mass distribution of seedlings, raising of Sandalwood plantations and encouraging growing of Sandalwood far and away from its natural habitat. Institute of Wood Science and Technology, Bengaluru, has been recognised as Centre for Advance Studies for Research on Sandalwood by Indian Council of Forestry Research and Education, Dehradun. The Institute has standardised methods for raising Sandalwood seedlings both in traditional and modern nursery practises. The Institute has been providing extensive training in this regard to State Forest Departments, Krishi Vignana Kendras, progressive farmers, private entrepreneurs and personnel from non-government organisations. Farmers and private entrepreneurs have started taking keen interest in Sandalwood cultivation. The plantations that are being grown would ultimately be the hub of activity for commercial utilisation as well as from research perspective. Therefore, the proactive participation of people in Sandalwood cultivation will not only conserve our precious Sandalwood, but also help in reviving the lost glory of Indian Sandalwood in the international market.

Acknowledgements We wish to thank Dr. H.Y. Mohan Ram, Professor (Retired), Delhi University, Dr. A. Seetharam, Project Coordinator, Small Millets (Retired), ICAR and Dr. Nataraja Karaba, Professor, Department of Crop Physiology, UAS Bangalore for their valuable inputs.

References

Azeez, A. S., Nelson, R., Prasadbabu, A., & Srinivas Rao, M. (2009). Genetic diversity of *Santalum album* using random amplified polymorphic DNAs. *African Journal of Biotechnology, 8*, 2943–2947.

Angadi, V. G., Jain, S. H., & Shankaranarayana, K. H. (2003). Genetic diversity between sandal populations of different provenances in India. *Sandalwood Research Newsletter, 17*, 4–5.

Annapurna, D., Rathore, T. S., & Somashekhar, P. V. (2005). Impact of clones in a clonal seed orchard on the variation of seed traits, germination and seedling growth in *Santalum album* L. *Silvae Genetica, 54*(4&5), 153–160.

Arun Kumar, A. N. (2005). *Evaluation and characterization of clonally propagated sandal (Santalum album L.) accessions of diverse origin with special reference to heartwood, oil content and other morphological characters* (Ph.D. thesis, Forest Research Institute, Dehra Dun).

Arun Kumar, A. N., Joshi, G., & Mohan Ram, H. Y. (2012). Sandalwood: History, uses, present status and the future. *Current Science, 103*, 1408–1416.

Arun Kumar, A. N., Srinivasa, Y. B., Joshi, G., & Seetharam, A. (2011). Variability in and relation between tree growth, heartwood and oil content in Sandalwood (*Santalum album* L.). *Current Science, 100*, 827–830.

Bagchi, S. K., & Veerendra, H. C. S. (1985). Study on intra tree and inter tree variations in leaves of *Santalum album*. *Myforest, 21*, 33–39.

Bhatnagar, S. P. (1965). *Studies in angiospermic parasites (No. 2) Santalum album—The Sandalwood tree*. Lucknow, India: National Botanical Gardens (Bulletin Number 112).

Brand, J., Kimber, P., & Streatfield, J. (2006). Preliminary analysis of Indian Sandalwood (*Santalum album* L.) oil from a 14-year-old plantation at Kununurra. *Western Australia. Sandalwood Research Newsletter, 21*, 1–3.

Burdock, G. A., & Carabin, I. G. (2008). Safety assessment of Sandalwood oil (*Santalum album* L.). *Food and Chemical Toxicology, 46*, 421–432.

Cameron, J. (1894). *Forest trees of Mysore and Coorg* (pp. 240–241). Bangalore: Mysore Government Central Press.

Dhanya, B., Viswanath, S., & Purushothaman, S (2010). Sandal (*Santalum album* L.) conservation in southern India: A review of policies and their impacts. *Journal of Tropical Agriculture 48*:1–10.

Jayappa, V., Nataraj, B. M., Shanbhag, P. K., Patil, K. B., & Srinivas, A. (1981). Regional variation in the yield and quality of sandal oil. A study by GLC. In *Proceedings of Second All India Sandal Seminar*. Salem: Tamil Nadu Forest Department.

Jones, C. G. (2008). *The best of Santalum album: Essential oil composition, biosynthesis and genetic diversity in the Australian tropical Sandalwood collection* (Ph.D. thesis, The University of Western Australia).

Kulkarni, H. D., & Srimathi, R. A. (1982). Variation in foliar charactersitcs in sandal. In P. K. Khosla (Ed.), *Biometric analysis in tree improvement of forest biomass* (pp. 63–69). Dehra Dun: International Book Distributors.

Mc Carthy, C. (1899). Progress report of forest administration in Coorg for 1898–1899.

Mitchell, J. E. M. (1941). Sandalwood problems, factors affecting heartwood and oil content in Sandalwood. In *Proceedings of 5th Silviculture Conference.* Dehra Dun.

MoEF. (2012). Report on the Committee headed by Shri A. K. Bansal, Addl. Director General (FC) on the regulatory regime regarding felling and transit regulations for tree species grown on non forests/private lands.

Nagaveni, H. C., & Ananthapadmanabha, H. S. (1986). Seed polymorphism and germination in *Santalum album* L. *Van Vigyan, 24*, 25–28.

Nageswara Rao, M., Ganeshaiah, K. N., & Uma Shaanker, R. (2007). Assessing threats and mapping sandal resources to identify genetic 'hot-spot' for *in-situ* conservation in peninsular India. *Conservation Genetics, 8*(4), 925–935.

Rai, S. N. (1990). Status and cultivation of Sandalwood in India. In L. Hamilton & C. E. Conrad (Eds.), *Proceedings of the Symposium on Sandalwood in the Pacific*, Honolulu, Hawaii US (pp. 66–71). Forest Service General Technical Paper PSW–122, April 9–11, 1990.

Rajan, B. K. C. (1992). *Ten forest products.* Bangalore: Jaya Publications.

Rama Rao, M. (1904). Notes on sandal. *Indian Forester, 30,* 248–267, 357–362, 394–403.

Ramalakshmi, S., & Rangaswamy, C. R. (1998). Physiological variation in seeds of provenances of sandal (*Santalum album*). In A. M. Radomiljac, et al. (Eds.), *ACIAR Proceedings Sandal and its Products* (Vol. 84, pp. 121–125).

Scot, J. (1871). Notes on horticulture in Bengal, No. 2, Loranthaceae, the mistletoe order, their germination and mode of attachment. *J. R.Hortic. Soc. India, 2*, 287.

Shashidhara, G., Hema, M. V., Koshy, B., & Farooqi, A. A. (2003). Assessment of genetic diversity and identification of core collection in Sandalwood germplasm using RAPDs. *Journal of Horticultural Science and Biotechnology, 78*(4), 528–536.

Singh, P. (1915). A further note on the oil value of some Sandalwoods from Madras. *Indian Forester, 41*, 123–129.

Srimathi, R. A., & Kulkarni, H. D. (1980). Preliminary findings of the heartwood formation in sandal (*Santalum album* L.). In *Proceedings of Second Forestry Conference* (pp. 1–13). Dehra Dun: F.R.I. and Colleges.

Suma, T. B., & Balasundaran, M. (2003). Isozyme variation in five provenances of *Santalum album* in India. *Australian Journal of Botany, 51*, 243–249.

Suma, T. B., & Balasundaran, M. (2004). Genetic diversity of eight *Santalum album* L. provenances of India based on RAPD analysis. *Indian Journal of Genetics and Plant Breeding, 64*(2), 167–168.

Swaminath, M. H., Hosmath, B. J., & Mallesha, B. B. (1998). The status of Sandalwood in India: Karnataka. In A. M. Radomiljac, et al. (Eds.), *ACIAR Proceedings Sandal and its Products* (Vol. 84, pp. 3–5). Bangalore: IWST & KSFD.

Veerendra, H. C. S., & Sarma, C. R. (1990). Variation studies in sandal (*Santalum album* L.) I: Time of emergence and seedling vigour. *Indian Forester, 116*, 568–571.

Climate Change and Sugarcane Productivity in Karnataka

Arun B. Chandran and K.N. Anushree

Abstract This study aims at understanding the relationship between climatic factors and its impact on sugarcane productivity in Karnataka. To understand the above objective we estimated the factors that affect sugarcane productivity. After appropriate statistical analysis, the strength of empirical results was checked through fixed effects panel regression model and Prais–Winsten model with panel corrected standard errors models. The data point has 592 observations corresponding to 16 districts of Karnataka, selected on the basis of the agro-climatic region to which they belong, with panel data for a period of 1966–2002. Since our prime interest was sugarcane productivity it was entered as a dependent variable in the models. Average rainfall, average maximum and average minimum temperature, area under irrigation and consumption of fertilizers are considered as explanatory variables. In order to capture the variability in the climatic factors it was decomposed into four seasons, viz., monsoon, summer, autumn and winter. The results conclude that though the average rainfall in different seasons has a direct positive impact, the rainfall in summer is significantly affecting sugarcane yield. The climatic variables of average minimum and maximum temperatures in different seasons have varied impacts on the yield of sugarcane. The maximum temperature in summer and the minimum temperatures in monsoon, winter and autumn have a significant positive impact on the sugarcane yield in Karnataka, whereas maximum temperature in autumn and minimum temperature in summer has negative relationship. The study suggests that adaptation strategies need special attention on technologies and management regimes that will enhance sugarcane tolerance to fluctuating temperatures.

Keywords Climate change · Sugarcane productivity · Karnataka

A.B. Chandran (✉) · K.N. Anushree
Institute for Social and Economic Change, Dr. VKRV Rao Road, Nagarabhavi, Bangalore 560072, India
e-mail: bchandran.arun@isec.ac.in

K.N. Anushree
e-mail: anushreekashyap91@gmail.com

S. Nautiyal et al. (eds.), *Climate Change Challenge (3C) and Social-Economic-Ecological Interface-Building*, Environmental Science,
DOI 10.1007/978-3-319-31014-5_23

1 Introduction

India is the second largest producer of sugar in the world economy. Sugar also holds an important position in Indian economy as it is not only providing employment for around 6 million farmer but it has also created employment opportunities in industrial sector, and is the second largest agro-based industry next to textile (Shrivastava et.al. 2011). The total area under sugarcane in India is around 5.01 million hectare and the estimated yield is on an average is around 69,838 kg/ha.[1] Sugarcane is an annual crop and grows in 12–18 months. The required ideal temperature for sugarcane cultivation is 27 °C and 100–175 cm of rainfall is ideal required for sugarcane production. Sugarcane is mainly grown in the tropical and subtropical regions in India.[2] Sugarcane regions in the tropical regions are Maharashtra, Karnataka, Kerala, Gujarat and Madhya Pradesh. UP, Bihar, Haryana and Punjab comes under the subtropical region. In India Uttar Pradesh is the major sugarcane growing state, contributing about 48 % of the area 40 % of the production, followed by Tamil Nadu, Maharashtra, Karnataka and Andhra Pradesh. In terms of productivity, Tamil Nadu is on first rank followed by Karnataka and Maharashtra. Coming to the state scenario, Sugarcane is one of the most important cash crops in Karnataka grown over an extent of 3.45 lakh hectares spreading in different ten agro-climatic zones. In the subsequent section, a brief review of the literature is presented. The third section deals with the data sources followed by econometric methodology, empirical results and conclusion.

2 Literature Review

This section is dealing with brief review of literature regarding climatic change and its impact on crop productivity in general and sugarcane in particular, and studies related to sugarcane yield at global, national and at state level. Study by Easterling et al., in 2000 reveals that there is increase in the average land and ocean surface temperatures globally and changing rainfall rates and changing pattern of rainfall. Chibo Onyeji (1993) suggests that the impact of climate change had decreased the agricultural productivity in Egypt. Yates and Strzepek (1998) assessed the integrated impacts of climate change using quadratic programming sector model and suggested that the change in climate scenarios had a minor impact on total economic welfare which is the total sum of consumer and producer surplus; with the major reduction of welfare approximately by around 6 %. Thus, the study concluded that the smaller food importing countries had a greater risk to climate change; while impacts can also be associated with changes in local, regional

[1] http://eands.dacnet.nic.in/.

[2] http://farmer.gov.in/.

biophysical systems and shifts in the national agricultural economy. Bosello et al. (2005) analysed the relationship between climate change and agriculture. This study suggests that the higher temperature had an impact on the production patterns. While Gbetibouo and Hassan (2005) studied the impact of climate change on major South African field crops. The research results suggest that the marginal changes in temperature had an impact on the production of field crops than changes in the precipitation. Deressa et al. (2005) used Ricardian approach to analyse the farmers' adaptation towards the impact of climate change on South African sugarcane production, using time series data for the period 1977–1998 pooled over 11 districts. The results suggest that the relationship between climate change and net revenue per hectare of sugarcane was nonlinear in nature with higher sensitivity to future increases in temperature than precipitation. Further, the study showcased that irrigation was not an appropriate substitute for mitigating climate change. Kabubo-Mariara and Karanja in 2007 used the same approach to examine the economic impact of climate change on Kenyan crop using a cross sectional data. The study suggests that there lies a nonlinear relationship between temperature and rainfall on revenue. Masters et al. (2010) studied the impact of climate change on agricultural commodities. The study revealed that the climate change has a significant negative effect on agriculture production. Marin et al. (2013) analysed the effects of climate change on sugarcane yield, water use efficiency in southern Brazil using two general circulation models and a sugarcane growth model.

Looking into home scenario, Seshu and Cady (1984) estimated the impact of climate variability on rice yield. The estimates reveals a decrease in rice yield at the rate of 0.71 tonne/ha with an increase in minimum temperature from 18 to 19 °C and a decrease of 0.41 tonne/ha with an average temperature increase from 22 to 23 °C Hingane et al. (1985) in their study observed an increase in the mean annual temperature in India by 0.4 °C for 100 years. In a simulation analysis done, Sinha and Swaminathan (1991) tested the vulnerability of rice and wheat production by increasing the temperature from 0.5 to 2 °C. The study results reveal that a 2 °C increase in the mean temperature would decrease the rice yield by 0.75 tonne/ha in high yield areas while in the low yield coastal region it is likely to decrease by 0.6 tonne/ha. Ramulu (1996) studied the impact of rainfall on sugarcane production in Andhra Pradesh for a period of 1973–1990 using Cobb–Douglas methodology. The estimates suggest that there lies no significant impact of rainfall on cane production in spite of sugarcane being highly water intensive crop. In another simulation analysis, Chatterjee (1998) found that a rise in temperature by 1–2 °C and carbon di oxide through stimulation on the yield of Maize and sorghum. The study reveals that increase in temperature by 1–2 °C would decrease the sorghum yields by 7–12 %. Study by Lal et al. (1999)on vulnerability of yield response of soya beans due to 10 % decline in rainfall, 3 °C and doubling of CO_2 suggests that soya bean yields could go up by as much as 50 % of the concentration of carbon dioxide in the atmosphere doubles. However, if the increase in carbon dioxide is accompanied by an increase in temperature then soya bean yields are expected to decline. Further, if the minimum and maximum temperatures go up by 1 and 1.5 °C, the gain in the yield is estimated to come down by 35 %. Kumar (2009) analysed the spatial differences on the impact of climate variation on Indian agriculture. The study made use of cross section data and used Ricardian approach for estimating the climate sensitivity on farm net revenue. The study estimates suggests that impact of change in climate can result in reduction of net farm revenue by 9 %. The estimates by Guiteras (2009) suggest that change in climate is has a negative impact on the combined yields of rice, wheat, jowar, bajra, maize and sugarcane. His study finds that the yield is likely to reduce by 4.5–9 % during (2010–39) due to impact of climate change. Study by Kumar (2011) reveals that the changes in weather condition has reduced the growing time of rice and sugarcane crops in UP and Uttarakhand. While understanding the sustaining sugar productivity in the presence of water depletion, Shrivastava et al. (2011) suggests that the cane productivity can only be sustained with appropriate technologies that economize water, by cultivating different varieties and also drought tolerant varieties of cane with relatively lesser water requirement will be advantageous in long run to sustain the cane productivity. Moorthy et al. (2012) examined the impact of climate change on five Indian major crops namely rice, wheat, sorghum, cotton, and sugarcane for a period of 50 years (1961–2010). This study was closely related to the other two studies done by Lobell et al. (2011) and Guiteras (2009). The estimates suggest that temperature and rainfall trends had no significant impact on major crop yields. Thus, the study concluded by emphasizing the role of error measurement in the prediction outcomes. While Kumar and Sharma (2014) estimated the climate change impact on sugarcane productivity using panel data methodology for a time period of 30 years (1980–2009). The study results suggest that actual rainfall, average maximum and minimum temperature were significant in determining sugarcane productivity. Also, the average maximum temperature in summer and in rainy season had a negative impact on the cane productivity. Though a plethora of empirical estimates are found in the literature pertaining to climate change and its impact on the productivity of the food crops in India, the empirical estimates pertaining to sugarcane is not consistent. This can be due to the employment of different methodologies, selection of appropriate variables, etc. For example, Ramulu (1996) did not include climatic variables related to temperature while estimating the impact of climate change on sugarcane productivity. The other reasons could be the combined analysis of climate change on different crops, which could have had different impacts (Krishnamurthy 2012; Sarker et al. 2012). For example, Guiteras (2009) study combined different crops which are likely to be impacted differently by climate change. In this regard the objective of this paper is to understand to what extent is the influence of climatic factors on sugar cane productivity in Karnataka.

3 Methodology

3.1 Data Description

The empirical data estimation of the study includes data from 1966 to 2002. Climatic and non-climatic variables from 16 districts of Karnataka, viz., Bangalore, Belgaum, Bellary, Bidar, Bijapur, Chikmanglur, Chitradurga, Dharwad, Gulbarga, Hassan, Kolar, Mandya, Mysore, Raichur, Shimoga and Tumkur. These districts compose more than 80 % of the area in six selected agro-climatic regions, viz., Central Dry Zone, North Eastern Transition Zone, Northern Dry Zone, Eastern Dry Zone, Southern Dry Zone and Southern Transition Zone of Karnataka. Such a selection provides us better insights into the climatic impacts in these regions. Climatic variables include season-wise average maximum temperatures (degree Celsius), average minimum temperatures and average annual rainfall (millimetres) whereas non-climatic variables considered includes the annual yield of sugarcane (of "raw sugar" (gur), in tonnes/ha), total annual fertilizer consumption (the total of nitrogen, phosphate and potash fertilizers used, in tonnes) and total annual sugarcane area under irrigation (in 000 ha). The climatic variables were divided into four different seasons—Summer, Winter, Monsoon and Autumn, according to the classifications made by Indian Meteorological Department. Summer includes the months from April to June; Monsoon season includes July to September; Autumn includes October and November and Winter season composes of December to March. These variables are identified in literature as to have substantial impact on yield of agricultural crops and especially on sugarcane (Deressa et al. 2005; Gbetibouo and Hassan 2005; Benhin 2006).

Data for all variables for the districts formed after 1966 have been given back to their parent district and removed from the database. There were 11 districts formed in Karnataka during this period and the data of these districts were traced back to the original districts that they belonged during 1966. A strongly balanced panel data of 592 observations (16 districts for a period of 37 years) was considered. A descriptive analysis of the data is given in Table 1.

3.2 Data Sources

1. **Agricultural Data**: The yield of sugarcane production were taken from Estimates of Area, Production and Yield of Principal Crops in Karnataka 1966/67–1989/90, Fully Revised Estimates of Principal Crops in Karnataka, 1990/91–1998/99, and Karnataka at a Glance, 1997/98–2007/08 published by Directorate of Economics and Statistics, Govt. of Karnataka.
2. **Climatic Variables**: The data for the rainfall was given by Report on Rainfall in Karnataka State of different years, by Directorate of Economics & Statistics, Govt. of Karnataka. The average maximum and minimum temperatures were collected from India Water-portal.

Table 1 A descriptive analysis of the data

Variable	Obs	Mean	Std. Dev.	Min	Max
Yield	592	8.439711	1.874944	3.25641	19.48148
Fertilizer	592	33884.51	29803.61	580	179806
Area_irrign	592	11.50725	20.42047	0	174.71
Max_winter	592	30.80219	1.798176	24.756	33.78075
Max_summer	592	32.96593	3.137419	24.947	39.64933
Max_monsoon	592	28.27522	2.100356	22.272	32.36633
Max_autumn	592	29.31305	1.754337	23.622	33.066
Min_winter	592	18.40842	.9610739	16.164	20.96225
Min_summer	592	22.64056	1.593508	18.69033	26.593
Min_monsoon	592	21.15136	1.240445	17.70833	23.33067
Min_autumn	592	19.85836	1.064107	16.798	22.3585
Rain_winter	592	5.600495	6.009835	0	40.52675
Rain_summer	592	152.5724	116.8271	12.62933	661.695
Rain_monsoon	592	244.4693	159.2002	41.517	827.376
Rain_autumn	592	98.74228	44.88098	3.9975	293.9725

3. **Non-Climatic Variables**: The fertilizer consumption data was taken from Fertilizer Statistics 1990–1995, 1998–2004, 2005–2009, by The Fertilizer Association of India. The Indian Agricultural Statistics Vol. II, 1966/67–1987/88, by the Directorate of Economics and Statistics, Govt. of India, Brochure on Irrigation Statistics in Karnataka 1980/81–1993/94, and Karnataka at a Glance 1995–96, published by Directorate of Economics and Statistics, Govt. of Karnataka gives data on the total sugarcane area irrigated.

4 Empirical Framework

Production function models, Ricardian cross sectional regression model, Agronomic-economic model, Agro-ecological zone model and integrated assessment model (United Nations Report 2011) are some of the commonly used econometric approaches in the existing literature to analyse the impact of climatic variables on yield of agricultural output. In this particular study, we use panel data regression estimation to analyse the impact of climatic and non-climatic variables on sugarcane yield.

A Hausman's Test (1978) specification test was used and fixed effects model of panel data estimation was identified as the best fit model. Finally, Prais–Winsten models with panels corrected standard errors (PCSEs) estimation are used for the proposed regression models to avoid the problems of heteroskedasticity, serial correlation, auto-correlation and serial auto-correlation in fixed effects regression model (Gupta 2012; Kumar and Sharma 2014).

Hausman test evaluates the consistency of an estimator when compared to an alternative, less efficient, estimator which is already known to be consistent. It is a χ^2 test based on the Wald criterion. Wald Criterion is as follows:

$$W = \left(b^{RE} - b^{FE}\right)' \left[var\left(b^{RE} - b^{FE}\right)\right]^{-1} \left(b^{RE} - b^{FE}\right) \sim \chi^2(k)$$

where RE represents random effects and FE represents fixed effects. k indicates the degrees of freedom and is given by the number of regressors in the Equation (13 in our study). b represents the coefficient values and var indicates the variance. If W is more than the critical χ^2 value at appropriate degree of freedom and at 5 % level of significance, the null hypothesis that the individual effects are uncorrelated with other regressors would be rejected and fixed effects model is used.

Fixed effects model is a statistical model used in panel data analysis that represents the observed quantities in terms of explanatory variables that are treated as if the quantities were non-random.

$$\text{Yield}_{it} = \beta_{1i} + \beta_2\text{Non-Climatic Variables}_{it} + \beta_3\text{Climatic Variables}_{it} + u_{it}$$
$$i = \text{districts} = 1, 2, 3\ldots16$$
$$t = \text{years} = 1, 2, 3\ldots37$$

Climatic variables are added season-wise and included into the model. In fixed effects model, we impose time independent effects for each entity that are possibly correlated with the regressors.

5 Results and Discussion

The results of the Hausman Specification test are illustrated in Table 3 in the Appendix. As shown in the results, the test has an asymptotic χ^2 distribution. The null hypothesis is rejected, concluding that the Random effects model is not appropriate because the random effects are probably correlated with one or more regressors. Hence, we use a fixed effects panel regression model based on this result. Prais–Winsten models are executed to avoid the problems of heteroskedasticity, serial correlation, auto-correlation and serial auto-correlation in proposed fixed effects regression model. The results of both these models are presented in Table 2.

According to the fixed effects model, fertilizer consumption, maximum temperature in autumn, minimum temperature in monsoon and autumn are statistically significant at 1 % level of significance; the rain in summer and the constant values are significant at 5 % level of significance and the area under irrigation is significant at 10 % level of significance. The Prais–Winsten model adds that after correcting for the heteroskedasticity and auto-correlation, the area under irrigation and the minimum temperature in summer is also statistically significant at 1 % level. The fertilizer consumption, area under irrigation, minimum temperature in monsoon and autumn and the rainfall in summer has a positive coefficient, whereas maximum

Table 2 The results of Fixed effects and Prais–Winsten Model

Variables	Fixed effects model	Prais–Winsten model
Fertilizer	8.46e−06***	8.69e−06**
Area under irrigation	0.0091654*	0.0114791***
Maximum temperature winter	−0.1474754	−0.281132
Maximum temperature summer	−0.9600607	0.6765891**
Maximum temperature monsoon	−0.8553481	−0.0056137
Maximum temperature autumn	−2.230395***	−0.839848***
Minimum temperature winter	0.1471964	0.5674326*
Minimum temperature summer	0.7787901	−1.054065***
Minimum temperature monsoon	2.162654***	0.5802172*
Minimum temperature autumn	1.787407***	0.3682153
Rainfall winter	0.0054152	0.0218034
Rainfall summer	0.0029156**	0.002192
Rainfall monsoon	0.0006462	0.0007406
Rainfall autumn	0.0014021	0.0036235
Constant	31.45268**	11.98454***
R^2	0.1178	0.2057

*** indicates significance at 1 % level of probability, ** indicates significance at 5 % and * indicates significance at 10 % level of probability

temperature in autumn and minimum temperature in summer has negative relation to the yield of sugarcane. Unlike shown by the earlier studies, like Ramulu (1996) for sugar production in Andhra Pradesh and Ajay Kumar et al. (2014) for a study of 13 states in India, the results shows that it may not be the rainfall in winter alone that is affecting the yield of sugarcane production in Karnataka, but the rainfall in summer is also significant and has a positive relationship to the sugarcane yield. The climatic variables with respect to minimum and maximum temperatures in different seasons have a very significant impact on the yield of sugarcane and this study points out the varying impacts of these variables in different seasons on the sugarcane productivity.

The non-climatic variables—fertilizer consumption and area under irrigation are both significant and has a positive relationship with sugarcane yield. Some studies direct us to the negative impact of fertilizers on sugarcane production (Ranuzzi and Srivastava 2012; Kumar and Sharma 2014), but this trend cannot be observed in the case of Karnataka and more fertilizer use in sugarcane production will increase the yield in the state.

Changes in the characteristics of seasons, a result of the climate change, has a profound impact on the yield of sugarcane in Karnataka. An increase of 1 °C in average maximum temperature in the autumn months (the months after the monsoon season), will decrease the sugarcane productivity by 0.9 tonne/ha. An increase in average minimum temperature in monsoon and autumn seasons has a positive impact on the yield of sugarcane, whereas an increase in average minimum temperature in summer has a negative relationship on the sugarcane yield.

It can be implied from the results that though the non-climatic strategies like fertilizers and improvement in irrigation positively impacts the sugarcane yield in Karnataka, the differential impacts of temperature and rainfall needs to be focused while making adaptation strategies. The varying temperatures in different seasons and fluctuations in the temperatures are a direct result of climate change. The future research should focus on making available sugarcane varieties which are less sensitive to fluctuations in temperature. Greater knowledge on the temperature impacts on sugarcane yield also needs to be disseminated. These results suggest a priority to intervention and adaptation strategies that target mitigation of fluctuating temperature impacts.

It should also be noted that the results of this study are based on only one crop and hence generalizations to the entire sugarcane or agricultural sector in the whole country cannot be made. The result also needs to be analysed with a data of greater time-horizon for generalized results.

Appendix

See Table 3.

Table 3 The results of the Hausman specification test

	Coefficients			
	(*b*)	(*B*)	(*b* − *B*)	sqrt(diag(*v*_*b* − *v*_*B*))
	Femodel	Remodel	Difference	S.E.
Fertilizer	8.46e−06	9.45e−06	−9.88e−07	1.61e−06
Area_irrign	0.0091654	0.0106159	−.0014505	0.0028576
Max_winter	−0.1474754	−0.3329382	0.1854628	0.5543859
Max_summer	−0.9600607	0.668375	−1.628436	0.5264949
Max_monsoon	−0.8553481	0.0329234	−.8882715	0.6034131
Max_autumn	−2.230395	−0.8796967	−1.350698	0.5908743
Min_winter	0.1471964	0.530713	−0.3835166	0.5481204
Min_summer	0.7787901	−1.027619	1.806409	0.4985125
Min_monsoon	2.162654	0.6518216	1.510832	0.5840345
Min_autumn	1.787407	0.3997172	1.38769	0.5616042
Rain_winter	0.0054152	0.0183406	−0.0129255	
Rain_summer	0.0029156	0.002446	0.0004696	0.0006801
Rain_monsoon	0.0006462	0.00071	−0.0000638	0.0004592
Rain_autumn	0.0014021	0.0030811	−0.0016789	

b = consistent under *Ho* and *Ha*; obtained from xtreg
B = inconsistent under *Ha*, efficient under *Ho*; obtained from xtreg
Test: Ho: difference in coefficients not systematic
chi2(13) = (*b* − *B*)'[(*v*_*b* − *v*_*B*)^(−1)](*b* − *B*) = 43.93
Prob > chi2 = 0.0000
(*v*_*b* − *v*_*B* is not positive definite)

References

Benhin, J. K. (2006). Climate Change and South African Agriculture: Impacts and Adaptation Option. *CEEPA Discussion Paper 21.*

Bosello et al. (2005, July). Retrieved August, 2015 from http://www.cepal.org/ilpes/noticias/paginas/1/35691/bosello_assesingclimatechangeimpacts.pdf

Chatterjee, A. (1998). *Simulating the impact of increase in temperature and CO_2 on growth and yield.* New Delhi: Indian Agricultural Research, M.Sc. Thesis (Unpublished).

Christensen, J. H. (2007). *Regional climate projections climate change, IPCC.*

Deressa, T., Hassan, R., & Poonyth, D. (2005). Measuring the impact of climate change on South African agriculture: The case of sugarcane growing regions. *Agrekon, 44*(4), 524-542.

Easterling, D. R., Meehl, G. A., Parmesan, C., Changnon, S. A., Karl, T. R., & Mearns, L. O. (2000). Climate extremes: Observations, modeling, and impacts. *Science.*

Gbetibouo, G. A., & Hassan, R. M. (2005). Measuring the economic impact of climate change on major South African field crops: A Ricardian approach. *Global and Planetary Change,* 143–152.

Guiteras, R. (2009). *The impact of climate change on Indian agriculture.* University of Maryland, College Park: Mimeo.

Gupta, S. S. (2012). Impact of Climate Change on Indian Economy: Evidence from Food Grain Yields. *I. Centre for Development Economics Working Paper 218,* Delhi.

Hausman, J. A. (1978). Specification Tests in Econometrics. *Econometrica, 46*(6), 1251–1271.

Hingane, L. S., Rupa Kumar, K., & Ramana Murty, B. V. (1985). Long-term trends of surface air. *Journal of Climatology,* 521–528.

Kabubo-Mariara, J., & Karanja, F. K. (2007). The economic impact of climate change on Kenyan crop agriculture: A Ricardian approach. *Policy Research Working Paper, 4334,* 1–40.

Krishnamurthy, C. K. (2012). The distributional impacts of climate change on Indian agriculture: A quantile regression approach. *Madras School of Economics Working Paper 69/2012.*

Kumar, A., & Sharma, P. (2014). Climate change and sugarcane productivity in India: An econometric analysis. *Journal of Social and Development Sciences, 5*(2), 111–122.

Kumar, K. S. (2009). Climate sensitivity of Indian agriculture does spatial effects matter. *SANDEE Working Paper,* 45-09, Kathmandu.

Kumar, V. S. (2011). Impact of climate change on the growth and production of *Saccharum offcinarum* and *Magnifera indica. International Journal of Science Technology and Management.*

Lal, M., Singh, K. K., Srinivasan, G., Rathore, L. S., Naidu, D., & Tripathi, C. N. (1999). Growth and yield response of soybean in Madhya Pradesh, India to climate variability and change. *Agricultural and Forest Meteorology, 93,* 53–70.

Lobell, D. B., Schlenker, W., & Costa-Roberts, J. (2011). Climate trends and global crop production since 1980. *Science, 333*(6042), 616.

Marin, F. R., Jones, J. W., Singels, A., Royce, F., Assad, E. D., Pellegrino, G. Q., et al. (2013). Climate change impacts on sugarcane attainable yield in southern Brazil. *Climatic Change.*

Masters, G., Baker, P., & Flood, J. (2010). Climate change and agricultural commodities. *CABI Working Paper,* (2).

Moorthy, A., Buermann, W., & Rajagopal, D. (2012). The impact of climate change on crop yields in India from 1961 to 2010.

Ramulu, M. (1996). Supply response of sugarcane in Andhra Pradesh. *Finance India, 10*(1), 116–122.

Ranuzzi, A., & Srivastava, R. (2012). Impact of Climate Change on Agriculture and Food Security. *ICRIER Policy Series 16.*

United Nations Report (2011). *An Assessment of the Economic Impact of Climate Change on the Agriculture Sector in Jamaica.* Subregional Headquarters for the Caribbean: Economic Commission for Latin America and the Caribbean.

Chibo Onyeji, S., & Fischer, G. (1993). An economic analysis of potential impacts of climate change in Egypt. *IIASA Working Paper, 93*(12).

Sarker, M. D., Rashid, A., Alam, K., & Gow, J. (2012). *A comparison of the effects of climate change on aus, aman and boro rice yields in Bangladesh: Evidence from panel data.* Paper presented at the 41st Australian Conference of Economists (ACE), Victoria University, Melbourne.

Seshu D.V., & Cady, F. B. (1984). Response of rice to solar radiation and temperature estimated from international yield trials. *Crop Science*, *24*, 649–654.

Sinha, S. K., & Swaminathan, M. S. (1991). Deforestation, Climate Change and Sustainable Nutrients. *Climatic Change, 16*, 33–45.

Shrivastava, A. K., Srivastava, A. K., & Solomon, S. (2011). Sustaining sugarcane productivity under depleting water resources. *Current Science*, 748–754.

Yates, D. N., & Strzepek, K. M. (1998). An assessment of integrated climate change impacts on the agricultural economy of Egypt. *Climatic Change, 38*, 261–287.

Dairy Farming and Organic Farming for Bio-resource Conservation and Livelihood Development in Tumkur District Karnataka State-India

N. Kumara, Nehal A. Farooquee and P.V.K. Sasidhar

Abstract A field study was conducted during 2013–2014 in Tumkur district to study the dairy farming and organic farming for bio-resource conservation and livelihood development. Further, the objective of the study is to (1) Measure the dairy farming generates year-round employment and substantial income to sustain their livelihood in rural. (2) Soil fertility and soil organic carbon and microbial growth in the soil. Studies have shown that dairying in rural areas has positively improved the life of those engaged in this business, directly or indirectly, bringing significant socio-economic changes by providing self employment. Dairy farming generates year-round employment and substantial income to sustain their livelihood. Majority (**86** %) were young age and middle-aged farmers contributing **75.67** % of milk yield to milk unions, which shows the employment opportunity provided by the dairy farming to small and marginal farmers and good source of income in turn improving the social and economic status of the rural farmers.Organic system significantly improved the organic matter (**1.96**) there by the soil quality and the sustainability index of the soil was maximum with organic compared to inorganic. There was improvement of soil structure, texture and microbial activity in organic farming. *Highlights* (a) Dairy farming has positively improved the life of those engaged in this business. (b) Dairy farming brought significant socio-economic changes by providing self employment. (c) Dairy farming has provided a year-round source of income for people. (d) Dairy farming generates year-round employment and substantial income to sustain their livelihood. (e) Organic farming increases microbial population and the organic carbon level. (f) Improvement of soil structure, texture and microbial activity in organic farming. (g) Bad effect of chemical fertilizers and pesticides were controlled in organic farming.

Keywords Farmers · Self employment · Socio-economic changes · Organic farming · Soil

N. Kumara (✉) · N.A. Farooquee · P.V.K. Sasidhar
School of Extension and Development Studies (SOEDS),
IGNOU, New Delhi, India
e-mail: nkumar278@gmail.com

S. Nautiyal et al. (eds.), *Climate Change Challenge (3C) and Social-Economic-Ecological Interface-Building*, Environmental Science,
DOI 10.1007/978-3-319-31014-5_24

1 Introduction

Dairying in India is by large in the hands of small/marginal landholders and agriculture laborers. The national average land holding is 1.68 ha per farm family and cattle and/or buffalo is a part of family. Eighty percent of 97.7 million farm families in India posses cattle and/or buffalo. Even agriculture labourers (11.5 % of 97.7 million) posses one or two dairy cattle/buffalo. Livestock keeping in general and dairying in particular, once a subsidiary enterprise, (Supplementary/ complementary) to agriculture, has now become the major enterprise economically and the crop production are becoming dependent on dairying (***Strategies for Sustainable Dairy Production in India***).

Scope for Dairy Farming and its National Importance; The total milk production in the country for the year 2008–2009 was estimated at 108.5 million metric tonnes and the demand is expected to be 180 million tonnes by 2020. To achieve this demand, annual growth rate in milk production has to be increased from the present 2.5–5 %. Thus, there is a tremendous scope/potential for increasing the milk production through profitable dairy farming.[1]

India derives nearly 33 % of the gross domestic population from agriculture and has 66 % of economically active population in agriculture. The share of livestock product is estimated at 21 % of total agriculture sector. The fact that dairying could play a more constructive role in promoting rural welfare and reducing poverty (www.allindiadairy.com).

Milk production alone involves more than 70 million producers, each raising one or two cows/buffaloes primarily for milk production. The dairy sector offers a good opportunity to entrepreneurs in India. India is a land of opportunity for those who are looking for new and expanding markets. Growth prospects in the dairy sector are very bright.[2]

Dairying contributes little more than 7 % to the national income; contribution of livestock sector to the national economy was in the range of Rs. 130,233 crores (1999–2000).[3] Milk output accounts for more than 17 % of India's agriculture production today as against 14 % in 1970–1971.

Organic farming is defined as a production system which largely excludes or avoids the use of fertilizers, pesticides, growth regulators, etc. and relies mainly on organic sources to maintain soil health, supply plant nutrients and minimise insects, weeds and other pests.

Organic farming is giving back to the nature what is taken from it. It is not mere non-chemicalism in agriculture; it is a system of farming based on integral relationship. Therefore, one should know the relationship among soil, water, plant

---

[1]Animal Husbandry-Dairy Farming-Serva Manav Vikas Samiti (An NGO) (All India Dairy Business Directory).

[2]Milk production A fact sheet—Dairy India (5th Edition).

[3]Dairy Year Book 05–06.

and microflora and overall relationship between plant and animal kingdom. It is the totality of these relationships which is the backbone of the organic farming (Funtilana 1990).

Climate change has become an important area of concern for India to ensure food and nutritional security for growing population. The impact of climate change is global, but countries like India are more vulnerable in view of the high population depending on agriculture. In India, significant negative impacts have been implied with medium-term (2010–2039) climate change, predicted to reduce yields by 4.5–9 %, depending on the magnitude and distribution of warming. Since agriculture makes up roughly 15 % of India's GDP, a 4.5–9 % negative impact on production implies a cost of climate change to be roughly up to 1.5 % of GDP per year.

Dairy farming is a livelihood for a large number of households in rural India and provides jobs to unemployed rural people, specially rural women, increase income of rural people, alleviate poverty in rural and less developed areas, help develop rural areas and eradicate social evils, provide food security, increase efficiency of agriculture sector, provide security against crop failure and increase export earnings (***Rural dairy farming and alleviation of poverty News letter Published in DAWN.com***). Dairy farming plays significant role in sustaining the rural livelihoods. In light of the above facts, to boost dairy farming for improving rural livelihoods in India as well as in the state of Karnataka, the present study was undertaken.

2 Methodology

A dairy farming and organic farming for bio-resource conservation and livelihood development in Tumkur district, Karnataka, India was carried out from March 2010 to April 2014.

The present study was undertaken in two talukas (Sira and pavagada) of Tumkur district Karnataka state, where Karnataka State Horticulture Department implemented the organic conversion programme on cluster base of 50 ha, consists of 20–30 small and marginal farmers having at least one or two cows, buffalos of 3–4 villages. 50 organic and dairy farmers were selected by proportionate random sampling method. Soil samples and yield of data of crop were collected yearly from 15 farmers and soil samples analyzed in government soil testing labs.

Data on dairy farming, milk yield and livelihood were collected by personal interview and observation method, also from secondary sources of information such as reports of Department of Animal Husbandry, Milk Unions. Discussions were held with officials of these departments, experts, executives, to elicit their views, ideas and opinion on the important issues pertaining to dairy farming and organic farming for bio-resource conservation and livelihood development. The data were analyzed using suitable statistical techniques such as average, frequency, percentages.

Table 1 Distribution of respondents according to their age

Sl. No	Category	Small farmers		Medium farmers		Big farmers		Total	
		Frequency	Percentage	Frequency	Percentage	Frequency	Percentage	Frequency	Percentage
1	Young age (<35 years)	09	39.13	09	42.85	01	16.66	19	38.00
2	Middle age (36–50 years)	11	47.82	10	47.61	03	50.00	24	48.00
3	Old age (>50 years)	03	13.05	02	9.50	02	33.33	07	14.00
	Total	23	100.00	21	100.00	6	100.00	50	100.00

3 Results and Discussions

The data revealed 23 small farmers (with 1 milch cattle), 21 medium farmers (2–3 milch cattle) and 6 large farmers (3 and above) (Reddy 1996). Out of 50 farmers 41 were male and 9 were female and 29 were organic and dairy farmers.

To understand the dairy farming improving, rural livelihood was analyzed. The findings from Table 1 showed that majority (48.00 %) of the respondents were middle aged followed by young age farmers 38 % and old age farmers were only 14 %.

It is implied from Table 2 that the small dairy farmers contributed majority (40.11 %) of milk yield to milk union followed by marginal/medium farmers 37.56 % and big farmers 22.33 %, respectively. Small and marginal farmers together contributed **75.67** % of milk yield to milk unions which shows the employment opportunity provided by the dairy farming to small and marginal farmers and good source of income in turn improving the social and economic status of the rural farmers (Table 3).

From the findings it can be concluded that small and marginal farmers together contributed 75.67 % of milk yield to milk unions which shows the employment opportunity provided by the dairy farming to small and marginal farmers and good source of income in turn improving the social and economic status of the rural farmers.

Table 2 Distribution of respondents according to their animals holding and milk yield

Sl. No	Dairy farmers	Number of farmers	Milk yield contribution per month (L)	Contribution (%)
1	Small farmers (1 animals)	23	6231	40.11
2	Marginal farmers (2–3 animals)	21	5838	37.56
3	Big farmers (3 and above animals)	06	3471	22.33
4	Total	50	15,540	100

Table 3 Organic matter of soil and yield parameters of areca nut

Sl. No	Data of soil and areca yield 2013–14	Treated	Control
1	Organic matter and soil quality and the sustainability index	1.96	0.81
2	Number spadix/plant	4.15	4.1
3	Number of bunches/spadix	12.30	12.05
4	Number of nuts/bunch	8.85	8.2
5	Weight of nut in g	32.00	30.65
6	Total yield/acre (480 plants)	6939	5960

It is evident from the study that there is a significant difference in organic matter and soil quality and the sustainability index. The organic matter and soil quality and the sustainability index gradually increasing in organically grown crops every year which was maximum (**1.96**) whereas in inorganic system due to application of chemical fertilizers microbial activity reduced and content of organic carbon is low (0.81). This is due to the increase in organic matter and soil quality and the sustainability index.

4 Conclusion

The study has clearly shown that organic farming increases microbial population and the organic carbon level; significant improvement of soil structure, texture and microbial activity. Bad effect of chemical fertilizers and pesticides were controlled. Dairy farming provided jobs to unemployed rural people, especially the small and marginal farmers, rural women to increase the income of rural people. The dairy sector holds high promises as a dependable source of livelihood for the vast majority of the rural poor in India. This in turn is a lesson for unemployed youths, landless labours small and marginal farmers of same villages as well as neighboring villages to opt dairy farming as occupation. Employment generation, especially the self employment generation in rural sector and adopt organic farming to bio-resource conservation. Dairy and organic farming are two faces of a single coin. Dairy provides employment to rural youths, women and organic farming improves soil texture, structure and enhance soil microbial activity.

Acknowledgment The authors greatly acknowledge the support rendered by Karnataka Milk federation, Milk Unions of Tumkur and Department of Agriculture.

References

Department of Animal Husbandry, Dairying and Fisheries. www.dahd.nic.in

Funtilana, S. (1990, March–April). Safe, in expensive, profitable and sensible. *International Agricultural Development* XI, 24.

National Dairy Development Board (NDDB). (2003). Website www.nddb.org

OrganicIndia

Reddy, R. V. K. (1996). *A critical analysis of dairy production technologies among the dairy farmers of East Godavari district of Andhra Pradesh* (Ph.D. thesis). Andhra Pradesh Agricultural University, Hyderabad.

Rao, D. L. N. (2005). *Soil microbial diversity in chemical and organic farming*. Paper presented at "National Seminar on Organic Farming-Current Scenario and Future Thrust" during April 27–28.

Part IV
Vulnerability, Sectoral Assessment, Models and Interfaces

Analysing Vulnerability to Climate Change in India with Special Reference to Drought Risk: Results from a Field Survey

Anu Susan Sam, Azhar Abbas, Muhammad Arshad and Harald Kächele

Abstract India faces considerable threat from changing climatic conditions locally and globally. Drought risk is one such threat which is projected to increase in frequency and intensity under such changing conditions. Many households are expected to suffer from these changes and associated risks such as drought. These risks may expose many vulnerable households in the absence of effective mitigation and adaptation measures. This study analyses climate change and drought risk vulnerability of marginal households exposed to such hazards. For this purpose, a field survey of 220 households in Balangir district of Odisha state in India was carried out using multistage random sampling technique. Using Climate Vulnerability Index (CVI) and other indicators of vulnerability to socioeconomic domains defined by IPCC, the current level of household's exposure and coping capacity is evaluated. Results indicate that vulnerability is a function of both climatic and non-climatic factors. The socio demographic characters like low literacy rates, resource availability and accessibility, social structure, etc., make people more vulnerable whereas access to social networks plays a significant role in uplifting the poor rural households. The research concludes that the impacts of climate change and drought risk expose those households disproportionately who are already vulnerable due to poverty, inequality and marginalisation. Moreover, a need to formulate policies based on regional and local estimates of vulnerability and coping capacity is emphasised for an effective mitigation to climate change and drought risk.

Keywords Adaptive capacity · Assessment · Exposure · Natural disaster · Risk · Sensitivity

A.S. Sam (✉) · A. Abbas · M. Arshad · H. Kächele
Leibniz Centre for Agricultural Landscape Research (ZALF),
Eberswalder Str. 84, 15374 Müncheberg, Germany
e-mail: anu.sam@zalf.de; rachelsusanrachel@gmail.com

S. Nautiyal et al. (eds.), *Climate Change Challenge (3C) and Social-Economic-Ecological Interface-Building*, Environmental Science,
DOI 10.1007/978-3-319-31014-5_25

1 Introduction and Background

Climate change is one of the most persistent threats to global stability in the future (Devkota et al. 2014). Even currently, the climate change effects have become widespread with increased intensity and frequency (IPCC 2007, 2012). Natural disasters and climate-related extreme events are such examples that are expected to increase as a result of relatively small changes in climate conditions (Adger et al. 2003; Huggel et al. 2015). A rise in these disasters, in turn will have specific effects of varying degrees on the environment, water resources, coastal zones, agricultural production, human livelihoods and biodiversity (Pant et al. 2014).

Among the climate-induced natural disasters, drought is one of the most frequently occurring extreme events with longer time scales and extensive geographical coverage. Changes in climate conditions such as variation in precipitation patterns and rising temperature can lead to increased frequency of drought worldwide (Dai 2013). Drought frequently causes sizeable losses to mankind as well as nature, especially in less developed countries but at the same time it is considered to be the most complex phenomenon affecting more people than by any other hazard (Lei et al. 2014). This is evident from the fact that nearly half of the countries in the world have suffered from drought over the past several decades (Wu et al. 2015a, b). IPCC has particularly declared drought as an important global challenge in the wake of its increasing frequency and severity (IPCC 2012). South Asian countries including India also face this challenge given highly diverse socioeconomic, geographical and hydrological regimes.

India is faced with immense challenges of climate change including drought. It is expected that it will face a greater seasonal variation in temperature given its diverse and changing climatic pattern (Christensen et al. 2007). It is shown that the country's temperature would rise to the tune of 0.57 °C per 100 years (Singh et al. 2001). This has started taking its effect while there is already an increase in the variability of average annual precipitation (Bhadwal 2003) which may lead to many natural disasters. As a result, recurrent drought has become a common phenomenon in different regions of the country, caused by a lack or a decreased amount of rain over a long period of time coupled with increasing amount of temperature (Pai et al. 2011). Subsequently, millions of people have become prone to drought and millions will be vulnerable to it in the future. This is evident from the fact that droughts have affected nearly 1061 million people and killed 4.25 million people in India during 1900–2015 (CRED 2015). Drought effects are wide ranging in the country that have taken their toll by adversely affecting national economy through decline in agricultural production, rise in rural unemployment, and fall in purchasing power and household food security (Kumar et al. 2005).

2 Defining Vulnerability to Climate Change

The IPCC defines vulnerability as the degree to which a system is susceptible to, and unable to cope with the adverse effects of climate change, including climate variability and extremes like drought, flood, cyclones and heat waves (IPCC 2007). Vulnerability encompasses three main aspects in terms of exposure, susceptibility and adaptive capacity (IPCC 2007). Exposure means the presence of people, environmental services and resources, infrastructure, or economic, social and cultural assets in places that could be adversely affected (Aleksandrova et al. 2014); susceptibility is the predisposition of a system to be negatively affected by climate variability or natural disaster (Birkmann et al. 2013) whereas adaptive capacity is the system's ability to absorb or recover from the effects of a hazardous event (IPCC 2012).

Vulnerability to climate change and climate extremes varies in time and space dimensions and is differed across groups and individuals (Maru et al. 2014). There is sufficient divergence between the extent of vulnerability in developed and developing nations. Developing countries are relatively more vulnerable to climatic vagaries mainly due to their weak coping (adaptive) capacity and unclear institutional frameworks (Yohe and Tol 2002). Other possible reasons for their increased vulnerability are: (i) relatively greater physical impacts given low level of preparedness and ad hoc nature of mitigation measures; (ii) heavy dependence of the majority of population on natural resources and primary production for livelihood and hence excessive exploitation leading to their degradation; and (iii) limited economic and technological capacity thus hindering adaptation process to rapidly changing climatic conditions (Gray and Mueller 2012).

Rapid changes in climate as well as the increased frequency and severity of natural disasters in several countries have caused a remarkable proportion of world population to suffer drastically (Leiserowitz et al. 2012). As a result, many development practitioners and climate change experts have extensively focussed the issue of vulnerability assessment and associated dimensions in order to come up with evidence-based solution particularly for developing world (Bardsley and Wiseman 2012). This assessment is highly relevant as it covers various domains and magnitudes of climate change and natural disasters on almost all aspects of households (Hahn et al. 2009).

3 Problem Statement

The vulnerability of different sectors towards climate change and natural disasters has serious implications for households and their livelihoods along with adverse effects on the economic and social development of the economy of human society (Calzadilla et al. 2013). The increased frequency and severity of such disasters have even disproportionate impacts on the resource dependent masses in the resource starved developing countries (Stern 2007). When faced with such a situation, people in these countries are forced to make choices regarding the necessary adjustments in social, economic, cultural and environmental contexts that are framed within the vulnerability of different systems constituted by both the human and natural environment (Bhattacharya-Mis and Lamond 2014). Henceforth, it is pertinent to analyse the magnitude and extent of the influence of such changes in a bid to effectively evaluate existing risk and coping potential of vulnerable communities. Many studies have focussed vulnerability to climate change and natural disasters but very little is known about the vulnerability to drought in South Asian countries in general and India in particular (Pandey and Jha 2011; O'Brien et al. 2007). Nonetheless, there exists a huge body of evidence which points towards increased risk of droughts in the country but very little is known about the vulnerability of local communities to such hazards. This gap is considered to lead to ineffective policy formulation and poor planning process thereof in mitigating drought risk. Moreover, a micro level household vulnerability analysis would be helpful in identifying the most vulnerable segments of the society to the natural hazards like drought (Aryal et al. 2014).

Therefore the aim of current study is to fill the above mentioned gap by quantifying the degree of vulnerability of the rural farming households that are prone to droughts in India.

4 Study Region

Odisha state, located in the eastern part of India, is selected for this research (Fig. 1). Odisha is one of the most drought affected states within India (Swain and Swain 2006) where 70 % of rural population depends on agriculture. On the other

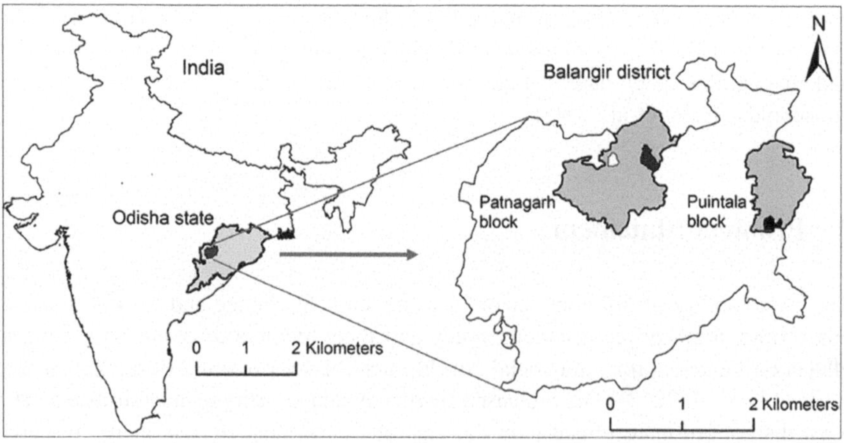

Fig. 1 Map of the study area and surveyed blocks in the state of Odisha, India

hand, the state has faced droughts or moisture stress for 22 years, during the period from 1950 to 2013 (Government of Odisha 2013). Even under normal conditions, agricultural production in Odisha is marked by low productivity, but its simultaneous susceptibility to climate change and droughts lead to wide fluctuations in output (Arora et al. 2015). Moreover, this state is one of the most underdeveloped states of India featured by higher poverty rate and lower growth rate. Under such conditions of lower social development with ever changing climatic conditions pose already marginalised remote regions of Odisha to higher drought risk.

5 Sampling and Data Collection

For the collection of primary data from the field, multistage random sampling procedure was employed for this research.

Balangir district was selected from the list of drought-prone districts of Odisha state. After the initial selection, two randomly selected blocks from Balangir were Patnagarh and Puintala (Fig. 1). At third stage, one panchayat from each selected block was randomly selected namely Tamian grama panchayat from the former and Mahimunda grama panchayat from the later. In the next step, two villages from each panchayat were selected and finally 55 households from each village were interviewed using face-to-face survey mode with the help of a pre-designed and pre-tested questionnaire. The selected villages were Aintalunga and Bagbahali (from Tamian) and Bilaikani and Sirabahal (from Mahimunda). As a result, the total sample size for this study was 220 households.

6 Analytical Framework

For measuring the vulnerability of rural households, the Climate Vulnerability Index (CVI) developed by Pandey and Jha (2011) was used. Major domains and sub domains were modified for this study which included only the sub domains that are relevant to rural communities of India. Vulnerability is a function of three dimensions, i.e. adaptive capacity, sensitivity and exposure (IPCC 2007) and it can be divided into six major domains, i.e. Demographic (Dem), Social (Soc), Economic (Eco), Physical (Phy), Drought (Dro) and Climate variability (CV).

Table 1 shows the organisation of these six domains in the IPCC framework. On the other hand, Table 2 presents the explanation about the sub domains.

The primary data collected from the study area were transferred into appropriate measurement units like counts, ratios, percentages and indices for each sub domain. In few cases, sub domains were case-specific and/or qualitative in nature. Since each of the sub domains was measured in different scale, each measure was standardised into an index using Eq. 1.

Table 1 Classification of major domains under the IPCC vulnerability framework

Dimensions of vulnerability	Major domains
Adaptive capacity	Demographic
	Social
Sensitivity	Physical
	Economic
Exposure	Drought
	Climate variability

$$\text{Index}_{sv} = \frac{S_{\text{v}} - S_{\min}}{S_{\max} - S_{\min}} \tag{1}$$

where

S_{v} is the original sub-domain for village v,

$S_{\min}$ and $S_{\max}$ are the minimum and maximum values, respectively, for each sub-domain which was determined using data from selected villages.

In addition, the index scores for major domains were calculated by taking the average of the standardised sub domains and then multiplied with their corresponding weights to get the weighted major domain score index. A balanced weighted approach was used in CVI assuming that each of the sub domains had an equal contribution to the overall index (Sullivan et al. 2002). The weights of each major domain were determined by the number of sub domain of which they are comprised. Once the weighted major domain index score was calculated, the index for exposure (Exp) (Eq. 2), adaptive capacity (Ada cap) (Eq. 3) and sensitivity (Sen) (Eq. 4) were calculated as follows:

$$\text{Exp} = \frac{W_{\text{Dro}}\text{Dro}_V + W_{\text{CV}}\text{CV}_V}{W_{\text{ND}} + W_{\text{CV}}} \tag{2}$$

$$\text{Adp Cap} = \frac{W_{\text{Soc}}\text{Soc}_V + W_{\text{Dem}}\text{Dem}_V}{W_{\text{Soc}} + W_{\text{Dem}}} \tag{3}$$

$$\text{Sen} = \frac{W_{\text{Eco}}\text{Eco}_V + W_{\text{Phy}}\text{Phy}_V}{W_{\text{Eco}} + W_{\text{Phy}}} \tag{4}$$

The indexed values for exposure, sensitivity and adaptive capacity were combined for calculating CVI as following:

$$\text{CVI}_{\text{v}} = 1 - \text{I}\left\{\frac{N_1\text{Exp} - N_2\text{Ada Cap}}{N_1 + N_2}\right\}\text{I} * \left\{\frac{1}{\text{Sen}}\right\} \tag{5}$$

where N_i is the number of major domains in the ith dimension of vulnerability.

The values of CVI range between 0 (least vulnerable) to 1 (most vulnerable).

Table 2 Major domains and sub domains of Climate Vulnerability Index (CVI)

Major domain	Sub domain	Explanation of sub domains
Demographic	Dependency ratio	Ratio of the population under 18 and over 65 years of age to the population between 19 and 64 years of age
	Female-headed households	Percentage of households where the primary adult is female. If a male head is away from the home for >6 months per year, the female is counted as the head of the household
	Illiterate household heads	Percentage of households whose heads have attended 0 years of school
	Backward caste	Percentage of households belonging to backward castes like OBC, SC and ST[a]
Social	Households seeking no assistance from NGOs/SHGs[c]	Percentage of households who have not received any assistance from NGO/SHG in the past 6 months
	Households with access to money lenders	Percentage of households who report to have access money lenders
	Households with no access to banks	Percentage of households who report no access to banks
	Households with no school access	Percentage of households who report no access to schools
Economic	Average livelihood diversification index[b]	The inverse of the number of livelihood activities +1 reported by a household
	Households depending solely on agriculture for livelihood	Percentage of households who have only agriculture as a source of income
	Households' average monthly income (US$)[b]	The inverse of average monthly income per household
	Average monthly food expenditure per household (US$)[b]	The inverse of average monthly food expenditure per household
Physical	Households depending on public/natural resources of water for household activities	Percentage of households who depend on public/natural source of water for household activities like cooking, washing, cleaning, bathing etc.
	Households depending on own farm for food	Percentage of households who get their food primarily from their own farms
	Households not depending on Public Distribution System (PDS)	Percentage of household who do not depend on PDS for subsided food items
	Households with no electricity connection	Percentage of households who lack electricity connection at house

(continued)

Table 2 (continued)

Major domain	Sub domain	Explanation of sub domains
Climate variability	Households who perceived a rise in average temperature	Percentage of household who reported any increase in the average temperature during the last few years
	Households who perceived variation in average rainfall	Percentage of households who reported any variation in the average rainfall during the last few years
	Households reporting an effect on agriculture due to variability in temperature and/or rainfall	Percentage of households who perceive that agriculture sector has been affected by the variability in temperature and/or rainfall during last few years
	Households experiencing any stress due to climate change	Percentage of household who reported any stress due to climate change or drought over the last few years
Drought	Households experiencing disease/death as a result of drought	Percentage of households who reported any disease/death in their family as a result of drought in the past six years (2009–2014)
	Households experiencing crop yield reduction/loss due to drought	Percentage of household reporting any yield reduction/loss of crops due to drought in the past six years (2009–2014)
	Households experiencing unemployment due to drought	Percentage of households reporting unemployment due to drought in the past six years (2009–2014)
	Household who faced food insecurity problems during drought periods	Percentage of households who faced food insecurity problems during drought period in the past six years (2009–2014)

[a]*OBC* Other Backward Castes, *SC* Schedule Castes, *ST* Schedule Tribe (backward caste classification of India)

[b]This study assumes that increase number/value of these sub domains will render households less vulnerable. Taking inverse of the crude value of such sub domains will yields a lower value for index scores for households with greater number/value of sub domains

[c]*NGOs* Non-Governmental Organisations, *SHG* Self Help Groups

7 Results and Discussions

The results of the analysed data are presented in two parts; the first part presents domain-wise findings along with their corresponding sub domains for the four study villages whereas the second part presents results about climate change vulnerability.

7.1 Domain-Wise Vulnerability

7.1.1 Demographic Domain

Several researchers argue that a better demographic profile can enhance the adaptive capacity of households to any shocks related to environment (e.g. Yohe and Tol 2002; Frank et al. 2011). The present study has identified specific sub domains related to demographic features that are likely to affect vulnerability to and coping potential for climate change and drought risk. Results in Table 3 show the indexed values for demographic domain with respect to various sub domains. Results indicate that, Bagbahali village is demographically the most vulnerable village (index score of 0.529) among the four surveyed villages. All the indicators have higher values in comparison with other villages including a higher dependency ratio. A high dependency ratio may increase the vulnerability of households (Block and Webb 2001).

Female-headed households are more vulnerable to the risk associated with the natural hazards in developing countries (Cutter et al. 2003) and our findings too support this notion. Although 90 % of households in rural India are headed by men (Chudgar 2009) but the households where female is the head, face severe risk from natural hazards such as drought as revealed by a relatively higher index values related to dependency ratio, illiteracy and backwardness. In the sampled households, most of the families were nuclear and men generally migrate to other states leaving behind women as the household heads in this village. The value for illiteracy index is also high for Bagbahali village (0.436), which explains a plight on the part of female-headed households and hence a relatively increased vulnerability to external risks as better education of household head has a negative association with natural disasters and climate change risks (Brody et al. 2008). The reason for lower level of education for female heads is the prevalence of several restrictions during the school going age thus confining them from going to school (Chudgar 2011).

The Indian caste system is historically an important rural dimension where people are socially differentiated through class, religion, region, tribe and language (Deshpande 2010). Castes continue to play a significant role in economic life in rural India, and specifically for those belonging to the backward castes are disadvantaged with respect to social and economic attainments (Rawal and Swaminathan 2011).

Table 3 Index values for demographic domain and its sub domains in the study area

Sub domains/major domain	Sirabahal	Bilaikani	Aintalunga	Bagbahali
Dependency ratio	0.158	0.115	0.159	0.206
Female-headed households	0.073	0.073	0.127	0.491
Illiterate household heads	0.236	0.236	0.273	0.436
Backward caste	0.909	0.891	0.673	0.982
Demographic	0.344	0.329	0.308	0.529

These castes would be prone to higher damages due to their constrained coping capacity in terms of their limited access to information, economic opportunities and seclusion from society. All four study villages have backward castes whereas Bagbahali is a tribal village where ninety 8 % of the sample households belong to *Gond* tribe.

7.1.2 Social Domain

Social networking and societal integration play an important role in motivating households to face natural disasters. As noted by Thomalla et al. (2006), the most vulnerable people to natural hazards are those with inadequate access to social capital (e.g. networks, information, relationships). Information related to social features linked with vulnerability to climate change is presented in Table 4. Results indicate that Sirabahal is the most vulnerable village with respect to social domain (index score of 0.7). As the access to social networks increase, households found themselves less vulnerable to shocks as their ability to cope with risks becomes high (Lokshin and Yemtsov 2001). The highest percentage of households who seek assistance from NGO/SHGs was in Bilaikani village. In Sirabahal village, there was no NGO/SHG to help the villagers.

Respondents in the study area have fairly easy access to informal credit system given by a higher value for its index whereas the access to banking system is somewhat constrained. In the absence of or constrained access to formal credit system, informal system is laden with some drawbacks such as higher interest rates. Credit surveys in many developing countries have concluded that non-institutional credit system like money lenders, landlords, traders, etc., charge very high rate of interest than those charged by institutional lenders like banks (Chakrabarty and Chaudhuri 2001). Bilaikani village had the highest index value for the percentage of household which had no access to banks. Most of the migrant households in the study villages had access to bank as the migrant family members send remittances to their families through banks. As Bilaikani had the least number of households with migrants, access to the banking system was found to be lower.

Bagbahali village had the least percent of households (29 %) with no access to school. People belonging to remote rural areas have meagre incomes. Children from

Table 4 Index values for social domain and its sub domains in the study area

Sub domains/major domain	Sirabahal	Bilaikani	Aintalunga	Bagbahali
Households seeking no assistance from NGOs/SHGs	1.000	0.327	0.727	0.745
Households with access to money lenders	1.000	1.000	1.000	1.000
Households with no access to banks	0.455	0.673	0.436	0.436
Households with no school access	0.345	0.636	0.327	0.291
Social	0.700	0.659	0.623	0.618

these families are generally unable to attend the school; instead they assist the earning member of the family to add-up some extra income. The primary and elementary schools are located within the villages whereas the high school and higher secondary schools are situated outside village boundaries. Most villages have poor connectivity from one place to another and as a result children have to walk miles mostly to reach the schools which often lead to demotivation among them. Access to education is differently perceived for male and female in these rural areas. If a family has to choose between educating a son or a daughter due to financial limitations, typically the son will be preferred for that purpose. Girls are often taken out of school to share the family responsibilities such as caring for younger siblings, etc.

7.1.3 Economic Domain

Highest vulnerability in case of economic domain was found for Aintalunga village with an index value of 0.405 (Table 5). Average livelihood diversification index is maximum for Aintalunga village and minimum for Sirabahal village. Diversification of livelihood strategies helps the households to choose protection strategies which in turn support them to survive during natural disasters (van den Berg 2010). The most common livelihood means in the study villages is agriculture. More than 80 % of households in all the study villages depend on agriculture as major income source. Apart from agriculture, people work as casual agricultural labours and also resort to migration. The highest percent of households who solely depend on agriculture was in Bilaikani village (60 %). As agriculture is highly dependent on nature and climate, a slight variation in rainfall or temperature will adversely affect its productivity. As a result, households engaged in agricultural production would be highly vulnerable to natural hazards such as droughts.

Apart from agriculture, wages from farm and non-farm sectors and remittances from migrant labour are the other sources of income in the study area. The average per month household income was the highest in Sirabahal village (US$ 87) and hence the index score was the least (0.085). The lowest per month household income was found in Bagbahali village (US$ 52) with index value of 0.275. Average household income and food expenditure have a positive relationship. Since

Table 5 Index values for economic domain and its sub domains in the study area

Sub domains/major domain	Sirabahal	Bilaikani	Aintalunga	Bagbahali
Average livelihood diversification index	0.436	0.648	0.764	0.576
Households depending solely on agriculture for livelihood	0.491	0.600	0.564	0.473
Households' average monthly income (US$)	0.085	0.157	0.122	0.275
Average monthly food expenditure per household (US$)	0.162	0.197	0.170	0.273
Economic	0.293	0.401	0.405	0.399

Sirabahal village had the highest average per month household income, the vulnerability associated with per month household food expenditure was least for this village (0.162).

7.1.4 Physical Domain

Table 6 presents the indexed values for physical domain and its sub domains. Index score for physical domain was the highest for Aintalunga village (0.797). As water is the most essential element which supports a wide range of domestic and productive needs of households, the access to own source of water was missing in case of the study area. It is evident from the fact that none of the household had his own source of water. All the households depend on public tube well (ground water) and natural sources for water needs (index score equal to 1). Problems of accessing safe quality water in ample quantity for drinking and domestic needs are found in the villages.

Considering access to food, the highest percent of households depending on own farm for food was in Bagbahali village (0.945). Farming on own farm is the primary source of food provision for many households in the study area. The surplus food after fulfilling household requirement is sold in order to repay loans and run other activities. Many a times, farmers are forced to retain only a small portion of the agriculture produce for own consumption for meeting urgent loan repayments which implies an overall fledgling financial standings of the households in the study area.

After consuming own-produced food items, households resort to Public Distribution System (PDS) for food provision on subsidised rates. The households having access to PDS are considered as less vulnerable. Only in Bagbahali village, all the surveyed households depended on PDS and the index score was equal to zero for this village. In this village insufficiency of food consumption was high and most of the families were below poverty line. Moreover, the access to electricity is also somewhat constrained. About 48 % households in Sirabahal village had no access to electricity. As observed by Barnet (2000), electricity can play an important role in improving quality of life, education for children, health while it

Table 6 Index values for physical domain and its sub domains in the study area

Sub domains/major domain	Sirabahal	Bilaikani	Aintalunga	Bagbahali
Households depending on public/natural resources of water for household activities	1.000	1.000	1.000	1.000
Households depending on own farm for food	0.764	0.709	0.836	0.945
Households not depending on Public Distribution System (PDS)	0.509	0.364	0.600	0.000
Households with no electricity connection	0.477	0.422	0.396	0.455
Physical	0.761	0.718	0.797	0.623

reduces the household workloads. The results of present study thus indicate lower coping potential augmented through limited access to amenities such as electricity.

7.1.5 Climatic Variability Domain

Climate variability is a transformative process that affects individuals and societies in different ways (O'Brien et al. 2007). For the present study area, the climate variability index was the highest for Bilaikani and the lowest for Sirabahal village (Table 7). The higher vulnerability to environmental changes thus expose the community in the study area in particular and in India in general to projected changes in annual mean surface temperature increase (ranging from 2.5 to 5 °C) by the end of century (IPCC 2001). The highest index value for the percent of household who reported increase in average temperature was for Bagbahali village (0.709). This is expected as extremes in maximum and minimum temperatures and rainfall were predicted to increase along with several years ago (Sathaye et al. 2006). Nevertheless, variability in the spatial patterns of the rainfall anomalies is seen all over India (Wang et al. 2001). This effect was also evinced by the present study and maximum anomaly in average rainfall was reported in case of Bilaikani village (0.764).

Subsequently, climate variability takes its toll by causing fluctuations in global food production in the arid and semi-arid tropical countries of the developing world (Sivakumar et al. 2005). This phenomenon is also observed by the respondents in the study area who reported a decline or variation in agricultural productivity brought about by changes in temperature and rainfall (index score value of 0.909). In addition, people reported that the drought and climatic variations occur every year and they had become part of their life (0.927).

7.1.6 Drought Domain

Between 2009 and 2014, Balangir district had faced drought for three years. The sub domains related to drought revealed that the households are still extremely

Table 7 Index values for climatic variability domain and its sub domains in the study area

Sub domains/major domain	Sirabahal	Bilaikani	Aintalunga	Bagbahali
Households who perceived a rise in average temperature	0.455	0.673	0.527	0.709
Households who perceived variation in average rainfall	0.509	0.764	0.582	0.673
Households reporting an effect on agriculture due to variability in temperature and/or rainfall	0.909	0.709	0.636	0.709
Households experiencing any stress due to climate change	0.927	0.709	0.655	0.727
Climate variability	0.700	0.714	0.595	0.705

Table 8 Index values for drought domain and its sub domains in the study area

Sub domains/major domain	Sirabahal	Bilaikani	Aintalunga	Bagbahali
Households experiencing disease/death as a result of drought	0.673	0.691	0.891	0.873
Households experiencing crop yield reduction/loss due to drought	0.764	0.818	0.873	0.945
Households experiencing unemployment due to drought	0.418	0.255	0.564	0.473
Household who faced food insecurity problems during drought periods	0.727	0.473	0.418	0.673
Drought	0.645	0.559	0.686	0.741

prone to drought. Results in Table 8 indicate that Bagbahali was the most vulnerable village to drought (0.741). Moreover, drought spells are accompanied by many diseases like hepatitis, dysentery, chicken pox, skin problems thus leading to protracted nature of hazardous impacts due to drought. The number of households affected by diseases is manifold during drought period than in normal years. This becomes evident from the fact that 89 % households caught diseases during drought in Aintalunga village.

Drought impacts are most prominent in the agricultural sector. Prolonged soil moisture deficits due to drought cause damage to crops which may ultimately lead to crop loss or yield reduction. A crop failure is the primary direct economic impact of drought within the agricultural sector which in turn causes negative supply shocks (Ding et al. 2011). The crop loss/yield reduction is very common phenomenon during drought in the study area too. Index score for yield loss/reduction was highest for Bagbahali village (0.95). Agricultural production is affected badly during drought period, leading to an increased unemployment which in turn increases the vulnerability of rural households (Singh et al. 2006). For the study area, the unemployment index was the highest for Aintalunga and the lowest for Sirabahal (Table 8). Drought lowers dietary choices and reduces overall food consumption which may lead to food insecurity problems and these problems are also reported to have taken place. As, most of the farming families in the study area have subsistence nature of agriculture, a drought spell casts a heavy influence on crop yields and hence reduces food availability and its consumption. The problem of food insecurity was reported by the respondents and the index score for food insecurity problem was maximum for Sirabahal (0.727) and minimum for Aintalunga (0.418).

7.2 *Climate Vulnerability Index*

Table 9 portrays results of CVI whereas Fig. 2 illustrates the index values of three IPCC dimensions of vulnerability for the study area. Scores for CVI were highest for Bilaikani village (0.861) and the lowest for Aintalunga village (0.84).

Table 9 Climate Variability Index of villages for the study area

Index	Sirabahal	Bilaikani	Aintalunga	Bagbahali
CVI	0.846	0.861	0.840	0.851

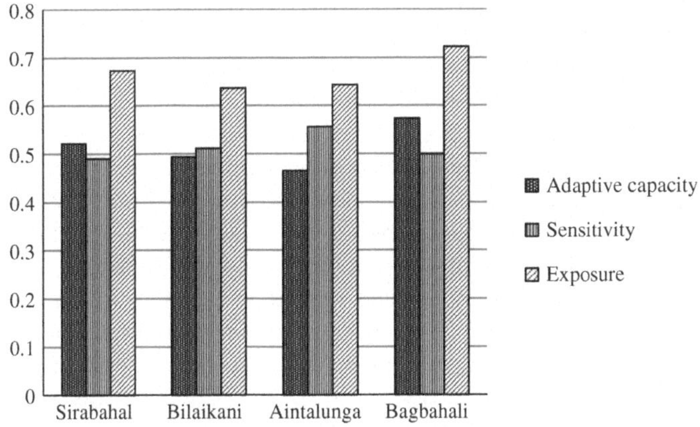

Fig. 2 Indexed values for IPCC dimensions of climate vulnerability in the study area

Results in Fig. 2 indicate that Bagbahali had the highest index value for adaptive capacity (0.574). The high dependency ratio, more number of female-headed households, more illiterate household heads and presence of backward castes in the study area might have led to this lower level of adaptive capacity in Bagbahali village. Sensitivity to climate variability was found to be maximum in case of Aintalunga village which mainly arises from higher dependency on own farm for food, weaker access to independent sources of water, low average household income and lower spending on food items seem to have made Aintalunga village highly sensitive to climate change and drought risk.

In case of exposure to the effects of climate change, its index value was maximum for Bagbahali village (0.723) followed by Sirabahal (0.6), Aintalunga (0.643) and Bilaikani (0.636) (Fig. 2).

8 Conclusions

India faces ever increasing challenge of climate change which may manifest itself in the form of increased number of drought spells with protracted durations as well as frequent flooding disaster. As this challenge keeps on mounting due to anthropogenic and socio demographic changes, more population is expected to suffer from its negative impacts. Proactive strategies would be required for an effective mitigation and adaptation process. In this regard, exact evaluation of the extent of exposure and vulnerability to such hazards is required to streamline and plan these proactive strategies especially for rural areas which are devoid of comprehensive system for *ex-ante* and *ex-post* dealing with such catastrophe. The current study quantifies the degree of vulnerability of rural households of and results indicate existence of many vulnerable households to climate change and drought.

The vulnerability of marginal households is further increased due to their inability to access pertinent information and their excessive dependence on natural resources for earning the livelihood. The climate change impacts in the form of drought would have a two-pronged consequence for such households through constrained agricultural production in the first place due to less availability of natural resource such as water and erosion of their coping capacity and social status brought about by dearth of financial autonomy. Moreover, lower levels of literacy, high dependency ratio, ethnic backwardness and weak housing structures may lead to increased levels of vulnerability. As a panacea to this situation, there is a felt need to re-focus local level coping and mitigation planning to natural disasters involving the local community. This would require an effective transmission of valuable information to at-risk communities along with the knowledge about anticipated changes and associated impacts as well as the benefits of timely preparedness. Regional mitigation plans and strategies also need to build on actual vulnerability indicators for specific region based on evidence in the context of socioeconomic setting for that region as well as its hydro-meteorological portfolio.

Acknowledgements We would like to thank the German Academic Exchange Service (DAAD) for the financial support for this research. The financial support by *Stiftung Fiat Panis* for the primary data collection is also highly acknowledged. We also appreciate the cooperation of our survey respondents in sharing their experience, household information and valid opinions data collection phase.

References

Adger, W. N., Huq, S., Brown, K., Conway, D., & Hulme, M. (2003). Adaptation to climate change in the developing world. *Progress in Development Studies, 3*, 179–195.

Aleksandrova, M., Lamers, J. P. A., Martius, C., & Tischbein, B. (2014). Rural vulnerability to environmental change in the irrigated lowlands of Central Asia and options for policy-makers: A review. *Environmental Science & Policy, 41*, 77–88.

Arora, A., Bansal, S., & Ward, P. S. (2015). *Do farmers value rice varieties tolerant to droughts and floods? Evidence from a discrete choice experiment in Odisha, India*. Agricultural and Applied Economics Association and Western Agricultural Economics Association Joint Annual Meeting, San Francisco, California.

Aryal, S., Cockfield, G., & Maraseni, T. N. (2014). Vulnerability of Himalayan transhumant communities to climate change. *Climatic Change, 125*, 193–208.

Bardsley, D. K., & Wiseman, N. D. (2012). Climate change vulnerability and social development for remote indigenous communities of South Australia. *Global Environmental Change, 22*, 713–723.

Barnett, A. (2000, June 29). *Energy and the fight against poverty*. Paper given as part of a series of Economic Research Seminars at Institute of Social Studies, The Hague.

Bhadwal, S., Bhandari, P., Javad, A., Kelkar, U., O'Brien, K., & Barg, S. (2003). *Coping with global change: Vulnerability and adaptation in Indian agriculture*. New Dehli: The Energy and Resources Institute (TERI).

Bhattacharya-Mis, N., & Lamond, J. (2014). Socio-economic complexities of flood memory in building resilience: An overview of research. *Procedia Economics and Finance, 18*, 111–119.

Birkmann, J., Cardona, O. D., Carreño, M. L., Barbat, A. H., Pelling, M., Schneiderbauer, S., et al. (2013). Framing vulnerability, risk and societal responses: The MOVE framework. *Natural Hazards, 67*, 193–211.

Block, S., & Webb, P. (2001). The dynamics of livelihood diversification in post-famine Ethiopia. *Food Policy, 26*(4), 333–350.

Brody, S. D., Zahran, S., Vedlitz, A., & Grover, H. (2008). Examining the relationship between physical vulnerability and public perceptions of global climate change in the United States. *Environment and Behavior, 40*(1), 72–95.

Calzadilla, A., Zhu, T., Rehdanz, K., Tol, R. S. J., & Ringler, C. (2013). Economywide impacts of climate change on agriculture in Sub-Saharan Africa. *Ecological Economics, 93*, 150–165.

Chakrabarty, D., & Chaudhuri, A. (2001). Formal and informal sector credit institutions and inter linkage. *Journal of Economic Behavior & Organization, 46*(3), 313–325.

Christensen, J. H., Hewitson, B., Busuioc, A., Chen, A., Gao, X., Held, I., et al. (2007). Regional climate projections, In S. Solomon, D. Qin, M. Manning, Z. Chen, M. Marquis, K. B. Averyt, et al. (Eds.), *Climate change 2007: The physical science basis. Contribution of working group I to the fourth assessment report of the intergovernmental panel on climate change*. Cambridge, United Kingdom and New York, NY, USA: Cambridge University Press.

Chudgar, A. (2009). The challenge of universal elementary education in rural India: Can adult literacy play a role? *Comparative Education Review, 53*(3), 403–433.

Chudgar, A. (2011). Female headship and schooling outcomes in rural India. *World Development, 39*(4), 550–560.

CRED. (2015). *India country profile of natural disasters, EM-DAT: The international disaster database*. Available via http://www.em-dat.net/disasters/Visualisation/profiles/countryprofile.php

Cutter, S. L., Boruff, B. J., & Shirley, W. L. (2003). Social vulnerability to environmental hazards. *Social Science Quarterly, 84*(2), 242–261.

Dai, A. (2013). Increasing drought under global warming in observations and models. *Nature Climate Change, 3*, 171–171.

Deshpande, M. S. (2010). *History of the Indian caste system and its impact on India today*. San Luis Obispo: Senior project Social sciences department, College of Liberal Arts, California Polytechnic State University.

Devkota, R. P., Cockfield, G., & Maraseni, T. N. (2014). Perceived community-based flood adaptation strategies under climate change in Nepal. *International Journal of Global Warming, 6*, 113–124.

Ding, Y., Hayes, M. J., & Widhalm, M. (2011). Measuring economic impacts of drought: A review and discussion. *Disaster Prevention and Management: An International Journal, 20*, 434–446.

Frank, E., Eakin, H., & López-Carr, D. (2011). Social identity, perception and motivation in adaptation to climate risk in the coffee sector of Chiapas, Mexico. *Global Environmental Change, 21*(1), 66–76.

Government of Odisha. (2013). *Disaster management plan for Odisha. Agriculture Sector*. Odisha, India: Department of Agriculture.

Gray, C., & Mueller, V. (2012). Drought and population mobility in rural Ethiopia. *World Development, 40*, 134–145.

Hahn, M. B., Riederer, A. M., & Foster, S. O. (2009). The livelihood vulnerability index: A pragmatic approach to assessing risks from climate variability and change—A case study in Mozambique. *Global Environmental Change, 19*, 74–88.

Huggel, C., Raissig, A., Rohrer, M., Romero, G., Diaz, A., & Salzmann, N. (2015). How useful and reliable are disaster databases in the context of climate and global change? A comparative case study analysis in Peru. *Natural Hazards Earth System Science, 15*, 475–485.

IPCC. (2001). Climate change 2001: The scientific basis. Contribution of working group I to the third assessment report of the intergovernmental panel on climate change. In J. T. Houghton, Y. Ding, D. J. Griggs, M. Noguer, P. J. van der Linden, X. Dai, et al. (Eds.), Cambridge, United Kingdom and New York, USA: Cambridge university Press.

IPCC. (2007). Climate change 2007: Impacts, adaptation and vulnerability. Contribution of working group II to the fourth assessment report of the intergovernmental panel on climate change. In M. L. Parry, O. F. Canziani, J. P. Palutikof, P. J. van der Linden, & C. E. Hanson (Eds.), Cambridge, UK: Cambridge University Press.

IPCC. (2012). Managing the risks of extreme events and disasters to advance climate change adaptation. A special report of working groups I and II of the intergovernmental panel on climate change. In C. B. Field, V. Barros, T. F. Stocker, D. Qin, D. J. Dokken, K. L. Ebi, et al. (Eds.), Cambridge, UK and New York: Cambridge University Press.

Kumar, R. H., Venkaiah, K., Arlappa, N., Kumar, S., Brahmam, G. N. V., & Vijayaraghavan, K. (2005). Diet and nutritional situation of the population in the severely drought affected areas of Gujarat. *Journal of Human Ecology, 18*(4), 319–326.

Lei, Y., Ja, Wang, Yue, Y., Yin, Y., & Sheng, Z. (2014). How adjustments in land use patterns contribute to drought risk adaptation in a changing climate—A case study in China. *Land Use Policy, 36*, 577–584.

Leiserowitz, A., Maibach, E., Roser-Renouf, C., & Hmielowski, J. D. (2012). *Extreme weather, climate and preparedness in the American mind*. Yale University and George Mason University Yale Project on Climate Change Communication, New Haven. CT: Yale Project on Climate Change Communication. Available via http://environment.yale.edu/climate/files/Extreme-Weather-Climate-Preparedness.pdf

Lokshin, M., & Yemtsov, R. (2001). *Household strategies for coping with poverty and social exclusion in post-crisis Russia*. World Bank Policy Research Working Paper (2556). Retrieved from http://ssrn.com/abstract=632623. Available via http://go.worldbank.org/0IX1HGHAG0

Maru, Y. T., Stafford, M. S., Sparrow, A., Pinho, P. F., & Dube, O. P. (2014). A linked vulnerability and resilience framework for adaptation pathways in remote disadvantaged communities. *Global Environmental Change, 28*, 337–350.

O'Brien, K., Eriksen, S., Nygaard, L. P., & Schjolden, A. N. E. (2007). Why different interpretations of vulnerability matter in climate change discourses. *Climate Policy, 7*, 73–88.

Pai, D. S., Sridhar, L., Guhathakurta, P., & Hatwar, H. R. (2011). District-wide drought climatology of the southwest monsoon season over India based on standardized precipitation index (SPI). *Natural Hazard, 59*(3), 1797–1813.

Pandey, R., & Jha, S. K. (2011). Climate vulnerability index-measure of climate change vulnerability to communities: A case of rural Lower Himalaya, India. *Mitigation and Adaptation Strategies for Global Change, 17*, 487–506.

Pant, L. P., Kc, K. B., Fraser, E. D. G., Shrestha, P. K., Lama, A. B., Jirel, S. K., et al. (2014). Adaptive transition management for transformations to agricultural sustainability in the Karnali mountains of Nepal. *Agroecology and Sustainable Food Systems, 38*, 1156–1183.

Rawal, V., & Swaminathan, M. (2011). Income inequality and caste in village India. *Review of Agrarian Studies, 1*(2). Available at http://ras.org.in/income_inequality_and_caste_in_village_india

Sathaye, J., Shukla, P. R., & Ravindranath, N. H. (2006). Climate change, sustainable development and India: Global and national concerns. *Current Science, 90*(3), 314–325.

Singh, M. B., Fotedar, R., Lakshminarayana, J., & Anand, P. K. (2006). Studies on the nutritional status of children aged 0–5 years in a drought-affected desert area of western Rajasthan, India. *Public Health Nutrition, 9*(8), 961–967.

Singh, R. S., Narain, P., & Sharma, K. D. (2001). Climate changes in Luni river basin of arid western Rajasthan (India). *Vayu Mandal, 31*(1–4), 103–106.

Sivakumar, M. V. K., Das, H. P., & Brunini, O. (2005). Impacts of present and future climate variability and change on agriculture and forestry in the arid and semi-arid tropics. *Climatic Change, 70*, 31–72.

Stern, N. (2007). *The economics of climate change. The stern review*. New York: Cambridge University Press.

Sullivan, C., Meigh, J. R., & Fediw, T. S. (2002). *Derivation and testing of the water poverty index phase 1*. Final Report, Department for International Development, UK.

Swain, M., & Swain, M. (2006). Drought induced changes in household income, expenditure and savings: Evidence from Bolangir district in Western Orissa. *The Journal of Income and Wealth, 28*(2), 49–73.

Thomalla, F., Downing, T., Spanger-Siegfried, E., Han, G., & Rockström, J. (2006). Reducing hazard vulnerability: Towards a common approach between disaster risk reduction and climate adaptation. *Disasters, 30*(1), 39–48.

van den Berg, M. (2010). Household income strategies and natural disasters: Dynamic livelihoods in rural Nicaragua. *Ecological Economics, 69*(3), 592–602.

Wang, B., Wu, R., & Lau, K. M. (2001). Inter annual variability of the Asian summer monsoon: Contrasts between the Indian and the Western North Pacific-East Asian monsoons. *Journal of Climate, 14*(20), 4073–4090.

Wu, Y., Bake, B., Zhang, J., & Rasulov, H. (2015a). Spatio-temporal patterns of drought in North Xinjiang, China, 1961–2012 based on meteorological drought index. *Journal of Arid Land, 7*, 527–543.

Wu, P., Ridley, J., Pardaens, A., Levine, R., & Lowe, J. (2015b). The reversibility of CO_2 induced climate change. *Climate Dynamics, 45*, 745–754.

Yohe, G., & Tol, R. S. J. (2002). Indicators for social and economic coping capacity-moving toward a working definition of adaptive capacity. *Global Environmental Change, 12*, 25–40.

An Assessment of Vulnerability of Livestock Farming to Climate Variability

G. Letha Devi, Dhirendra Varma and Mukund A. Kataktalware

Abstract The livelihood vulnerability index (LVI) was developed to estimate climate change vulnerability in the Chitradurga and Kolar Districts of Karnataka, India. 120 households were surveyed, each district collected data on socio-demographics, livelihoods, social networks, health, food and water security, natural disasters and climate variability. Data were aggregated using a composite index and differential vulnerabilities were compared. Results suggest that Chitradurga may be more vulnerable in terms of water resources, while Kolar may be more vulnerable in terms of socio-demographic structure. This pragmatic approach may be used to monitor vulnerability, program resources for assistance, and/or evaluate potential program/policy effectiveness in data-scarce regions by introducing scenarios into the LVI model for baseline comparison.

1 Introduction

Fallout of climate change on environment, social ecology and livelihood need serious attention. It threatens to deepen vulnerabilities, erode hard own gains and seriously undermine prospects of sustainable development. Vulnerability to climate change varies across regions, sectors and social groups. Understanding the regional and local dimensions of vulnerability is essential to develop appropriate and targeted adaptation efforts. We must recognize that climate change impacts will not be felt in isolation, but in context of multiple stresses.

G. Letha Devi (✉)
National Institute of Animal Nutrition and Physiology, Bangalore 560 030, India
e-mail: lethaayur@gmail.com

D. Varma
Indian Institute of Science, Bangalore, India

M.A. Kataktalware
Southern Regional Station, National Dairy Research Institute, Bangalore 560 030, India
e-mail: mtalware@gmail.com

S. Nautiyal et al. (eds.), *Climate Change Challenge (3C) and Social-Economic-Ecological Interface-Building*, Environmental Science,
DOI 10.1007/978-3-319-31014-5_26

The Intergovernmental Panel on Climate Change (IPCC) in its fourth assessment report (2007) indicated that many of the developing countries tend to be especially vulnerable to extreme climatic events and adverse impacts of a gradual climate change as they largely depend on climate sensitive sectors like agriculture and forestry. India has a geographic disadvantage as it is already in the warmer part of the world. Nearly two thirds of the Indian population is rural mostly living in harsh climatic regions of mountains, deserts and river deltas, which are more susceptible to climate change.

In the case of livestock, global warming and climate change are likely to impact negatively on production and health. Increase in physiological reactions at high temperatures will elevate heat loads of animals resulting into a decline in productivity of meat, wool, milk and draught power (Upadhyay et al. 2008).

Livestock is an essential component of the dry land ecosystem. Grazing animals in dry lands contribute to a healthy soil through manure and seeding. It improves the soil's physical properties (porosity, water holding capacity, drainage) and its fertility (dung has high carbon: nitrogen ratio and urine is rich in 'nitrogen' and 'potassium'). Small ruminants spread the manure and urine very evenly. It is reported that a flock of 1000 sheep and goats, when spent five nights in the field can very well manure 1.32 acres of land. Almost eighty percent of milk in India is produced in integrated mixed crop-livestock farming systems. A well managed integrated crop livestock system has the potential to create a win-win situation for both the farmers and the environment. However, there are two areas where livestock in India has likely negative contributions on the environment. They are (a) methane emission and (b) degradation of common lands. These are issues that need to be addressed through technical and policy interventions. But valuing livestock only from one perspective (environmental) in a context where it plays many roles might lead to making expensive mistakes. This is because a preponderant number of farmers in the marginal lands in countries like India are surviving only because of livestock. In these countries, livestock is kept by people not just for production but because of its multiple (livelihood, social, environmental) contributions. When there was agrarian distress in India, farmers committed suicide. But no suicides were reported in areas where livestock was prominent as there was something to fall back. Livestock is found to be positively egalitarian. Therefore, a more detailed and holistic assessment is required before drawing conclusions (Pasha 2001).

The poor livestock keepers depend heavily on common property resources for their survival. At the same time, there are few organized efforts for the development of common lands and its sustainable management. This is likely to have a negative impact on the land. This scenario is not inevitable, provided the poor are offered alternative options that will reduce their dependency on the common resources and that will regulate the use, enhance the regeneration and raise the productivity of common property resources. Therefore, any change in the status and productivity of common property resources directly influences the economy of the rural poor. Jodha (1992) rightly suggested certain key elements of such an approach. Some of them are (1) introduction of technological investments and creation of economic incentives to conserve such resources while raising their productivity and (2) regulation of common resource use with the involvement of user groups and mobilization of a community strategy that complements state interventions with the essential participation of local people.

An effort was made to analyze the impact of climate change on livestock farming, in Chitradurga and Kolar districts of Karnataka state, through a combination of quantitative and qualitative methods, with the objective to assess the vulnerability of Indian livestock farming to climate variability and socio-economic impact of climate variability and to study the coping strategies of Indian livestock farmers to impacts of climate vulnerability.

2 Methodology

Vulnerability index was developed which includes social vulnerability, infrastructure development, biophysical conditions, climate, agriculture and livestock and transportation. These were combined into vulnerability indices and mapped as vulnerability profiles. Five case studies were undertaken in depth to understand their coping strategies to impacts of climate vulnerability. It was observed that there is a shift in cropping pattern and farmers themselves evolve their strategies to minimize the economic losses due to changes in climate and market changes (Fig. 1).

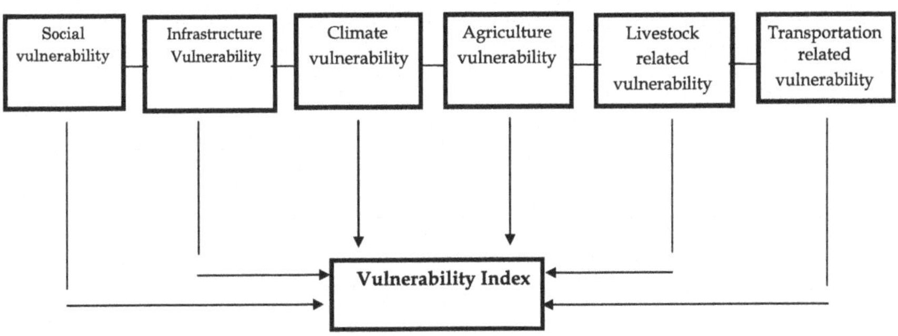

Fig. 1 Construction of vulnerability index

3 Components of Vulnerability Index

The following components were selected for composite index, after consulting the literature and experts in the field.

- Social vulnerability (social participation, access to common facilities),
- Infrastructure development (roads, health care, common facilities),
- Climate (rainfall, temperature, humidity),
- Agriculture (crops, pattern of cropping),
- Livestock,
- Transportation.

4 Development of Vulnerability Index

There were three major steps as described in detail below:

(a) Determination of scale values
(b) Selection of items/indicators of components of vulnerability index
(c) Computing the composite index of vulnerability.

4.1 Normalized Rank Method by Guilford (1954) for Determining Scale Values

- Different components of VI were ranked by a group of judges according to their perceived significance in determining the vulnerability to climate change.
- Rankings were obtained from 36 judges who were experts in the fields of social science and climate change research.
- The rankings were then tabulated and a frequency distribution denoted by 'f_{ji}' was worked out for each component.
- For different numbers of stimuli ranked 'n' corresponding 'c' values were made available from the statistical table.
- Then 'f_{ji}' multiplied by the corresponding 'c' value and summated over all the ranks gave a total score '$\sum f_{ji}c_i$' for each component, which was further divided by the total number of judges and the consequent value 'R_i' was subjected to linear transformation to arrive the scale value 'R_c' using the formula $R_c = 2.357R_j - 7.01$ (Table 1).

Table 1 Scale values of components of vulnerability index

Components	Scale value
Climate	11.53
Infrastructure development	11.44
Social/financial vulnerability	10.16
Agriculture	9.45
Livestock	9.36
Transportation	8.13

4.2 *Selection of Items/Indicators of Components of Vulnerability Index*

Items under each component of vulnerability index were selected through expert consultation and the literature scan. Content validity of items ascertained. Thirty-eight items covering the different components were selected initially. The selected items were pretested on a sample of thirty farmers in a non-sampling area. The items, which were found irrelevant, were dropped. Thirty-two items covering the different components constituted the final indicators for vulnerability index.

4.3 *Computing the Composite Index of Vulnerability*

Each component of index consisted of different number of items/indicators and hence their range of scores was different. Therefore, the scores of all the seven components were converted into unit scores by using simple range and variance as given underneath.

$$U_{ij} = \frac{Y_{ij} - \text{Min}y_i}{\text{Max}y_i - \text{Min}y_i}$$

where,

U_{ij} Unit score of the *i*th respondent on *j*th component
Y_{ij} Value of the *i*th respondent on the *j*th component
Max y_j Maximum score on the *j*th component
Min y_j Minimum score on the *j*th component.

The score of each component ranged from 0 to 1 i.e. when y_{ij} is minimum, the score is 0 and when y_{ij} is maximum the score is 1.

5 Vulnerability Index

The unit scores of each respondent were multiplied by respective component scale values and summed up. The scores thus obtained were divided by the total scale value and multiplied by 100 to get the vulnerability index for each respondent.

Different components were combined into vulnerability index

$$VI_i = \frac{\sum U_{ij} \cdot S_j^{i=1-240, j=1-6}}{\sum S_j} \times 100$$

where,

VI_i Vulnerability index of *i*th respondent
U_{ij} Unit score of the *i*th respondent on *j*th component
S_j Scale value of the *j*th component

Subsequently, mean index for each group of respondents was calculated. Reliability of the index was tested using 'test–retest' method and the '*r*' value 0.813 was found to be highly significant.

6 Findings of the Study

6.1 Vulnerability to Climatic Variability in Karnataka

Vulnerability is often reflected in the condition of the economic system as well as the socio-economic characteristics of the population living in that system. Assessment of current vulnerability can be done using a variety of socio-economic indicators that capture the exposure of the population in concern. The socio-economic status of a group is closely linked to the adaptive capacity of that particular group. Many factors contribute to social and economic vulnerability including rapid population growth, poverty and hunger, poor health, low levels of educations, gender inequality, social exclusion, fragile, marginal and/or hazardous location, resource degradation, and lack of access to infrastructure, resources and services, including knowledge and technological means. The exposed population has a limited capacity to protect themselves from natural hazards, especially from extreme events such as storms, droughts and floods. They bear the brunt of the consequences of large-scale environmental change, including land degradation, biodiversity loss and climate change, which affect the welfare of the most vulnerable populations. Over the long term, vulnerable populations have to learn to cope with the effects of climate change on their production systems (Table 2).

The sample was stratified into three groups viz., crop farming system, livestock farming system and integrated farming system based on the crop or livestock enterprise which they were practicing. It was observed that all the villages were vulnerable

Table 2 Comparison of vulnerability index among groups (n = 240)

Village	Vulnerability index (crop farming system)	Vulnerability index (livestock farming system)	Vulnerability index (integrated farming system)	*F*-value
Adilaur	58.10	57.50	56.74	0.57
Bharamapura	55.70	54.64	53.65	1.04*
Kumbarakatte	56.96	55.28	54.58	1.01*
Shahpur	56.12	54.98	54.73	0.81
Nanadanahalli	56.91	53.21	54.74	2.07**
Yaranghatta	57.12	55.58	54.64	1.99**

*Significant at 5 % level and **Significant at 1 % level

to the climate change and economic change and in general, it was observed that the district Chitradurga was much vulnerable as compared to Kolar. Further, it was found that groups with single livestock species (Cattle) were highly vulnerable to climate vagaries. Integrated farming system with a few cattle, sheep/goat and livestock was found to be the least vulnerable system. Significant difference was found among integrated farming system and livestock farming system.

Most of the households reported that they suffered crop loss and loss of animals due to extreme climatic conditions. The major loss was due to field crops, which are much sensitive to climatic conditions. Vegetable crops also faced the similar issues. The most vulnerable livestock species was found to be cross bred cattle as compared to sheep and goat, which can withstand extreme weather conditions.

Animal Feeding and management were the worst affected in case of climate vagaries. In extreme climate affected situations, livestock was the first option to encash, followed by cash crops and trees. Shelter, food and basic sustenance were the most essential needs in case of climate vagaries, both for human and animal. 95 % of respondents reported that meeting out the water requirement of animals was the challenge during drought periods. There is very little compensation received for the loss of livestock, due to natural disasters, as is the case of crops. 98 % were not able to repay the agricultural loans during climate disasters. 79 % of the total respondents changed their livelihood pattern, as coping strategy to climate changes (species of crops and livestock, management practices, housing of animals, etc.).

6.2 *Case Studies*

Five case studies were conducted in the study area, to find out the micro-level adaptation strategies for coping the impact of climate variability. In general, six measures were zeroed in as coping behaviour by the households to hedge against the shocks. They were (i) interest-free loans from friends and relatives, (ii) relief from government, (iii) selling of land, (iv) additional loans, (v) selling household assets and (vi) selling livestock.

The findings are summarized in the following points:

- Households belonging to BPL, households which own the land and those receiving medical or any other kind of aid are least likely to opt for selling of livestock.
- Most of women-headed households opted for sale of household assets, whereas male-headed households were not opting it as coping strategies.
- Female-headed households were not opting for additional loans to cop up with losses due to climate vagaries. Similar is the case with that receiving government relief for natural disasters.
- Respondents belonging to OBC and above middle class were not receiving the government relief for natural disasters, and had to resort to other coping strategies.
- Money received from friends and relatives were used mainly to cover up short fall in supply of food and other necessary items.

7 Discussion

The areas which were vulnerable in the earlier time continued to be more vulnerable. But the study focused changes during the past 5 years only, as collecting information through recall method is difficult for beyond 5 years period. Majority of households were affected by climate extremes and most of them carved out their own adaptation strategies, especially in case of livestock rearing. There is still lack of initiatives in the climate adaptation policies from the government agencies and most of the households believe that there has to be intervention from the government, especially in natural disasters, with respect to crops and livestock. Insurance sector also can play a vital role in this regard.

The trend in crop-livestock production shows that milk production is less susceptible to draft conditions, compared to crop production. Farmers have developed their own adaptation strategies to vagaries of nature by shifting from field crops to cash crops, adoption of scientific management practices for crops and livestock. It was observed that milk production can be sustained even under stress if population is optimized along with proper management of available feed-fodder resources. This is more important for feed-fodder deficit states like Karnataka.

8 Suggestions and Implications

The study brings out a few suggestions for helping to cop up the vulnerability of climate variability for the farmers. It requires efforts at various levels of stakeholders such as farmers, line department officials, scientists, policy makers, etc.

1. Farmers need to be trained in various aspects of coping with climate variability.
2. Awareness has to be developed at the level of line department officials working closely with the farmers.
3. Modern Information and communication tools should be utilized appropriately, for timely and proper dissemination of climate-related information.
4. Research organizations should work in close participation with farmers so that technologies can be developed for coping the climate challenges.
5. Policies need to be framed, keeping emphasis on challenges posed by the changing climate.

9 Conclusions

Policy makers face a multiple issues like poverty, illiteracy, infant mortality, malnutrition, environmental pollution, provision of basic amenities, etc. so that broader issues such as climate vagaries and natural disasters get sidelined. It can be concluded that poor population find it challenging to cop up with impacts of climate vagaries. The study results may help in policy decisions in formulating effective strategies for coping up climate variability, especially in livestock sector, which contributes to more than 60 % of livelihoods of rural population in India.

Acknowledgement The authors are highly grateful to ICAR-NIANP for providing the necessary support for carrying out the study. The financial support provided by ICSSR, New Delhi is greatly acknowledged.

References

Guilford, J. P. (1954). *Psychometric Methods*. McGraw-Hill: New York.

IPCC. (2007). Climate change 2007: The physical science basis-summary for policy makers. *Fourth Assessment Report of the Intergovernmental Panel on Climate Change. IPCC Secretariat, Switzerland.*

Jodha, N. S. (1992). *Common property resources and dynamics of rural poverty in India's dry regions*. India: New Delhi.

Pasha, S. A. (2001). *Economic and ecological dimensions of livestock economy*. India: New Delhi.

Upadhyay, R. C., Singh, S. V., & Ashutosh. (2008). Impact of climate change on livestock. In *Proceedings of 36th Dairy Industry Conference, Varanasi, India.*

Assessment of Carbon Sequestration in Soils of Different Land Use Land Cover in Honnaver Taluka of Uttara Kannada District

A.G. Koppad and Pavan Tikhile

Abstract The study was conducted in Honnaver taluka (14° 8′ 0″ N to 14° 25′ N Latitude and 74° 25′ E to 74° 45′ E Longitude) of Uttar Kannada district, Karnataka India to assess the carbon sequestration in soils of different land use systems. The IRS P6 LISS-III imageries of the study area was procured from NRSC, Hyderabad and different land use systems in Honnaver taluka were identified with the ground truth data collected from GPS and processed in ERDAS software. The land use land cover (LULC) classes, viz., dense forest, sparse forest, agriculture and open land were identified. The total area in each class was assessed through supervise classification. The soil samples at 1 m depth were drawn at grid point in flat land and along the profile in sloppy land in different land use system. The SOC was estimated using Walkley and Black rapid titration method. The total area in four land use classes is 68520 ha with SOC pool of 12.112 million tonnes. Among different classes dense forest covers highest area (44053 ha) and highest SOC pool (8.82 million tonnes). Among the different land use classes, higher SOC was sequestered in dense forest (200.10 t/ha) followed by sparse forest (166.89 t/ha). The SOC in open land and agriculture land is 145.78 and 82.79 t/ha, respectively. The carbon mitigation potential of dense forest is 2.42 times higher compared to agricultural land followed by sparse forest (2.02 times).

Keywords Soil organic carbon · Imageries · LULC · Forest

1 Introduction

Global warming has become a problem for the society due to carbon emission in the wake of modernization and urbanization. Global surface temperatures have increased by 0.8 °C since the late nineteenth century, and 11 out of the 12 warmest years on record have occurred since 1995 (IPCC 2007). Carbon sequestration is the

A.G. Koppad (✉) · P. Tikhile
Department of Natural Resource Management, College of Forestry, Sirsi, Karnataka, India
e-mail: koppadag2007@rediffmail.com

S. Nautiyal et al. (eds.), *Climate Change Challenge (3C) and Social-Economic-Ecological Interface-Building*, Environmental Science,
DOI 10.1007/978-3-319-31014-5_27

process of capture and long-term storage of atmospheric carbon dioxide (Sedjo and Sohngen 2012) in soil. Soil can be considered as the largest pool of terrestrial organic carbon in the biosphere storing more CO_2 (2200 Pg) than plants (Batjes 1996). Forest and forest soil play an important role in the global carbon balance (Pareta and Pareta 2011 and Jha et al. 2003). Soil stores 2.5–3.0 times higher as that stored in the plants and two to three times more than the atmospheric CO_2 (Davidson et al. 2000). The forest soils are one of the major carbon sink in earth having higher content of organic carbon (Dey 2005).

Through the process of photosynthesis, plants assimilate carbon and return to the soil as litter and gets stored as soil organic matter (Negi and Gupwta 2010). Accurate estimation of soil organic carbon pool helps for climate change study and prediction of variation in climate change and its effect on local and global climate as well as biodiversity. The remote sensing technologies are being used for preparation of land use land cover (LULC) map and area estimation under different land use practices. Geographical information system (GIS) technology is useful for spatial analysis and preparation of SOC pool map. The present study is an attempt to estimate the carbon sequestration stock in forest soils of various land use system in Honnaver taluk of Uttar Kannada district (Karnataka).

2 Material and Method

The study was conducted in Honnaver taluka of Uttar Kannada district, Karnataka, India. The study region lies between 14° 8′ 0″ N to 14° 25′ N Latitude and 74° 25′ E to 74° 45′ E Longitude covering a surface area about 68520 ha. The study area is shown in Fig. 1.

The toposheets covering the study area with scale 1:50,000 were procured from SOI Bangalore. IRS LISS-3 satellite image path 097 row 063 dated 22 January 2010 spatial resolutions with 23.5 × 23.5 m procured from NRSC Hyderabad. The image was atmospheric and corrected to obtain actual reflectance value of the earth surface and also topographic normalization procedure was performed to reduce hill shade effects on the images using ERDAS IMAGINE 2011 software. The toposheet was georeferenced with projected coordinate system (WGS84). All georeferenced toposheet were clipped and mosaic to cover geographical area of the study region. Digitization of the study region boundary was done on mosaic toposheet and the same boundary was used to subset study area from the satellite image. The images were classified based on GPS data point using supervised calcifications method. Totally six classes were classified, i.e. dense forest, sparse forest, open land, agriculture land, water body and settlement. Out of these six LULC classes dense forest, sparse forest, open land and agriculture land were selected for the estimation of soil organic pool.

Soil samples were collected from different land use land class area, viz., dense forest, sparse forest, open land and agriculture land. While taking the samples the land terrain was considered. In sloppy land, the number samples were taken along the slope from top to bottom and in case of flat land grid at equidistance were

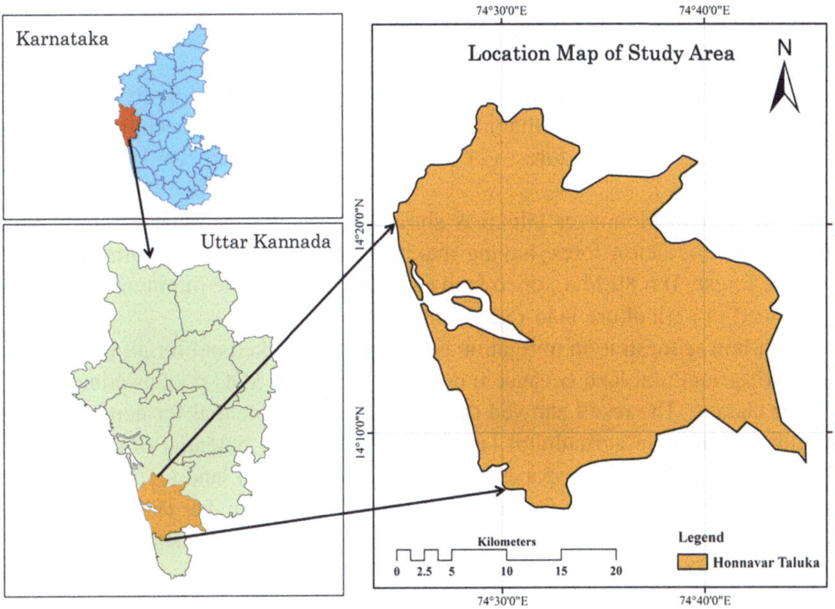

Fig. 1 Location map of study area

followed and representative soil sample was collected. The total 200 soil samples were collected from different land use. The soil sample to the depth of 1 m was taken using soil screw auger and core sampler was used to collect the soil core at different depth for estimating the bulk density. SOC was determined using Walkley and Black rapid titration method (Walkley and Black 1934). The % of SOC value obtained from the WB method was multiplied by standard correction factor of 1.32 (De Vos et al. 2007) to obtain the corrected SOC.

$$\mathbf{SOC\ \%} = (\mathrm{BTV} - \mathrm{STV}) \times 0.5\,\mathrm{N\ FAS} \times 0.003 \times 100/\mathrm{wt.of\ soil}$$

$$\mathbf{Total\,SOC(tones)} = \mathrm{SOC\ \%}/100 * \mathrm{BD}\ \left(\mathrm{t/m^3}\right) * \mathrm{area}\left(\mathrm{m^2}\right) * \mathrm{depth\,of\,soil(m)}$$

$$\mathrm{BTV} = \mathrm{Blank\,titrated\,value},\quad \mathrm{STV} = \mathrm{Sample\,titrated\,value},$$

$$\mathrm{FAS} = \mathrm{Ferrous\ Ammonium\ Sulphate}$$

$$\mathrm{BD} = \mathrm{Bulk\,density.}$$

The results were tested using different statistical techniques such as F-test and correlation among all samples for the all LULC.

3 Result and Discussion

The LULC map and SOC pool is shown in Fig. 2 and 3, respectively. The image classification results indicated that about 64.29 % of the area is occupied by dense forest. Followed by sparse forest (21.32 %), agriculture (7.09 %), least was with open land (4.56 %).

The SOC pool Honnaver taluka is shown in Fig. 3 and Table 1. The results indicated that the dense forest having maximum SOC pool (200.10 t/ha) followed by sparse forest (166.89 t/ha), open land (145.78 t/ha), and the minimum SOC pool was noticed in agriculture land (82.79 t/ha).

The carbon sequestration mitigation potential was worked out for all LULC with respect to agricultural land because it contains minimum SOC pool as compared to all other classes. The result showed that the mitigation potential for dense forest is 2.42 times more than agricultural land (1.00). Mitigation potential of dense forest was maximum followed by sparse forest (2.02 times), open land (1.76 times). The mitigation potential of dense forest is higher mainly due to higher leaf litter fall and microbial activities which creates more soil organic content (Conanat et al. 2001).

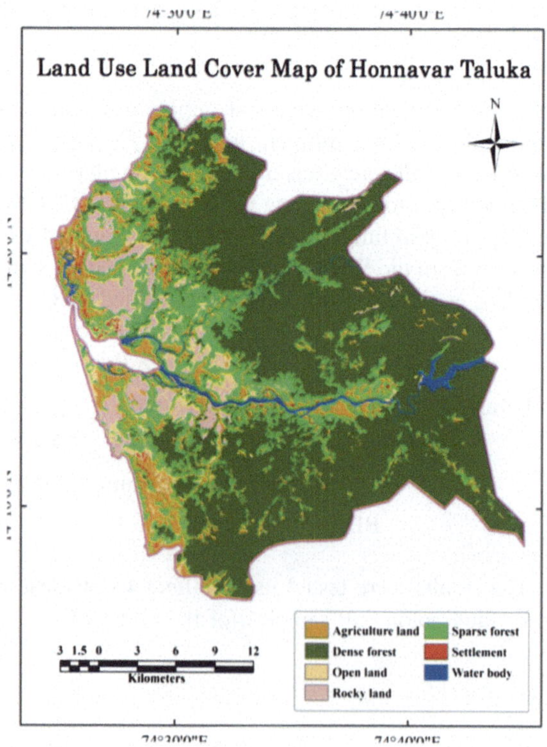

Fig. 2 LULC classification map of Honnaver taluka

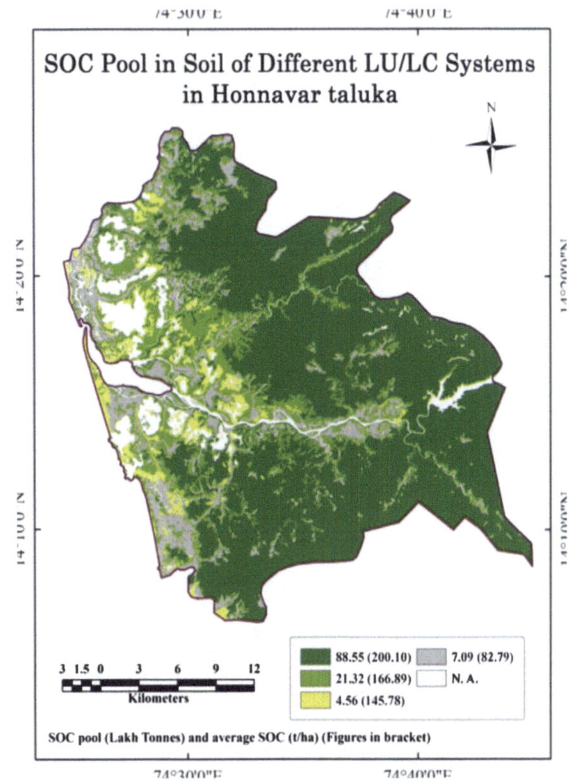

Fig. 3 SOC pool in Honnaver taluka

Table 1 Total SOC (Lakh tonnes) from different forest land use in Honnaver taluka of UK District

Sl. no	LULC	Area (ha)	Area (%)	SOC (%)	BD (t/m^3)	Total SOC (Lakh tonnes)	SOC (t/ha)	Mit.pot LU wise
1	Dense forest	44053	64.29	1.38	1.45	88.15	200.10	2.42
2	Sparse forest	12775	18.64	1.12	1.49	21.32	166.89	2.02
3	Open land	3128	4.57	0.96	1.52	4.56	145.78	1.76
4	Agriculture	8564	12.50	0.60	1.38	7.09	82.79	1.00

Sl No	LULC	SD	H Mean	Variance	Skewness
1	Dense forest	0.55	1.17	0.30	0.39
2	Sparse forest	0.44	0.96	0.20	0.29
3	Open land	0.37	0.83	0.13	0.41
4	Agriculture	0.20	0.48	0.04	−0.12

Sl No	Variables	Correlation	Covariance	F Value
1	Dense forest Sparse Forest	0.064	0.015	0.246
2	Dense forest Open land	−0.045	−0.009	0.078
3	Dense forest Agriculture	0.185	0.021	0.000
4	Sparse forest Open land	−0.122	−0.017	0.709
5	Sparse forest Agriculture	0.170	0.013	0.002
6	Open land Agriculture	0.389	0.022	0.005

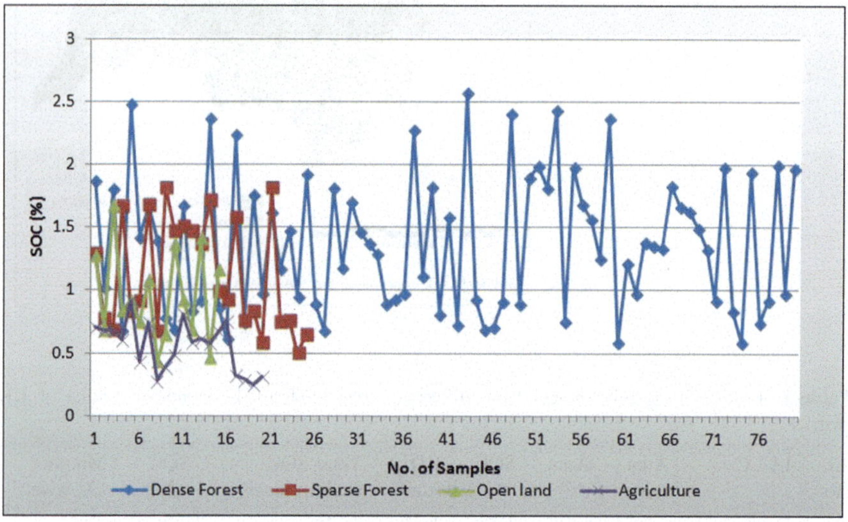

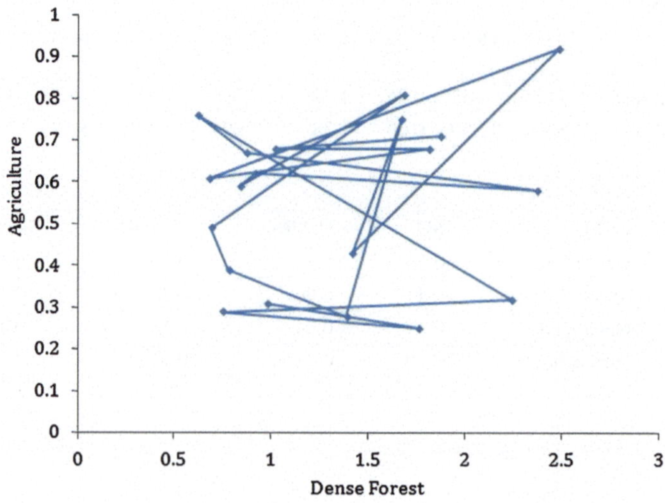

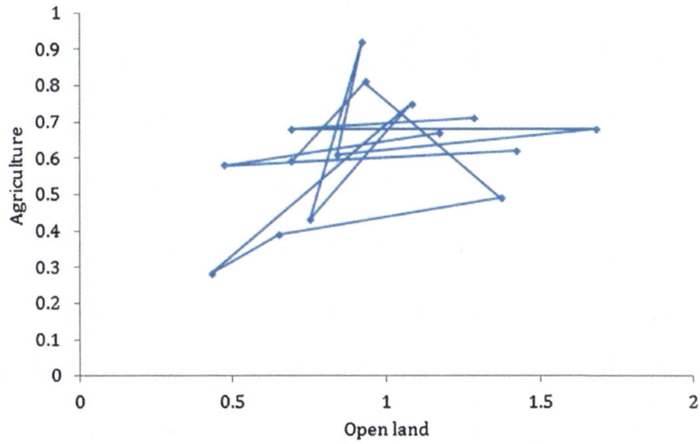
Agriculture
Open land
0
0.1
0.2
0.3
0.4
0.5
0.6
0.7
0.8
0.9
1
0.5
1.5
2

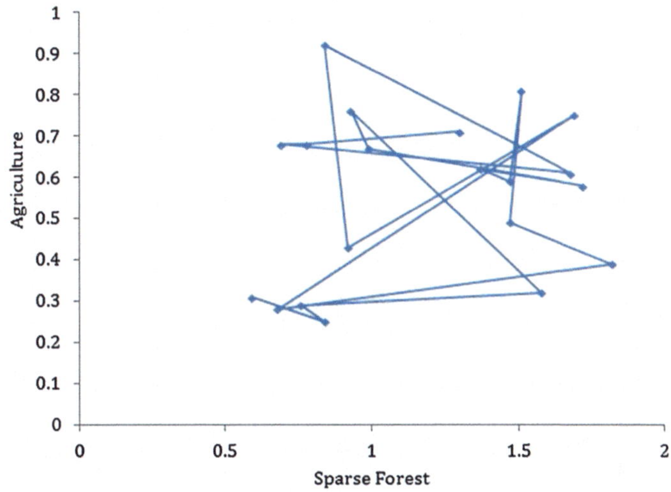
Agriculture
Sparse Forest
0
0.1
0.2
0.3
0.4
0.5
0.6
0.7
0.8
0.9
1
0.5
1.5
2

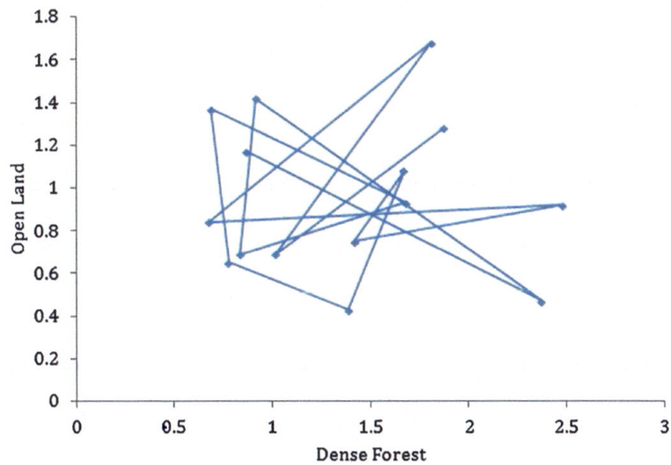
Open Land
Dense Forest
0
0.2
0.4
0.6
0.8
1
1.2
1.4
1.6
1.8
0.5
1.5
2
2.5
3

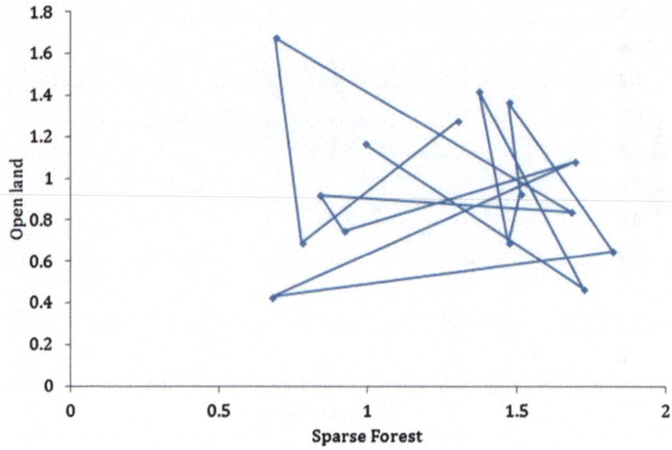

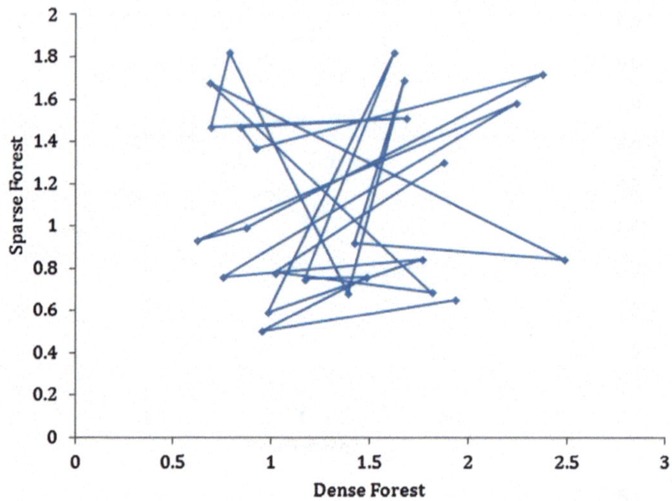

4 Conclusions

The study implies that the dense forest in Honnaver taluka sequester more carbon and CO_2 mitigation potential of dense forest is 2.42 times more than agricultural land. Dense forest helps to reduce concentration of CO_2 in the atmosphere. The second best land use system to sequester more carbon in soil is spare forest followed by open land.

Acknowledgement Thanks to Department of Science and Technology, New Delhi for providing the financial support for conducting research work.

References

Batjes, N. H. (1996). Total carbon and nitrogen in the soils of the world. *Soil Science, 47*(2), 151–163.

Conanat, R. T., Paustian, K., & Elliott, E. T. (2001). Grassland managment and conversion into grassland: Effect on soil carbon. *Ecological Applications, 11*, 343–355.

Davidson, E. A., Trumboe, S. E., & Amudson, R. (2000). Soil warming and organic carbon content. *Nature, 408*, 789–790.

De Vos, B., Lettens, S., Muys, B., & Deckers, J. A. (2007). Walkley-Black analysis of forest soil organic carbon: Recovery, limitation and uncertainty. *Soil Use and Management, 23*, 221–229.

Dey, S. K. (2005). A preliminary estimation of carbon stock sequestrated through rubber (Hevea brasiliensis) plantation in North Eastern regional of India. *The Indian Forester, 131*(11), 1429–1435.

IPCC Climate Change. (2007). Climate change impacts, adaptation and vulnerability. In *Working Group II 2007*. Geneva, Switzerland: IPCC.

Jha, M. N., Gupta, M. K., Saxena, A., & Kumar, R. (2003). Soil organic carbon store in different forest in India. *The Indian Forester, 129*(6), 714–724.

Negi, S. S., & Gupta, M. K. (2010). Soil organic carbon store under different land use systems in Giri catchment of Himachal Pradesh. *The Indian Forester, 136*(9), 1147–1154.

Pareta K., & Pareta, U. (2011). Forest carbon management using satellite remote sensing techniques: A case study of Sagar district. *E-International Scientific Research Journal, III*(4), ISSN-2094-1749.

Sedjo, R., & Sohngen, B. (2012). Carbon sequestration in forests and soils. *Annual Review of Resource Economics (Annual Reviews), 4*, 127–144. doi:10.1146/annurev-resource-083110-115941.

Walkley, A., & Black, I. A. (1934). An examination of Degtjareff method for determining organic carbon in soil: Effect of variation in digestion condition of inorganic soil constitution. *Soil Science, 63*, 251–263.

Lack of Environmental Planning Exacerbating Climate Change Risks in Urban India: *Experiences from Megacities of Bengaluru and Mumbai*

S. Manasi and Nidhi Jamwal

Abstract The world is urbanizing fast and climate change impacts are already being felt in urban agglomerations across the globe. The United Nations Report has projected that India will add 404 million urban inhabitants by 2050 (World Urbanization Prospects 2014). Because of high population densities and increased industrial activities, cities contribute to climate change and also face serious risks due to impacts of climate change. Being engines of growth, cities are a hub of economic centres, thus, economic losses due to climate changes are expected to be significantly high in urban areas making them extremely vulnerable to disasters. In spite of growing evidence of climate change impacts on Indian cities, both the state authorities and the civic bodies are unable to wed urban planning exercise with environmental planning. Current plans and policies are not holistic, and fail to meet the future escalating demands of urban population and possible urban disasters. This paper explores these issues through two case studies of Indian megacities—Bengaluru in Karnataka and Mumbai in Maharashtra. Former focuses on problems faced by urban residential communities located in low-lying areas of the city. Poor urban planning and mindless construction mania has accentuated cases of urban flooding in Bengaluru. The second case study of Mumbai looks at the July 2005 floods when the entire metropolis was marooned due to unprecedented heavy rainfall. It puts in perspective various scientific studies that show impacts of climate change in Mumbai and the associated risks. Both the cases re-emphasise the fact that urban planning is not a top-to-down approach and it cannot be done undertaken in isolation. In wake of growing evidence of climate change impacts in urban areas, it is imperative that a holistic "environmental urban planning" is undertaken to help Indian cities adapt to climate change.

S. Manasi (✉)
Centre for Research in Urban Affairs, Institute for Social and Economic Change, Dr. VKRV Rao Raod, Nagarabhavi, Bangalore 560072, India
e-mail: manasi@isec.ac.in

N. Jamwal
Environment Journalist, Mumbai, India
e-mail: nidhijamwal@gmail.com

S. Nautiyal et al. (eds.), *Climate Change Challenge (3C) and Social-Economic-Ecological Interface-Building*, Environmental Science,
DOI 10.1007/978-3-319-31014-5_28

1 Background

The world is urbanizing fast and climate change impacts are already being felt in urban agglomerations across the globe. At present, 54 % of the world's population lives in cities, which is likely to increase to 66 % by 2050. The United Nations Report has projected that the concentration of urban population will be more in Asia and Africa, and India is projected to add 404 million urban inhabitants by 2050 (World Urbanization Prospects 2014). Because of high population densities and increased industrial activities, cities both contribute to climate change and also face serious risks due to impacts of climate change. Being engines of growth, cities are a hub of economic centres, thus, economic losses due to climate changes are expected to be significantly high in urban areas.

Increase in India's urban population means an increased pressure of providing basic amenities, such as drinking water, housing, sewerage, etc., to the urban people. Increased rate of urbanization causes pressures on the city's ability to provide necessary infrastructure and transportation requirements for the increasing population. This has a direct bearing on the natural resources (high consumption of sand, wood, minerals, etc.) and the urban ecology (increase in air pollution, water contamination, groundwater depletion, etc.). Coupled with these are the impacts of climate change, such as rise in temperature, extreme rainfall events, sea level rise, and increase in vector-borne diseases, urban heat island effect, which make urban centres extremely vulnerable to disasters, such as the July 2005 floods in Mumbai.

Flood related problems and uncertainties have been occurring in many South Asian cities and nearly half of the global flood fatalities over the last quarter century occurred in Asia, of which South Asia was affected more of deaths and economic loss. Besides the subcontinent's large coastal and deltaic cities are the ones who get flooded, cities located inland have also been affected with flood risks to a large extent, which may be attributed to resulting effects of social, spatial, and political economic drivers of urban flood risk and uncertainty (Ranganathan 2014). Urban flooding is increasingly experienced in cities and towns post 2000. Even small downpours disturb the city life. Besides, cities continue to face water scarcity and inadequacy while water markets thrive (Ramachandra 2012).

In spite of growing evidence of climate change impacts on Indian cities, both the state authorities and the civic bodies are unable to wed urban planning exercise with environmental planning. Current plans and policies are not holistic, and fail to meet the future escalating demands of urban population and possible urban disasters. Worse, the current practices of urban development are only making our cities more vulnerable to natural calamities.

Our paper explores these issues through two case studies of Indian megacities—Bengaluru in Karnataka and Mumbai in Maharashtra. Former focuses on causes and implications faced due to urban flooding and related concerns reflecting poor urban planning and mindless construction mania that has accentuated cases of urban flooding in Bengaluru, particularly the encroachments of wetlands that acted as natural drainage channels for the city. The second case study of Mumbai looks at the July 2005 floods when the entire metropolis was marooned due to unprecedented heavy rainfall. It puts in perspective various scientific studies that show impacts of climate change in Mumbai and the associated risks; and argues that unless environmental planning is made an integral part of the city planning exercise, Mumbai will witness more flooding and huge loss of property. It also talks about adaptation measures being undertaken by the civic administration and if these are good enough.

2 Bengaluru's Flooding Woes

Bengaluru, known to be the hi-tech city with the IT boom, is one of the fastest growing cities in the world. Bengaluru meteoric growth into a globally integrated network of trade besides highly competitive framework has aggravated profound changes in the settlement patterns impacting strain on the infrastructure. City's built up area has increased by almost 600 % in the last forty years. Increased migration from rural areas have increased the number of slums portraying urban poverty, the official data indicates 597 slums in Bengaluru, which may be much more, since the data is based on the survey conducted more than three decades ago. The city did not witness obvious instances of urban flooding until the recent past. Given the topographical conditions of being above 3000 ft the sea level, well planned wetland system and no rivers close by, floods did not affect the city for decades. Study by T. V. Ramachandra et al. have reiterated that urban heat island phenomenon is evident across localities with higher local temperatures showing that pattern of growth and its implications on local climate, besides, decline in natural resources—loss of vegetation by 76 % and decline in water bodies by 79 %), poor drainage network and encroachment of flood plains have added to the problems, which have caused urban flooding problems.

The urbanization process and the hydraulic insufficiency of drainage systems are seen as the two most major causes of urban flooding in Bengaluru. Ad hoc urbanization has altered the drainage characteristics of natural catchments resulting in increase in the volume and rate of surface run-off. The current drainage systems do not have the capacity to cope with the increased volume of water, further blocked with solid waste and debris. Encroachment of wetlands, floodplains have obstructed floodway's causing loss of natural flood storage. Studies have indicated the gravity of the situation highlighting the fact that the city has lost more than half of its water bodies and two thirds of its green space in the past four decades (Sridhar et al. 2012). Bengaluru does not have water resources of its own other than its aquifers and ground water reservoirs. These tanks were gravity-fed and excess water was redirected through canals located at a lower gradient, as it has a natural elevation of 920 m indicating that water can percolate on its own down the slopes, thus sustained through careful networking of numerous water bodies like lakes[1] and tanks since several decades and forms the cascade of tanks. Bengaluru's undulating terrain was suitable to such irrigation system. These wetlands offer important services of providing drinking water requirements of the city, act as flood storage and climate stabilizer, meet fodder requirements besides filtering sediments and nutrients from surface water. Cauvery river[2] is the major source that meets the water demands of the city.

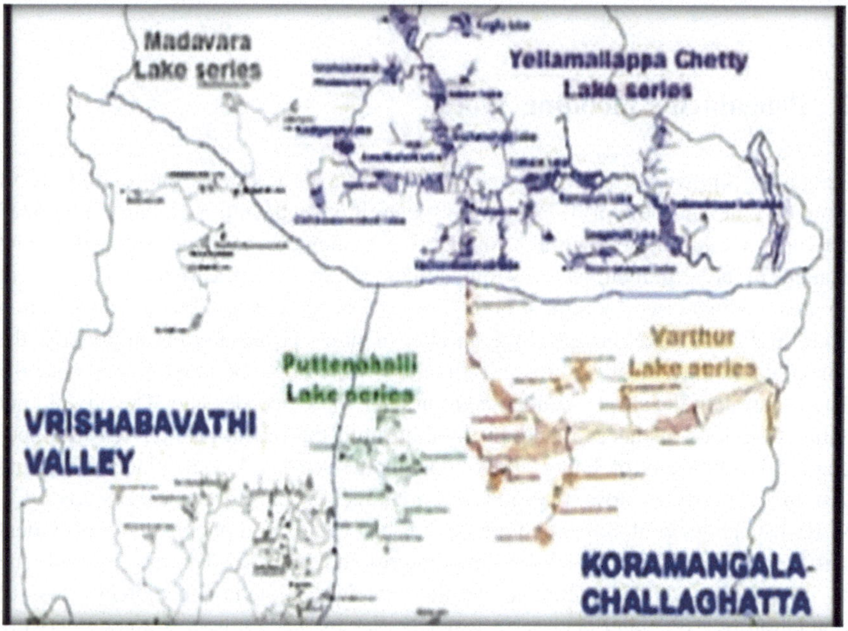

Cascade of tanks. *Source* Lake development authority

There were about 370 tanks in Bengaluru, but have declined gradually due to urbanization and encroachments to build various establishments, both private and government resulting in urban flooding (Table 1).

Decline in the water bodies may be attributed to lack of proper legislation and absence of management structure. Study by Indian Institute of Science, 2015

[1]Fernando (2008) paper titled 'Disappearance and privatisation of lakes in Bengaluru' mentions that Bengaluru has only 117 lakes. The disappearance of Banglaore lake started from 1980s but has increased due to rapid urbanization process. While there were 262 lakes in 1961, but official statistics mentions presence of only 117 lakes, but only 33 lakes are still more or less visible on satellite imagery. (http://base.d-p-h.info/en/fiches/dph/fiche-dph-7689.html).

[2]The city of Bengaluru is getting around 950 MLD of water from Cauvery and Arkavathy sources as against the present demand of 1350MLD for a population of around 9 million. As such, there is a short fall of 400MLD (www.bwssb.org).

Table 1 Decline in water bodies of Bengaluru City

Bengaluru city			Greater Bengaluru	
Year	Number of water bodies	Area (in ha)	Number of water bodies	Area (in ha)
SoI	58	406	207	2342
1973	51	321	159	2003
1992	38	207	147	1582
2002	25	135	107	1083
2007	17	87	93	918

Source Survey of India, topographic maps (published in 1973), Ramachandra and Uttam Kumar 2009

indicates that many lakes are encroached for construction of illegal structures up to 54 % while 66 % of lakes are sewage fed, 14 % surrounded by slums, 72 % show loss of catchment area. Several of the lake catchments are dumped with solid waste and debris. The encroachments by construction of multistoreyed buildings in lake beds intervened with the natural catchment flow leading to sharp decline in the catchment yield and deteriorating quality of water bodies.

2.1 *Development Characteristic Over Bengaluru with the Erstwhile City Corporation Limits, Greater Bengaluru Region and Bengaluru Metropolitan Area*

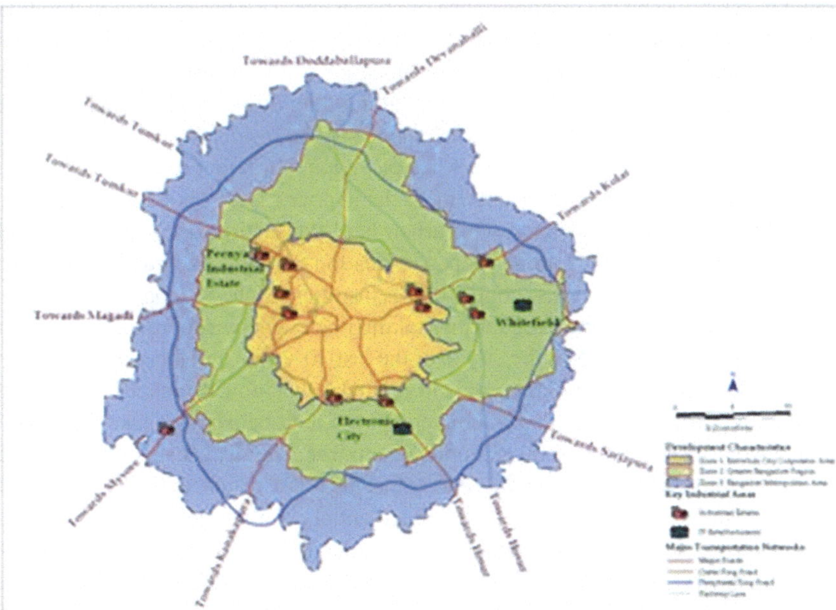

Bengaluru has an advantage of having nearly 70 rainy days spread over the year. However, Bengaluru drainage system (infrastructure) of 180 km long primary and secondary storm water drainage system, can handle only 30 mm of rainfall in one hour. So whenever there is a heavy down pour, the city faces problems. Lack of open space has led to flooding low-lying areas during downpour. BBMP has listed five main areas as flood prone and also identified 134 low-lying areas in Bengaluru, prone to flooding. The East is most vulnerable with 108 areas, West has 31 areas and Southern areas of Bengaluru are all prone to floods, while Northern parts of Bengaluru are not an exception. A study by IISc has indicated that the fragile storm water drain network in the 73 km^2 area between IISc and GKVK campus in Jakkur has rainwater run-off exceeding the drain capacity. The reasons for flooding may be attributed to excess flow into the conduits, silt deposition due to heavy flow of water run-off and weakening of conduits or velocity of heavy run-off could damage conduits, breaching them and flooding the areas. Changing the conduit width, slope and roughness of material lining of the conduit considering conduit erosion and deposition issues were possible solutions suggested (Gouri 2015).

The natural valleys are unable to pump out the excess rain and drain water. All the valleys are choked with debris and heavily silted besides being encroached. In August 2000, rains caused havoc in most areas of Bengaluru and BBMP had to pump out one crore litres of water in the City market area. The operation went on for three weeks. Since the water could not be drained out naturally, it had to be done manually. Urban flooding is a new trend affecting Bengaluru, given its natural gradient, the city did not face this problem earlier. The rainwater infiltration has declined as the soil exposed to this purpose has gradually decreased over time. Besides, the construction of tar/cement roads has increased leading to the decrease of ground water table and the disappearance of open wells. This has resulted in flooding of the city roads during heavy rainfall. The side drains and the shoulder drains filled with mud and other garbage adds to the blocking of the natural water flow. The low-lying areas in the city are prone to annual flooding due to overflowing of water from the drains causing immense inconvenience to the residents.

The role of global economic players has played an important role in aspects like utilization of land in specific. Bengaluru's land prices have shot up immensely. Some of the problems associated with urban land transactions include an unusual escalation in the demand for urban property, creation of bogus documents, '*benami*' and deceitful land transactions, coupled with insufficient checks/control at various official levels for monitoring irregularities taking place. There is more than 500 % of increase in the built up area of the city over the last 30 years, and there is some form of violation of planning, building or land-related legislation. Demand on housing with booming real estate has led to conversions of tank beds/major lakes into residential layouts. There are several examples of environmentally sensitive lands being violated and not getting protected by the authorities. For instance, there are several cases of urban conflicts in low-lying urban residential communities due to ineffective urban planning in metro Bengaluru. For instance, an apartment complex of 300 plots was built on the gradient of Shinivagulu Tank Bed in 1989 by the Bengaluru Development Authority at Koramangala. A storm water drain of about 40 feet passes through the layout. During heavy rains water enters the households causing inconvenience to the residents. Hence, a secondary drain was built to divert flood waters into storm water drains solving the problem partially. The natural gradient of the Koramangala valley of which Shinivagulu Tank bed is a component facilitates water to drain through the wetlands into the Bellandur lake affecting natural flow and causing flooding along the belt. The residents filed a writ petition in the High Court against the nuisance of the drain asking for coverage. Federation—Forward 68 (group of 12 Residents Welfare Associations) filed a Public Interest Litigation and resolved it after a long legal battle. There are several such instances of violation who have not found solutions as yet, in the city (Manasi and Smitha 2012).

2.2 *Flooding Damages Property*

Flooded roads/homes, traffic jam, power cut, sanitary water entering the homes are the common situations encountered in Bengaluru in the recent past. Urban slums and settlements are more vulnerable as they live in dwellings that do not have drainage. Since, several of the slums are located along side drainage lines, sewage drains, low lying, congested and unplanned areas; they are susceptible to receive the onslaughts of flooding. Houses constructed are of poor quality using available local materials that do not protect them. Slums lack basic facilities in sanitation and hygiene aggravating the problems further. Newspapers reports results of rain havoc report instances of houses falling and people getting killed, trees falling and disrupting traffic, damaging vehicles and at times causing death, electric poles getting uprooted. The congested areas where the poor reside are flooded, making it impossible for them to move out of their dwellings.

Photo Congested lanes, Gulbarga slum, Peenya and Summanahalli slum

BWSSB's survey in Bengaluru has shown that flooding was more in the central part of the city, 7 % of houses faced flooding, 10 % of households of urban poor have been encountering flooding for more than a decade. Among the respondents surveyed, flooding had lasted between one and three days 2/3rds indicated while

3 % indicated that it lasted for a week, 2/3rds indicated blockage in the sewerage, 12 % indicated flow of water into their houses, lower income households and slums faced flooding caused by blocked drains. All households suffered damage to assets, loss of working days. These were people who were mostly daily wage labour or petty businessmen affecting their household economy.

2.3 Safety at Stake

Bengaluru stands low on the Walkability Index[3] (WI), an indicator that tells us how walker friendly a city is. It scores WI score of 46 out of 100 as per the research conducted by Clean Air Initiative in Asia has found that Bengaluru is among the four cities where roads and pavements are not pedestrian friendly. Many pavements are shrinking while some vanishing, many are encroached by hawkers and shopkeepers. Skywalks constructed are not user friendly as they do not have lifts and are not user friendly to senior citizens and physically challenged. Thus, safety of citizens is more at risk as flooding aggravates it. There are several instances of fatalities as people have got washed away in open storm water drains, causing anguish and fear among citizens during rainy days, particularly elderly. Such instances were never experienced in Bengaluru prior to 2000. Newspapers have captured painful stories of a three member family travelling in a car being washed away in an open drain in JP Nagar in Feb 2002, two-year-old boy on SJP road in 2002, 67-year-old merchant in VV Puram, Newly married couple at Prakash Nagar, in October 2003, three year old at Guddadahalli in April 2011, nine year girl in Oct 2014 in Bilekahalli and 12-year-old child drowned falling into an uncovered pit filled with rainwater in the vicinity of Government Arts and Science College.

2.4 Flooding Affects Toilet Access for the Urban Poor

Providing slums with access to toilet, construction of toilets by itself does not ensure a positive usage. In fact, toilet usage is determined by several factors like space, lighting, water facilities, etc. During floods, people accessing public toilets/ community toilets/shared toilets gets complicated, women, elderly and children suffer acutely during such times. Besides, already prevailing poor quality of toilets gets messier making it impossible to use them. Blockage and overflowing of drainage lines becomes worse and frequent during instances of flooding. In the case of pit latrines, pit collapsing and overflowing are common problems encountered (Manasi et al. 2015).

[3]Walkability Indix measures several factors—easy and safe access to mass transit and other public transport modes, good and continuous footpaths, easily crossable roads, disabled friendly pavements with ramps and absence of obstructions like trees, electricity poles and street hawkers.

Photo Urban flooding adds to the menace of poor quality toilets and blockage causing leakage in manholes, Ventateshwara slum, Bengaluru

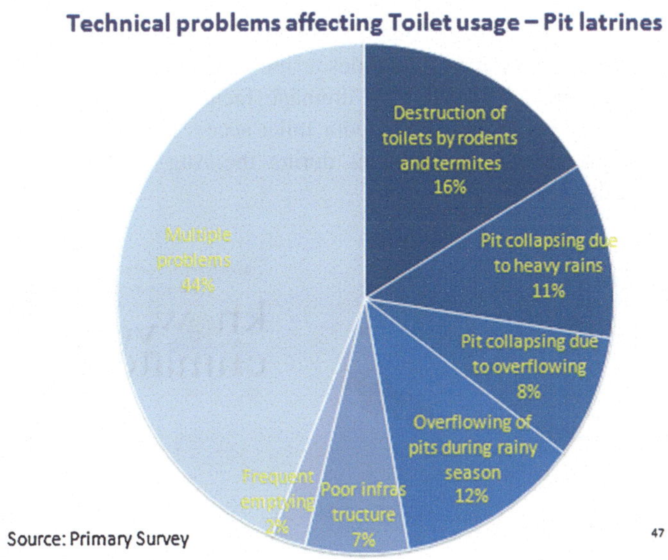

2.5 *Health Ailments Rampant*

Allergies and skin ailments were common among the slum populace. Although, health ailments may be attributed to several other factors related to hygiene, contamination during flooding was also observed as one of the major reasons for spreading of vector-borne and waterborne diseases.

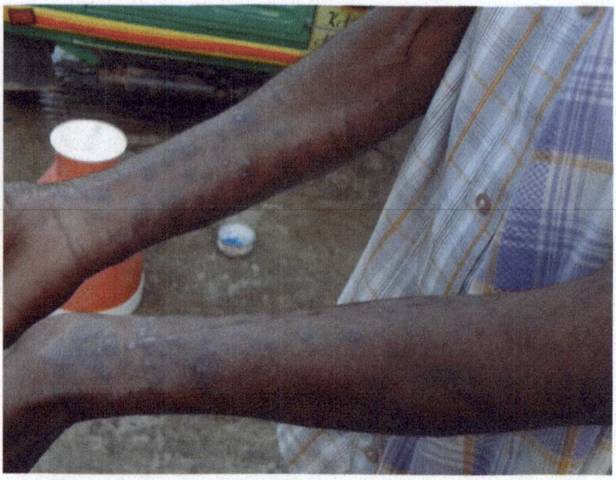

Photo Skin allergies common among slum residents, Nayandahalli slum

Study on Sanitation based on 20 slums in Bengaluru slums reveal that 43 % of the slum dwellers consider multiple reasons (a combination of poor hygiene and sanitation, poor ventilation, poor drainage, mosquitoes and fatigue due to intensive physical work) as responsible for their present health condition, while 19 % believe poor hygiene and poor sanitation facilities as the main cause and 17.5 % attributed it to mosquitoes and 5.75 % to poor drainage facilities prevailing in the slums which in turn, may be attributed to a poor toilet access (Manasi et al. 2015).

An interesting analysis of problems during the Aug–Sep 2011 downpour is represented in the fig.

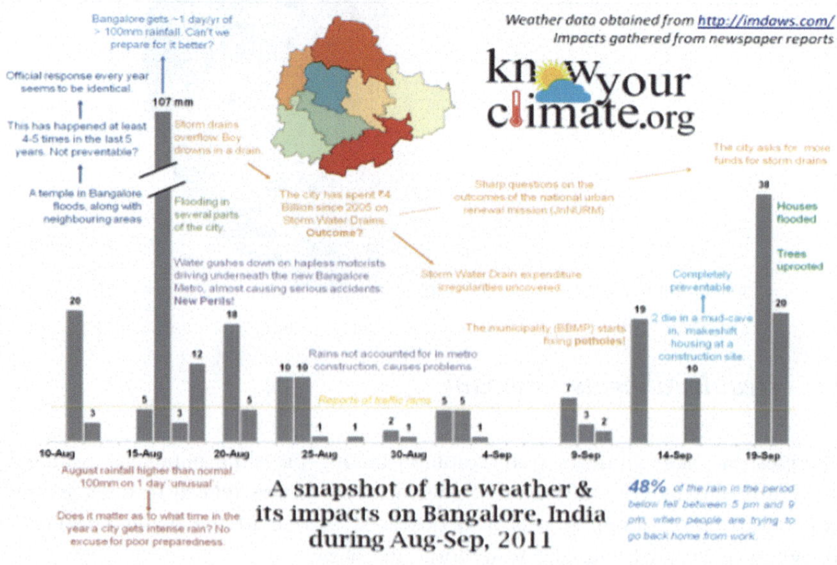

Source http://blog.knowyourclimate.org/2012/11/weather-impacts-bangalore-2011-graphic/

2.6 Response Through Demolitions Spree of Encroached Buildings on Lake Bed

Although demolitions were the only option left to reclaim lake beds, given the contexts, it is also important to address the issue of prevention of such encroachments in the future. Besides, it is important to address the issues concerning citizens who are victims in the process since their investments of life time would be changed into rubble in the most unexpected way. The demolition was a resultant of a public interest litigation filed by People's Campaign for the Right to Water four years ago. In August 2014, Karnataka High Court's order on clearing the encroachments was taken up after intimating the occupants. People were served multiple notices and court order was given three times before commencing the drive. People are gullible as the plots were purchased more than two decades ago, leaving them in anguish and despair making them homeless instantly for no fault of theirs. People blame the officials to have provided them sanctions of services—water, sewerage lines and power, means, officials who cleared them were to be questioned as well. This calls for the need for effective implementation of Urban Property Ownership Records initiative execution effectively that ensures land ownership in the real sense.

Demolition drive in Bengaluru brought out several encroached buildings on tank beds. The urban district administration has launched an operation, a 10 day demolition drive in April 2015 to clear lake beds that are encroached to construct illegal buildings, both housing and commercial. 200 buildings that included schools, medical centres, houses were demolished, the one that occupied Sarakki lake. Several buildings in Banaswadi on the lake bed and South Bengaluru on Puttenahalli lake were demolished. Several surveys of the lake area had indicated that 34 acres of the 84 acres of the lake area across three villages Sarakki, Puttenahalli and Jaraganahalli was encroached by private layouts/commercial establishments and estimated to cost around Rs. 2000 crores in the current open market. Flats at Yamare village, Anekal Taluk built on Raja Kaluve (storm water drain) were demolished as the builder had encroached 700 m of storm water drain to construct a residential project of 140 flats, extending up to 2 acres, that valued Rs. 12 crores. Encroached land in Iblur lake was also revived. Iblur lake was spread across 18 acres and 6 guntas but had got reduced to 7 acres and 32 guntas. 3500 BDA plots also were to be taken over.

3 Mumbai Floods—That Sinking Feeling

Being a heavily populated coastal city, Mumbai is highly vulnerable to the impacts of climate change. Extreme weather events, flooding, sea level rise, missing winter, urban heat island effect, some of these effects have already started to show up. Whereas the municipal corporation has undertaken projects to adapt the city to threats of climate change, most of these projects are disjointed and lack a comprehensive plan.

Mumbai is no stranger to flooding. Ever year during the monsoon season, low-lying areas of the city get submerged in knee-deep water. Mumbaikars are habituated to such '*regular*' flooding. However, July 26, 2005 is a watershed in the history of Mumbai. On this fateful day, the metropolis received an unprecedented rainfall of 944 mm in 24 h. The entire city was flooded, bringing life to a standstill. The horror of flood caught both the people and the government unaware. Public transport, including local trains, halted. Airport was shut down. A large number of Mumbaikars were marooned overnight on the roads. Some remained locked up inside their cars and died of suffocation. A few were electrocuted, many drowned and innumerable lives were washed away by the gushing water. Those who remained at their workplace were considered fortunate. A large number of children spent the night at higher floors of their schools.

As telephone lines went dead, people were unable to find out whereabouts of their loved ones. Both water and electricity supply was snapped across the city. This extreme weather event killed 914 people and lead to a conservative financial loss of Rs. 4,500,000,000.[4] The night of July 26–27, 2005 is rightly considered the darkest night of Mumbai. Whereas the government pushed the blame on 'natural' disaster, God's fury and one-in-a-hundred-years episode, experts claim the coastal city of Mumbai has every reason to expect more such extreme weather events (*see box 1*: *More floods, no winter*). Climate change is expected to increase the severity and frequency of flooding (see Table 2: *Estimated economic losses due to the impact of climate change in Mumbai*). As per a 2007 study conducted by A.S. Unnikrishnan and D. Shankar of Goa-based National Institute of Oceanography, mean sea level rise trends along the Indian coasts are about 1.30 mm/year. Mumbai is witnessing 0.77 mm/year sea level rise based on last 113 years of tide-data gauge.[5]

"It is a well documented fact that events such as 2005 floods will not remain once in a 50 years episode. They will happen more frequently. Thus, the government and the local administration need to undertake urgent steps to adapt the city to climate change," said Anand Patwardhan, professor at Shailesh J. Mehta School of Management of IIT Bombay.[6] Patwardhan has researched on policy responses to climate change, including adaptation and mitigation. Strengthening city's infrastructure to better resist flood risks is not an easy task. This is because of several reasons—scarce land, surrounded by the sea on three sides, several low-lying areas, dilapidated infrastructure, rapid urbanization, very high population density, huge informal settlements, public transport bursting at the seams, and poor building

[4]http://www.indiawaterportal.org/sites/indiawaterportal.org/files/why_mumbai_must_reclaim_its_mithi_gautam_kirtane_orf_2011.pdf.

[5]http://envfor.nic.in/sites/default/files/Unni%20-%20Coastal%20India_0.pdf.

[6]Anand Patwardhan 2013, professor, Shailesh J Mehta School of Management, IIT Bombay, September 28, *personal communication*.

Table 2 Estimated economic losses due to the impact of climate change in Mumbai

Type of impact	Type of costs and period of impact	Cost in rupees (crore)
Dislocation due to extreme events of flooding of low-lying areas—every five years till 2050	Cumulative costs over the period 2005–2050	407.6
Material damage to low-lying areas due to extreme events—every five years till 2050	Cumulative costs over the period 2005–2050	6413
Mortality costs due to extreme events of flooding—every five years till 2050	Cumulative costs over the period 2005–2050	3050
Disability adjusted life years (DALYs) lost due to diseases like malaria, diarrhoea and leptospirosis	Cumulative costs over the period 2005–2050	3153
Building foundation damages for the period till 2050 due to sea level rise	Single-cost estimate for the year 2050	1,501,725
Tourism loss: less number of tourists visiting Mumbai	Single-cost estimate for the year 2050, as compared with the base year 2005	1,963,500

Source Sibananda Senapati and Vijaya Gupta 2012, Presentation on *Dying fisheries in a changing environment: A study on livelihood strategies of fishery communities in Mumbai, India*, National Institute of Industrial Engineering, Mumbai

regulations. However, adaptation can be efficient to reduce some climate change impacts. For instance, Netherlands is experiencing a sea level rise of 0.2 m a century. After the 1953 great flood, institutional and legal innovations were implemented to manage future risks. Climate change is now naturally taken into account, and flooding risks are monitored and managed on regular basis.

Keeping in mind the present condition of Mumbai's infrastructure, which is dilapidated and set for an overhaul, it is crucial to factor in climate change and then make long-term investments in sectors, such as water management; sewerage; energy production and distribution infrastructure; transportation infrastructure; urbanism, housing and architecture; and natural disaster protections. This will help the city adapt and deal with the challenges of climate change.

Box 1: More Floods, No Winter

In its 2009 report, *Environmental degradation, disasters and climate change*, the India Meteorological Department's (IMD) regional team at Mumbai analyzed 100 years of weather data from 1901 onwards and found a rise of 1.62○C in the average maximum temperature of Mumbai. Winters are already giving Mumbai a miss, claimed this report. IMD also noted that from 2001 onwards rainfall over Mumbai has increased. This could be due to a change in the lapse rate or the rate of change of any meteorological element (temperature in this case) with altitude. The lapse rate determines the growth of clouds. The lapse rate is increasing with the cooling of air in the upper atmosphere and warming of the lower atmosphere. With such a steep lapse rate, cloud formation will be rapid and there will be more thunderstorms. There will also be heavy precipitation, warned the IMD report.[7]

More recently, UK-based Maplecroft's *Climate Change and Environmental Risk Atlas 2013* has ranked Mumbai eighth in the global list of total 50 cities (chosen for their current and future importance to global business) that face a range of risks due to climate change. The 2013 atlas clubs Mumbai under "high risk" category in the climate change vulnerability index (CCVI). The CCVI has been developed by Maplecroft to identify risks to populations, company operations, supply chains and investments in 197 countries. It evaluates exposure to climate related natural hazards; the sensitivity of populations; development; natural resources; agricultural dependency; research and development; government effectiveness and education levels.[8]

3.1 Mumbai's Challenge of Staying Afloat

Mumbai has a unique topography, whereas the island city is made of merger of seven islands, the suburbs consist of four island and some hilly areas. Much of the city lies *just above* sea level, with average elevations ranging from 10 to 15 m. A large part of the city sits upon reclaimed land (see map 1: *Mumbai, an amalgamation of islands*). Railway lines are typically about 10 m above low tide level and the subways are very close to high tide level. Mumbai's gravity drainage system was designed in 1920s, and is affected by low and high tides. During high tide, flood gates are closed to stop ingress of sea water. Thus, if high tide and heavy rains occur at the same time, there is no drainage of storm water and large parts of the city get flooded. The storm water drainage network of the city is over 100 years old, and is designed for a rainfall of 25 mm/h and a run-off coefficient of 0.5 only (The percentage of rainfall that appears as storm water run-off from a surface is called the ***run-off coefficient***). However, under present circumstances, when the city is overdeveloped with major changes in land use pattern, the run-off coefficient has increased to 1. The storm water drains do not have the capacity to drain out excess water from the city.

Interestingly, both Mumbai and Singapore receive about 2500 mm rainfall per annum. But, Singapore's rainfall is better distributed over the year, whereas Mumbai receives majority of rainfall during the two months of July and August. Analysis by an ex-officer of Municipal Corporation of Greater Mumbai (MCGM)

[7]http://www.downtoearth.org.in/node/3471.

[8]http://maplecroft.com/about/news/ccvi_2013.html.

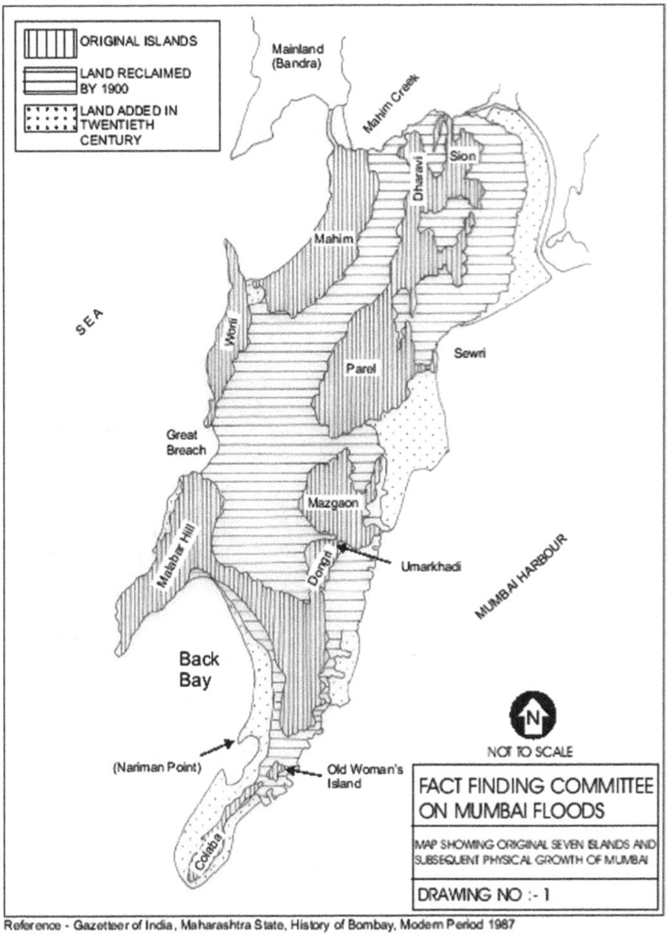

Map 1 Mumbai, An Amalgamation of Islands

shows that the ratio of number of flooding days with rainfall of more than 75 mm have increased from 1:7 in the 1940s to 1:1 in the 1970s to 1.5:1 in the 2000s.

3.2 Mumbai's Little-Known River System

Till July 26, 2005, both the authorities and the residents of Mumbai did not know there were rivers criss-crossing the metropolis and draining excess water into the sea. To them Mithi river was a mere stinking *nullah*, which over a period of time

Map 2 Mumbai's river systems. *Source* http://upload.wikimedia.org/wikipedia/commons/6/69/River-Geography-Mumbai.png

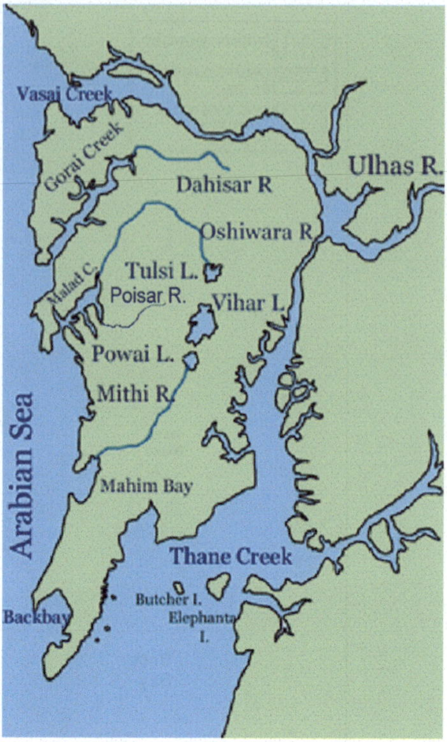

had been encroached, both legally and illegally, to construct buildings and other infrastructure projects. The flood of 2005 brought to the fore the fact that Mithi is an integral part of city's drainage network. And so are the other rivers—Oshiwara, Poisar and Dahisar (*see box 2: Mumbai's Unknown Rivers, and map* 2: *Mumbai's river systems*).

The floodplains of these rivers have completely disappeared and are thickly populated. The course of some rivers has been diverted, such as Mithi river near the airport, to facilitate construction activities. "The administration does not understand that rivers are not *nullah*s. They have their own course and cannot be channelized like gutters. If we tinker too much, the rivers will fight back with vengeance, as Mithi did on July 26 when it breached the airport wall and flooded the runway," warned Girish Raut, an advocate and environment activist based in Mumbai. "The reason why rivers have floodplains is because rivers are expected to flood. Mumbai has eaten into floodplains of all the rivers," added Patwardhan.

Box 2: Mumbai's Unknown Rivers

Mithi river: It originates from the overflow of Vihar Lake and also receives the overflows from Powai Lake about 2 km later. It flows for a total of 15 km before meeting the Arabian Sea at Mahim Creek. It is a natural drainage channel for Mumbai suburbs carrying excess waters during the monsoons.

Oshiwara river: It begins in the Aarey Milk Colony, cuts through the Goregaon hills, before emptying into the Malad Creek. On its way, it picks up industrial effluents and sewage from various colonies, slums and industrial estates. It is a highly polluted and encroached upon river. For instance, most of the call centres in Malad area of Western Suburbs have been built on reclaimed ground at the mouth of river.[9]

Dahisar river: It is a river in the northern suburbs near Dahisar and originates from Tulsi Lake in Sanjay Gandhi National Park. The river is currently highly polluted with the dumping of industrial effluents from workshops, and sewage from slums and storm water drains. The river empties into Manori Creek.[10]

Poisar river: This river also begins in the Sanjay Gandhi National Park, but empties into the Marve Creek. It is a small stream, which is highly contaminated because of industrial effluents and sewage.[11]

3.3 *Slew of Projects*

"Post 2005 flood, the MCGM has undertaken various initiatives/projects—adaptation strategies—to prepare the city better and adapt it to various impacts of climate change. Most of these projects are in accordance with the recommendations of the Chitale committee," said Rajiv Jalota, additional municipal commissioner, MCGM.[12] Chitale committee, headed by Madhav Chitale, was a fact-finding committee set-up by the state government to investigate the reasons behind the July 26 flood. This committee submitted its exhaustive 283-page report in 2006. Before Chitale committee, there were few more committees that had warned the city administration about mindless development and neglect of the local water systems. But, all these reports were ignored (*see box 3: Forewarnings ignored*).

[9]http://en.wikipedia.org/wiki/Oshiwara_River.

[10]http://en.wikipedia.org/wiki/Dahisar_River.

[11]http://en.wikipedia.org/wiki/Poisar_River.

[12]Rajiv Jalota 2013, additional municipal commissioner, MCGM, Oct 3, *personal communication*.

The officials at MCGM claim that they have now finally woken up to the challenge. "We are following a variety of measures, such as widening and deepening of *nallah*s, setting up pumping stations, widening of underground drainage network in island city, widening of open drains in the suburbs, contour mapping, etc. Some of these activities have been finished, whereas others are ongoing," informed L S Vhatkar, director (ES&P), MCGM.[13] Vhatkar is responsible for the entire drainage and sewerage network of Mumbai. Mithi river, the main channel for draining excess water from the city, has come to the centre stage. Every year before the onset of monsoon, tenders are floated for desilting of the river to increase its capacity to carry storm water and run-off. This has become a business in itself.

Box 3: Forewarnings Ignored

i. Natu Committee Report, 1975
ii. Report on model studies on the effect of proposed reclamation in Mahim Creek (Bandra-Kurla Complex) by the Central Water and Power Research Station (CWPRS), 1978
iii. Dharavi Storm Water Drainage System—Detailed Project Report by Shah Technical Consultants (STC), 1988
iv. Paranjape Committee for development of a gated barrage across the mouth of the Mithi River in 1988
v. Brihanmumbai Storm Water Drainage Project (BRIMSTOWAD) Report, 1993
vi. The Mithi River Water Pollution and Recommendations for its Control —Klean Environment Consultants, 2004

Source Making a sewer … river again, ORF Mumbai, http://www.indiawaterportal.org/sites/indiawaterportal.org/files/why_mumbai_must_reclaim_its_mithi_gautam_kirtane_orf_2011.pdf

3.3.1 Desilting and Widening of Nullahs

According to a presentation, *Adapting to climate change in Asia's coastal megacities*, by Kishore Gajbhiye, ex-additional municipal commissioner of MCGM, in phase-I, 5.68 lakhs cubic metre (m^3) silt has been removed from Mithi. Another 3.70 lakh m^3 was removed in phase-II after March 2007. Apart from this, 2652 residential and 1148 commercial structures have been removed. About 1769 residential and 349 commercial structures are rehabilitated.

[13]L.S. Vhatkar 2013, director (ES&P), MCGM, October 3, *personal communication*.

However, activists and environmentalists allege that there is lack of coordination between MCGM and the Mumbai Metropolitan Region Development Authority (MMRDA), both of which are responsible for desilting the river and depositing the silt at Kanjurmarg and Mulund dumping grounds. For instance, in May this year, politician and activist Kirit Somaiya filed a PIL in the Bombay High Court alleging that the removed silt was being dumped along the river and would be washed back into the river.[14]

According to Vhatkar, as of October 2013, Mithi river has been desilted until the lakes, from where the river starts its journey. It has also been widened wherever land was available on the sides of the river. However, questions are being raised on the efficacy of desilting operations. "Both Kanjurmarg and Mulund dumping grounds are not scientific landfills. They are not in accordance with the municipal solid waste management rules and are a major cause of environmental pollution …. As far as adaptation strategies are concerned, the government lacks holistic approach. In order to solve one problem, it ends up creating few more," said Rishi Aggarwal, a civic activist and member of Observer Research Foundation (ORF) Mumbai.

Apart from desilting, the corporation is also widening the existing *nullahs*. The length of major *nullahs* across Mumbai is 340 km. A large number of *nullahs* are at present 3 m wide. These are being widened as per the available space. In the island city, the underground arch drains are over 100 years old and were built by the British. These are 8 × 4 m wide and run 10 feet below the ground surface. "There is not much scope to widen the underground arch drains …. A total of 30 km arch drains had to be rehabilitated. We have already completed 20 km stretch. The rest 10 km will be done by May-June 2014," assured Vhatkar.[15] This is expected to take care of drainage of excess water from the island city.

In case of the suburbs, there are only open drains and no underground drains. Thus, there is a scope for widening these drains. Chitale committee report had recommended widening of 4 m drains to at least 6 m. According to Vhatkar, these open drains are being widened. In some places, there is no scope for widening because of heavy construction along the drains. The corporation is digging additional drains in such areas.

Drains that have less than 1.5 m width are called minor *nullahs* and fall under the local ward offices' jurisdiction. These drains are 200 km in length. Apart from major and minor drains, Mumbai also has road side drains that run for 2000 km. These are also being desilted and widened by the corporation.

---

[14]http://news.oneindia.in/2006/05/19/mmrda-and-bmc-to-remove-silt-from-mithi-bankshc-1148049513.html.

[15]L.S. Vhatkar 2013, director (ES&P), MCGM, October 3, *personal communication*.

3.3.2 Two Decades Post Brimstowad

The most crucial adaptation strategy for Mumbai—the much-hyped BRIMSTOWAD (Brihanmumbai Storm Water Drainage) project—still remains incomplete. Way back in 1989, MCGM appointed M/s Watson Hawksley India Pvt. Ltd. to study the storm water drainage system of the entire city. The consultant submitted its final report in 1993, 20 years ago, with a series of recommendations.

Apart from other measures, the report recommended increasing the capacity of storm water drains considering rainfall of 50 mm/h. The present capacity of Mumbai's storm water drains is 25 mm rainfall/h. "As per the latest estimates, the rainfall has increased and we need to prepare the drainage system for 55 mm/h capacity. Municipal corporation is already working on it," said Jalota.

In 1993, the total cost of the project was Rs. 616.30 crore. However, the projects could not be undertaken systematically due to financial constraints and encroachments on various sites. Twenty years later, the cost of the project has jumped up to Rs. 1200 crore (see Table 3: *Proposed category of works and fund requirement*). Further, the cost of rehabilitating project affected persons is pegged at Rs. 600 crore. The MCGM is still struggling to implement the recommendations of the BRIMSTOWAD report.

Post 2005 flood, projects under BRIMSTOWAD received a fillip. The Prime Minister sanctioned a special grant of Rs. 1200 crore outside of JNNURM (Jawaharlal Nehru National Urban Renewal Mission) as 100 % subsidy. The work under this grant involves widening of drains, construction of pumping stations and rehabilitating of slums up to 01.01.2000. Following projects are being carried out under BRIMSTOWAD project with aid of the Central government:

- GIS-based urban Storm Water Drain management system
- Hydro mechanism of Storm Water Drain System.
- Flood warning system
- Rehabilitation/upgradation of *nullahs*/storm water system.

"There are some areas in the city where widening of drains is not possible. In such areas, BRIMSTOWAD report has recommended setting up of pumping

Table 3 Proposed category of works and fund requirement

Categories	Unit	Rs. in crore			
		City area	Western suburb	Eastern suburb	Total
Rehabilitation of old SWD system	24.45 km	203.62			203.62
Widening and training of nullahs	102.95 km	153.33	272.31	283.27	708.91
Pumping stations	8 number	148	111.00	29.00	288.00
		504.95	383.31	312.27	1200.53

Source Kishore Gajbhiye 2009, additional municipal commissioner MCGM, A presentation on *Adapting to climate change in Asia's coastal megacities*

stations to throw out water in the sea. The 1993 report has recommended 8 such pumping stations—5 in the island city and 3 in the suburbs," informed Vhatkar. "Of these 8 pumping stations, two have already been constructed at Haji Ali and Irla. Work is under progress for two more stations at Love Grove in Worli and Cleveland Bunder in Reay Road. For the last 2 pumping stations at Khar Danda and Britannia, tender process is on," he added. According to Vhatkar, the two proposed pumping stations at Mahul and Mogra have run into rough weather due to presence of mangroves. "Our proposal to set up pumping stations at these two sites was shot down by the Union ministry of environment and forests. However, we have modified the plan asking for replantation of the mangroves at an alternate site. We are waiting for the ministry's approval," said Vhatkar.

In a nutshell, under phase-I of BRIMSTOWAD report, 20 works were listed. Of these, 15 have already been complete, whereas five are facing difficulties due to the problem of encroachment. Under phase-II, 58 works were listed. Of these 34 have been started, two completed and 32 are under progress. Implementation of BRIMSTOWAD project has already been delayed by 20 years. And, it seems it will take couple of more years before the corporation can complete all the works. Till then, Mumbai will remain on tenterhooks. According to MCGM officials, the corporation faces two key hurdles while implementing BRIMSTOWAD projects; First, encroachment on the site and rehabilitation of the people, Second, taking care of utilities, such as cables, gas pipelines, etc., while carrying out the projects. Both these factors slow down the pace of project implementation.

3.3.3 Flood Warning System and Emergency Response

Flood warning system consists of automatic rain gauges to study exact pattern of the rain all over the island city and suburb. "Since fire brigade is the first respondent and is on 24 h alert, MCGM has installed rain gauges at 26 fire brigade stations—across Mumbai. This rain gauges data is transmitted to control room every one hour," said Jalota. The rain gauges can be further calibrated to give alarm at prefixed rainfall intensity (above 20 mm rainfall). This data can serve as warning and evacuation signal if required. The approximate cost of equipment/system is Rs. 2 crore.

MCGM has also set-up an Emergency Operations Centre (EOC), which is equipped with an array of communication systems—satellite phones, ham radios, TETRA (Terrestrial Trunk Radio Access); video conferencing set-up; water supply and ration stocks, etc. "MCGM is the only corporation in India that has undertaken an aerial survey of the entire city and has prepared the detailed contour maps. This task was completed in 2009–2010 and is more accurate than a GIS map. This map can help predict the extent and intensity if area-wise flooding during rainfall. It can thus be used to evacuate people in case of extreme weather events," informed Vhatkar.

3.3.4 Disaster Preparedness

MCGM has put in place a robust disaster management system, which includes 196 temporary pumps to discharge water; zone-wise six search and rescue teams under the fire brigade; 500 additional BEST buses. Apart from this, 288 retaining walls in 74 places have been undertaken by the Slum Improvement Board. The state government has also set-up a State Disaster Management Sub-Committee for Mumbai, which is headed by ACS (Home). There are ward-wise Disaster Management Committees headed by assistant commissioners of respective wards.

MCGM is undoubtedly trying to reduce risks that Mumbai faces due to climate change. However, all these projects are disjointed and lack a comprehensive plan. For instance, drainage is looked after by one department headed by Vhatkar. Warning system and disaster management is the responsibility of disaster management department of the corporation. Making city more pedestrian-friendly and promoting public transport system is looked after by the roads department. Dilapidated buildings and building codes fall under development plan (DP) department.

3.4 Will Mumbai Swim or Sink?

According to the findings of a 2010 OECD study, '*Flood Risks, Climate Change Impacts and Adaptation Benefits in Mumbai*', by 2080 s, an upper bound climate scenario could see the likelihood of a 2005-like event more than double. Financial losses could also triple compared with the 2005 situation. Rapid urbanization will further increase the risk level.

However, all is still not lost. In June 2011, MCA4climate published a case study titled '*Flood risks, Infrastructure Resilience and Climate Change Adaption in Mumbai, India*' to illustrate the evaluation of climate adaptation policy to increase infrastructure resilience in the coastal city of Mumbai. It came out with five broad policy options to increase infrastructure resilience. These include public transport, building codes, warning systems and emergency response, insurance, and enforced retreat. It also listed and discussed instruments for each of these policy options (see Table 4: *Policy options to increase infrastructure resilience in Mumbai*) (Table 4).

According to Patwardhan, in order to develop adaptation strategies for any city, it is important to understand the link between hazard, exposure and response/capacity. "Hazard signifies parameters such as how much rainfall a city receives, on which local people/state governments/administration has no control. Exposure stands for urban land use, where we have built, what we have built and how much we have built. This is something that is a result of human activities. Lastly, response shows what capacity we have to deal with a disaster or extreme weather event. We cannot control hazard, hence we have to focus on exposure and response, thus these two elements define the adaptation strategy for a city".

Table 4 Policy options to increase infrastructure resilience in Mumbai

Policy	Instrument	Description
Public transport	Infrastructure development would be funded through a combination of taxation and public investment	Build the planned transport link on stilts as opposed to underground. Significant new infrastructure would be required to complete the project, which would bring the opportunity to link to the wider transport network (monorail and roads) creating better city wide mobility
Building codes	Command and control type regulatory instrument	Amend existing building regulations and introduce new regulations where necessary to ensure that in 20 years' time all floodplain buildings are on stilts, and earthquake proof. Unauthorized colonies must be prevented and regulations enforced
Warning systems and emergency response	A combination of public investments and information-based instruments (targeted education and communication)	The formation and coordination of a 'disaster management' cell—including government, emergency services, logistics and weather forecast services. It would also involve the building of designated safe shelters, as well as targeted communication and education to relevant sectors of the population
Insurance	Public investment since this is a government financed scheme	A government scheme aimed at low-income households and the informal sector. The insurance would be compulsory. The aim of this policy option is to achieve 80 % coverage within ten years
Enforced retreat	A command and control regulatory instrument	The definition of high-risk priority areas, the enforced movement of people living in those areas, the location and development of alternative settlements and the restoration of vacated areas. The target would be up households and small businesses. Although this would be government supported policy, the opportunity for public/private partnerships would be promoted in the development of new settlements

Source Stephane Hallegatte and Valerie Belton 2011, *Flood risks, Infrastructure Resilience and Climate Change Adaption in Mumbai, India*, MCA4climate and UNEP, June

The present adaptation strategies are mainly focused on population that is either public property or is covered by insurance. For instance, during the 2005 flood, a large number of poor people were affected. However, their financial losses were never factored in while arriving at a total loss figure. This must change. “A team of IIT Bombay conducted over 1000 surveys in various low-lying areas of Mumbai and realized the loss of poor was far greater than the public sector loss. These poor people have devised their own adaptation strategies to deal with flooding, such as raising the toilet seats, installing electricity metrer transformers at a higher level, etc. These are all private adaptations. Unless these converge with the larger long-term adaptation plan of Mumbai, they will remain a failure and will only add to the financial burden of the poor,”. Insurance can be of much help in disasters such as July 26 flood, Insurance model works well when there is low probability of an event, as the risks are pooled together. Flooding in Mumbai is a chronic problem and a large chunk of population is affected by it at the same time (Discussion with Patwardhan 2013).

According to Aggarwal, Mumbai needs to have a comprehensive adaptation plan that not only focuses on drainage, but also on other elements such as solid waste management, urban land use, public transport, encouraging pedestrians, etc. “At present, the exercise of preparing a 20 years development plan for Mumbai is going on. It is a big opportunity for the corporation to factor in climate change, encourage people’s participation in planning, and then come up with a comprehensive plan.” The Development plan (DP 2014–2034) for Greater Mumbai is being prepared by the development plan department of the corporation in consultation with M/S Groupe SCE (I) Ltd. The existing land use (ELU) maps and reports have already been prepared and put up on corporation’s website for comments from the public. Mumbai-based think tank, Urban Design Research Institute (UDRI) is actively campaigning to ensure the DP 2014–2034 is comprehensive, robust and people-centric. This may just be the last chance for Mumbai to stay afloat.

4 Conclusion

Both the cases re-emphasise the fact that urban planning is not a top-to-down approach and it cannot be undertaken in isolation. In wake of growing evidence of climate change impacts in urban areas, it is imperative that a holistic “environmental urban planning” is undertaken to help Indian cities adapt to climate change. Infrastructure development must take into account the natural water channels and urban water bodies. Urban forests need to be created and protected to address the rising temperature in urban cities and increased heat waves. Local level disaster preparedness and management needs to be strengthened. Capacities of our civic institutions need to be built to address challenges of climate change. Above all, citizens must be made part of the "environmental urban planning". This will lead to creation of smart cities in India. This is the key to prevent flooding and water logging of drains and roads, inundation of low-lying areas and overflow of sewage onto to the streets. Bengaluru/Mumbai must put a price on the "free" drainage services its rivers/water bodies provide to keep the city afloat against all odds. Only then will it learn to protect these natural water channels and make them an integral part of city planning exercise.

Urban water harvesting is another important aspect that would improve water demands and reduce urban flooding. Unless the natural outflows are cleared completely for smooth flow of rain and waste water away from Bengaluru, floods will continue to occur. Environmental protection is an integral part of any developmental process supported by constitutional provisions.

A rational urban planning must ensure holistic development and effective land use and land management practices which ensure citizens right to safe environment and protection of natural resources. Watershed areas need to be protected not only from water resource point of view, but also for other vital environment and hydrological norms that directly affect lives of people. Bengaluru metropolitan area's most potential and live ground water reservoirs lie in southeastern part of South Bengaluru. Frequent land use changes, and haphazard occupation of tank beds as a consequence of new layouts especially by blocking natural drainage would inevitably destroy most prolific and promising aquifers and naturally formed water harvesting areas in urban centres. This would inevitably constitute denial of citizens of their basic right to enjoy vital natural resources like water. Citizens are entitled to a healthy and productive life in harmony with nature as enshrined in Article 21 of the constitution of India. Ineffective urban planning infringe upon collective rights of citizens.

The main focus of city development plan must aim towards achieving rational land use and land management, which ensures that people enjoy a clean and safe environment and are not subjected to improper planning that denies them or inhibits use of natural resources. As land rights in India are not well-defined, there have been implications in terms of conflicts, misuse and inequitable distribution of resources. Need and emphasis on well-defined land rights system has been stressed time and again by researchers and policy makers. Land security can be attained with a well-defined land rights system in place. There are various institutions (Department of Survey Settlement and Land Resources, Bengaluru Development Authority, Private Land Developers, Cooperative Societies, etc.) involved in processing of records related to property in terms of issuing and approving which have been largely misunderstood in relation to property rights besides poor maintenance of land records. It is crucial that there is a comprehensive understanding of roles, responsibilities and coordination amongst these institutions in land management and service delivery. It is also important that appropriate legal framework should be structured to correspond and support the varied institutions. ST bed is one of the cases of such violation in urban development; several such cases are witnessed in the city with the demand for land space increasing enormously. Other studies (Helen, Yeo 2013) have reiterated the fact that that there is lack of awareness among government officials on climate change, limited institutional coordination regarding disaster assistance, government agencies have implemented preventive measures over the years but the residents do not acknowledge them, there is relationship between income levels and vulnerability towards floods.

References

Helen, S. N. H., & Yeo, W. C. (2013). Lee Kuan Yew School of Public Policy National University of Singapore, April 2013.

http://archive.deccanherald.com/deccanherald/oct24/i5.asp, Friday, October 24, 2003.

http://epaperbeta.timesofindia.com/Article.aspx?eid=31806&articlexml=Heavy-rainfall-can-flood-north-Bluru-IISc-study-12052015005026

http://m.newshunt.com/india/english-newspapers/deccan-herald/our-city/apartment-complex-faces-demolition_35202551/c-in-l-english-n-deccan-ncat-OurCity

http://ramuBengaluru.blogspot.in/2013/06/the-first-flooding-of-Bengaluru.html

http://timesofindia.indiatimes.com/city/bengaluru/Apartment-owners-wake-up-to-a-demolition/articleshow/31443989.cms

http://timesofindia.indiatimes.com/city/bengaluru/Bangalore-scores-low-on-walkability-score/articleshow/30921100.cms

http://www.bangaloremirror.com/bangalore/civic/Encroachments-take-a-pounding/articleshow/46950730.cms

http://www.ibtimes.co.in/bengaluru-demolition-drive-encroached-lake-beds-expands-over-3500-bda-plots-face-threat-631602

http://www.ksndmc.org/articles.aspx?Arctileid=11

http://www.ndtv.com/bangalore-news/demolition-drive-aims-to-reclaim-encroached-lake-in-bengaluru-756040

http://www.pacindia.org/uploads/default/files/publications/pdf/5d2e2897b6e7d46ccf7035d9a7be709d.pdf

http://www.thehindu.com/news/cities/bangalore/bengalurus-largest-encroachment-eviction-drive-begins-at-sarakki-lake/article7108731.ece

Malini, R., https://lsecities.net/media/objects/articles/Bengaluru-high-tech-and-the-monsoon/en-gb/

Manasi, S., Latha, N., & Poornima (2015). Pathways to sanitation: Growing challenges and access to urban poor—A study of Bengaluru City, deliverable 2, ISEC project report submitted to HSMI.

Manasi, S., & Smitha, K. C. (2012). Case study—floods in Bengaluru: No respect for water bodies and water courses. In Ekalavya Prasad, K.J. Joy, Suhas Paranjape, Shruthi Vispute (Eds.), *Agony of floods—Flood induced water conflicts in India*. Forum for Policy Dialogue on Water Conflicts in India, October.

Ranganathan, M., (2014). 'Mafias' in the waterscape: Informality and everyday public authority in Bangalore. *Water Alternatives, 7* (1): 89–105

Ramachandra, T. V., Aithal, B., & Uttam Kumar (2012). Conservation of wetlands to mitigate urban floods. *Journal of Resources, Energy and Development, 9*(1), 1–22.

Ramachandra, T. V., & Mujumdar, P. Urban floods, case study of India.

Ramachandra, T. V., & Uttam Kumar (2009). Land Surface temperature with land cover dynamics: Multi-resolution, spatio-temporal data analysis of Greater Bangalore. *International Journal of Geoinformatics, 5*(3), 43–53.

Sridhar, K., & Kumar, S. (2012). India's urban environment: Macro-economic context, challenges and policies.

World Urbanization Prospects (2014). http://www.un.org/en/development/desa/news/population/world-urbanization-prospects-2014.html

Climate Change at Peri-urban Contexts: A Case Study of Jadigenahalli Gram Panchayat, Bangalore Rural District

S. Manasi, K.V. Raju, S. Poornima, K.P. Rashmi and B.R. Hemalatha

Abstract This paper is an attempt to observe the possibilities of climate change at the micro level. Accordingly, a trend analysis of climate change variability and its implications for various issues like the livelihood of people, migration, changes in the cropping pattern, and land use at the local level have been carried out based on meteorological and spatial data and perceptions of the local communities regarding the implications of climate change as well as local environmental changes. The analysis has been carried out at three levels: the Macro level (An overview of climate change), the micro level (implications of Climate change), and finally the micro level implications for the macro level. An examination of meteorological data, while showing no change in climate at the macro level, points to the local-level environmental changes which have led to socioeconomic environmental problems like a greater stress on water availability, changes in cropping pattern and livelihoods. Interestingly, an analysis of the community perceptions regarding these issues shows some conformity with scientific claims.

1 Introduction

The Stockholm Conference in the year 1972 was the first international recognition and manifestation of the urgency to address climate change, affecting both the developed and developing countries. Ever since, there have been several national and international efforts directed toward combating the effects of climate change. In response to the international endeavor, India has come out with a "National Action Plan on Climate Change" (NAPCC). Increased economic activities and changes in lifestyle

S. Manasi (✉)
Centre for Research in Urban Affairs, Institute for Social and Economic Change, Bangalore 560 072, India
e-mail: manasi@isec.ac.in

K.V. Raju · S. Poornima · K.P. Rashmi · B.R. Hemalatha
Centre for Ecological Economics and Natural Resources, Institute for Social and Economic Change, Bangalore 560 072, India

S. Nautiyal et al. (eds.), *Climate Change Challenge (3C) and Social-Economic-Ecological Interface-Building*, Environmental Science,
DOI 10.1007/978-3-319-31014-5_29

patterns have caused serious environmental damage to the earth. It is observed that 'climate change' is the most serious environmental issue that the modern world is going to face in the twenty-first century. The earth's climate has been steadily changing, leading to the degradation of biodiversity, water and soil resources, desertification, coastal erosion, decrease in agricultural productivity and much more.

Climate change mainly refers to a statistically significant variation in either the mean state of the climate persisting for an extended period, usually decades, or in its variability (GOI 2008). A change in climate is attributed directly or indirectly to human activity that alters the composition of the global atmosphere in addition to a natural climate variability observed over comparable time periods. The United Nations Framework Convention on Climate Change (UNFCCC) thus makes a distinction between climate change attributable to human activities altering the atmospheric composition and climate variability attributable to natural causes (UNFCCC 2011).

India is a large country with 15 agro climatic zones, diverse seasons, crops, and farming systems. For a majority of the people in India, agriculture is the main source of livelihood. Agriculture is also most vulnerable to climate change, because it is inherently sensitive to climate variability. Climate change can have a significant impact on Indian agriculture in various direct and indirect ways besides affecting the lives and livelihoods of millions of Indians (Ninan and Satyasiba 2012). Agriculture and allied activities, such as livestock rearing and fisheries, constitute an important component of India's Gross Domestic Product, while providing employment opportunities to two-thirds of the population. Climate change can also have an economic impact on agriculture, including changes in farm profitability, prices, supply, demand, trade, and regional comparative advantages (Khan et al. 2009).

Several studies have predicted a changing weather pattern in the coming decades with a warmer climate, depletion in rainfall, and increase in drought situations necessitating the adopting of mitigating options. Changes in seasons and season length are an indicator, as well as an effect, of climate change. Seasonal changes can profoundly affect the balance of life in ecosystems besides impacting the essential human activities such as agriculture and irrigation (Anandhi 2010). Given this backdrop, our study was an attempt to observe the possibilities of climate change at the micro level in a semi-arid region, peri urban area of Bengaluru.

2 Objectives and Methodology

1. To develop a methodology for detection and assessment of climate change using spatial data.
2. To understand the effect of climate change at the micro level in a semi-arid region of Karnataka, India.

Criteria for selection of Jadigenahalli Gram Panchayat (JDGP):

- Near to metropolitan city and highly influenced by urbanization.
- High water scarcity seen in the villages (depletion in ground water levels).

3 Methodology

A methodology has been developed for understanding and accomplishing the objectives of this chapter in two separate parts as seen below.

3.1 *Part 1: Methodology for the First Objective*

The methodology for detection and assessment of climate change in Jadigenahalli comprises the following major steps:

1. To develop a spatially linked indicator database for detecting climate change evidence.
2. To form an index for measuring the extent of climate change.

Figure 1 shows the methodology adopted for the exercise.

Base map creation: Satellite images were geo-referenced with survey of India ground control points identified. Features like water bodies, hillocks, quarries, and settlements were extracted to form a geo-database on GIS platform. Revenue maps were also used to append the features. The cadastral maps collected from the DSLR were geo-referenced and features such as administrative boundaries, revenue

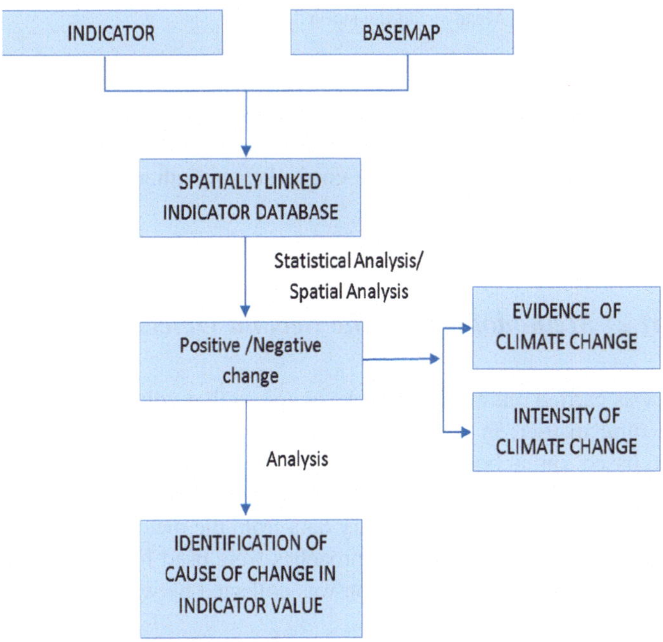

Fig. 1 Methodology adopted for first objective

boundary, taluk boundary, existing roads, and natural features were digitized. Cartography has been used for all these features as per the KTCP act 1961 for producing a base map.

Formation of a spatially linked indicator database: Quantitative indicators were identified pertaining to the categories of 'land use–land cover,' 'demography,' 'economy,' and 'climatology.' Data pertaining to the indicators were collected from the relevant sources (census, IMD, KSRSAC) and an indicator database formulated as per the formulae/methodology indicated in Table 1. The database was then linked to the villages in the geo-database forming a spatially linked indicator database.

The following table presents a list of 15 quantitative indicators, the data required for forming the indicators and the sources of the same.

The indicators were measured across three points of time 1991, 2001, and 2014 to assess the increase or decrease in the indicator values. Based on the range of values for the first year of assessment, the data was classified as 'low,' 'medium,' and 'high.' A similar classification for all the years helps us trace an increase or decrease in the indicator values.

Mean and standard deviation of the data range was calculated with 'high,' 'medium,' and 'low' classes classified as per the table presented here.

Class	Values
High	Mean + Std deviation
Medium	(Mean + Std deviation) − (Mean − Std deviation)
Low	Mean − Std deviation

Mapping of indicator data: The indicator values derived were linked to the village units to spatially represent the indicators in the form of maps. The indicator values superimposed with land use/land cover data and other field data helped infer results.

3.2 *Part 2: Methodology for the Second Objective*

The study was carried out at the micro level across five villages of Hoskote Taluk, Bangalore rural district, in a semi-arid tropical region of Karnataka, India. This study was based on a bottom-up approach for understanding climate change implications for the village-based people in terms of their perceptions, experiences at the local level in addition to secondary data from the meteorological department.

Both qualitative and quantitative approaches were used for eliciting information at various levels. Qualitative information was collected by way of conducting focus group discussions (FGDs).

Table 1 Indicator database for climate change assessment in JDGP

Sl	Category	Indicators	Methodology	Source
1	Land use/land cover	Percentage of area under agriculture *	% agricultural area: (total irrigated area + total un-irrigated area)/total area * 100	CENSUS
2		Percentage of area under irrigation	% area irrigated = total area under irrigation/total area * 100	CENSUS
3		Percentage of area under waste land	% area waste land = (cultural waste + area not available for cultivation)/total area * 100	CENSUS
4	Climatological indicators	Percentage change in annual precipitation	[(Precipitation in year 2) − (precipitation in year 1)/ (precipitation in year 1)] * 100	IMD
5		Percentage increase in annual max temperature	[(Annual max temp in year 2) − (annual max temp in year 1)/(annual max temp in year 1)] * 100	IMD
6		Percentage increase in annual min temperature	[(Annual min temp in year 2) − (annual min temp in year 1)/(annual min temp in year 1)] * 100	IMD
7	Demography	Population growth	Growth = (total population current decade − total population previous decade)/total population previous decade * 100	CENSUS
8		Population density	Density = total Population/total Area	CENSUS
9		Percentage of SC/ST population	% SC − ST population = (total SC population + total ST population)/total population * 100	CENSUS
10		Sex ratio	Sex ratio = no. of females/no. of males * 1000	CENSUS
11		Literacy rate	Literacy = total literate population/total population with above 6 years age * 100	CENSUS
12	Economy	Work participation	Work participation = (total main workers + total marginal workers)/total population * 100	CENSUS
13		Percentage of nonworkers	Non work participation = Total non workers/total population * 100	CENSUS
14		Percentage of nonagricultural workers	Non agriculture workers = Total non agriculture workers/total main workers * 100	CENSUS
15		Percentage of agricultural workers	% agricultural workers = (total main culti. + total agri. workers)/total main workers * 100	CENSUS

A total of 24 FGDs were held covering different socioeconomic groups along with a few personal interviews with informers in all the villages to understand farmer's knowledge and awareness about changing climate in the rural context and its implications for the local environment, livelihoods, migration, cropping pattern, land use pattern related to forest area, grazing lands, and tanks and water availability in the region.

Quantitative information was collected from secondary sources like government offices at the local level like Nada Kacheri (village-level government office), Gram Panchayat, Revenue department, Karnataka State Agricultural department, and Meteorological Department.

Based on the primary and secondary level data, a trend analysis as well as perception analysis was carried out at the micro level across all the five villages for assessing the implications of climate change variability.

4 Analysis and Discussion

Based on categorized indicators as presented in Table 2, analysis and certain inferences have been drawn.

4.1 Direct Indicators

(a) **Climatological indicators**

A description of data used in this assessment is as follows: Rise in temperature, changes in precipitation patterns, sea level rise, melting of snow cover and mountain glaciers, coastal erosion, and occurrence of health hazards and disaster

Table 2 Indicators used

Direct indicators	Indirect indicators			
Climatological indicators	Demographic indicators	Economic indicators	Land use and land cover indicators	
			Agricultural sector	Society and ecosystem sector
Temperature	Population growth	Work participation	Agricultural area	Change in land use
Rainfall	Population density	Nonworkers	Area under irrigation	pattern of Forest area
Relative humidity	Sex ratio	Agricultural workers	Food production	Grazing land
Drought	Literacy rate	Nonagricultural workers	Cropping pattern	Wasteland
			Livestock	Change in livelihood and migration

Source IPCC—IV report

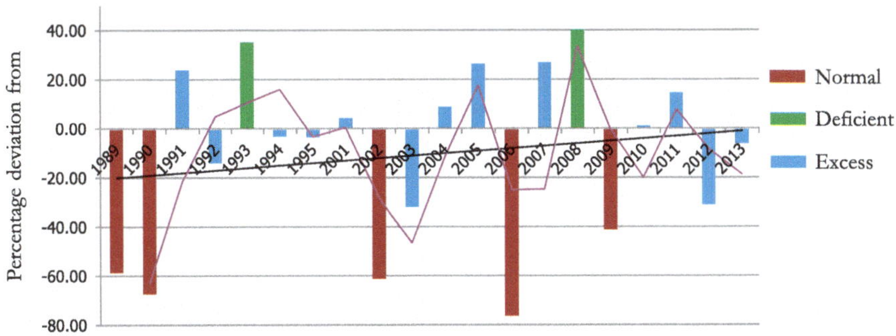

Fig. 2 Annual variability of rainfall: 1989–2013. *Source* Indian Meteorological Department

events are perceived as the visible impacts of climate change.[1] Rainfall and temperature data, provided by the Indian Meteorological Department (IMD) for Hoskote and Jadigenahalli, has been used in this assessment. The average annual rainfall data from 1989 to 2013 as also the mean annual temperature data from 1971 to 2013 compiled from the weather reports of the IMD have been analyzed.

Indicator 1: Annual Rainfall

Rainfall variability is a recognized indicator of climate change.[2] The annual rainfall data from 1989 to 2013 (data from 1996 to 2000 was missing) shows a mean rainfall of 750 mm and a standard deviation of 238 mm from it (see Fig. 2).

Rainfall and Climate Change: "Farmers occupying mainly rain fed farming land are more suspected to be prone to Climate change effects like loss of income and livelihood" (IPCC, 4th Report).

In this region, farmers posses mainly rainfed farming and according to people's perceptions there has been a reduced rainfall in the region over the years due to which lakes and tanks have dried up with a severe scarcity of water being experienced and observed. But this view is not supported by meteorological data recorded and as the average rainfall pattern in the study area hovers around 750 mm and the deviation is due to region receiving more than average rainfall in 1993 and 2008 (1200 mm) and a reduced rainfall in 1989, 2002, and 2009 (420, 430, and 560 mm, respectively) as shown in Fig. 3.

People perceived that 30 years ago the rainfall pattern was more than the normal and tanks filled (1993), but people also observed that despite the years 1993 and 2008 witnessing a fairly high rainfall, the water scarcity continued to persist. The reason was reduction in the number of lakes/water storage tanks in the region, as before three decades there were five lakes/water bodies (one in each village) but

[1]Statistics related to climate change, Ministry of Statistics and Programme Implementation, Govt. of India.

[2]Climate Change and India; A 4 × 4 assessment, INCAA report, Ministry of Environment and Forests Govt. of India.

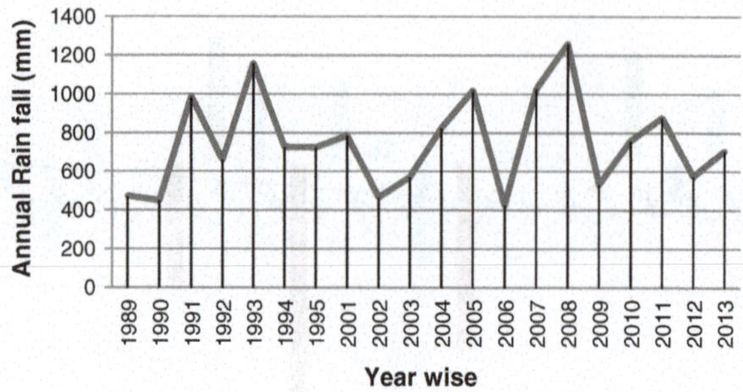

Fig. 3 Rainfall pattern of JDGP (mm). *Source* Indian Meteorological Department. *Asterisk* Data from 1995–2000 not recorded and maintained with the department

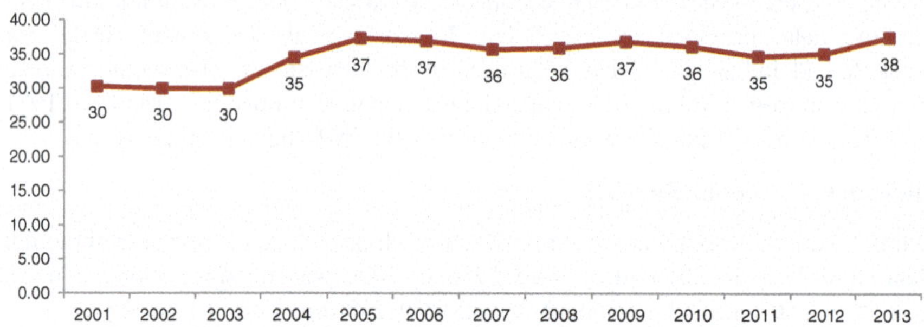

Fig. 4 Variations in max annual temperature. *Source* Master Plan 2031 for Hoskote LPA

now only one lake/water body exits in Govindapura, which dramatically brought down the water holding capacity in the region and also water recharge capacity. Along with this there was a decreased moisture content retention capacity of the soil system and heavy/overexploitation of ground water for agricultural activities.

This shows less awareness among villagers about water usage and conservation (each farmer, on an average, owns 6 bore wells) in terms of cultivating more crops, leading to a depletion of ground water levels. Further, de-silting of lakes for brick making industrial units (there are 500–600 brick making units spread across Jadigenahalli to Malur) has reduced water storage in lakes which cannot be attributed to climate change.

Indicator 2: Annual 'Max' Temperature

The recent global and Indian temperatures have been considered a standard indicator for assessing climate change (see Foot note 2). The mean 'Maximum Annual Temperature' in Bangalore for the period 2001–2013, as availed from the IMD, was 34 °C (as shown in Figs. 4 and 5).

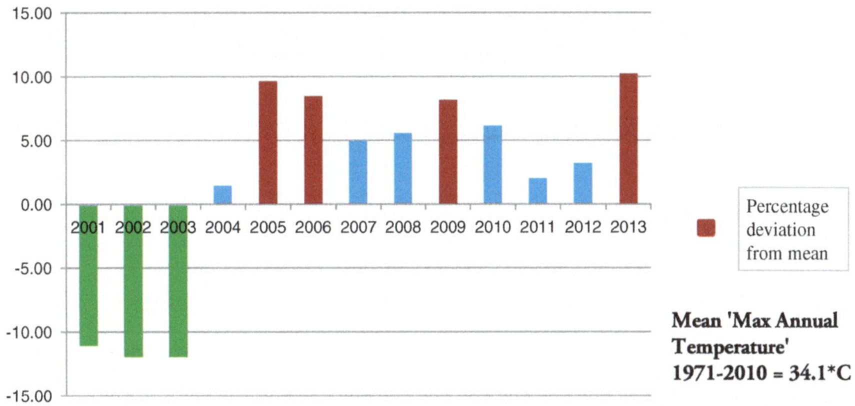

Fig. 5 Variations in max annual temperature. *Source* Indian Meteorological Department, Master Plan 2031 for Hoskote LPA

Indicator 3: Annual 'Min' Temperature

The mean 'Minimum Annual Temperature' in Bangalore from 2001–2013, as availed from the IMD, amounts to 15 °C with a standard deviation of 0.5 °C (as shown in Fig. 6).

The data availed from the IMD data for Bangalore as shown in the above graphs tells us that there is a variation in the temperature among different years but there has been no such increase of temperature.

The maximum average temperature recorded in the study area from 2001 to 2010 amounts to 29 °C and the average minimum temperature to 18 °C with a variation of 1 °C in the mean maximum temperature and 0.9 °C mean minimum temperature (as shown in Fig. 7); this shows there is a very slight variation, but not an increase in the temperatures over the last decade. Except a drought occurrence 53–60 years ago, there are no extreme events like hot waves recorded in the region and as the above graph also shows temperatures being moderate over the said decade, it can be surmised that there is no climate change in the region.

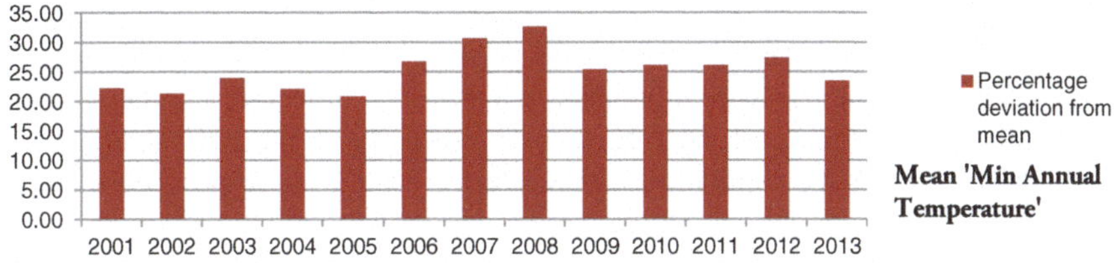

Fig. 6 Variations in min annual temperature. *Source* Indian Meteorological Department, Master Plan for Hoskote LPA

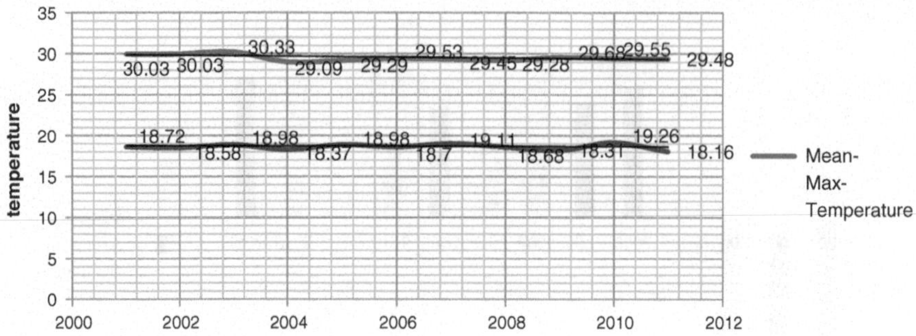

Fig. 7 Mean maximum and mean minimum temperature in Jadigenahalli gram panchayat. *Source* Indian Meteorological Department

Inferences from Climate-Related Indicators

In Asia and the pacific region, there is evidence of a prominent increase in the intensity of extreme events such as heat waves, tropical cyclones, prolonged dry spells, intense rainfall, tornadoes, avalanches, thunderstorms, and sever dust storms. The region is highly subjected to natural hazards like 2004 Indian Ocean Tsunami, 2005 Pakistan Earthquake, and 2006 landslides in Philippines (IPCC, 4th Report), but in the case of the semi-arid region, the villages belonging to Jadigenahalli Gram panchayat are not found affected to such an extreme extent by climate vulnerability, as the region is located 900 m above (average) from the sea level. With respect to temperature and rainfall there is a variation in it but not an increase.

4.2 *Indirect Indicators*

(a) **Demographic indicators**:

Social vulnerability to climate change is a function of exposure, sensitivity, and adaptive capacity.[3] Based on the availability of data, population, sex ratio, and literacy have been considered as demographic indicators.

Indicator 4: Population Growth

Data pertaining to population shows 6208 persons residing in the study area in 2011. There has been a growth in population in the area; however, a higher growth of 54 % has been observed for 1991, as compared to 19 % for 2001 and 4 % for 2011. Table 3 shows the population statistics (refer Table 3) in respect of the study area.

[3]Indicators to assess community-level social vulnerability to climate change, SocMon and SEM-Pasifikaregional socioeconomic monitoring guidelines, CRISP and IUCN, 2011.

Table 3 Population statistics

Village name	Years						
	1981	1991		2001		2011	
	Persons	Persons	Growth (%)	Persons	Growth (%)	Persons	Growth (%)
Jadigenahalli	1573	1707	8.52	1877	9.96	1983	5.33
Haralur	109	553	407.34	692	25.14	832	20.23
Vadigehalli	97	101	4.12	120	18.81	103	−14.17
Karibeeranahosahalli	410	419	2.20	532	26.97	583	9.59
Govindapura	6	602	9933.33	810	34.55	724	−10.62
Total	2195	3382	54.08	4031	19.19	4225	4.81

Source Census of India

Over the last two decades, there has been a steady decrease in the growth of population in Vadigehalli and Govindapura, indicating high out migration on a daily basis.

Indicator 5: Population Density

Population density data related to the study area shows no distinguishable increase over the last three decades. Population density (see Table 4) is very low across the entire study area ranging from 2.66 to 3.32 pph.

Indicator 6: Sex Ratio

Sex ratio is defined as number of females per 1000 males. Sex ratio in the study area is higher than the national average of 943, highlighting out migration of male population from the villages (see Table 5). The sex ratio in Vadigehalli is abnormally high at 1512 females per 1000 males, while that of Govindapura is also intriguing at 1069 females per 1000 males.

The data can also be corroborated by that of land use data. Vadigehalli and Govindapura are the two villages with decreasing agricultural land holdings. A high daily out migration is commonly attributed to the lack of employment opportunities in the native area. In the present case, the decrease in agricultural land holdings could be attributed to a decrease in employment opportunities.

Table 4 Population density

Village name	Population density (persons/ha)		
	1991	2001	2011
Jadigenahalli	4.07	4.48	3.23
Haralur	4.26	5.32	6.40
Vadigehalli	0.61	0.73	0.63
Karibeeranahosahalli	2.31	2.92	3.20
Govindapura	3.29	4.43	3.96
Total	2.66	3.17	3.32

Source Census of India

Table 5 Sex ratio

Village name	Sex ratio		
	1991	2001	2011
Jadigenahalli	969	945	961
Haralur	1048	983	917
Vadigehalli	1405	1449	1512
Karibeeranahosahalli	1024	1094	990
Govindapura	1034	1061	1069
Total	1011	1005	984

Source Census of India

Table 6 Work participation

Village name	Work participation (%)		
	1991	2001	2011
Jadigenahalli	38.72	41.13	41.75
Haralur	38.52	49.57	33.89
Vadigehalli	42.57	36.67	31.07
Karibeeranahosahalli	42.48	40.41	46.66
Govindapura	38.04	53.46	35.77
Total	39.15	44.83	39.60

Source Census of India

Indicator 7: Literacy Rate

Literacy data related to the study area shows a steady increase in the literacy level. However, overall, the literacy rate of 71 % is still lower than the national average of 73 % and the state average of 75 %.

(b) **Economic indicators**

Indicator 8: Work Participation

Work participation is considered an indicator of demographic vulnerability to climate change.[4] The work participation rate in the study area shows a steady decrease over the years (see Table 6). The work participation rate is found highest in Karibeeranahosahalli and lowest in Vadigehalli.

Indicator 9: Nonworkers

Data pertaining to workers shows a higher percentage of nonworkers in the study area. The nonworking population share is highest in Vadigehalli followed by Haralur (see Table 7).

---

[4]The Demography of Adaptation to Climate Change, United Nations Population Fund, International Institute for Environment and Development & El Colegio de México, 2013.

Table 7 Nonworking population

Village name	Nonworkers (%)		
	1991	2001	2011
Jadigenahalli	60.28	58.87	58.25
Haralur	56.24	50.43	66.11
Vadigehalli	57.43	63.33	68.93
Karibeeranahosahalli	57.52	59.59	53.34
Govindapura	61.96	46.54	64.23
Total	59.49	55.17	60.40

Source Census of India

Table 8 Agricultural workers

Village name	Agricultural workers (%)	
	2001	2011
Jadigenahalli	59.07	27.29
Haralur	90.96	63.83
Vadigehalli	38.64	43.75
Karibeeranahosahalli	37.67	90.44
Govindapura	87.99	16.22
Total	69.01	42.32

Source Census of India

Indicator 10: Agricultural Workers

The share of agricultural workers in relation to the total workforce is directly proportional to the agricultural activity in the area. The data pertaining to agricultural workers (see Table 8) availed from the census shows a steady decrease over the last two decades. The decrease is highly prominent in the villages of Jadigenahalli, Haralur, and Govindapura.

Indicator 11: Nonagricultural Workers

In tandem with the trend observed in the 'agricultural workers' category, there has been a steady increase in the share of 'nonagricultural' workers in the total workforce as shown in Table 9.

There is an overall increase of 47 % in the share of nonagricultural workers. The increase is more discernible in the villages of Govindapura, Haralur, and Jadigenahalli. The trend can be attributed to the growth of industrial units in Haralur village and real estate activity with the villagers getting employed as laborers. As our field observations and FGDs reveal, villagers also travel to the nearby towns of Hoskote and Malur, K R Puram, Bangalore in search of employment.

Table 9 Nonagricultural workers

Village name	Nonagricultural workers (%)	
	2001	2011
Jadigenahalli	25.00	55.43
Haralur	9.04	35.11
Vadigehalli	13.64	34.38
Karibeeranahosahalli	19.07	9.56
Govindapura	12.01	77.99
Total	17.87	47.64

Source Census of India

4.3 Land Use Land Cover Indicators

Land use and land cover data published by the Census of India in the years 1991 and 2001 have been used for assessment. The aim of the assessment is to find evidence of a decreased agricultural activity and agricultural production in relation to climate change factors.

Indicator 12: Agricultural Area

Area under agriculture with respect to the study villages shows an overall 5 % increase in agricultural area. Karibeeranahosahalli, Govindapura, and Jadigenahalli show a slight increase, while Vadigehalli shows an actual decrease in agricultural area by 1 % (refer Table 10). However, the increase is found highest in respect of Haralur.

The data collected through field visits and on-site interviews with the village residents reveals that most of the lands previously under agricultural use have now been converted into eucalyptus plantations. Hence, even though 'land' is still under 'agricultural use,' the actual pattern of agriculture has changed considerably. This is further substantiated with the data availed for the following indicators.

Table 10 Change in the coverage of land under agricultural use

Village name	Percentage of area under agriculture (%)		Percentage change
	1991	2001	
Jadigenahalli	73.33	75.48	2.15
Haralur	71.79	76.05	4.25
Vadigehalli	62.32	63.68	1.36
Karibeeranahosahalli	49.50	68.57	19.07
Govindapura	70.40	77.37	6.98
Total	56.72	61.79	5.06

Source Census of India

Indicator 13: Cropping Pattern and Agricultural Productivity

Agricultural productivity is assumed to be directly proportional to the area under various crops. As per data collected from the district handbook, there has been a decrease in land under agricultural crops over the last two decades as shown in Table 11.

The total area at present under food grains comes to 14,308 Ha as compared to 18,874 ha in 2000. There has been a decrease in the area under cereals, minor millets, pulses, food grains, and commercial crops. The area and production of sericulture have also decreased over the last decade as shown in Fig. 8. However, the decrease is merely due to the availability of high yielding varieties of crops and the introduction of fruits and vegetables as major crops.

Table 11 Agricultural area in Hoskote Taluk

Agriculture crops	1995–96	1999–2000	2010–11	Remarks
	Area (ha)			
Paddy	1762	1600	566	Decrease
Ragi	14,226	12,231	11,460	Decrease
Maize	542	584	204	Decrease
Total	16,530	14,415	12,230	Decrease
Total cereals and minor millets	16,610	14,415	12,229	Decrease
Total pulses	4529	4459	2079	Decrease
Total food grains	–	18,874	14,308	Decrease
Total commercial crops	20	7	5	Decrease
Fruits	–	1228	4085	Increase
Vegetable	–	1607	4084	Increase
Area under mulberry	2798	3551	1852.43	Decrease
Cocoon production (tonnes)	4763	2694	2913.56	Decrease
Total area sown	34,520	32,660	36,278	Increase

Source District handbook

Fig. 8 Area under cereal crops (Ha) in Hoskote Taluk. *Source* Agriculture census

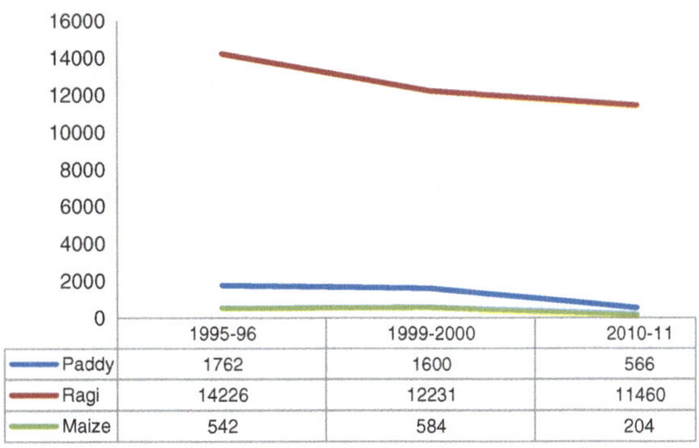

In the absence of Jadigenahalli Gram Panchayat data, Hoskote Taluk data is used here because Jadigenahalli Gram Panchayat comes under Hoskote Taluk and is 9 km off from Hoskote. From secondary data (see graph), it can be observed that the area under maize (542–204 ha) and paddy (1762 to 566 ha) has drastically decreased as compared to ragi (14,226–11,460 ha) mainly because of a shift in the cropping pattern; as ragi is the staple food, instead of leaving the land fallow, many are cultivating ragi under rainfed area.

Indicator 14: Area Under Irrigation

The availability of irrigation facilities is a major factor affecting agriculture. Data pertaining to 'Irrigated land,' as published by the Census of India, shows a 10 % increase in irrigated land over the last decade. Haralur, however, shows a 0.1 % decrease in irrigated area (see Table 12). The percentage of land irrigated in the villages is, however, low, ranging from 0 to 4 %, excepting Govindapura village. The table below shows the statistical details pertaining to 'irrigated land' across the study villages.

The data on source-wise irrigation collected from the District Handbook of Hoskote shows (see Table 13) a steady decrease in the tank irrigation and a marked increase in the bore well irrigation in Hoskote Taluk.

Similar trends have been seen across the study area. Although a large number of bore wells are present, 64 % percent of them are non-functional. A total of 143 bore wells and 17 open wells are present in the study area. Table 14 shows the statistical details of dry bore wells, failed bore wells, dry open wells, and bore wells with water.

Table 12 Area under irrigation

Village name	Percentage of area under irrigation (%)		Percentage change
	1991	2001	
Jadigenahalli	14.83	17.54	2.71
Haralur	19.00	19.17	0.17
Vadigehalli	20.50	22.08	1.58
Karibeeranahosahalli	7.03	10.84	3.80
Govindapura	14.02	77.37	63.57
Total	12.49	23.27	10.77

Source Census of India

Table 13 Irrigation sources—Hoskote Taluk

Irrigation	1995–96	1999–2000	2008–09
Canal	0	0	0
Tank	1632	864	209
Open well	82	228	3528
Bore well	5440	3843	7921
Total	7154	4935	11,658

Source District hand book of Hoskote Taluk

Table 14 Number of wells

Bore well status	Number
Dry bore wells	76
Failed bore wells	16
Functional bore wells	51
Total	143

Source Authors, 2014

Fig. 9 Open wells in JDGP. *Source* Authors, 2014

As observed from a primary survey (refer Table 14), the depth of bore wells dug in the recent years goes beyond 1300 ft, showing a considerable depletion of ground water resources. However, a better water situation has been observed in Govindapura, where surface wells and tanks with water are present, which can be attributed to a higher percentage of irrigated area in the village.

It can also be seen from the 'Ground water potential' map and 'Bore well map' that majority of the bore wells (see Fig. 9) are clustered in the villages of Jadigenahalli, Vadigehalli, and Karibeeranahosahalli which are areas with a relatively high water potential. Cluster of bore wells is also seen in Haralur village, which lies in the zone with a lower water potential. However, all the clusters show the presence of dry and failed bore wells, signaling an overexploitation of water resources.

In the study area it was observed that due to scarcity of water, people who are growing vegetables and flowers have adopted to drip irrigation as shown in Fig. 10. Earlier, agriculture farming was based on open well, tank, and rainfed system, but now due to overexploitation of ground water, it leads to adaptation of new technologies to conserve water (Fig. 10).

Indicator 15: Area Under Wasteland

Data pertaining to 'area under waste land' shows an overall decrease of waste lands by 5 %. The change in waste land corresponds to the change in agricultural land as presented in Table 15. Vadigehalli, which shows a decrease in agricultural land, also shows a corresponding increase in waste land, while all other villages with an increase registered in agricultural land show a decrease in waste land.

The percentage of waste land in the study area is considerably high, ranging from 20 to 35 %.

Fig. 10 Drip irrigation system for vegetable and flower cropping. *Source* Authors, 2014

Table 15 Area under waste land

Village name	Percentage of area under wasteland (%)		Percentage change
	1991	2001	
Jadigenahalli	26.67	24.52	−2.15
Haralur	28.21	23.95	−4.25
Vadigehalli	37.68	36.32	−1.36
Karibeeranahosahalli	50.50	31.43	−19.07
Govindapura	29.60	22.63	−6.98
Total	28.00	22.96	−5.04

Source Census of India

5 Agricultural Sector

Changes in agricultural production can have corresponding impacts on incomes and well-being of rural people. (IPCC report, 2013). Indian agriculture is facing challenges from several factors such as increased competition for land, water and labor from nonagricultural sectors, and increasing climatic variability. Food production in India is sensitive to climate change such as variability in monsoon rainfall and temperature changes within a given season. Small changes in temperature and rainfall can have a significant impact on quality and quantity of production (refer Table 16).

FGDs revealed that most of the farmers are not able to express their perceptions regarding climate change directly, but they express themselves through effects or changes that have occurred relative to the earlier years or based on the experiences of the elderly residing in the village. The study also reveals that most of the respondents are not aware of climate change; rather they readily talk of rainfall- and warming-related aspects.

The area under total pulses as shown in Fig. 11, total cereals and minor millets, has also got reduced, i.e., (see graph) pulses (from 4529 to 2079 ha), cereals, and minor millets (16,610 to 12,229 ha). This reduction is due to the fact that people have diverted their cropping pattern from the cultivation of food grains to eucalyptus, vegetables, and flowers, as these villages are located near to Hoskote and Bangalore and also because of water scarcity, labor problem, high input cost, and low profit and this also led to change in cultivable land as shown in Table 17.

India has a variety of crops grown under irrigated and rainfed conditions. In the study area, many farmers have shifted from cultivating local varieties to hybrid varieties in respect of all crops like ragi, pulses, and some others, leading to an increase in the per acre production of ragi. While local ragi is sown at a spacing of 9 * 4, hybrid variety is sown at 2 * 2 spacing thus increasing the number of seedlings sown per acre, and in addition, gypsum application also helps to attain the required moisture level, uniform filling of seeds, and increase in yield by 8–10 %.

Table 16 Hybrid varieties used in Jadigenahalli Gp (2013)

Crops	Hybrid
Paddy	KRH2, Rashi, Telahana, IR64, MTU 1001, MTU 1010, JGI 1798
Ragi	GPU 28, MR1, M1365
Maize	Hybrid variety
Popcorn	Popcorn variety
Minor millets	Arka, Navane, Samee
Red gram	Hyderabad 3c, TTB 7, BRG1, BRG2
Cowpea	C15, TVX 944,
Bengal gram	TMV2, JC24
Sesame and Niger	Local

Source Department of Agriculture, Hoskote, 2014

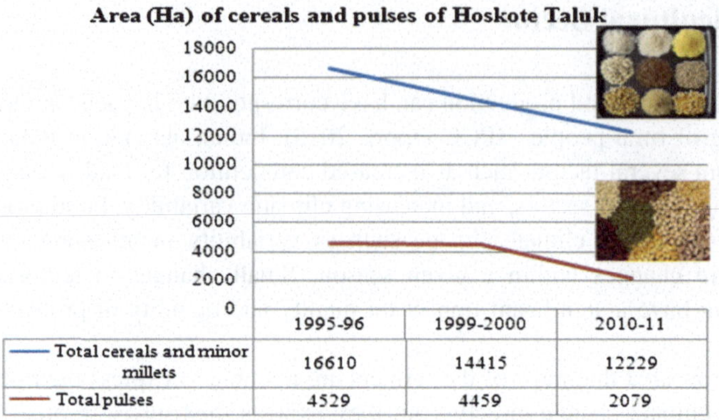

Fig. 11 Area under cereals and pulses in Hoskote Taluk. *Source* Agriculture census

Table 17 Cultivable and uncultivable land in Hoskote Taluk

Land not available for cultivation	1995–96	1999–2000	2010–11
Nonagriculture	1049	9438	10,942
Barren	9008	1049	1049
Other uncultivable land			
Cultivable waste	750	978	498
Permanent pasture	660	500	356
Tress and Groves	6226	6245	1584
Area sown (ha)			
Net	28,220	27,910	34,130
More than once	6300	4750	2148

Source Annual season and crop report, various issues, District Handbook

The graph below shows per hectare yield of rainfed ragi from 2006 to 2010. It is evident that there is a decline in the yield of ragi for 2008 (1408 kg) and 2009 (1389 kg), respectively. The reason could be that there was a below average rainfall in 2009 (532 mm) (Fig. 12).

Ragi is still the staple food and several farmers cultivate ragi as it is rainfed and instead of leaving the land barren, they prefer to cultivate ragi. Besides, changing cropping pattern from cultivating food grains to eucalyptus plantations, vegetables, and flowers, due to factors like demand from nearby town Hoskote and Bangalore city, combined with challenges of water scarcity, labor problems, high input costs, and low profits. This has also effected the changes in the coverage of agricultural land as shown in Table 18.

Fig. 12 Per hectare yield of rainfed ragi (kg) *Source* District Handbook, Hoskote Taluk, Bangalore Rural District

Table 18 Changes in the coverage of land under agricultural use

Village name	Percentage of area under agriculture (%)		Percentage change
	1991	2001	
Jadigenahalli	73.33	75.48	2.15
Haralur	71.79	76.05	4.25
Vadigehalli	62.32	63.68	1.36
Karibeeranahosahalli	49.50	68.57	19.07
Govindapura	70.40	77.37	6.98
Total	56.72	61.79	5.06

Source Agriculture census

6 Livestock: Local to Hybrid

It has been observed from the FGDs that earlier all households were rearing local breeds (Hallikar), but now they are replaced with hybrid varieties like Holstein Friesian, Jersey (Imported Cattle breed), Jamunapari, and Bijapur varieties (Goat). From the secondary data it is evident (see graph) that Holstein Friesian (cow) population in all the villages has increased (from 348 to 375) as compared to other livestock. This is due to more milk yield from hybrid varieties (15–25 l/day/cow) as compared to local cultivars (2–3 l/day/cow). As a result, villagers are able to earn additional income from milk sale.

Interestingly, the population of ox shows a rapid decrease; in Vadigehalli and Haralur there is no presence of ox pairs, but in Jadigenahalli there are 3–5 pairs and in Govindapura and KB Hoshalli 2 and 4 pairs of ox, respectively. And also the cost has increased from Rs 40,000 to 80,000. This is due to urbanization and mechanization of agriculture and many are dependent on tractors for plowing, as a way of time saving. This could also be due to a variety of socioeconomic factors like low returns from agriculture, high input cost, changing occupational preferences, and so on.

Another important species in India are buffaloes. The morphological and anatomical characteristics of buffaloes make them well suited to hot and humid

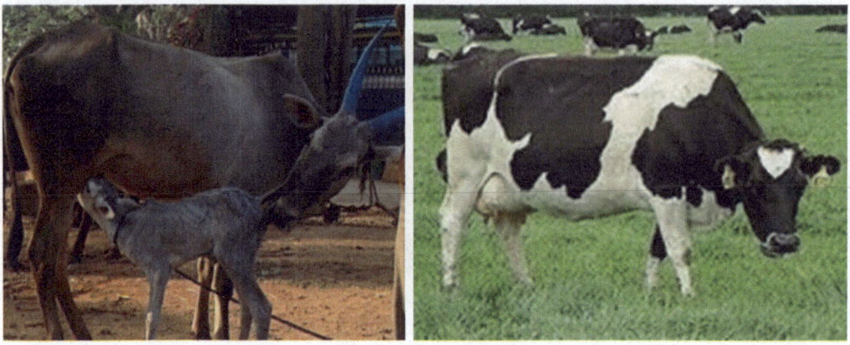

Fig. 13 Shift in livestock from local to hybrid varieties. *Source* Authors, 2014

climates, but heat stress has a detrimental effect on the reproduction of buffaloes (Tailor and Nagda 2005). From the FGDs, it becomes evident that the number of households with buffalo has decreased overtime. This is due to scarcity of fodder and labor, water, and grazing lands which have been converted into agricultural lands.

And in all the study villages, the cost of livestock has increased, 30 years ago, the cost of local cows (local variety) was from Rs. 500/- to 1000/-, and for hybrid from Rs. 50,000 to Rs. 80,000/- (because of their high milk yielding capacity). For buffaloes, earlier the cost range was from Rs. 300 to Rs. 700, but now it is from Rs. 15,000 to Rs. 25,000.

Indigenous varieties were replaced by livestock. This is because of short duration, high yielding, and pest- and disease-resistant varieties. In case of livestock because of high milk yield people have shifted to hybrid varieties, as shown in Fig. 13.

7 Decrease in Sheep and Goat Rearing

The population of sheep (from 803 to 612) and goats (from 479 to 354) has decreased in all the study villages and according to villagers, earlier, there were many groups (in one group there were 50–100) of sheep and goats in a family, but now, due to urbanization, scarcity of space and labor, many people have stopped rearing sheep and goats. Also earlier, villagers used to take their livestock to *Gomals*[5] for grazing, but now these grazing lands have been allotted to landless and SC people for agriculture purpose. In the surrounding waste land many factories have come up, while previously, there were a large number of trees (*casuarina equisetifolia)* used as fodder for goat and sheep; but now, due to eucalyptus plantation, *casuarina equisetifolia* has totally disappeared from this region, which is also quoted as one of the reasons for decrease in livestock (see Fig. 14).

[5]Government allocated land for grazing.

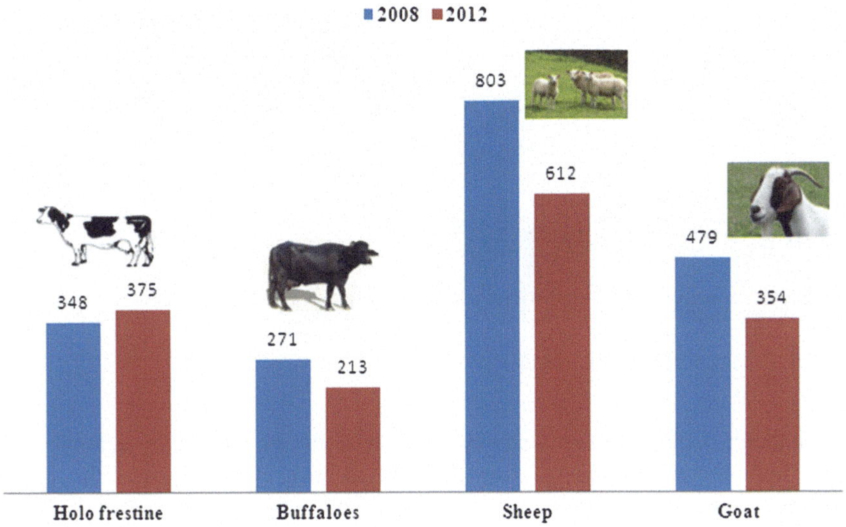

Fig. 14 Livestock population—JDGP (2008 and 2012). *Source* Livestock census 2008 and 2012 (Draft), Jadigenahalli Gp, Veterinary Department

8 Water Relations in Eucalyptus Plantation

Eucalyptus is a controversial tree species globally, due to its merits as well as notoriety. Its merits like fast-growing nature, quick adaptation to a wide range of ecological situations, several industrial applications, and as a means of livelihood for the unprivileged have elevated it to one of the most desirable tree species introduced in afforestation, farm forestry, and social forestry programs. Nevertheless, eucalyptus is also known to cause a number of environmental hazards like depletion of groundwater, dominance over other species through allelopathic effects, loss of soil fertility, and negative impacts on local food security issues (Mukund and Palanisami 2011).

Tree planting in the study area is mainly meant for fuel wood, income generation, and construction purposes. No respondent has reported that trees are planted for environmental conservation. As per people's perceptions, eucalyptus is the main reason for a scarcity of water in the recent years and also no crop will grow in the surrounding lands once it is uprooted because it absorbs moisture from the surrounding land. In this respect, Engel (2005), in a study on hydrological consequences of eucalyptus in Argentine pampas, finds that it utilizes ground water (67 % of its total water use) as well as water from upper vadose zone (unsaturated zone), the main source of supply to ground water. This corroborates the earlier study carried out by Thorburn and Walker (1993) who also find evidence of eucalyptus drawing water with a greater suction from the surrounding areas. The proportion of groundwater used by eucalyptus trees from distant places from the plantation increased from 40 to 63 % with the surface soil drying up.

Fig. 15 Eucalyptus plantation and its marketing. *Source* Authors, 2014

Although depletion of groundwater resources can be caused by a number of other hydrological factors or overextraction for irrigation purposes, the above evidences suggest that eucalyptus plantations can aggravate the water depletion much earlier than expected by other reasons. Respondents observed that eucalyptus was introduced in their village's way back in the 1980s by the government with seedlings provided by the Forest Department. Earlier, a very few households were cultivating eucalyptus in all villages, but currently majority of households have resorted to eucalyptus cultivating due to less cost of cultivation. Eucalyptus is harvested once in 3 or 6 years with costs varying from Rs. 4000/- to 4500/- per acre (see Fig. 15).

9 Change in Livelihood

Livelihoods are sustainable when it enables people to cope with and recover from shocks and stresses (such as natural disasters and economic or social upheavals) and enhance their well-being and that of future generations without undermining the natural environment[6] or resource base (International Federation of Red Cross and Red Crescent Societies).

Indicators used for understanding changes in livelihood include social sector–population density, caste, gender ratio, literacy rate, and access to transport. Economic indicators[7] like contribution of agriculture to livelihood and contribution of off-farm income to livelihood have been considered. Data that supports social and economic indicators is shown in Table 19.

Change in Livelihoods, Agriculture Food habits: The cost incurred on agriculture has increased more than ever before with one-third of income obtained from cultivation of any crop going towards water purchasing water. The income obtained

[6]http://www.epa.gov/climatechange/impacts-adaptation/health.html.

[7]http://www.iisd.org/casl/caslguide/indicatorchecklist.htm.

Table 19 Occupational details

Villages	Agriculture dependent families (2001)	Agriculture labor-dependent families (2001)	Business dependent families	Private employment	Government employment	Total
Jadigenahalli	130	80	12	03	15	262
Karibeeranahosahalli	38	60	03	05	03	120
Govindapura	44	39	12	67	04	101
Vadigehalli	07	04	00	00	00	15
Haralur	42	50	04	20	10	130
Total	457	285	37	115	32	890

Source Jadigenahalli Grama Panchayat, 2008

from ragi cultivation amounts to 9100 rupees, the cost incurred on water purchasing to 2250 rupees per yield per 3 months, and the same is the case with other crops. Although income obtained from cultivating ragi amounts to 2000 rupees per yield, still they grow it asbecause it is their staple food. Thus, income obtained from cultivation has decreased and so also the contribution of agriculture to livelihood has come down over the years, forcing people to seek off-farm occupations. According to respondents, earlier everybody were practicing agriculture barring a few households, but now there is decline in the agricultural activities. In Jadigenahalli, 70–90 households are engaged in agriculture out of 380 households, and similarly in other villages. It could be because, earlier, there was no option other than agriculture, agriculture labor, and rearing livestock, but now, due to high literacy rates, awareness, urbanization, and easy access to transportation and communication, many people are able to find alternative ways of earning income like business, driving, eucalyptus tree cutting, working in brick making units, and garment industries. And this also proves that region is not even vulnerable to the change in weather condition in the region as it is next to a metropolitan city.

Table 20 indicates a good access to basic needs of villagers like schools, health care centers, bank, communication, and transportation.

Changes in Livelihood and Water Scarcity: Another major reason for change in livelihoods is acute scarcity of water in the region (see Fig. 16). According to people, reduced water availability, soil moisture loss, loss of soil fertility, and low ground water tables are responsible for this situation. Ground water has depleted from 30 to 1650 ft still with water not being available. In 1993, tanks were full, but thereafter no tanks are getting filled up; currently, among six tanks only one contains a little water. This high scarcity of water is mainly due to overexploitation of ground water resource and unscientific and excess use of fertilizers and pesticides, leading to soil fertility loss.

Improved transportation has helped in supply of drinking and domestic water to villagers during high water stress months. Besides this, technologies like drip irrigation and mulching have helped in agricultural cultivation.

Change in livelihood and migration: The above issues related to change in livelihood have led to an increase in the migration rate due to good access to communication and transportation. Most of the people migrate daily to nearby cities

Table 20 Basic infrastructure available to the community

Indicators	Jadigenahalli	Vadigehalli	Govindapura	Haralur	K.B. Hosahalli
Schools	✓	X	✓	✓	✓
Health center	✓	X	X	X	X
Veterinary center	✓	X	X	X	X
Communication (telephone exchange, newspaper, TV)	✓	✓	✓	✓	✓
Grocery and daily needs shop	✓	X	✓	✓	✓
Markets	X	X	X	X	✓
Transportation	✓	X	✓	✓	✓
Bank	✓	X	X	X	X
Electricity (domestic and agriculture)	✓	✓	✓	✓	✓
Agriculture information centers	✓	X	X	X	X
Community hall	✓	X	X	X	X
Safe drinking water	✓	✓	✓	✓	✓
Medical shop	✓	X	X	X	X

Source Authors, 2014

Fig. 16 Change in irrigation water tanks in Hoskote Taluk *Source* District handbook

and towns for education and job opportunities—private/government/daily wages. They migrate mostly to Bangalore, Hosakote, Malur, and Vagata. The migration level of villagers in Govindapura is high; 40 % of the people are dependent on agriculture with most of them going out to work, and the same is the case with Govindapura colony where scheduled caste people reside. Most of them are landless and as there is an easy access to Malur, they migrate daily in search of wage employment.

According to the villagers, 3–6 households from each village migrate permanently. This is due to changes in the occupational pattern, while in-migration of 20–25 HH has been observed in Jadigenahalli and Haralur due to the presence of five factories in the region. People have immigrated from different states like Rajasthan, northeastern India in search of employment over the past 5–6 years.

Women and livelihoods: The occupational patterns have changed for women in JDGP. Earlier (before 30 years), they were engaged in as daily wage labor since they were practicing agriculture, but now, due to increase in floriculture and vegetable cultivation women are making a living by making flower garlands and selling vegetables. Women also collect pongamia seeds in and around villages during summer, dry them, and sell it to small-scale oil refineries located in Hoskote and Malur.

10 Key Findings

- **Weather and climate**: In Asia and the pacific region, there is evidence of a prominent increase in the intensity of extreme events of climate change like cyclones, tsunamis, and earthquakes, but in this study there is no change observed in climate at the micro level across all the five villages in a semi-arid region of Karnataka as it is located 900 m above from the mean sea level and it does not fall under climate change vulnerability index and the extreme events are comparatively nil.
- **Rainfall**: The average rainfall in the study area is 750 mm even though it was more than the average in 1993 and 2008 (1200 mm) and less than the average in 1989, 2002, and 2009 (420, 430, and 560 mm, respectively). However, the rainfall has not deviated much from the normal rainfall pattern over the past three decades but due to loss of water storage tanks and high ground water extraction, people face water scarcity in the region.
- **Temperature**: There is a variation in the temperature in different seasons and across the last three decades but there is no increase in the temperature, so this semi-arid region is not affected by climate change.
- **Shift in the cropping pattern, Decrease in area under agricultural crops and livestock rearing**: Earlier, people were growing crops like paddy, ragi, pulses, turmeric, onion, vegetables, and groundnut as main crops, but now ragi and eucalyptus have become main crops. And also people have shifted from cereals and pulse crops to vegetables (beans, potato, tomato, cowpea, bitter gourd) and flower crops, because Hoskote is nearby where these vegetables and flowers can be easily marketed. This shift is mainly due to changing needs of people and demand for eucalyptus tree. The area under pulses (from 4529 to 2079 ha), cereals, and minor millets (from 16,610 to 12,229 ha) has decreased overtime. There has been a decline in livestock rearing, because earlier, people used to take their livestock to Gomalas, or waste lands with grass and shrubs for grazing, but now, they have been converted into agriculture lands. Interestingly, the population of ox and buffaloes is less compared to cows, sheep, and goats.
- **Change in varieties**: Earlier, villagers were cultivating local varieties, but now people have shifted to hybrid varieties of short duration. This is a normal phenomenon seen in most of the villages due to invention of new hybrid varieties which give more yield and income.

- **Society and ecosystem**: Since Jadigenahalli is just about 40 km away from the metropolitan city center, with a good communication and transportation network with NH4 (National Highway) connection and also as it possesses a forest area of 750 acres with natural resource access, the region is less vulnerable to climate change effects (as compared to the remote region/villages).
- **Changes in livelihood and its impact on Migration**: Now with only 25–26 % of the villagers dependent on agriculture as their major occupation and the remaining 60–70 % of the people on garments, factories, nonagriculture labor, and livestock rearing as major alternative sources of their livelihood, which are less vulnerable to climate change. In view of this change, 60 % of the people migrate daily to other cities in search of employment and most of the youngsters migrate daily in pursuit of education, and thus changes in livelihoods and migration are not due to climate change in this region.

11 Conclusion

Climate change poses a serious threat to poor farmers who live in remote areas, marginal, dry lands, and deserts with a limited access to natural resources, communication, and transportation. On the other hand, Jadigenahalli Gram Panchayat located in semi-arid region, with analyzing all the above indicators, reveals that it is not vulnerable or has any drastic climatic variation in the region. However, there is scope for government and local bodies to work toward sustainable natural resources management to avoid further deterioration and possibilities of future climate change implications.

References

Aavudai, Anandhi. (2010). Assessing impact of climate change on season length in Karnataka for IPCC SRES scenarios. *Journal of Earth System Science, 119*(4), 447–460.

Census of India. (1991–2011). Government of India.

Census, All Indian Latrine Data. (2001–2011). Government of India.

Climate Change, Ministry of Environment and Forests. (2008). New Delhi: Government of India. Retrieved October 2008, from http://envfor.nic.in/cc/what.htm

Data on temperature and Rainfall, Indian Meteorological department, Government of India.

Engel, Vic. (2005). Hydrological consequences of Eucalyptus afforestation in the Argentine Pampas Water Resources Research. Vol 41, W 10409, p. 14

Inter Governmental Panel on Climate Change (IPCC). (2007). Fourth Assessment Report: Climate change.

Khan, A. Shakeel, Sanjeev Kumar, Hussain, M. Z., & Kalra, N. (2009). Climate Change, Climate Variability and Indian Agriculture: Impacts Vulnerability and adaptation Strategies.

Mukund Joshi and Palanisami, K. (2011). 21st International Commission on Irrigation and Drainage, (ICID), 15–23 October 2011, Tehran, Iran.

Ninan, K. N., & Bedamatta, S. (2012). Climate change, agriculture, poverty and livelihoods: A status report. Working paper, Institute for Social and Economic Change, Bangalore.

Revenue department report and Gram Panchayat records, Hoskote, Karnataka, India.

Tailor, S. P., & Nagda, R., K. (2005). Conception rate in buffaloes maintained under sub humid climate of Rajasthan.*Ind J Dairy Sci*, *58*(1):69–70.

Thorburn, P. J., & Walker, G. R. (1993). The source of water transpired by Eucalyptus camaldulensis: soil, groundwater, or streams? In: Stable isotopes and plant carbon water relations, Pub: CAB International. pp. 511–527.

Umesh Babu, M. S., & Puttaiah, E. T. (2012) Issues and constraints (climate change) of water resource management in Tungabhadra River Basin in India. *Research Journal of Agriculture and Forestry Sciences*, *1*(1), 17–26.

Energy Generation by Use of Crop Stubble in Punjab

Parmod Kumar

Abstract There is growing interest in alternative uses of field-based residues for energy applications. The use of agriculture waste for generating power is an innovative idea as biomass is a carbon neutral fuel, which helps in reducing the greenhouse effect and global warming. The advantages of utilizing crop residue over and above conventional resources are that the residue is renewable, readily available, and can be used successfully by burning in boilers with an efficiency of 99 %. Further, they are available at low cost relative to coal and their ash contents are also much less as compared to coal. Biomass-based power plants of 10–20 MW capacity setup in a group of villages or every *Taluka* can meet energy needs of villagers and employ thousands of people. Following the initiatives started by Ministry of New and Renewable Energy, Government of India, Punjab Energy Development Agency (PEDA) has allocated 30 sites/tehsils for setting up plants with a total generation capacity of 337 MW out of which 2–3 plants are already commenced and working successfully. Since the Punjab State Electricity Board (PSEB) is already planning to install additional power generating units based on paddy straw, steps need to be taken to ensure the availability of paddy straw in the respective areas.

This paper is mere compilation of information based on the inputs received from Punjab Energy Development Agency (PEDA) Chandigarh, Punjab State Council for Science and Technology, Chandigarh, Punjab Pollution Control Board, Patiala, Department of Agriculture, Punjab Government Chandigarh.

P. Kumar (✉)
Agriculture Development and Rural Transformation Centre (ADRTC), Institute for Social and Economic Change, Nagarabhavi P.O, Bangalore 560072, India
e-mail: pkumar@isec.ac.in

S. Nautiyal et al. (eds.), *Climate Change Challenge (3C) and Social-Economic-Ecological Interface-Building*, Environmental Science,
DOI 10.1007/978-3-319-31014-5_30

1 Introduction

Punjab made a commendable progress in the production of food grains in the post-green revolution period. Food grain production underwent a big jump from 3.16 million tons in 1960–1961 to 28.90 million tons in 2013–4. The green revolution also known as the new agricultural strategy was marked with the arrival of new high-yielding varieties (HYV) of wheat, rice, maize, and *bajra* (millet) and package of other inputs like chemical fertilizers, insecticides, pesticides, and assured irrigation facilities. Focusing on popularizing modern inputs and practices in the productive areas where the likelihood was more for the high-yielding seeds to show results was the most important feature of this new strategy. Punjab with requisite irrigation and infrastructure facilities became a major beneficiary of this national strategy and has been shown as a showpiece of India's successful green revolution strategy. Short duration and HYV of rice and wheat were introduced in Punjab to boost up the production of food grains. During the first decade of the green revolution, the technology was confined only to the wheat crop. A remarkable growth rate of 5 % was achieved by the state's agricultural sector since the beginning of the green revolution in the mid-1960s.

The decade from the mid-1970s to mid-1980s was characterized by the extension of new seed fertilizer technology from wheat to rice crop. Due to the input and output price structure and superior yields of rice and wheat as compared to other crops, Punjab agriculture has virtually become a rice–wheat monoculture. During 1966–67, total area under rice was 0.29 million ha, which increased to 2.85 million ha by 2013–14. There was also a substantial increase in the average rice productivity, which increased from 1186 kg/ha in 1966–67 to 3952 kg/ha by 2013–14 (Fig. 1). During the same period, the area under wheat crop increased from 1.61 to 3.51 million ha and productivity from 1544 to 4848 kg/ha. Punjab has achieved a crop intensity of 191 % as against 138 in the country as a whole. The present level

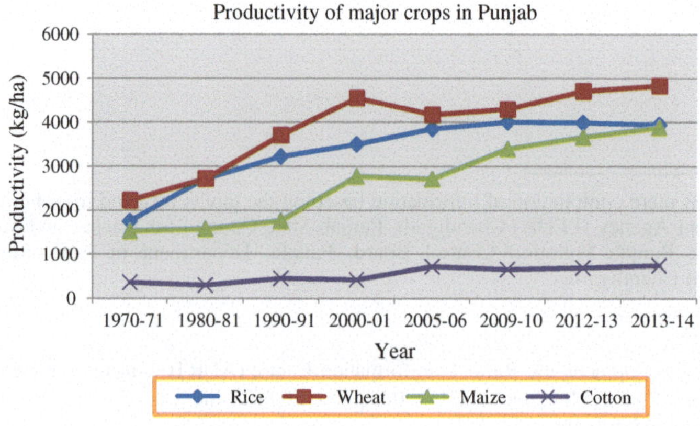

Fig. 1 Productivity of major crops in Punjab

of consumption of fertilizer (NPK) is 249 kg/ha as compared to the Indian average of 142 kg/ha. Similarly, Punjab has 98 % of high-yielding variety coverage, which is the highest among the Indian states. About 18 % of the total tractors in India are in Punjab. Production is supported by about 97 % irrigation coverage with 970,139 tube wells.

2 Agricultural Residue Burning

Under the rice–wheat cropping pattern, rice has to be harvested early in order to accommodate the wheat crop. This means, a very little time is left in the hands of the farmers to turn around for planting the wheat crop. Within this period, the farmer has to get rid of the rice stubble and prepare the land for sowing the wheat crop. The previous varieties of rice and wheat crops were of long duration and could fit rice–wheat rotation only in small areas. But with the availability of photo-period non-sensitive short-duration varieties of wheat as well as rice it became possible to grow high-yielding 120–130 days rice crop, i.e., June–July to October–November followed by a high-yielding 110–120 days wheat crop, i.e., November–December to March–April. With the adoption of these varieties rice–wheat crop rotation was practiced in areas which formerly produced only wheat or rice but not both in the same field in any one farming year. The major constraint in the rice–wheat cropping system is the available short time between rice harvesting and sowing of wheat and any delay in sowing adversely affects the wheat crop. Preparation of the field also involves removal or utilization of rice straw left in the field.

Various modern inputs were introduced in Punjab to harvest the rice crop within such a short period of time. One such input which has become the most popular implement in the rice–wheat cropping system is the use of the combined mechanized harvester. The use of the combined harvester has increased at a tremendous rate in Punjab. Almost 80 % of the rice crop is harvested using this implement in Punjab. However, the use of the combined harvester has in reality exacerbated the problem of crop residue management. The use of combined harvesters leaves behind a large amount of rice residue to be burnt in the open fields. The combined harvester spreads the rice residue in the fields which is difficult to collect. It is widely perceived that farmers find it the easiest and the most economical way of getting rid of the rice stubble through burning it. Also, the shortage of time for sowing the wheat crop, after the rice crop harvest, leaves farmers with no other option but to burn it.

Thus, burning has emerged as the standard method of rice residue/stubble management in the combine harvested rice–wheat cropping system that is practiced on a broad scale in the state of Punjab in northwest India. Every year almost 15 to 20 million tons of paddy straw are generated in Punjab. Of this, according to various estimates, on an average, almost 7–8 million tons of rice residue are set on fire in open fields.

Rice residue burning results in extensive impacts both on farms, e.g., losses in soil nutrients, soil organic matter, production and productivity, air quality, biodiversity, and water and energy efficiency, and on human and animal health. In India, air pollution from residue burning can be severe, with impacts on human health by directly causing or exacerbating a range of health hazards and contributing to the incidence of traumatic road accidents through significantly reduced visibility. One of the recognized threats to the rice–wheat cropping system sustainability is the loss of soil organic matter as a result of burning. The straw collected from the fields is of great economic value as livestock feed, fuel, and industrial raw material. In Northern India, wheat straw is preferred while in Southern India paddy straw is fed to livestock (Hegde 2010). The residue generated from the rice–wheat cropping system can be put to many uses, but this is possible if the residue is separated from the grain and carried out of the field. Burning reduces the availability of straw to livestock, which is already in short supply by more than 40 % (Table 1). However, in the case of combine harvesting, most of the residues are left in the field for burning adversely affecting overall sustainability of the rice–wheat cropping system (Thakur 2003). Zero tillage after stubble burning is now being adopted by many farmers. In 2005–06, around 10 % of the total area sown under wheat was using zero till machines. Apparently, less than one percent of farmers incorporate the paddy straw because in the case of incorporation more tillage operations are

Table 1 State-wise percentage of short fall of crop residue and greens

States/UTs	Availability	Requirement	Shortfall (%)	Availability	Requirement	Shortfall (%)
	Crop residues (million tons)			Greens (million tons)		
Andhra Pradesh	15.69	31.71	50.52	4.88	16.91	71.14
Arunachal Pradesh	0.47	1.00	53.00	1.57	0.53	−196.23
Assam	5.82	12.39	53.03	0.95	6.61	85.63
Bihar	16.23	23.49	30.91	0.81	12.53	93.54
Chhattisgarh	9.93	14.93	33.49	2.83	7.96	64.45
Goa	0.13	0.15	13.33	0.05	0.08	37.50
Gujarat	10.61	22.32	52.46	14.48	11.9	−21.68
Haryana	8.75	9.95	12.06	6.57	5.31	−23.73
Himachal Pradesh	2.3	4.60	50.00	1.98	2.45	19.18
J & K	2.53	6.79	62.74	0.64	3.62	82.32
Jharkhand	4.10	13.59	69.83	0.88	7.25	87.86
Karnataka	14.59	20.66	29.38	3.55	11.02	67.79
Kerala	0.71	2.91	75.60	0.39	1.55	74.84
Madhya Pradesh	24.3	37.41	35.04	11.65	19.95	41.60
Maharashtra	22.21	33.68	34.06	25.12	17.96	−39.87
Manipur	0.36	0.72	50.00	0.00	0.38	100.00
Meghalaya	0.31	1.17	73.50	0.4	0.62	35.48
Mizoram	0.15	0.06	−150.00	0.5	0.03	−1566.67

(continued)

Table 1 (continued)

States/UTs	Availability	Requirement	Shortfall (%)	Availability	Requirement	Shortfall (%)
	Crop residues (million tons)			Greens (million tons)		
Nagaland	0.56	0.74	24.32	0.3	0.4	25.00
Orissa	12.25	22.27	44.99	2.46	11.88	79.29
Punjab	**13.71**	**10.58**	**−29.58**	**7.38**	**5.64**	**−30.85**
Rajasthan	21.67	33.53	35.37	33.53	17.88	−87.53
Sikkim	0.23	0.25	8.00	0.01	0.13	92.31
Tamil Nadu	7.01	16.46	57.41	3.7	8.78	57.86
Tripura	0.53	1.09	51.38	0.19	0.58	67.24
Uttar Pradesh	42.07	57.19	26.44	15.73	30.5	48.43
Uttarakhand	2.05	4.90	58.16	1.73	2.61	33.72
West Bengal	13.77	30.30	54.55	0.51	16.16	96.84
A & N Islands	0.02	0.11	81.82	0.00	0.06	100.00
Chandigarh	0.00	0.04	100.00	0.00	0.02	100.00
Dadra & Nagar H.	0.04	0.80	95.00	0.20	0.40	50.00
Daman Diu	0.01	0.10	90.00	0.00	0.00	–
Delhi	0.09	0.43	79.07	0.10	0.23	56.52
Lakshadweep	0.00	0.10	100.00	0.00	0.00	–
Pondicherry	0.06	0.11	45.45	0.01	0.06	83.33
India	**253.26**	**415.83**	**39.10**	**142.82**	**221.63**	**35.56**

The bold figures indicate the Punjab state that is the subject of the paper and All India which is sum total of all states.

Source Lok Sabha Unstarred Question No. 726, dated on 24.11.2009

required than after burning (Singh et al. 2008). The options for crop residue management may include developing systems to plant residue into bailing and removal for use as animal feed or for industry. Enhanced decomposition of machine-harvested straw to improve nutrients in the soil can be useful. The use of microbial sprays that can speed decomposition of residue is also an option. The option of planting into residue needs further investigation of inorganic nitrogen and its adverse effect due to nitrogen deficiency.

Rice straw is a major field-based residue that is produced in large amounts in Asia. In fact the total amount of 668 tons could produce theoretically 187 gallons of bioethanol if the technology were available (Kim and Dale 2004). However, an increasing proportion of this rice straw currently undergoes field burning. This waste of energy seems inapt, given the high fuel prices and the great demand for reducing greenhouse gas emissions as well as air pollution. As climate change is extensively recognized as a threat to development, there is growing interest in alternative uses of field-based residues for energy applications.

There are primarily two types of residues from rice cultivation that have potential in terms of energy—straw and husk. Although the technology of using rice husk is well established in many Asian countries, rice straw is rarely used as a source of renewable energy. One of the principal reasons for the preferred use of husk is its easy procurement, i.e., it is available at the rice mills. In the case of rice straw, however, its collection is a tedious task and its availability is limited to harvest time. The logistics of collection could be improved through baling but the necessary equipment is expensive and its purchase is uneconomical for most rice farmers. Thus, technologies that enable straw to be used for energy generation must be efficient to compensate for the high costs involved in straw collection.

3 Agricultural Residues for Power Generation

Punjab produces around 23 million tons of rice straw and 17 million tons of wheat straw annually (Agrawal et al. 2006). More than 80 % of paddy straw (18.4 million tons) and almost 50 % wheat straw (8.5 million tons) produced in the state are burnt in fields annually. Almost all of paddy straw, except Basmati paddy, is burnt in the field to enable early sowing of next crop. Recently, the farmers have extended this practice to wheat crop also. Although part of the wheat straw is used as dry fodder for cattle, the remaining straw is usually burnt for quick disposal.

Biomass, such as agricultural residue, bagasse, cotton stalks, rice husk, etc., is emerging as a viable source of power. Direct burning of such waste is inefficient and leads to pollution. When combusted in a gasifier at low oxygen and high temperature, biomass can be converted into a gaseous fuel known as producer gas. This gas has a lower calorific value compared to natural gas or liquefied petroleum gas, but can be burnt with high efficiency and without emitting smoke.

The advantages of utilizing crop residue over and above conventional resources are that the residue is renewable, readily available, and can be used successfully by burning in boilers with an efficiency of 99 %. Further, they are available at low cost relative to coal and their ash contents are also much less (as compared to 36 % ash content of coal).

The calorific values of both coal and paddy straw are comparable, i.e., 4200 and 3590 kcal/kg, respectively. Additional income to the farmers from the sale of straw is an added advantage. At the same time, state agencies involved could also take advantage of carbon credit policy setup under the UNFCCC (United Nation Framework Convention on Climate Change). The policy involves the allocation of credits to programs which help in curbing global warming. The Indian government should encourage private parties/agencies to take advantage of this carbon credit policy of UNFCCC.

Biomass-based power plants of 10–20 MW capacity setup in a group of villages or every *Taluka* can meet energy needs of villagers and employ thousands of people. Kirangatevalu village in Karnataka is a prime example in this regard. The initial electrification of Kirangatevalu meant supply of power to a few homes and farms for 4–5 h a day. The transformation of the village is the result of an initiative taken by a private firm that has set up a power plant using agricultural waste such as sugarcane refuse and coconut fronds that are plentiful in the area. Villagers sell their agricultural waste to the plant and get access to reliable power at commercial rate. A supply chain to procure agricultural waste from villages in a radius of 10 km has been established to ensure the supply of agricultural waste throughout the year. The waste that was burnt in open fields has now become a source of energy and also income for farmers. The 4.5 MW power plant set up by Malaballi Power Plant Private Limited supplies electricity to 48 villages inhabited by 120,000 people in the Mandya district in Karnataka.

In Punjab in the 1980s Punjab State Electricity Board (PSEB) had set up a 10 MW power plant based on rice straw at village Jalkheri, district Fatehgarh Sahib in which 250–3000 tons per day (TPD) of fuel is burnt in a boiler furnace of steam generation capacity of 50 tons per hour (TPH). The plant earlier used rice straw but due to clinkerisation of boiler, rice straw was replaced with rice husk, cow dung, and other agro wastes. This plant has since been leased out by PSEB to M/S Jalkheri Power Private Limited. Now these plants will be using improved technology and M/S Punjab Biomass Power Limited has signed two agreements with PSEB for setting up 12 MW rice straw-based power plants at Baghaura in Rajpura Tehsil and Sawai Singh village in Patiala Tehsil. A total amount of 0.1 million tones of paddy straw would be collected from a command area of 25 km^2 around each unit and a barter system of providing electricity will be worked out with the farmers. The units will be run on BOO (Build Own Operate)[1] basis. Detailed project reports (DPR) have been prepared and land is being purchased.

4 Energy Technologies

The transportation of biomass is one of the key cost factors for its use as a source of renewable energy. Decentralized energy provides an opportunity to use biomass to meet local energy requirements in terms of electricity. In contrast to straw, the use of rice husk for energy has been realized faster. Rice mills can use husk to serve their internal energy requirements. As an alternative, rice millers could sell the husk to a power plant operator. The propagation of rice husk use for energy was accelerated by energy providers, who deal with a relatively small number of rice millers for supplying husk, which is an easier task than dealing with thousands of farmers supplying rice straw. Millers can also produce electricity themselves and then sell to a power grid. This setup appears to be the most promising option in terms of logistics and transportation for energy generation.

Transportation costs of straw are a major constraint to its use as an energy source. As a rule of thumb, transportation distances beyond a 25–50 km radius (depending on local infrastructure) are uneconomical. For longer distances, straw

[1]In a BOO project ownership of the project remains usually with the Project Company. Therefore the private company gets the benefits of any residual value of the project. This framework is used when the physical life of the project coincides with the concession period. A BOO scheme involves large amounts of finance and long payback period.

could be compressed as bales or briquettes in the field, rendering transport to the site of use a viable option. Nevertheless, the logistics of a supply chain is more complicated in the case of straw. Although five different energy conversion technologies seem to be applicable for rice straw in principle, only combustion technology is currently commercialized and the other technologies are at different stages of development. As a general rule for energy use, each step in the chain consumes a certain amount of energy and thus reduces the net energy supplied at the final stage.

5 Thermal Combustion

Rice straw can either be used alone or mixed with other biomass materials (the latter is called co-firing or co-combustion) in direct combustion. In this technology, combustion boilers are used in combination with steam turbines to produce electricity and heat. In thermal combustion, air is injected into the combustion chamber to ensure that the biomass is completely burned in the combustion chamber. Fluidized bed technology is one of the direct combustion techniques in which solid fuel is burned in suspension by forced air supplied into the combustion chamber to achieve complete combustion. A proper air-to-fuel ratio is maintained and, in the absence of a sufficient air supply, boiler operation encounters various problems.

In straw combustion at high temperatures, potassium is transformed and combines with other alkali earth materials such as calcium. This in turn reacts with silicates, leading to the formation of tightly sintered structures on the grates and at the furnace wall. Alkali earths are also important in the formation of slags and deposits. This means that fuels with lower alkali content are less problematic when fired in a boiler (Jenkins et al. 1998). The by-products are fly ash and bottom ash, which have an economic value and could be used in cement and/or brick manufacturing, construction of roads and embankments, etc.

The National Biomass Assessment Project of Ministry of New and Renewable Energy, Government of India, conducted a biomass study in which 29 tehsil were surveyed from 1995 to 2001. A total of 36 Talukas were included from different districts. The power generating potential was estimated at 342 MW (see details in Table 2). Following these developments, PEDA facilitated setting up of a 6 MW biomass power project at village Gulabewala in district Muktsar. This was undertaken as a joint project with M/s Malwa Power Pvt. Ltd., at a cost of Rs. 21.50 crore. The project was completed and commissioned in May 2005 and is operating satisfactorily.

Encouraged by the success of Gullabewala, PEDA is collaborating with the private sector for the establishment of power projects based on agricultural residues. A total of 20 projects with a total installed capacity of 320 MW have been allotted on BOO basis to private developers. These projects are being set up by the private developers with the state-of-the-art technologies such as biomethanation, combustion, etc. The plants are designed to receive mixed waste such as paddy straw, cotton stalks, and other agro residues available in the state. Two projects of 8 and 14.5 MW

Table 2 Availability of surplus biomass in Punjab—National Biomass Assessment Project Studies

Taluka	District	Availability of major surplus biomass (Tons)	Current utilization (Tons)	Estimated power generating potential (MW)
Nakodar	Jalandhar	733027.3	539293.6	12
Batala	Gurdaspur	228898.03	1,533,899	10
Mansa	Mansa	215557.2	246604.4	20
Muktsar	Muktsar	171545.2	218264.1	10
Ferozepur	Ferozepur	2,533,630	562,931	5
Malerkotla	Sangrur	4,423,495	17,242,469	20
Patiala	Patiala	7,500,000	816,000	10
Ludhiana	Ludhiana	1,071,000	870,000	15
Derabassi	Patiala	520,000	592,000	8
Nawanshahar	Nawanshahar	1,038,553	4,411,210	10
Fazilka	Ferozepur	445,160	534317.11	
Ajnala	Amritsar	143,155	342939.2	10
Barnala	Sangrur	622285.41	853061.82	10
Patti	Amritsar	359,295	459,267	10
Anandpur Sahib	Roopnagar	72,349	385,641	5
Nihal Singh Wala	Moga	70,377	1,559,985	3–5
Samrala	Ludhiana	133,079	577,012	5
Raikot	Ludhiana	174,137	771,348	7–14
Talwandi Sabo	Bathinda	75,077	784,927	6
Garhshankar	Hoshiarpur	166,553	877,917	10
Jaito	Faridkot	249,202	297,680	13
Abohar	Ferozepur	146066.10	288351.05	5
Sultanpur Lodhi	Kapurthala	177886.34	316778.25	9
Tarn Taran	Amritsar	293,000	298,000	12
Jagraon	Ludhiana	160064.6	302057.1	10
Sunam	Sangrur	1539707.62	–	31
Fatehgarh Sahib	Fatehgarh Sahib	290,700	198,700	11
Amloh	Fatehgarh Sahib	157,400	107,400	6
Bathinda	Bathinda	1458328.8	–	13.5
Phagwara	Kapurthala	675,809	566,766	5
Shahkot	Jalandhar	659,564	443,251	10
Gurdaspur	Gurdaspur	365,264	317,815	1–2
Faridkot	Faridkot	2,281,680	2,013,887	7–9
Kharar	Ropar	547,803	478,597	1.5–2
Hoshiarpur	Hoshiarpur	374,180	327,260	2.25–4.5
Phillaur	Ludhiana			10
Aggregate				342

Source National Biomass Assessment Programme, Ministry of New and Renewable Energy, Government of India were commissioned in 2009. Land was already allocated to the developers for another five plants and work commencement started during March–April 2009.

According to information provided by PEDA, biomass power projects are planned to be allocated in the following phases:

- Phase I—agreements reached with two companies—M/s Turbo Atom TPS and M/s Green Field Energen Pvt. Ltd., in New Delhi and Chandigarh, respectively for two tehsils, Ferozepur and Patti with a total capacity of 56 MW (for details see Annexure 1).
- Phase II—three companies were there for Abohar, Sunam, and Ajnala, respectively. The two companies of Sunam and Ajnala are canceled having 41 MW. With the capacity of 8 MW the company M/s Dee Development in Abohar Tehsil is commissioned (see Annexure 1 for details).
- Phase III—six companies with different feed stocks. The M/s Green Plant of Chandigarh is based on paddy straw which is planned in 14 tehsils with total 146.5 MW of capacity, out of which Garhshankar with 10 MW capacity is likely to begin. M/s Universal Biomass of Muktsar which is mostly based on cotton stock with 14.5 MW in Malout Tehsil is commissioned. The Malwa Power Ltd., in the village Gulabewala in the district of Muktsar, was started before PEDA took over with 6 MW. Other three companies have total capacity of 65 MW (for detail, see Annexure 1).

PEDA has so far allocated 30 sites/tehsils for setting up plants with a total generation capacity of 337 MW. Annexure 1 gives detailed status of biomass-based power projects.

6 Conclusion

The use of agriculture waste for generating power is an innovative idea as biomass is a carbon neutral fuel, which helps in reducing the greenhouse effect and global warming. It can also be source of valuable carbon credits. Each project will benefit around 10–15 thousand farmers in their command areas. According to one estimate, farmers can earn around Rs. 1500–2000 per acre for their stubble waste that is otherwise wasted at the moment. The farmer will have access to free power in exchange of rice straw. Locally generated and distributed power will also reduce wasteful transmission and distribution losses. Moreover, controlled combustion in a modern boiler with electrostatic precipitators will be far less polluting than the open burning of straw in the field as is being practiced at present. The ash generated during the combustion can also be used as fertilizer in the field.

Since the PSEB is already planning to install additional power generating units based on paddy straw, steps need to be taken to ensure the availability of paddy straw in the respective areas. Consideration should be given to providing incentives to the private sector to set up such units all over the state. Appropriate balers need to be developed to collect the paddy straw during the harvesting season to reduce the cost of collection and transportation. Incentives could also be given to the farmers for storing paddy straw for its regular supply to the power plants throughout the year to fulfill need of the plant in the off-season as well.

Annexure 1

Biomass power projects commissioned in the state by PEDA: (62.5 MW)

Sl. No	Name of the company	Site	Capacity (MW)	Month of commissioning	Remarks
01	M/s Malwa Power Ltd.	Vill. Gulabewala, Distt. Muktsar	6	May, 2005	First project allocated by PEDA
02	M/s Dee Development Engineers Pvt. Ltd	Vill. GaddaDhob, Tehsil. Abohar Distt Ferozepur	8	Feb., 2009	Project was allocated under Phase-II
03	M/s Universal Biomass Energy Pvt. Ltd.	Vill. ChannuTeh. Malout, Distt. Sri Muktsar Sahib	14.5	Oct., 2009	Project was allocated under Phase-III
04	M/s. Punjab Biomass Power Pvt. Ltd.	Distt. Patiala	12	June, 2010	Project allocated by PSPCL
05	M/s. Green Planet Energy Pvt. Ltd.	Binjon, Distt. Hoshiarpur	6	March, 2012	Project was allocated under Phase-III
06	M/s. Green Planet Energy Pvt. Ltd.	Bir Pind, Distt. Jalandhar	6	Feb., 2013	Project was allocated under Phase-III
07	M/s Viaton Energy Pvt. Ltd.	Khokhar Khurd Distt. Mansa	10	July, 2013	Project was allocated under Phase-III
		Total	62.5		

Source http://peda.gov.in/eng/Bio-mass%20Power.html; accessed on February 2, 2016

Detail of biomass projects under implementation

Sl. No.	Name of the company	Name of site	Capacity (MW)	Project status
1	M/s Green Planet Energy Pvt. Ltd.	Vill. Binjon, Tehsil Garhshankar, Distt. Hoshiarpur	4	Under implementation
2		Vill. BirPind, Tehsil Nakodar, Distt. Jalandhar	4	Under implementation
3		Vill. Manuke Gill, Tehsil Nihal Singh Wala, Distt. Moga	6	Under implementation

Source http://peda.gov.in/eng/Bio-mass%20Power.html; accessed on February 2, 2016

References

Agrawal, S., Trivedi, C., & Sengupta, R. (2006). Air pollution due to burning of residues. In *Proceeding of the Workshop on Air Pollution Problems Due to Burning of Agricultural Residues, held at PAU, Ludhiana organized by the Indian Association for Air Pollution Control in collaboration with the Punjab State Pollution Control Board Patiala and the Central Pollution Control Board*, New Delhi.

Hegde, N. G. (2010). Forage resource development in India. In *Souvenir of IGFRI Foundation Day*, November, 2010. www.baif.org.in

Jenkins, B. M., Baxter, L. L., Miles, T. R, Jr, & Miles, T. R. (1998). Combustion properties of biomass Fuel process. *Technol, 54*, 17–46.

Kim, S., & Dale, B. E. (2004). Global potential bioethanol production from wasted crops and crop residues. *Biomass and Bioenergy, 26*, 361–375.

Punjab Energy Development Agency, Government of Punjab. www.peda.gov.in

Singh, R. P., Dhaliwal, H. S., Sidhu, H. S., Manpreet-Singh, Y. S., & Blackwell, J. (2008). Economic assessment of the Happy Seeder for rice–wheat systems in Punjab, India. In *Conference Paper, AARES 52nd Annual Conference*, Canberra, Australia: ACT.

Thakur, T. C. (2003). *Crop residue as animal feed: Addressing resource conservation issues in rice–wheat systems of south Asia, a resource book*. Rice Wheat Consortium for Indo-Gangetic Plains (CIMMYT), March, 2003.

Adaptation and Mitigation Strategies for Sustainable Dairy Production Under Changing Climate Scenario

Mukund A. Kataktalware, S. Nazar, G. Letha Devi and K.P. Ramesha

Abstract In India, dairy sector plays vital role in the national economy and in the socio-economic development of the country by providing gainful employment and income generating avenues for small, marginal farmers and landless labourers. Apart from the existing challenges like increased occurrence of emerging and re-emerging animal diseases, vulnerability to exotic diseases, perennial shortage of feed and fodder resources and an urgent need to increase production as well as productivity to meet ever increasing demand for animal products, etc. now livestock sector is confronting a very serious challenge of 'climate change'. Changes in climate which encompasses decrease in minimum temperature, increase in maximum temperature, erratic rainfall, higher incidence of extreme weather events, etc. may negatively affect water availability, pasture and fodder crop quality and quantity, animal immune response, epidemiological pattern of vector-borne diseases, productive and reproductive efficiency, which in long term would affect the livelihood and food security of millions of farmers' dependent of livestock. Adaptation strategies like selection of animals for thermal tolerance, alteration in herd composition, microclimate modification, access to cool clean drinking water, changing feeding frequency and time of feeding and changing ingredients, e.g. addition of dietary fat to increase energy density, etc. could be adopted to maintain dry matter intake during stressful environmental conditions. The important strategies for mitigation of greenhouse gas (GHG) emissions from dairy animals include diet manipulation, direct inhibitors, feed additives, propionate enhancers, methane oxidisers, probiotics, defaunation, hormones, vaccination for reduced activity of rumen protozoa, etc. Improving waste management by adopting waste-to-energy concepts like biogas production, electricity generation, etc. would reduce GHGs such as CH_4, N_2O significantly. The existing adaptation and mitigation strategies have the potential to significantly reduce the likely impact of climate change on dairy production system. Therefore, it is high time that concerted efforts are made to

M.A. Kataktalware (✉) · S. Nazar · K.P. Ramesha
SRS, ICAR-NDRI, Adugodi, Bengaluru 560 030 India
e-mail: mtalware@gmail.com

G. Letha Devi
ICAR-NIANP, Adugodi, Bengaluru 560 030 India

S. Nautiyal et al. (eds.), *Climate Change Challenge (3C) and Social-Economic-Ecological Interface-Building*, Environmental Science,
DOI 10.1007/978-3-319-31014-5_31

enhance the resilience of the farms and animals through dissemination of adaptation and mitigation strategies in a sustained manner so that, the production and productivity levels of dairy animals are not only maintained but also improved even in challenging environmental conditions.

1 Introduction

Since time immemorial agriculture and livestock production are inherently linked, interdependent and critical for overall food security of the country. Animal husbandry is an integral part of India's agricultural economy and performs diversified role in providing livelihood security to the rural population. As per the 19th Livestock Census, 2012, India has the largest livestock population (512.05 million) in the world. According to estimates of the Central Statistics Office (CSO), the value of output from livestock sector at current prices was about Rs. 537,535 crore during 2012–2013 which is about 25.63 % of the value of output from total agricultural, fishing and forestry sector at current price and 26.02 % at constant prices (2004–2005). Livestock sector also contributes significantly in gainful employment, asset creation, coping mechanism against crop failure and social and financial security. It is estimated that about 75 million rural households own one or the other species of livestock. Several measures initiated by the Government resulted in increasing the milk production significantly from the level of 17.0 million tonnes in 1950–1951 to 137.7 million tonnes in 2013–2014 and thus making India the largest producer in the world. In India, dairy sector plays vital role in the national economy and in the socio-economic development of the country by providing gainful employment and income generating avenues for small, marginal farmers and landless labourers. Apart from the existing challenges like increased occurrence of emerging and re-emerging animal diseases, vulnerability to exotic diseases, perennial shortage of feed and fodder resources and an urgent need to increase production as well as productivity to meet ever increasing demand for animal products, etc. now livestock sector is confronting a very serious challenge of 'climate change'. The probability of extreme high temperature events (heat waves) would increase as mean temperature increases, and an increasing number of heat waves could significantly increase the negative impact of global warming, especially on livestock (Klinedinst et al. 1993). The climate variability which encompasses decrease in minimum temperature, increase in maximum temperature, erratic rainfall, higher incidence of extreme weather events, etc. are likely to negatively affect water availability, pasture and fodder crop quantity and quality, animal immune response, epidemiological pattern of vector-borne diseases, productive and reproductive efficiency, which in long term would affect the livelihood and food security of millions of farmers' dependent on livestock. Even though the change in climate is a global phenomenon, its adverse effects are more severely experienced by poor people in developing countries who rely heavily on the natural resource base for their livelihoods (IFAD 2009).

2 Climate Change

Climate can be described as long-term average weather that is, in part, determined by the greenhouse effect (IPCC 2007). Climate change refers to the statistical variations in the properties of the climate system such as changes in global temperatures, precipitation, etc. due to natural or human drivers over a long period. The greenhouse effect refers to the process where long-wave radiation released from the Earth's surface because of the absorption of solar radiation is partially trapped by greenhouse gas (GHG), thereby warming the lower atmosphere. The greenhouse effect allows the average surface temperature of the earth to be 14 °C, whereas, without the greenhouse effect, the global mean surface temperature would be −19 ° C (IPCC 2007). Change in climate is a natural phenomenon, however, after the industrial revolution the rate of change has accelerated, and in last couple of decades it is found to be alarming. The Summary for Policymakers submitted in 2013 by the IPCC, it is concluded that it is now more certain than ever before that human-caused climate change is real, and GHG emissions are causing changes to the planet that could possibly trigger dangerous consequences by the turn of the century (The Hindu 2013). The increased emissions of anthropogenic (human-caused) GHG have led to atmospheric concentrations of carbon dioxide, methane and nitrous oxide that are unprecedented in at least the last 800,000 years (AR5, IPCC 2014). Climate change could severely change the distribution and quality of natural resources which may adversely affect the livelihood security of millions of people worldwide.

3 Heat Stress

Animals are adaptable and can maintain productive performance in a relatively broad range of environments. Environmental stressors beyond threshold limits for coping and compensatory mechanisms could compromise health, growth and productive and reproductive performance of an animal. Intensive selection of dairy animals for increased production is making these animals susceptible to hot and humid environments. Selection for milk yield reduces thermoregulatory ability in the face of heat stress (Berman et al. 1985) and magnifies the seasonal depression in fertility caused by heat stress (Al-Katanani et al. 1999). Heat stress can be defined as the sum of forces external to a homeothermic animal that acts to displace body temperature from the resting state (Yousef 1984). Heat stress occurs when any combination of environmental factors (temperature, relative humidity, solar

Table 1 Thermo-neutral zone for dairy animals

Animal/breed	TNZ	UCT	Source
Dairy cows	24 °C	27 °C	Johnson (1965)
Dairy cattle	4–25 °C		McDowell (1972)
Lactating Holstein cows	4–24 °C		Hahn (1981)
Dairy cows		25 to 26 °C	Berman et al. (1985)
Lactating dairy cows	−0.5 to 20 °C		Johnson (1987)
Zebu/Indigenous cows	32–34 °C		Thomas and Sastry (1991)
Indian dairy cows	27 °C		Dutt et al. (1992)
Temperate-region adult cattle	5–15 °C		Hahn (1999)

radiation, air movement, and precipitation) cause the effective temperature of the environment to be higher than the animal's thermo-neutral zone (TNZ). Several researchers have reported different ranges of the TNZ and upper critical air temperature (UCT) for dairy animals which is summarized in Table 1.

4 Impact of Climate Change

Direct effects of environmental stress on livestock include temperature-related sickness and death, and the morbidity of animals. Indirectly it influences emergence of pathogens, distribution of vector-borne diseases, host immunity, scarcity of feed and water, etc. which affects animals' health. The acclimation of the animals to meet the environmental stressors leads to reduced dry matter intake (DMI) and adjustments in many physiological functions which are associated with impaired health ultimately compromising productive and reproductive efficiency. If exposure to high air temperature is extended, lower feed intake is followed by a decline in the secretion of calorigenic hormones (growth hormone, catecholamines and glucocorticoids in particular) in thermogenic processes of digestion and metabolism, and metabolic rate (Johnson 1980) which could lead to impaired metabolism of lipids, protein and mineral including liver function. A higher incidence of mastitis in dairy cows during periods of hot weather has also been reported by several researchers worldwide. Heat stress compromises oocyte growth in cows by altering progesterone, the secretion of luteinizing hormone and follicle-stimulating hormone and dynamics during the oestrus cycle (Ronchi et al. 2001). De Rensis and Scaramuzzi (2003) also reported that in cattle, the heat stress has negative effect on oestrus expression due to a reduced estradiol secretion from the dominant follicle developed in a low luteinizing hormone environment, embryo development and survival (Fig. 1). Roy and Prakash (2007) reported that prolactin and progesterone profiles during the summer and winter months are directly correlated with the reproductive performance of buffaloes, and that hyperprolactinaemia may cause acyclicity/infertility in buffaloes during the summer months due to severe heat stress. Semen concentration, number of

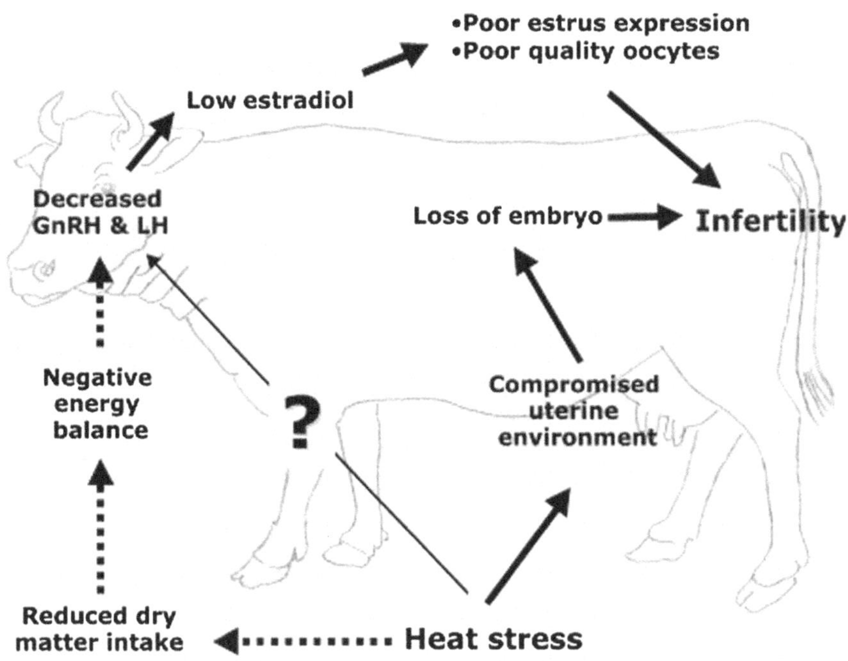

Fig. 1 Effect of environmental stress of reproduction performance of dairy cows. *Source* De Rensis and Scaramuzzi (2003)

spermatozoa and motile cells per ejaculate of bulls are lower in summer than in winter and spring (Mathevon et al. 1998). Nichi et al. (2006) reported higher percentage of major sperm defects during summer than the winter in Simmental and Nellore bulls. The increase in milk yield increases sensitivity of cattle to thermal stress and reduces the "threshold temperature" at which milk losses occur (Berman 2005) indicating that high yielding animals are more vulnerable to environmental stress.

Indian perspective: The negative impact of direct temperature rise on total milk production of cattle and buffaloes for India has been estimated about 1.6 MT in 2020 and more than 15 MT in 2050 (Upadhyay 2014). Climate change has a substantial effect on water availability which could cripple the dairy farming like never before. Sirohi and Sirohi (2010) attempted to quantify the potential losses that could accrue to the milk producers due to climate change induced decline in productivity of dairy animals (Table 2). The States with higher coastal areas are more vulnerable to milk production losses due to high relative humidity for longer duration. The enormity of the cost to the milk producers could be even greater after accounting for the losses accruing from incidence of diseases, extreme events, etc.; the high magnitude of direct production losses has implications for the potential investments in climate change adaptation strategies for the livelihood security of millions of milk producers in the country (Sirohi and Sirohi 2010).

Table 2 Annual loss in milk production in 2030 from elevated heat stress (Sirohi and Sirohi 2010)

States	Mean annual THI increase = 1 unit				Mean annual THI increase = 2 unit			
	Crossbred cows	Local cows	Buffaloes	Total annual loss (million liters)	Crossbred cows	Local cows	Buffaloes	Total annual loss (million liters)
	(liters/animal/year)				(liters/animal/year)			
Andhra Pradesh	697.0	15.0	52.9	433.0	856.5	18.4	65.0	532.0
Assam	640.0	20.8	73.5	109.9	786.6	25.6	90.3	135.1
Bihar	716.3	27.6	97.3	514.6	880.2	33.9	119.6	632.4
Chhattisgarh	567.7	10.0	35.4	38.4	697.6	12.3	43.5	47.2
Gujarat	740.2	26.7	94.4	467.5	909.6	32.8	116.0	574.4
Haryana	665.0	36.0	127.3	339.9	817.2	44.3	156.4	410.3
Jharkhand	745.5	25.6	90.3	95.6	916.0	31.4	110.9	117.4
Karnataka	319.0	1.3	4.5	169.6	391.9	1.6	5.6	208.4
Kerala	838.2	10.2	36.0	499.2	1030.0	12.5	44.3	613.5
Madhya Pradesh	618.9	16.2	57.1	283.5	760.5	19.9	70.2	348.3
Maharashtra	532.4	8.3	29.3	421.1	654.2	10.2	36.0	517.4
Orissa	817.1	27.3	96.5	271.8	1004.1	33.6	118.6	334.0
Punjab	610.0	18.2	64.1	488.3	749.6	22.3	78.8	600.0
Rajasthan	659.6	39.6	139.7	808.3	810.6	48.6	171.7	993.3
Tamil Nadu	888.4	20.4	72.1	1475.3	1091.6	25.1	88.6	1812.8
Uttar Pradesh	616.5	20.8	73.6	979.5	757.5	25.6	90.4	1203.6
Uttarakhand	224.3	4.1	14.4	19.3	275.6	5.0	17.7	23.7
West Bengal	828.7	35.6	125.7	443.3	1018.3	43.7	154.5	544.8
India				7852.0				9648.6

5 Adaptation to Climate Change

Adapting appropriate strategies would ensure sustainability of livestock production system under changing climate scenario. An advanced planning of production management systems is required, with an understanding of animal responses to thermal stress and ability to provide management options to prevent or mitigate adverse consequences (Nienaber and Hahn 2007). Alleviation of environmental stress in livestock warrants a multi-disciplinary approach. The environmental stress could be ameliorated by modifying the microclimate, nutritional manipulation and selection of animals for thermal tolerance. Another alternative could be to change the species, e.g. small ruminants like goat instead of cattle or buffalo. Provision of an animal shelter is one of the most easily implemented and economical strategies to minimize the impact of environmental stress. The microclimate modification strategies that would facilitate improved heat exchange between an animal and its environment are use of shade, misters, foggers or pad cooling; and improving the ability of the animal to dissipate body heat by increasing sensible heat or increasing evaporative heat loss, e.g. using sprinklers to wet animals. However, sensible and evaporative heat losses are mutually dependent, which makes using one without the other limits its utility. The other livestock management strategies could include reduction of livestock numbers—a lower number of more productive animals leads to more efficient production and lower GHG emissions from livestock production (Batima 2007); changes in livestock/herd composition (selection of large animals rather than small); improved management of water resources through the introduction of simple techniques for localized irrigation (e.g. drip and sprinkler irrigation), accompanied by infrastructure to harvest and store rainwater, such as tanks connected to the roofs of houses and small surface and underground dams (IFAD 2009). Round the clock access to cool clean drinking water to meet potential peak demands is critical for reducing evaporative heat loss and maintaining DMI by the lactating animals. Nutritional manipulations like changing feeding frequency and time of feeding, and changing ingredients, e.g. addition of dietary fat to increase energy density, or additional roughage added to ration to reduce heat increment could be adopted to maintain DMI during environmental stress conditions. Breeding goals should appropriately be reset taking into account abiotic stress factors, poor quality feed and fodders and incidence of various infectious diseases. Species and breeds which are evolved in similar geo-climatic conditions and well adapted to such conditions should be chosen over their exotic counterparts. The long-term sustainability of production system should be the goal rather than short-term gains from high yielding exotic breeds alien to local climatic conditions. The milk producers should be made aware about the modern agronomic practices for the production and conservation of fodder which would improve the supply of animal feed reducing undernourishment and mortality in herds.

6 Mitigation of GHG Emissions

The major sources of GHG emissions in the livestock production system include the enteric fermentation of animals, manure disposal and production of feed and forage. The feed ingredients influence enteric fermentation and methane emission from the rumen. There are a number of nutritional technologies for improvement in rumen efficiency like, diet manipulation, direct inhibitors, feed additives, propionate enhancers, methane oxidisers, probiotics, defaunation and hormones (Moss 1994). The methane mitigation from molasses urea supplementation was 8.7 % (Srivastava and Garg 2002) and 21 % from use of feed additive monensin (De and Singh 2001). Methane emission from animals may be less on feeding leguminous forages than feeding of grasses (Malik et al. 2013). Malik and Singhal (2008) documented a significant reduction in methane production on the inclusion of limited quantity of lucerne fodder (30 %) to the wheat straw-based ration. Plant secondary metabolites such as tannin, saponin, essential oil, etc. showed promising results under in vitro conditions and need to be evaluated further for their action in animal systems by looking at the nutrient availability and productivity of animals (Malik et al. 2013). The most noteworthy achievement to reduce CH_4 production in livestock farms is the development of a vaccine containing an antigen derived from methanogenic bacteria and an immunogenic preparation which reduces the activity of rumen protozoa (Varma 2013). Improving animal waste management like covered storage facilities, biogas production, electricity generation, etc. would reduce GHGs like CH_4, N_2O significantly.

7 National Innovations on Climate Resilient Agriculture (NICRA)

The Prime Minister's National Action Plan on Climate Change has identified agriculture as one of the eight National Missions. Therefore, the Indian Council of Agricultural Research (ICAR) has launched the NICRA project during 2010–2011 with the objectives to enhance the resilience of Indian agriculture covering crops, livestock and fisheries to climatic variability and climate change through development and application of improved production and risk management technologies, to demonstrate site specific technology demonstration packages on farmers' field for adapting current climate risks and to enhance the capacity of scientists and other stakeholders in climate resilient agriculture research and its application. The ICAR-National Dairy Research Institute (NDRI), Karnal is the core institute dealing with the strategic research on adaptation strategies in livestock (cattle and buffalo) to thermal stress through nutritional and environmental manipulations. Under this project, major cattle and buffalo breeds are being subjected to detailed investigation with respect to their unique climate adaptability traits which would help to prepare a broad road map based on scientific evidence for making Indian dairying climate resilient.

8 Conclusion

In India, dairy sector plays an important role in the national economy and in the socio-economic development of the country. Changes in climate are likely to impact farm animals by means of compromised performance, health status and immune response which would not only threaten the sustainability of dairy production system and livelihood of millions of milk producers but also food and nutritional security of the country. The existing adaptation and mitigation strategies have the potential to significantly reduce likely impact of climate change on dairy production system. Therefore, it is high time that concerted efforts are made to enhance the resilience of the farms and animals through dissemination of adaptation and mitigation strategies in a sustained manner so that, the production and productivity levels of dairy animals are not only maintained but also improved even in challenging environmental conditions.

References

Al-Katanani, Y. M., Webb, D. W., & Hansen, P. J. (1999). Factors affecting seasonal variation in 90-d nonreturn rate to first service in lactating Holstein cows in a hot climate. *Journal of Dairy Science, 82*, 2611–2616.

Batima, P. (2007). Climate change vulnerability and adaptation in the livestock sector of Mongolia. *Assessments of impacts and adaptations to climate change, International START Secretariat*. Washington DC, US.

Berman, A. (2005). Estimates of heat stress relief needs for Holstein dairy cows. *Journal of Animal Science, 83*, 1377–1384.

Berman, A., Folman, Y., Kaim, M., Marnen, M., Herz, Z., Wolfensen, D., et al. (1985). Upper critical temperature and forced ventilation effects for high-yielding dairy cows in a subtropical climate. *Journal of Dairy Science, 68*, 1488–1495.

De Rensis, F., & Scaramuzzi, R. J. (2003). Heat stress and seasonal effects on reproduction in the dairy cow—a review. *Theriogenology,* 60, 1139–1151.

De, D., & Singh, G. P. (2001). Monensin enriches UMMP supplementation on in vitro methane production in crossbred calves. In: *Proceedings of the Tenth Animal Nutritional Conference (Abstract papers),* (p. 161). Animal Nutrition Society of India, ICAR-NDRI, Karnal, India.

Dutt, T., Taneja, V. K., Singh, Avtar, & Singh, A. (1992). Comfort zone for maximal milk production in crossbred cattle. *Indian Journal of Dairy Science, 45*(3), 119–122.

Hahn, G. L. (1981). Housing and management to reduce climatic impacts on livestock. *Journal of Animal Science, 52*, 175–186.

Hahn, G. L. (1999). Dynamic responses of cattle to thermal heat loads. *Journal of Animal Science, 77*(Suppl.2), 10–12.

IFAD (International Fund for Agricultural Development) (2009). Livestock and climate change. Available in: https://www.ifad.org/documents/10180/48b0cd7b-f70d-4f55-b0c0-5a19fa3e5f38.

IPCC (Intergovernmental Panel on Climate Change), (2007). Climate Change: Synthesis Report; Summary for Policymakers. Available in:http://www.ipcc.ch/pdf/assessment-report/ar4/syr/ar4_syr_spm.pdf.

Johnson, H. D. (1965). Environmental temperature and lactation (with special reference to cattle). *International Journal of Biometerology, 9*, 103–116.

Johnson, H.D. (1980). Depressed chemical thermogenesis and hormonal function in heat. In M.K. Yousef (Ed.) *Environmental Physiology: Ageing, Heat and Altitude* (pp. 3–9). New York, USA: Elsevier North Holland, Inc.

Johnson, N. N., McGowan, M. R., McGuigan, K., Davison, T. M., Hussain, A. M., Juniewicz, P. E., Johnson, B. H., & Bolt, D. J. (1987). Effect of adrenal steroids on testosterone and luteinizing hormone secretion in the ram. *Journal of Andrology, 8,* 190–196.

Klinedinst, P. L., Wilhite, D. A., Hahn, G. L., & Hubbard, K. G. (1993). The potential effects of climate change on summer season dairy cattle milk production and reproduction. *Climate Change, 23,* 21–36.

Malik, P. K., & Singhal, K. K. (2008). Influence of lucerne fodder supplementation on enteric methane emission in crossbred calves. *Indian Journal of Animal Sciences, 78,* 293–297.

Malik, P. K., Bhatta, R., & Prasad, C. S. (2013). Nutritional and biological interventions for enteric methane mitigation. In M.A. Kataktalware, et al. (Eds.), *Management strategies for sustainable livestock production against impending climate changes* (pp. 54–60). Model Training Course, Southern Regional Station, ICAR-NDRI, Bengaluru, India.

Mathevon, M., Buhr, M. M., & Dekkers, J. C. M. (1998). Environmental, management and genetic factors affecting semen production in Holstein bulls. *Journal of Dairy Science, 81,* 3321–3330.

McDowell, R. E. (1972). *Improvement of Livestock Production in Warm Climates*. Freeman, California, USA: San Francisco Press.

Moss, A. R. (1994). Methane production by ruminants – literature review of I. Dietarymanipulation to reduce methane production and II. Laboratory procedures for estimatingmethane potential of diets. *Nutrition Abstracts and Reviews Series B, 64,* 786–806.

Nichi, M., Bols, P. E., Zuge, R. M., Barnabe, V. H., Goovaerts, I. G., Barnabe, R. C., & Cortada, C. N., (2006). Seasonal variation in semen quality in Bos indicus and Bos taurus bulls raised under tropical conditions. *Theriogenology, 66,* 822–882.

Nienaber, J. A., & Hahn, G. L. (2007). Livestock production system management responses to thermal challenges. *International Journal of Biometeorology, 52,* 140–157.

Ronchi, B., Stradaioli, G., Verini Supplizi, A., Bernabucci, U., Lacetera, N., Accorsi, P. A., Nardone, A., & Seren, E. (2001). Influence of heat stress and feed restriction on plasma progesterone, estradiol-17β LH, FSH, prolactin and cortisol in Holstein heifers. *Livestock Production Science, 68,* 231–241.

Roy, K. S., & Prakash, B. S. (2007). Seasonal variation and circadian rhythmicity of the prolactin profile during the summer months in repeat-breeding Murrah buffalo heifers. *Reproduction, Fertility and Development, 19,* 569–575.

Sirohi, S., & Sirohi, S. K. (2010). Vulnerability of milk producers to climate change: Technological and policy options for livelihood security. In: R.C. Upadhyay, et al. (Eds.), *Climate Change & Livestock Productivity in India* (pp. 143–152). National Symposium ICAR-NDRI, Karnal, India.

Srivastava, A. K., & Garg, M. R. (2002). Use of sulfur hexafluroide tracer technique for measurement of methane emission from ruminants. *Indian Journal of Dairy Science, 55,* 36–39.

The Hindu. (2013). Despite the hiatus, global warming is unequivocal. Available in: http://www.thehindu.com/sci-tech/energy-and-environment/despite-the-hiatus-global-warming-is-unequivocal/article5175117.ece.

Thomas, C. K., & Sastry, N. S. R. (1991). *Dairy bovine production.* New Delhi: Kalyani Publication.

Upadhyay, R. C. (2014). Physiological determinants of climate resilient animal production and future global challenges. In: P.S. Yadav, et al. (Eds.), *Physiological determinants of climate resilient and sustainable animal production* (pp. 31–39). National Symposium ICAR-CIRB, Hisar.

Varma, G. G. (2013). An introduction to livestock meteorology and events regulating climate change. In: M.A. Kataktalware, et al. (Eds.), *Management strategies for sustainable livestock production against impending climate changes* (pp. 1–7). Model Training Course, Southern Regional Station, ICAR-NDRI, Bengaluru, India.

Yousef, M. K. (1984). Stress physiology: Definition and terminology. In: Yousef, M.K. (Ed.) *Stress Physiology in Livestock* (pp. 3–7). CRC Press, Boca Raton, FL.

Climate Change and Its Impact on Milk Production in India

K. Sreenivasaiah

Abstract In the world today 7.08 billion people inhabit on the earth presently to feed everyone adequately the world needs 910.10 million tons of milk, against which global production is only 720.98 million tons. This discrepancy in need and production left over 868 million. People undernourished worldwide and 850 million people of them live in developing countries. Climate change poses a formidable challenge to the development of livestock sector in India. The anticipated rise in temperature between 2.3 and 4.8 °C over the entire country together with increased precipitation resulting from climate change is likely to aggravate the heat stress in dairy animals, particularly hybrid animals, adversely affecting their productive and reproductive performance, and hence reducing the total area where high yielding dairy cattle can be economically reared. Milk is an important component of food that is significantly increasing in demand. Increased heat stress associated with global climate change may, however, cause distress to dairy animals and possibly impact milk production. The Temperature- Humidity Index was used to relate animal stress with productivity of milk of buffaloes, crossbred, and local cows.

Keywords Dairy · Climate change · Milk production · Live stock

1 Introduction

> The cows are able to pay for their feed food, my bank loan monthly instalments, feed for non-milking cows, the maintenance of calves, the salary of workers and the running of my household; I can't ask for anything else. I am in dairy farming because I love cows and they are not putting any additional financial burden on me. It is very important that seasonally taking care is very important for all cows.

K. Sreenivasaiah (✉)
Economics Department, Government First Grade College for Women's Balmatta, Mangalore, Karnataka, India
e-mail: economicssrini@gmail.com

S. Nautiyal et al. (eds.), *Climate Change Challenge (3C) and Social-Economic-Ecological Interface-Building*, Environmental Science,
DOI 10.1007/978-3-319-31014-5_32

Jayaram Kumar, Karnataka, INDIA co-owner, who is very happy with the financial returns from the Dairy farm.

Milk production and reproductive functions of milk animals are negatively impacted by temperature rise during summer and also by the sharp temperature decline in winter. The high temperature causes stress due to increased body heat leading to low heat dissipation from the body surface. High heat load in lactating cross breed cows reduces their milk production and shorten duration of lactation periods. Information on milk production of cross breed cows with specific emphasis on climate change is not available therefore an attempt has been made to find out the sensitivity of lactating cross breed cows to changes in temperature during extreme summer. The impact of temperature rise and decline has been analysed on the milk production and reproductive functions. From the macro perspective, a systematic research study is required to examine the impact of climate change an animal production system in India; however, it is appropriate to highlight some important points. The livestock development policy in India has emphasised crossbreeding of Indian cattle with exotic sires since the inception of planning in India in early fifties. The goal was increased milk production; as a result, about 17.5 % of the dairy cattle in the country are now crossbred, starting from a negligible proportion three decades ago it shows that the crossbred animals, on account of their higher milk yield, are economically most suitable for farmers. However, cross-sectional data shows that, in general, the productivity of crossbred cattle is lower in areas where the mean annual temperature is higher.

2 Literature Review

The studies indicated that India loses 1.8 million tons of milk production at present due to climatic stresses in different parts of the country. Global warming will further negatively impact milk production by 1.6 million tons by 2020 and more than 15 million tons by 2050. High producing crossbred cows and buffaloes will be impacted more than indigenous cattle. Northern India is likely to experience a greater impact of global given the vulnerability of India to rise in sea level, the impact of increased intensity of extreme events on the livestock sector would be large and devastating for the low-income rural areas. The predicted negative impact of climate change on Indian agriculture would also adversely affect livestock production by aggravating the feed and fodder shortages. The livestock sector, which will be a sufferer of climate change is itself a large source of methane emissions, an important greenhouse gas (GHG). In India, although the emission rate per animal is much lower than the developed countries, due to vast livestock population the total annual methane emissions are about 9–10 Tg from enteric fermentation and animal wastes. Estimates of the current contribution of livestock to anthropogenic climate change, expressed in carbon dioxide equivalents, range from 8.5 to 18 %. This includes carbon dioxide itself, mainly due to land use changes; methane emissions through enteric fermentation by ruminants; and nitrous oxide emissions, mostly from manure-handling practices. Total emissions of all three GHGs are likely to increase in the coming decades: it has been estimated that emissions of methane due to livestock and nitrogen dioxide due to agriculture will increase by up to 60 % by 2030: East Asia and sub-Saharan Africa are expected to increase steeply, driven by increasing numbers of ruminants. In response, there will be increased pressure to increase efficiency of livestock production: more milk, meat, and eggs with fewer inputs and decreased GHG emissions per unit of production. Shifting to fewer, more productive animals of more productive breeds is one way to do this although doing so would require enhanced access to breeding, with ideal climate animal health and feed services will be enhanced there by productivity per animal will increases, with this we can gain increasing in income, employment and finally it will be resulted in knowledge environmental conservation by milk farmers. Such an approach also provides an opportunity for "triple-win solutions" of anthropogenic interference with the climate system through GHG emissions has led to worldwide research on assessing the impacts that might result from potential climate change associated with GHG accumulation. As the ecosystems are sensitive to changes in climate, it is necessary to examine the likely impact of climate change on various sectors within the ecosystems to be able to comprehensively understand the effects of climate change. Global food security is high on the development agenda. Some estimates anticipate that a 50–70 % increase in food production will be needed by 2050 to feed an additional two billion people. This is especially crucial for developing countries, where the problems of feeding poor people have been highlighted by recent food price shocks: the expectation is for more and sustained rises in food prices. East Asia and sub-Saharan Africa are expected to increase steeply, driven by increasing numbers of ruminants. From the above studies, it is clearly state that, the livestock sector is contributing negatively on climate change it required to increase production efficiency among the live stock animals instead of increasing the number of animals. Since dairy sector contribute lot on human development and economic development.

3 The Methodology of This Study

This paper has tried, to examine dairy industry with seasonal variation of climate to milk production, Climate Change and its Impact on Milk Production in India. This study is based on secondary data.

4 The Objectives of the Study

The three main objectives of the study were as follows:

(i) To Identify how climate change effects animal health.
(ii) Quantify the climate change impacts on the production of milk.
(iii) To understand and analyse growth of milk production with per capita availability.

5 Climate Change and Milk Production

According to International food policy report By 2050, the decline in calorie availability will increase child malnutrition by 20 % relative to a world with no climate change. Climate change will eliminate much of the improvement in child malnourishment levels that would occur with no climate change. The greatest threats and effects of climate change are on food security and the impact on dairy sector on this issue. Ensuring food security at the national level is therefore a high priority. The issue is especially important because it provides the link between production and availability of milk on the one hand and potential use of the other. It is a primary objective in the livelihoods and aspirations of farmers should be able to produce milk for the growing demand, and provide for a better tomorrow. Developing countries may experience a decline of between 9 and 21 % in overall potential agricultural productivity as a result of global warming at the world level. The anticipated rise in temperature between 2.3 and 4.8 °C over the entire country together with increased precipitation resulting from climate change is likely to aggravate the heat stress in dairy animals, particularly hybrid animals, adversely affecting their productive and reproductive performance, and hence reducing the total area where high yielding dairy cattle can be economically reared. The temperature level as depicted over the decade.

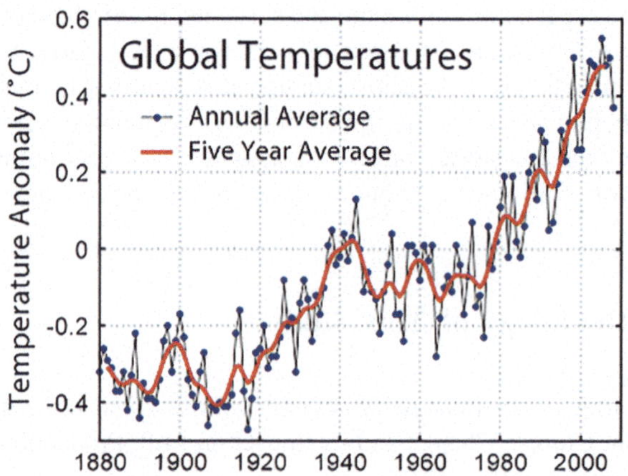

The livestock industry is likely to be impacted greatly by the effects of climate change. Some climate forecasters predict that will experience increasing average temperatures and humidity in the future, and more hot spells. A 'hot spell' means seven days or more consecutive days where temperatures are above 40 °C that will impact on the negative on milk production.

Scientists use a kind of scale, called the **temperature-humidity index (THI)**, to measure heat stress on cattle, and this can also measure how productive the cattle are. When cattle are under heat stress, farmers can expect their stock to have:

- Reduced grazing time (because the animals might be seeking shade).
- Reduced feed intake.
- Increase in body temperature.
- Increased sweating and panting.
- Weight loss.

5.1 Temperature Humidity Index (THI)

Temperature Humidity Index (THI) is a measure that has been used since the early 1990s. It accounts for the combined effects of environmental temperature and relative humidity, and is a useful and easy way to assess the risk of heat stress among the animals.

$$\mathrm{THI} = \mathrm{Tdb} - [0.55 - (0.55 \times \mathrm{RH}/100)] \times (\mathrm{Tdb} - 58)$$

- When the THI exceeds 72, cows are likely to begin experiencing heat stress and there in calf rates will be affected.

- When the THI exceeds 78, cow's milk production is seriously affected.
- When the THI rises above 82, very significant losses in milk production are likely, cows show signs of severe stress and may ultimately die.

A number of important points should be made about the THI:

- A THI of 72 may under-estimate heat load in high-yielding Holstein-Friesian cows—increasing milk yield increases cows' sensitivity to heat stress.
- Recent research shows that increasing milk production from 35 to 45 l/day reduces the threshold temperature for heat stress by 5 °C.
- THI does not account for solar radiation or air movement—those two factors, along with air temperature and relative humidity, determine the heat gained and lost between the cow and the environment.
- THI does not enable you to measure the accumulation of heat load over time, e.g. after several days. Despite these limitations, THI is still a useful and easy way to assess and predict the risk of heat stress; however, it is wise to be

conservative. If you have a herd of high-producing Holstein-Friesian, it is better to overestimate the risks of heat stress using a lower THI than get caught out. It is strongly being realized that gradual stress induced by global warming and climate change is also influencing fertility and productivity of livestock directly, and, through reduction in fodder production indirectly. Milk production of high producing European, crossbreds and buffaloes is reduced when the THI of the region exceeds 75. On an average THI exceeds 75 at 75–80 % places in India throughout the year. The Majority of places in India observes THI > 75 and more than 85 % places in India experience moderate to high heat stress during April, May and June and THI range of 75–85 at 2.00 p.m. This situation is likely to worsen in the years to come. All THI maps for baseline and 2030 indicates a temperature rise and change in THI in Uttar Pradesh, Madhya Pradesh, Gujarat, Rajasthan and other states of India based on the regional climate model PRECIS A1B scenario (source NDDB).

5.2 *Effects of Heat Stress on Fertility Among the Cows*

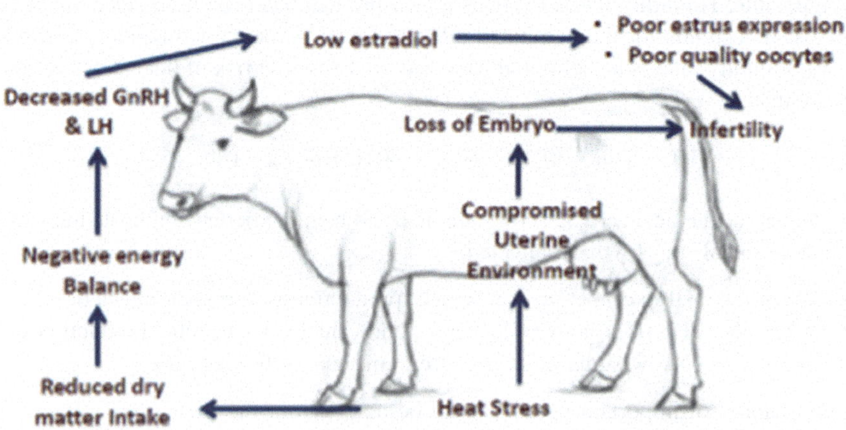

Fertility of postpartum dairy cows generally decreases during the summer when compared to winter seasons, so therefore it is necessary to understand message to be sent all milk farmers understand the situation and to act accordingly. The important factor behind the decrease in fertility is increase in ambient temperature that results in reduced levels of the fertility hormones especially LH (Luteinizing Hormone) and estradiol due to direct effect of heat stress on the hypothalamus in the brain. Heat stress also acts on ovary and uterine environment, which creates unfavorable condition to the oocytes and embryo. In addition, heat stress also causes a decrease in dry matter intake, which in turn leads to negative energy balance. This also affects the levels of fertility hormones. To overcome these situations, measures like providing cooling system for animals, providing adequate drinking water and providing high quality forage and feed may help and improve the fertility in animals. India is a diversified climatic condition country to propagate such knowledge to set up whether portals needed. Effects of heat Stress on fertility among the cows will be resulted decrease in the fertility rate in, turn it becomes uneconomical for milk farmers to feed dry cows and non milch cows.

5.3 *Intensive Livestock*

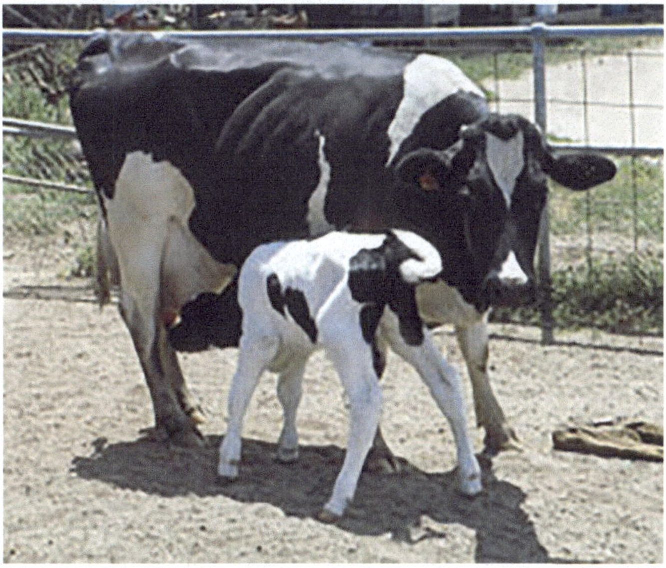

Holstein-Friesian cow and calf

In dairy cows, heat stress reduces the amount of milk produced, reduces milk fat and protein content, and decreases reproduction rates. High-producing dairy cows is the most susceptible to increases in the THI. Heat stress day's where THI > 80 lead to a substantial effect on the reproduction of dairy cows, particularly of the Holstein-Friesian breed. A change in average temperatures over the hot dry period and a change in the number of 'extreme' days will both likely lead to changes in dairy production in understanding the climate change capability to manage climate change and seasonal risk, and equip them to meet the challenges associated with reducing GHG emissions.

Shelter from the sun provided for cattle at a feedlot

Beef cattle in feedlots subject to heat stress can experience reduced health and a reduction in Reduced milk production due to heat stress is attributable only partly to decrease in feed intake. Actually 35 % of reduced milk production is due to decreased feed intake while remaining 65 % is attributable to direct effect of heat stress. Other factors resulting reduced milk production during heat stress are decreased nutrient absorption, the effect in rumen function and hormonal status and increased maintenance requirement resulting reduced net energy supply for production.

Milk production in cow has been found to be reduced when ambient temperature and temperature, humidity index increases above critical threshold. Heat stress during 60 days prepartum period negatively affects postpartum milk production and cows perpetrated during summer produce less milk as compared to other season. Similarly, the quantity of milk protein and solid not fat (SNF) have been found to be reduced during heat stress in dairy cattle. Mallonee reported 20 % less milk yield in cattle kept in sun than milk yield in cattle kept in shed. Similarly, Roman-Ponce found 10.7 % higher milk production in cows kept in shed than that in cows kept in the sun during hot weather.

6 Milk Federations and Whether Study Department

The important key strategy where all farmers can adapt to the dairy development which intern helps industry with the help of milk federations and these federations has to establish whether stations to guide farmers.

- A diverse feed base provides flexibility to take advantage of rainfall when it occurs as opposed to a reliance on one or two species limited to a particular growing season. This may include a shift towards annual crops and pastures where pasture persistence is a problem and livestock.
- Shade, sprinklers, timing of activities can all help to reduce heat stress. The Cool Cows program provides information on managing heat stress.
- Securing supplementary feeds through the use of forward contracts.
- Ensuring your farm is water efficient, and you have sufficient and secure supplies of water available to meet your needs and those of their stock. Shade can help significantly here as it reduces daily stock drinking requirements.

A range of case studies is available that demonstrate farmers taking action to adapt to climate change.

It is important for farmers to

- Know the risks, plan ahead and allow time for implementation, for varying breeds rather than needing to undertake poorly thought-out actions or local cows.
- Assess any adaptation strategy in context of their own personal and farm business goals, and
- Focus on what they can control.

7 Average Milk Production and Temperature in Selected Cities

Name of the district	Avg. temperature (in °C)	Avg. milk production of cross breeds (in l)
Mangalore	29.74	3.8
Kurnool	29.5	3.02
Ananthpur	28.61	3.21
Chennai	28.27	4.65
Ahmedabad	27.86	5.56
Cuttack	27.72	3.15
Rajkot	27.14	5.28
Tiruvanthapuram	27.10	5.65
Jodhpur	27.09	3.27
Kolkata	26.98	5.48
Balasore	26.88	3.57
Nagpur	26.85	3.29
Hyderabad	26.10	4.86
Patna	25.58	3.50

(continued)

(continued)

Name of the district	Avg. temperature (in °C)	Avg. milk production of cross breeds (in l)
Jaipur	25.17	5.45
Aurangabad	25.04	3.78
Lucknow	25.01	4.00
Agartala	24.86	5.50
Bhopal	24.85	6.01
Pune	24.79	6.28
Ranchi	24.33	6.00
Guwahati	24.21	5.85
Bangalore	23.80	6.34
Ludiyana	23.50	7.58
Ambala	23.45	6.44

Source (1) Indian Meteorological Department (2) Directorate of Animal Husbandry

The above table shows Avg. milk production of different cities in different temperature levels. It is evident from the table that the overall average increase in temperature leads decrease in milk production of cross bred cows Seasonal variation in milk production was observed and it was found to be 1500, 1800 and 1080 l production of milk in rainy, winter and summer seasons, respectively. The category-wise milk production came to 3255.50 l on medium farms, 3508.50 l on small farms, 4057.16 l on marginal farms and 5152.83 l in landless families. In small and marginal farmers, they are the better taking care of animals.

8 Milk Production in India

Year	Production (million tons)	Per capita availability (g/day)
2001	80.6	220
2005	92.5	233
2010	116.4	273
2011	121.8	290
2012	127.9	290
2013	132.4	295
2014	144.1	298

Source NDDB

From the above table it is clear that, milk production of India as increased from 80.6 million tons to 144.1 in the span of 14 years with 6 % an annual increase in milk production. Currently, India is facing a challenge of producing adequate food from shrinking natural resource base for the ever-increasing population.

Intensification of agricultural activities through enhanced productivity and efficient resource use is the only option available as competition for land and water is increasing from non-farm sectors. In other words, more production is needed with reduced natural resources under a variable climate. India also needs to take steps towards a carbon and energy efficient economy. All this calls for a Climate Resilient Agriculture (CRA) leading to sustainable food security through integrating innovations, technologies, efficient resource use, sound public policies, establishment of new institutions, and development of infrastructure. The Indian agricultural production system faces the daunting task of having to feed 17.5 % of the global population with only 2.4 % of land and 4 % of the water resources at its disposal. With the continuously degrading natural resource base compounded further by global warming and associated climate changes resulting in increased frequency and intensity of extreme weather events, "business as usual" approach will not be able to ensure food and nutrition security to the vast population as well as environmental security (the need of the hour). The challenge is formidable because more has to be produced with reduced carbon and water footprints.

9 Scope of Indian Dairy

The economic development of the country is largely linked with its rural development because majority of her population live in the villages. The rural people depends directly or indirectly on agriculture for live hood. The growth strategy, as recommended by the National Commission on Agriculture (1976) in its report, seeks to reserve a major share of the dairy industry for the weaker of farmers and to adopt an integrated area development approach mainly based on a system of producers' co-operatives. The All India Rural Credit Review Committee has also emphasized the need for providing subsidiary occupations to the peasants. Hence, the governments through the Departments of Agriculture and Animal Husbandry have to encourage subsidiary and allied occupations to agriculture like dairy, fishery, poultry, sheep-rearing, etc. A dairy industry is a sub-sector of agriculture economy of India is most important.

Animal Husbandry and Dairying may be regarded as a source to create the employment in rural areas all round the year. Indian Agriculture is mainly dependent on the monsoon and hence agriculture field faces certain bottlenecks to provide employment during such periods. On an average Agriculture sector may provide 200 days employment to the rural persons. This means they have to find alternate source of employment for income during the rest of the year. Dairy farming, sheep and goat rearing, poultry production, pig farming, rabbit rearing are the alternate sources of mixed farming. It may be possible to generate the employment for the farmers as well as landless labourers who can do this job themselves, or it may be possible to employ young and the old family persons as a side business. Many of the operations in Animal Husbandry and Poultry Farming can be done by the rural women. It is estimated that on an average 35 million human years/annum employment generation has been potential through this sector. There are significant growth prospects. Although India is one of the world's leading milk producers, yet the northern parts of the country are facing acute shortage of milk supply. States of the north-east, West Bengal, Delhi and Kerala are dependent on milk powders. In fact, the Delhi government is keen to procure milk from Karnataka Milk Federation (KMF) via a rail tanker and the supply is awaiting clearance from the Indian Railways. There are also opportunities for development of low-fat and sugarless milk products, paneer and cheese variants. India's real economic inclusiveness is with livestock with the adapting automatic weather stations breedable female bovines (cattle and Buffalo) by ear tagging (polyurethane) with laser printed unique numbers. To provide a detailed pass book with details for the animals to enable unique identification and collection of data, the following benefits are expected

- Tracking of animals from the point of origin without unnecessary burden to producers and other stake holders.
- Animal movement data can be maintained and readily accessed when necessary.
- Respond to intentionally or unintentionally introduced animal disease outbreaks quickly and effectively.

To provide the capability to identify all animals and premises that have had a direct contact with a disease of concern, identify beneficiaries for implementing various schemes by the govt. The milk producing ability of animals is related to the expression of various genes and proteins in the mammary gland apart from the hormonal regulation. Similarly, the reproduction potential of the farm animal is dependent on the expression of genes and proteins in the reproductive systems. (1) Proteomics is an upcoming area wherein the detailed proteome analysis of the systems as a whole, will help us understand the gene network pathway of production and reproduction traits. (2) Bovine genome has been recently annotated and the annotation of buffalo genome is under progress by various groups. Many of the annotated genes have been categorized as pseudo and hypothetical genes. Therefore there is a need to annotate the genes at the protein level by proteomic approaches. (3) Generation of voluminous data from genomics and proteomics approaches from farm animals needs further validation in terms of their structure at molecular level. Therefore, there is a need to characterize the proteins at molecular level by X-ray crystallography to make a data base of animals' proteins leading to structural genomics of farm animals from there also need to understand climate change on milk production.

10 Dairy Calves

The challenging issue in dairy production in India is what to do with male dairy calves. Many states have banned the slaughter of dairy cattle, and beef is not a popular meat due to cultural and religious factors. As a result, male dairy calves have a very low value. Therefore, all milk farmers expecting female calves which will become future milch cows and is also is economically benefited to them because at market price one milch cow cost more than Rs. 50,000 this will be expensive to small and marginal farmers. They are often not allowed colostrum from their mother after birth, which is essential to building immunity against infections and would provide them with a good start in life. Some of these animals are destined for 'gaushalas' (or retirement homes for cattle), but the size of the national dairy herd in India means it is not easy to provide every single animal with adequate care. Sexed semen is a recent innovation that can help to avoid this challenge, but its high cost does not make it a feasible option for many farmers in India. Based on economically feasible option is more advisable.

11 Limitations

The estimates of livestock production are available for three major products, namely, milk, eggs and wool, although the time lag of these estimates is of the order of two years. The utilisation rates of milk in fluid form or after being converted into desi ghee, butter, khoa, cream, and similar other items, within the households, are not available. In the absence of this data, estimates of the value of output of milk are being prepared as if milk is consumed as such. In the case of meat, estimates of yield are not generally based on any scientific techniques, although some States are generating the estimates using the Integrated Sample Survey results. Besides, reliable estimates of the number of animals slaughtered in places other than the slaughterhouses like unorganised slaughter houses and religious slaughters are not available. Because of this limitation there is huge scope for Indian dairy industry. Indian Zebu cows are very fertile. They get easily pregnant and calving is cake walk for them. Believe me Indian cows give birth without any assistance and infant mortality rate is almost nil. Even for bigger breeds (gir, tharparkar, kankrej, ongole etc.) calving is easy because calves at birth are very small. Where as western breed calf are generally bulky and bigger creating complication even to the mother cow (holstein friesian semen when used on a small jersey or on a hybrid cow normally creates this problem). Because of this quality farmers have large herd or number of animals born in a few years that too without any hassle (no antibiotics or vet assistance is required). Where as hybrid or western breeds struggle and mortality is very high. Most of the time farmers are at loss because of this phenomena in the long run so therefore it is to understand for all cross breed cows infant mortality rate to be decreased.

12 Cleanliness Milk Cows

Cleanliness systems for dairy cows—where the herd or a representative proportion of the herd is scored on the basis of how much muck and dirt adhere different body parts—can be used to broadly measure the standard of various aspects in the management of dairy cows, particularly with regard to controlling the incidence of infectious diseases or monitoring cow comfort levels, with this there is positive impact on milk production. Over the years several different systems have been used but they share common similarities. The scoring process can take place at any time of the year, but may be particularly useful during housing or when grazing during particularly wet conditions. Most obviously, cleanliness method is used to monitor how clean cows are being kept by the system in which they are housed in terms of environmental mastitis prevention, but it can also indicate nutritional and general health issues. **In particular, cleanliness scoring can help to indicate how the incidence of lameness in the herd affects levels of cow cleanliness, and, perhaps more importantly, how the levels of cleanliness themselves, such as poor bedding management or insufficient passageway scraping, may be factorial in causing hoof conditions which can lead to mobility problems, such as** digital dermatitis **and erosion. On** an individual cow basis. Dirtiness may be related to the health of the cow and her nutrition, but can also be due to the particular habits of the cow when walking, lying down or feeding, or housing management issues.

13 Recommendations

- Need to find out cross breed cows according to weather.
- Establish automatic weather station.
- Develop specialized, state of art, climate control facilities (CO_2, temperature, water).

- More area for cow sheds.
- More area during pregnancy of animals.
- Implement immediate emission reduction targets with the aim being to reduce net emissions to zero as soon as practicable. The goal should be to achieve 95 % of power from renewable sources by 2020, with a 90 % cut in overall emissions by 2030. Introduce annual reduction targets.
- Enhance national capacity on decision support systems.
- Training on cow rearing for different seasons.
- Intensify efforts for increasing climate literacy among all stakeholders of agriculture, including students, researchers, policy planners, science administrators, industry as well as farmers.
- Weather watch groups.
- Initiate further international treaty negotiations aimed at getting all countries to agree to a global target of 90 % emissions reductions on 1990 levels by 2030.
- Balanced food feeding for fragment animals.
- Explore feasibility of establishing feed, fodder, and seed banks.
- Increase farm insurance coverage using weather derivatives.
- Enhance climate literacy.
- Start the transition to a zero-waste economy. Engage workers in industry, with technical experts, to redesign their products and jobs sustainably.
- End industrial farming based on fertilisers, pesticides and fuel sourced from petroleum. Restrict farming areas to ensure that riverine, forest and other indigenous ecosystems return to healthy states. Encourage new farming practices including organic and urban farming.

14 Conclusion

Atmospheric temperature of the earth has been increased due to cumulative effects of GHGs in the atmosphere emitted from different industrial and agricultural activities of human. There is a serious threat of climatic changes (in the form of severe droughts, floods, intense rainfall, and landslides) undermining development programmes and millennium development goals aimed at reducing poverty. Climate induced disasters directly affect the livelihood of the farmers. Since livelihood of the farmers is based on agriculture and animal husbandry, all of the respondents said that the decrease in the animal agricultural production weakened the economic condition. Warming of the climate system of the earth has multifaceted effects on animals. Intensification and increase frequency of thermal stress is the most prominent impact of global warming in dairy cattle resulting in different physiological, metabolic and production disturbances. Importance of thermal stress has been increased to the dairy farmers in tropical, subtropical and even in temperate region of the world due to atmospheric warming. Increased area for cow shed and between the animals which will bring down the temperature in the cowshed. It also suggested maintain water fountains and musical system nearby cow sheds. Responding to the challenge of climate change requires formulation of appropriate long term adaptation strategies and mitigation options for the livestock sector. Factors affecting variability in enteric CH_4 Production Requires urgent attention and efforts to decrease the uncertainty in GHG emission inventories. It is very essential to identify viable GHG reduction strategies. That is much more clarity is needed concerning the benefits of livestock, the negative impacts they can have on greenhouse-gas emissions and the environment, and the effects of climate change on livestock system. Although the reduction in GHG emissions from livestock activities are seen as high priorities, strategies for reducing emissions should not reduce the economic viability of enterprises. The importance of such a Movement will be obvious considering the fact that 60 % of India's population of 1.2 billion depend upon agriculture for their livelihood. In addition, India has to produce food and feed for over 1.2 billion human, and over a billion farm animals. Scaling up the findings of the current project is therefore the pathway for a sustainable food, water and livelihood security system in rural India.

Bibliography

Bajagai, Y. S. (2011). Global climate change and its impacts on dairy cattle. *Nepalese Veterinary Journal, 30*, 2–16.

CAST. (2004). *Climate change and greenhouse gas mitigation: Challenges and opportunities for agriculture* (p. 120). Task Force Report 141, Council for Agricultural Sciences and Technology, USA.

Clifford, B. C., Davies, A., & Griffith, G. (1996). UK climate change models to predict crop disease and pest threats. *Aspects of Applied Biology, 45*, 269–276.

Crutzen, P. J., Aselmann, I., & Seiler, W. (1986). Methane production by domestic animals, wild ruminants, other herbivorous fauna and humans. *Tellus, 38B*, 271.

CSO. (2000). *Compendium of environment statistics* (p. 306). Central Statistical Organisation, Ministry of Statistics and Programme Implementation, Govt. of India.

De, D., & Singh, G. P. (2001). Monensin enriches UMMP upplementation on in vitro methane production in crossbred calves. In *Proceedings of the X Animal Nutritional Conference (Abstract papers), Karnal, India*. Animal Nutrition Society of India, National Dairy Research Institute, 161.

Dinar, A., Mendelsohn, R., Evenson, R., Parikh, J., Sanghi, A., Kumar, K., et al. (Eds.). (1998). *Measuring the impact of climate change on Indian agriculture* (p. 266). World Bank Technical Paper No. 402, World Bank, Washington, DC.

Dutt, T., Taneja, V. K., & Avtar Singh, S. A. (1992). Comfort zone for maximal milk production in crossbred cattle. *Indian Journal of Dairy Science, 45*(3), 119.

Evenson, R. E. (1999). Global and local implications of biotechnology and climate change for future food supplies. *Proceedings of the National Academy of Sciences of USA, 96*(11), 5921.

Gaylean, M. L., & Hubbert, M. E. (1995). Effects of season, health and management on feed intake by beef cattle. In *Symposium: Intake by Feedlot Cattle* (pp. 226–234). Stillwater, OK: Oklahoma Agricultural Experiment Station, Oklahoma State University.

Hahn, G. L., Klinedinst, P. L., & Wilhite, D. A. (1992). *Climate change impacts on livestock production and management*. Paper 92-7037. ASAE, St. Joseph, MI.

Henderson-Sellers, A., & McGuffie, K. (1997). Climate models. In R. D. Thompson, & A. Perry (Eds.), *Applied climatology–Principles and practice* (pp. 36–50). New York: Routledge.

Intergovernmental Panel on Climate Change (IPCC). (1995). *Climate change 1995: The Science of climate change*. Cambridge: The Cambridge University Press.

Mader, T.L., Harrington, Jr. J. A., Hahn, G. L., Davis, M. S., & Nienaber, J. A. (2001). *Potential climate change effects on warm-season production of livestock in the United States*. ASAE Paper Number: 01-3042.

Ramarao, D. (1988). Seasonal indices and meteorological correlates in the incidence of foot-and-mouth disease in Andhra Pradesh and Maharashtra. *Indian Journal of Animal Sciences, 58*(4), 432–434.

Reilly, J. (1996). Agriculture in a changing climate: impacts and adaptation. In R. T. Watson, M. C. Zinyowera, & R. H. Moss (Eds.), *Climate Change 1995: impacts, adaptations, and mitigation of climate change: Scientific-technical analyses* (pp. 429–467). Contribution of Working Group II to the Second Assessment Report of the Intergovernmental Panel on Climate Change. Cambridge: Cambridge University Press.

Rosenzweig, C., & Iglesias, A. (1998). The use of crop models for international climate change impact assessment. In G. Y. Tsuji, G. Hoogrnboom, & P. K. Thorton (Eds.), *Understanding options for agriculture production* (pp. 267–292). Dordrecht, The Netherlands: Kluwer.

Saseendran, S. A., Singh, K. K., Rathore, L. S., Singh, S. V., & Sinha, S. K. (2000). Effect of climate change on rice production in the tropical humid climate of Kerala, India. *Climate Change, 44*, 495–514.

Scheehle, E. (2002). Emissions and projections of non CO_2 greenhouse gases from developing countries 1990–2020 (p. 73).

Sharma, S. K., Singh, G. R., & Pathak, R. C. (1991). Seasonal contours of foot-and-mouth disease in India. *Indian Journal of Animal Sciences, 61*(12), 1259–1261.

Sheppard, J. C., Westberg, H., Hopper, J. F., Ganesan, K., & Zimmerman, P. (1982). Inventory of global methane sources and their production rates. *Journal Geophysical Research, 87*, 1305–1312.

Shinde, S., Taneja, V. K., & Singh, A. (1990). Association of climatic variables and production and reproduction traits in crossbreds. *Indian Journal of Animal Sciences, 60*(1), 81–85.

Singh, G. P. (1998). *Methanogenesis and production of greenhouse gases under animal husbandry system*. Final report A.P. Cess fund project, National Dairy Research Institute, Karnal, India.

Upadhyay, R. C., Singh, S. V., Kumar, A., Gupta, S. K., & Ashutosh, A. (2010). Impact of Climate change on Milk production of Murrah buffaloes. *Italian Journal of Animal Science, 6*(2s), 1329–1332.

Traditional Health Foods Security as Corporate Social Responsibility for Pregnant Women Health Monitoring

B.N. Lakshmi and T.S. Indumathi

Abstract The changing climate activities have outlined several health risks generated by a warming and a more variable climate. Traditional foods in India are receiving greater attention in market and as a healthy food preventive for centuries. The term malnutrition is generally better understood than intelligence, behaviour, or mental development. There is a severe malnutrition prevailing among the low income groups of population comprising of adolescent girls, lactating and pregnant women. The present era is providing women a high rate of success in all fields, but yet women face problems in terms of nutrition health during the period of pregnancy. The period of pregnancy is one where women of all ages suffer from acute nutritional health problems and health abnormalities due to variations in a few physiological parameters like blood pressure (BP), blood glucose level (BGL) and weight. The magnitude and implications of the food security and nutritional problems in pregnant women call for urgent actions through corporate social responsibility in community development. Several research and evaluation studies conducted in the country reported that a multi-pronged approach including increased traditional food production, its equitable distribution, and nutrition intervention and nutrition education with respect to pregnant women need to be considered as the most important focal points. Traditional health foods help largely to counterpart this issue of malnutrition in pregnant women. Climate change can affect problems that are already huge, malnutrition lowers resistance to diseases resulting in morbidity, apathy, lethargy and reduction of working efficiency. Due to the lack of effective communication, many technologies to overcome malnutrition are not reaching to the vulnerable groups. Therefore, nutrition intervention and nutrition education are the most important focal points, and in viewpoint of achieving food security traditional foods have the potential.

B.N. Lakshmi (✉) · T.S. Indumathi
PG Research Center, Visvesvaraya Technological Univeristy, Mudenahalli, Belgaum, India
e-mail: keerthisri.20@gmail.com

T.S. Indumathi
e-mail: drindumathits@gmail.com

S. Nautiyal et al. (eds.), *Climate Change Challenge (3C) and Social-Economic-Ecological Interface-Building*, Environmental Science, DOI 10.1007/978-3-319-31014-5_33

1 Introduction

The changing climate activities have outlined several health risks generated by a warming and a more variable climate. Traditional foods in India are receiving greater attention in market and as a health food preventive for centuries, anti-diabetic potential, hypocholesterolemic activity, control of cholesterol, gall stones, anti-inflammatory, arthritic aspects and anticancer potency have been brought out in recent studies. Further, tailor made foods are needed to meet the various situations through optimization of dietary energy sources in low calorie diets with reference to fat and proteins to prevent degradation of body proteins. Further, Cognitive process as learning, memory development, problem-solving abilities, creative thinking and psychological factors such as depression, anxiety, aggression and behaviour aspects do play an important role in traditional health food design. Further, problem of micronutrient deficiency (hidden hunger) is gaining widespread attention through corporate social responsibility in community development. Malnutrition is an ecological problem that does not occur alone. Its consorts are poverty, disturbed family structure ignorance and despair. Although, the need for dietary supplements is important the problem of malnutrition is larger than diet alone. Malnutrition is so often found enmeshed in the circumstances of a deprived environment that it is impossible to eradicate it completely without substantial social and economic reform. The term malnutrition is generally better understood than intelligence, behaviour or mental development. It has been defined in various ways. There is a severe malnutrition prevailing among the low income groups of population comprising of adolescent girls, lactating and pregnant women. Deficiencies of micronutrients lead to growth retardation morbidity, mortality, brain damage, eye damage leading to blindness and affect the cognitive and working capabilities. This drains on the country's human resource due to the problem of malnutrition and micronutrient deficiencies is unnecessary and is tragic. The government of India has drawn up many action plans to reduce the gravity and enormity of this chronic malnutrition. Malnutrition and micronutrient malnutrition are matters of serious concern in our country as it affects more than 33 % of India's population. Women are more likely to suffer from nutritional deficiencies than men are, for reasons including women's reproductive biology, low social status, poverty and lack of education. Socio-cultural traditions and disparities in household work patterns can also increase women's chances of being malnourished. Globally 50 % of all pregnant women are anaemic, and at least 120 million women in less-developed countries are underweight which hinders women's productivity and can lead to increased rates of illness and mortality. Micronutrient deficiencies are a matter of major public health concern. Recognizing the international, regional and national resources coordination and support required, governments, in collaboration with international agencies, NGOs, the private sector/industry, other expert groups and the community, should adopt an appropriate combination of strategies under changing climate. Climate change is a continuous process and health effects on human beings due to climate change are inevitable. Although climate change is a global phenomenon, its consequences will not be evenly distributed. Scientists agree that developing countries and small island nations will be the first and hardest hit. Warmer temperatures, shifting rainfall patterns and increasing humidity affect the transmission of diseases by vectors like mosquitoes. They are quite sensitive to changes in temperature and rainfall and are among the first organisms to extend their range when environmental conditions become favourable Thus, higher temperatures could influence the incidence of diseases such as malaria, dengue fever, yellow fever and several types of encephalitis. Cold temperatures are often the limiting factor in mosquito survival, so any increase in minimum winter temperatures would likely extend mosquito ranges into temperate regions or higher altitudes where they do not survive. However, there is an urgent need to mitigate the ill effect of climate change on human health. The present era is providing women a high rate of success in all fields, but yet women face problems in terms of nutrition health during the period of pregnancy. The period of pregnancy is one where women of all ages suffer from acute nutritional health problems and health abnormalities due to variations in a few physiological parameters like blood pressure (BP), blood glucose Level (BGL) and weight. The nutritional health problems and health abnormalities faced by pregnant women can be counterpart by consuming traditional health foods. These traditional health foods are rich in nutrients and hence help largely in reducing problems and abnormalities in health faced by women during pregnancy. However, women of this era require a system/medical practitioner to provide the necessary advice regarding traditional health foods that can be consumed during pregnancy, and this very system is the nutritional health monitoring system that provides required advice and information about traditional health foods necessary for pregnant women, to reduce the problems and abnormalities in health that occur during pregnancy. Further, availability of abundant resources to pregnant women is not an important criteria but how efficiently these resources are utilized is mainly required. Henceforth, here entrepreneurship and skill development among pregnant women is required for the improving efficiency of utilisation and better health maintenance. There is a growing need for entrepreneurship and skill development among pregnant women to decide upon the need for mobile nutritional monitoring system that provides healthy nutritional advice and traditional health foods for efficient health maintenance during pregnancy for women and this must become corporate social responsibility for community development.

1.1 Objectives and Methodology

Objectives of the study are to know infectious diseases and chronic diseases, to understand the five major health consequences, to ascertain food security and nutritional problems of pregnant women and strategies to prevent malnutrition. The respondents were selected at random and used personal interview method. The study was conducted in and around Bangalore and randomly selected 60 pregnant women in Bangalore North Taluk, Karnataka constituted population for the study. The respondents were interviewed using pretested structured interview schedule and conducted during 2013.

2 Results

2.1 *Infectious Diseases and Chronic Diseases in Changing Climate*

The Intergovernmental Panel on Climate Change (IPCC) projections of increased temperature and precipitation suggest the emergence of more disease-friendly conditions in regions that did not previously host diseases or disease carriers. Climate change accelerates the spread of disease primarily because warmer global temperatures enlarge the geographic range in which disease-carrying animals, insects and microorganisms—as well as the germs and viruses they carry—can survive. In addition to changing weather patterns, climatic conditions affect diseases transmitted via vectors such as mosquitoes (vector-borne disease) or through rodents (rodent-borne disease). The climate-sensitive diseases among the largest global killers are diarrhea, malaria and protein-energy malnutrition. The extreme events—like floods, storms, droughts, and uncontained fires can be devastating for health and can be solved through corporate social responsibility in community development. Floods spread bacteria, viruses, and chemical contaminants, foster the growth of fungi, and contribute to the breeding of insects. Prolonged droughts interrupted by heavy rains, favour population explosions of both insects and rodents. Extreme weather events have been accompanied by new appearances of harmful algal blooms. In India, deaths due to chronic diseases were 3.78 million in 1990 (40.4 % of all deaths) and are expected to reach 7.63 million in 2020 (66.7 % of all deaths). In medicine, a chronic disease is a disease that is long-lasting or recurrent. The term *chronic* describes the course of the disease, or its rate of onset and development. A chronic course is distinguished from a recurrent course; recurrent diseases relapse repeatedly, with periods of remission in between. As an adjective, *chronic* can refer to a persistent and lasting medical condition. *Chronicity* is usually applied to a condition that lasts more than three months. Examples of chronic diseases include: asthma, chronic fatigue syndrome, chronic osteoarticular diseases: rheumatoid arthritis, osteoarthritis, chronic respiratory diseases: chronic obstructive pulmonary disease, asthma, pulmonary hypertension, Chronic renal failure, diabetes mellitus, chronic hepatitis, autoimmune diseases, such as ulcerative colitis, lupus erythematosus, Crohn's Disease and Coeliac Disease, cardiovascular diseases: heart failure, ischemic cardiopathy, cerebrovascular disease, epilepsy, neoplastic. Over the past few decades, chronic diseases have their roots in unhealthy lifestyles or adverse physical and social environments. Risk factors like unhealthy nutrition over a prolonged period, tobacco use, physical inactivity, excessive use of alcohol and psychosocial stress are among the major lifestyle issues. The entire population is at risk because of mass elevated risk factors in which individual susceptibility is enhanced by culture, economic factors and the environment. India is the second largest producer and consumer of tobacco in the world, consequently huge rates of cancer, particularly oral cancer in the world. This costs India, for the individuals affected (treatment costs for tobacco related diseases) about US $7.2 billion just for the year 2002–2003. In diabetes, we had an estimated 19.3 million in 1995. The projected number is 57.2 million by 2025. Shifting dietary patterns, a decline in energy expenditure associated with a sedentary lifestyle, an ageing population. Changing lifestyles and stressed workplace environment have given rise to many chronic diseases in India.

2.2 *Five Major Health Consequences and Nutritional Problems in Changing Climate*

(1) The agricultural sector is extremely sensitive to climate variability. Rising temperatures and more frequent droughts and floods can compromise food security. Increases in malnutrition are expected to be especially severe in countries, where large populations depend on rain-fed subsistence farming. Malnutrition, much of it caused by periodic droughts, is already responsible for an estimated 3.5 million deaths each year. (2) More frequent extreme weather events mean more potential deaths and injuries caused by storms and floods. In addition, flooding can be followed by outbreaks of diseases, such as cholera, especially when water and sanitation services are damaged or destroyed. Storms and floods are already among the most frequent and deadly forms of natural disasters. (3) Both, scarcity of water which is essential for hygiene, and excess water due to more frequent and torrential rainfall will increase the burden of diarrhoeal disease, which is spread through contaminated food and water. Diarrhoeal disease is already the second leading infectious cause of childhood mortality and accounts for a total of approximately 1.8 million deaths each year. (4) Heatwaves, especially in urban "heat islands", can directly increase morbidity and mortality, mainly in elderly people with cardiovascular or respiratory disease. Apart from heat waves, higher temperatures can increase ground-level ozone and hasten the onset of the pollen season, contributing to asthma attacks. (5) Changing temperatures and patterns of rainfall are expected to alter the geographical distribution of insect vectors that spread infectious diseases. Of these diseases, malaria and dengue are of greatest public health concern. In short, climate change can affect problems that are already huge, largely concentrated in the developing world, and difficult to combat. However, these problems can be solved through corporate social responsibility in community development. The magnitude and implications of the nutritional problems call for urgent actions.

Several research and evaluation studies conducted in the country reported that a multi-pronged approach including increased traditional food production, its equitable distribution and nutrition intervention and nutrition education need to be considered as the most important focal points. Micronutrients are the nutrients which are needed for the body in minute quantities for growth, development and maintenance. Malnutrition is a condition resulting from the deficiency or excess intake of calories and one or more nutrients among vulnerable groups in India. Micronutrient and malnutrition are the terms commonly used to refer to vitamin and mineral nutritional deficiency diseases. However, malnutrition lowers resistance to diseases, resulting in morbidity, apathy, lethargy and reduction of working efficiency. These lead to low income, low standard of living, poverty and infection among vulnerable groups. The experience of past two decades indicated that there is an improvement in the nutritional status of the population in developing societies which is mainly because of eradication of poverty, application of measures to overcome socio-economic disabilities. Although there is an improvement in the nutritional status and the micronutrient rich foods may be both available and consumed, often they are not consumed in sufficient quantities to prevent deficiencies and they may not be consumed by all vulnerable groups. Further, education and information for behavioural changes in the eating habits are being increasingly recognized as the key of interventions in the improvement of health and nutrition. However, functional illiteracy, lack of infrastructure, and expensive technologies often mean that efforts in healthy and nutritional communication at the community level are minimal, difficult and disappointing. Hence, there is need for food security.

2.3 Food Security, Nutritional Problems and Traditional Health Foods for Pregnant Women

The magnitude and implications of the nutritional problems in pregnant women call for urgent actions through corporate social responsibility in community development. Several research and evaluation studies conducted in the country reported that a multi-pronged approach including increased traditional food production, its equitable distribution, and nutrition intervention and nutrition education with respect to pregnant women need to be considered as the most important focal points. Micronutrients are the nutrients, which are needed for pregnant women in minute quantities for growth, development and maintenance. Malnutrition is a condition resulting from the deficiency or excess intake of calories and one or more nutrients among pregnant women in India. Micronutrient and malnutrition are the terms commonly used to refer to vitamin and mineral nutritional deficiency diseases in pregnant women. However, malnutrition lowers resistance to diseases, resulting in morbidity, apathy, lethargy and reduction of working efficiency among pregnancy women. These lead to low income, low standard of living, poverty and infection. The experience of past two decades indicated that there is an improvement in the nutritional status of the women during pregnancy in developing societies which is mainly because of eradication of poverty, application of measures to overcome socio-economic disabilities. Although there is an improvement in the nutritional status and the micronutrient rich foods may be both available and consumed, often they are not consumed in sufficient quantities to prevent deficiencies and they may not be consumed by all women during the period of pregnancy. Further, education and information for behavioural changes in the eating habits are being increasingly recognized as the key of interventions in the improvement of health and nutrition among pregnant women. However, functional illiteracy, lack of infrastructure and expensive technologies often mean that efforts in healthy and nutritional communication at the community level are minimal, difficult and disappointing. To overcome all the nutritional problems faced by women during the period of pregnancy and provide them a healthy pregnancy and post pregnancy period there is an overwhelming requirement for traditional healthy foods. The traditional healthy food intake by women during pregnancy safeguards them and their foetus from malnutrition during and post pregnancy. Hence, traditional healthy foods are particularly needed for women during pregnancy. Traditional foods in India are receiving greater attention in market as a health food for pregnant women for centuries. Further, traditional health foods are having anti-diabetic potential, hypocholesterolemic activity, control of cholesterol, gall stones, anti-inflammatory, arthritic aspects and anticancer potency have been brought out in recent studies that indicated the need for corporate social responsibility in community development. However, tailor made foods are needed to meet the various situations through optimization of dietary energy sources in low calorie diets with reference to fat and proteins to prevent degradation of body proteins in pregnant women. The cognitive process as learning, memory development, problem-solving abilities, creative thinking and psychological factors such as depression, anxiety, aggression and behaviour aspects can be enhanced through the intake of traditional health foods by women during pregnancy. The problem of micronutrient deficiency (hidden hunger) is gaining widespread attention and can lead to growth retardation morbidity, mortality, brain damage, eye damage leading to blindness and affect the cognitive and working capabilities. This drain on the country's human resource due to the problem of malnutrition and micronutrient deficiencies is unnecessary and is tragic, hence, the government of India has drawn up many action plans to reduce the gravity and enormity of this chronic malnutrition. Malnutrition and micronutrient malnutrition is a matter of serious concern in our country as it affects more than 33 percent of India's population. Women are more likely to suffer from nutritional deficiencies than men are, for reasons including women's reproductive biology, low social status, poverty, and lack of education. Socio-cultural traditions and disparities in household work patterns can also increase women's chances of being malnourished. Globally 50 percent of all pregnant women are anaemic, and at least 120 million women in less developed countries are under weights, this hinders women's productivity and can lead to increased rates of illness and mortality. Micronutrient deficiencies are a matter of major public health concern. Traditional health foods help largely to counterpart this issue of malnutrition in pregnant women and fetus.

2.4 Strategies to Prevent Malnutrition in Pregnant Women Through Food Security

The following strategies are suggested to overcome malnutrition and micronutrient deficiency through corporate social responsibility in community development using traditional foods are like (1) Encourage the concerned international agencies and NGO's to provide assistance in combining all aspects of micronutrient deficiency problems, through monitoring and surveillance, research, production and consumption of micronutrient rich-traditional-foods for pregnant women. (2) Ensure that nutrition education and training programmes are implemented at the community, school and national levels to provide information on proper traditional food preparation, nutritional value and bioavailability and other factors that affect micronutrient status, especially of the pregnant women, and to promote consumption of traditional foods that are rich in micro nutrients. (3) The sustainable traditional food-based strategies are given first priority particularly for pregnant women deficient in vitamin A and iron, favouring locally available traditional foods and taking into account local food habits. Supplementation of intakes with vitamin A, Iodine and iron may be required on a short-term basis to reinforce dietary approaches in severely deficient pregnant women, utilizing possible primary health care services. Supplementation should be progressively phased out as soon as micronutrient rich traditional food-based strategies enable adequate consumption of micronutrient. (4) Plan and implement programmes to correct micronutrient deficiencies and prevent disease occurrence by promoting and dissemination of nutrition information and giving priority to sustainable traditional food-based approaches that encourage dietary diversification through the production and consumption of micronutrient rich-traditional-foods, including appropriate processing and preservation techniques allowing the conservation of micro nutrients should be promoted among pregnant women. (5) Implement the most appropriate combination of the measures for improved traditional food availability, traditional food preservation, nutrition education and training, dietary diversification, food fortification, supplementation and pertinent public health message such as primary health care, promotion of safe drinking-water. (6) Develop sustainable institutional capacities along with human resources, including training of professionals, and community leaders, in order to achieve the goals of micronutrient deficiency control and prevention of diseases in pregnant women. (7) There is need for more and more research on the role of micronutrients in pregnant women health and disease, on the development of inventories and traditional food consumption tables of existing and potentially significant traditional food sources of micro nutrients, including inter alia green and yellow vegetables and fruits, palm oil and other locally available traditional food sources of micronutrients, on weaving foods, on factors affecting the bioavailability of nutrients in traditional food, on indigenous methods of traditional food processing and preparation affecting micro nutrient availability, etc. and on the improvement of existing techniques for the assessment and correction of micronutrient deficiencies through corporate social responsibilities in community development.

2.5 Nutritional Interventions on Pregnant Women Health Through Food Security

Interventions aim at changing diet and nutrition of pregnant women that include educating individuals, changing the environment, modifying the traditional food supply, undertaking corporate social responsibility community interventions, and implementing economic policies. In most cases, quantifying the effects of the intervention is difficult, because behavioural changes may take many years and synergies are potentially important but hard to estimate in formal studies. Substantial nihilism often exists regarding the ability to change diets or behaviours of pregnant women, but major changes are possible over extended periods of time since changing behaviours related to diet and lifestyle require sustained efforts and long-term persistence. However, opportunities exist that do not require individual behavioural changes among pregnant women, and these can lead to more rapid benefits (1) **Eat a Healthy Traditional Diet, i.e.** Medical experts have long recognized the effects of traditional diet on health of pregnant women. Efforts to change diets, physical activity patterns, and other aspects of lifestyle during pregnancy have traditionally attempted to educate women through schools, health care providers, worksites, and general media. Hence, eating a traditionally healthy food during pregnancy helps women largely to have a healthy lifestyle pre, during and post pregnancy. (2) **Worksite Interventions, i.e.** worksite interventions can efficiently include a wide variety of health promotion activities because working pregnant women spend a large portion of their working hours and eat a large percentage of their food there. Interventions can include educating pregnant employees; screening them for behavioural risk factors; offering incentive programs to walk, ride a bicycle, or take public transportation to work; offering exercise programs during breaks or after work; improving the physical environment to promote activity; and providing traditionally healthier foods in cafeterias. Worksite health promotion can result in a positive return on investment through lower health costs and fewer sick days. (3) **Interventions by Health Care Providers,**

i.e. studies of traditional dietary counselling by physicians indicate that even brief messages about traditional food nutrition can influence behaviour and that the magnitude of effect is related to the intensity of intervention. Identifying pregnant women who are overweight or obese, or who are gaining weight but are not yet overweight, is an initial step in preventing and treating overweight. However, many physicians are not well trained to measure and calculate BMI and identify weight problems. (4) **Fortifying Traditional Health Food, i.e.** food fortification has eliminated iodine deficiency, pellagra, and beriberi in much of the world. In regions where iodine deficiency remains a serious problem, fortification should be a high priority. Folic acid intake is suboptimal in many regions of both developing and developed countries. Fortifying traditional health foods with folic acid is extremely inexpensive and could substantially reduce the rates of several chronic diseases. Grain products—such as flour, rice, and pasta—are usually the best traditional foods to fortify, and in many countries, they are already being fortified with other B vitamins. (5) **Increasing the Availability and Reducing the Cost of Healthy Traditional Foods, i.e.** policies regarding the production, importation, distribution, and sale of specific foods can influence their cost and availability. Policies may be directed at the focus of research and the types of production promoted by health services. Policies often promote grains, dairy products, sugar, and beef, whereas those that encourage the production and consumption of fruits, vegetables, nuts, legumes, whole grains, and healthy oils would tend to enhance rather than pregnant women, i.e. almost every national effort to improve nutrition incorporates the promotion of traditional health food choices, such as fruits, vegetables and legumes. Ideally, such efforts are coordinated among government groups, retailers, professional groups and non-profit organizations, and investment in such efforts should include the careful testing and refining of social-marketing strategies. Another strategy is to protect pregnant women from aggressive marketing of unhealthy traditional foods. Producers spend billions of dollars a year encouraging pregnant women to consume foods that are detrimental to their health. Hence, there is need for food security.

3 Level of Communication at Community

The results of the study indicated that lack of effective communication at the community level, many technologies to overcome malnutrition and micronutrient deficiencies are not reaching to the pregnant women and vulnerable groups. Hence, there is a need for a strategy with effective communication to overcome malnutrition and micronutrient deficiencies. Knowledgeable malnutrition and micronutrient deficiencies resource scientist is the person who starts the process of communication. The key factor influencing communication is the communicator himself.

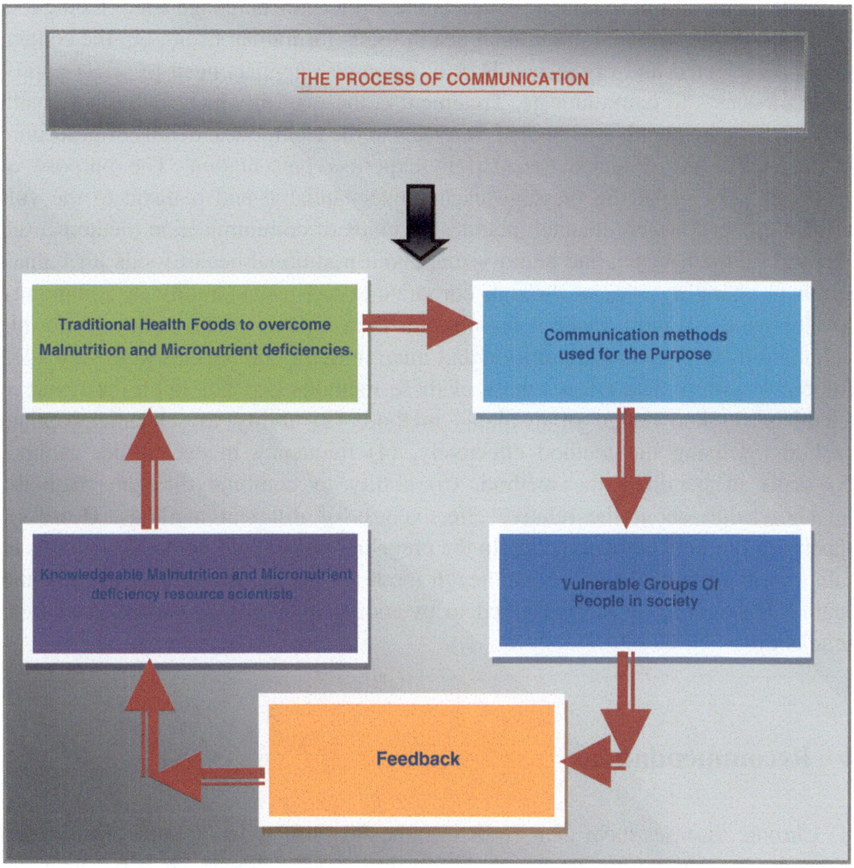

The factors affecting the knowledgeable resource scientists' effectiveness in communicating traditional health foods for human health in changing climate are like (1) Credibility in the eyes of vulnerable groups, i.e. expertise and trustworthiness (2) cultural and language compatibility (3) degree of contact with vulnerable groups (4) communication skills (5) Attitude towards the available technologies and vulnerable groups (6) role perception and (7) Empathy of resource scientist. The message is traditional health foods for human health can overcome malnutrition and micronutrient deficiencies. Further, the traditional health foods should be in line with the objective to be attained, clear and should be in line with the mental, social, economic and physical capabilities of the vulnerable group specific, simple accurate and timely. However, appealing applicable, adequate and manageable by the vulnerable groups supported by the factual information, covering both advantages and disadvantages. The resource scientists wish his vulnerable groups to receive the technologies to prevent malnutrition and micronutrient deficiencies, understand, accept and act upon the message. However, three things need to be considered, i.e. code, content and treatment. Code is the symbol of language used, Content is the details of message, i.e. traditional health foods for human health but the content has to be selected and organized. Hence, resource scientists need to select appropriate messages to communicate. Treatment is the way in which traditional health foods for human health are handled. It relates to the techniques, details of procedure or manner of performance essential to expertise presentation. The purpose of treatment is to make the message clear, understandable and realistic to the vulnerable group. Further, channel includes all kinds of communication methods used like oral, written, visual and audio-visual about traditional health foods for human health in changing climate through corporate social responsibility in community development. All these make it possible for the resource scientists to transmit his technologies to prevent malnutrition and micronutrient deficiencies to the intended vulnerable group. The effectiveness of these methods depends upon the resource scientists (1) knowledge of available method, (2) ability to select appropriate method (3) using the method effectively, (4) frequency in use of the method, (5) skills in handling the method, (6) ability to combine different methods, (7) knowledge about the relative effectiveness of different methods. However, effective communication should help the pregnant women and vulnerable groups to gain a clear view of the traditional health foods for human health and quantity and quality of nutrients to be consumed to overcome malnutrition and micronutrient deficiency.

4 Recommendations

(1) Chronic diseases have their roots in unhealthy life styles, facilitating to overcome diseases as corporate social responsibility. (2) Climate change can affect problems that are already huge and largely concentrated and difficult to come back. (3) Malnutrition lowers resistance to diseases resulting in morbidity, apathy, lethargy and reduction of working efficiency.(4) Lack of effective communication many technologies to overcome malnutrition are not reaching to the vulnerable groups.(5) Nutrition intervention and nutrition education are the most important focal points and traditional foods are receiving greater attention.(6) Nutrition education and training to correct micronutrient deficiency at community level to prevent malnutrition.(7) Effective food security in traditional health foods as corporate social responsibility for health monitoring of pregnant women.

Research Approach to Analyze Climate Change Impacts in Rural Regions of India and to Explore Potential Adaptation Strategies for Biodiversity Conservation and Livelihood Development

Sunil Nautiyal and Ruediger Schaldach

Abstract This article contains the methodological and theoretical approaches related to assessing the impacts of various driving forces including climate on the diverse socio-ecological systems. The detailed description is given to formulate the appropriate approach in interdisciplinary/transdisciplinary research for producing successful outcome that can be implemented for resource management in various regions having contrast setup in the ecology, climate, and socio-economy. The scientific framework and integrated interdisciplinary approaches described at this juncture are toward working for complex problems in real world. As a result, realistic outcomes can be obtained for proposing the solutions so that the goals of sustainable social and ecological development under changing environment can be achieved.

Keywords Climate change · Biodiversity conservation · Land use land cover · Agroecological regions · Remote sensing and GIS · Landshift model · Interdisciplinary modelling

1 Introduction

Climate change is the biggest environmental threat faced by humanity. Climate change and its manifestations, particularly through rising temperatures, changing rainfall, sea-level rise and increasing droughts and floods have the potential to

S. Nautiyal (✉)
Centre for Ecological Economics and Natural Resources (CEENR), Institute for Social and Economic Change, Dr. V.K.R.V. Rao Road, Nagarabhavi, Bangalore 560 072, India
e-mail: nautiyal_sunil@rediffmail.com

R. Schaldach
Center for Environmental Systems Research (CESR), University of Kassel, Kassel, Germany
e-mail: schaldach@usf.uni-kassel.de

S. Nautiyal et al. (eds.), *Climate Change Challenge (3C) and Social-Economic-Ecological Interface-Building*, Environmental Science,
DOI 10.1007/978-3-319-31014-5_34

adversely impact natural ecosystems (such as forests, grasslands, rivers, and oceans) and socioeconomic systems (such as food production, fisheries, and coastal settlements). India is one of the countries likely to be the most adversely impacted. There are efforts at global, national, state and even at local levels to address climate change through mitigation and adaptation. Addressing climate change requires policy formulation, research, technology transfer and diffusion, financing and enhancing adaptive capacity of the poor.

The Government of India has adopted the National Climate Change Action Plan with eight missions aimed at mitigation, adaptation, and research. So far, climate change has remained largely an international issue and within India it has remained largely at the National level and has not yet been seriously considered at the micro level. However, a few state governments have initiated policies and programmes aimed at addressing climate change. Addressing climate change requires serious attention of policymakers not only at international and national level, but also, at the regional and local.

India, with a huge diversity in land, topography, climate and socioeconomic conditions, is divided into 15 agroecological zones. Further, to help developing location-specific research and development strategies at the micro level, a total of 127 subzones (agro-climatic subregions) have been identified in India (NARP 2009). In defining zonal boundaries several indicators (such as water availability, soil types, rainfall and pattern of rainfall, edaphic factors, land use and land cover, area under irrigation and rainfed, crops and cropping pattern, etc., were taken into consideration (Sharma 2007). Therefore, research on climate change and its impact only at the national level may not be a sound approach toward adaptation and mitigation activities at the micro level. Thus, a micro-level research needs to be undertaken that might help us understand climate change impacts on the landscape, i.e., biodiversity, health, natural resource management, land use and land cover development, adaptation and the development of socio-ecological systems (Nautiyal 2010).

The United Nations Framework Convention on Climate Change's Conference of the Parties says that the vulnerable groups particularly in developing countries and whose livelihood is land use practices are the most victims in present world as in most of the cases their activities are driven by climate. Therefore, solving climate dilemma through mitigation process and scientific research is an ethical concern. The climate change harms the vulnerable people and this is not mere inconveniences, but in some cases catastrophic losses of life or the ability to sustain life (Brown 2010). This is important to undertake in-depth research on land use and climate modeling (ex-post; ex-ante) at local level as so far orchestrated land use modeling work in India is very limited. The research should help in developing the better strategies in the perspective of ongoing climate change and corresponding limit of sustainable land use—socio-ecological development in India.

2 Land Use Research and Science Policy-Interface

The development of better land use policy (such as land use development at regional, state or national level) builds on the sectoral (micro study) perspectives. These depends on the perceived risks of uncertainty in production due to variety of factors—such as climate change, socioeconomic, ecological cultural characteristics of the regions involved, and the technical feasibility of policy measures (Klabbers et al. 1996). For effective implementation, the scientific and technological research should support the policy making processes. When science and policy differ then the outcome in the form of communication is often problematic. The requirement is —that the scientific information is as per the requirement of policy demand and should be easily accessible to policy makers and decision takers. The integrated modeling approach strongly supports this viewpoint (Kaechele and Dabbart 2002; Schaldach et al. 2011a, b; Nautiyal et al. 2010, 2011). On the other, policy makers and decision takers should formulate requisite information in such a way that is easily understandable for researchers to provide available scientific information in their deliberations (van den Hove 2007). The integrated modeling approach should be able to deliver the outcome toward evaluating different policy scenario at various spatial and temporal scales. Thus, the research needs to help in making effective science policy recommendations for climate change, land use development (i.e., impact, causes, effects, adaptation, and mitigation) at local, regional level, which help to support better policy formulations. The research endeavors should focus on developing recommendations for micro levels depicting various geo-climatic regions in country.

There is a need for not only for developing strategies at regional or national level which is largely a top-down approach but more importantly, to develop a bottom-up approach to address climate change mitigation and adaptation. This research approach therefore should aim at adopting a bottom-up approach for sustainable land use development and to develop mitigation and adaptation strategies, programmes and projects and subsequently after the fruitful research results, to undertake pilot mitigation and adaptation projects in various agro-climatic regions of India.

3 Problems Intended to Be Addressed by Proposed Research Project

The consequences of climate variability could be manifested in several ways, i.e., floods, droughts, unprecedented rains, inconsistencies in seasonal temperature etc. This poses a challenge to developing world due to its dependence on climate-sensitive economic activities and predominantly in practicing rain-sustained agricultural activities (Shisanya and Khayesi 2007). About 60 % of the total population in India are engaged in practicing land use related activities. In rural ecosystems in India, the land use (agriculture), forests, and animal husbandry are strongly interlinked together. Therefore, un-sustainability in any one of the branch leads to unsustainable development and consequences appear in terms of unsustainable resource flow among the different branches of the ecosystems, pollution, overexploitation of the natural resources, biodiversity/habitat loss, and migration from rural to semi-urban and urban areas. Hence, it can be foreseen that the high demand for resource will deteriorate the ecosystem quality and reduction of resource supply required for sustainable land use. To maintain the equilibrium between sustainable land use development and resource availability, additional scientific and technical efforts require toward developing adapted cultivation strategies and techniques to keep the crop yield in a balance with the growing population, depleting resources and vulnerability of land use-related practices to climate change (Nautiyal 2010). It has been stated that unfavorable climatic conditions are one of the most serious hurdles to sustainable land use development in India (Swaminathan 1999). Research on the identification of issues related to climate change and consequently the evaluation of current land use ecosystems and resource use efficiency are therefore necessary to provide a conceptual framework for actions in this respect.

Developing countries like India already face social, economic and environmental stresses and resource constraints that limit their ability to adapt to climate change (IPCC 2001, 2007; WBGU 2009). In contrast producers from developed countries have the physical, technical, and economic resources to moderate or adapt to, the impacts of climate variability on food production system (Fischer et al. 2005; Challinor et al. 2007). In country like India consistent production from the agriculture sector is inherently sensitive to variability in climate as largely the production-related activities depend directly on rainfall. Therefore, climate is key driver to food security and irregularity in climate which certainly influencing the production practices, promoting the migration (mostly the male population) from rural to urban areas to seek employment/jobs to fulfil their family requirement (Findley 1994; Henry et al. 2004; Nautiyal and Kaechele 2008, 2009). Forced migration hinders development in many ways, i.e., by increasing pressure on urban infrastructure and services, by undermining economic growth; by increasing risk of conflicts for the resources, (Brown 2008) and migration also has consequences on gender-related issues as this pose a tremendous work pressure on women in rural sector which ultimately influences the health of the rural people (Kalkstein and Smoyer 1993; Meze-Hausken 2000) and education of the children in rural societies. Therefore, a study that will focus on impact of climate change on socio-ecological setup and predict the land use and climate scenario is important to undertake in different agro-ecological zone of India for better policy formulation, innovations for land use resource management.

There are several studies/reports available at national and global level (McCarthy et al. 2001; IPCC 2001, 2007; Klein et al. 2004; Warren et al. 2006; Stern 2006; Haberl et al. 2006; He et al. 2006; Fuessel 2007 Halsnæs and Verhagen 2007; Hellden 2008; Li et al. 2008; Vavruset et al. 2008; Nautiyal and Nayak 2009, 2010; Arndt et al. 2010; Bauer and Scholz 2010; Nath and Bahera 2011) that deals with climate change and its impact. However, in India it is important to understand the trends of localized land use change and its impact on rural livelihood systems at micro level. The micro-level investigations can help to identify what are trends of changes in the climatic variables in the region, how these changes are impacting the existing livelihood options and land use? Who are the vulnerable sections of the society? What are impacts, how the affected communities are adapting to these changes? How effective are the formal meso-level adaptation measures? and in this endeavor what solution we would provide based on the researchers' own research? Thus, the attempt has been made to explore the above questions by intensive empirical field studies in some of the geo-climatic regions of the country.

4 Need Attention to Create Proactive Response to Climate Change and Associated Land Use Development

Climate change can be viewed as one of the most critical environmental problems ever to confront us as it is most immediately and inextricably linked to well-being, development, and economic growth. Thus, the response to it cannot be left to the confines of the environment but needs to seek clarity and consolidate its response relating the agendas and interests of the multiple constituencies (WWF online). Therefore, climate change assessment has to be done in viewpoint of impact, vulnerability, and adaptation in different agroecological regions of India. The adaptation strategies are important to be evaluated for future sustainability.

5 Climate Change Assessment: Impact, Vulnerability, and Adaptation

Impact assessment: Impact of climate change has to be assessed using the following approach:

- Climate variability: Assess current climate variability at micro level.
- Obtain climate data at regional level for all the study regions: temperature, rainfall, droughts, floods etc.
- Select the sectors: rain fed and irrigated agriculture, evergreen and deciduous forests, arid regions, river valley systems, and water resources availability etc.
- Models for impact assessment: Methodological perspective of the climate model that need to be followed for this aspect.
- Collect secondary and primary data required for modeling from forests, farms, streams, rivers etc.
- Research results make projections of climate impacts on biophysical and socioeconomic indicators.

Vulnerability assessment: Biophysical and socioeconomic systems' vulnerability need focus of following issues

- Assess the impacts of projected climate change on the selected biophysical or socioeconomic systems.
- Identify indicators for assessing the vulnerability.
- Develop vulnerability indices: agricultural vulnerability, water vulnerability, livelihood vulnerability, biodiversity vulnerability, drought and flood vulnerability.
- Conduct field studies to generate data on parameters and indicators for vulnerability indices.
- Develop a composite vulnerability after normalization.
- Rank the areas/villages according to vulnerability: biophysical and socioeconomic factors.

Adaptation strategies: Adaptation requires approach as described below

- Assess the impacts of projected climate change on the selected biophysical or socioeconomic and socio-cultural systems of all three study regions.
- Identify the vulnerable natural ecosystems and socioeconomic systems: rainfed and irrigated agriculture, evergreen and deciduous forests, coastal fisheries, river valley systems, and water resources availability (ground and surface water).
- Identify practices that reduce the vulnerability and enhance resilience: soil and water conservation, irrigation, multiple cropping, drought resistant varieties, biodiversity enhancement, anticipatory planting of tree species, flood protection barriers etc.
- Identify strategies to promote adaptation: halting forest fragmentation, expansion of irrigation, forest conservation, diversification of farm incomes etc.
- Develop pilot adaptation projects for implementation in other agroecological regions of the country.
- Model-based testing and evaluation of adaptation strategies.

6 Transdisciplinary Research Approach

Varied agroecological regions being selected for the study on climate change, should have diverse pattern in rain fall, soil quality, natural resource availability, agricultural practices, and consequent agricultural outputs (for example Fig. 1). The climate change has a major impact on these factors which finally go on to affect the livelihood and vulnerabilities of farmers as well as the local population. For impact and vulnerability assessment and adaptation strategies to climate change the possibility is to concretely develop science policy-interface for sustainable socio-ecological systems for different agroecological regions. However, this requires sound methodological

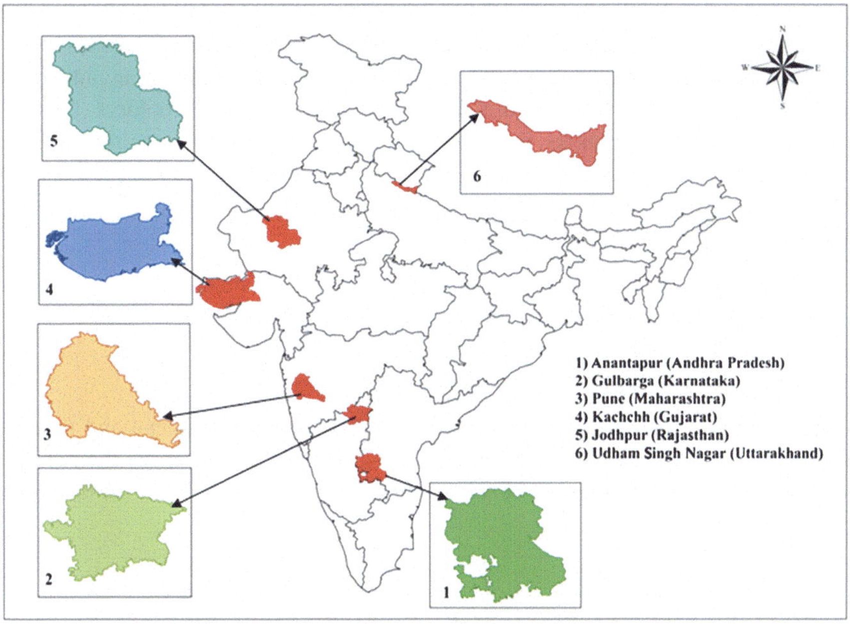

Fig. 1 Selection of study areas in various agroecological regions

perspective from the overall understanding which will emerge stratagem for adaptation. The research approach in this endeavor is described in following sections.

7 Structure of an Integrated Interdisciplinary Land Use Modeling Approach in Indian Perspective for Sustainable Socio-ecological Development

Land use management and sustainable socio-ecological development is becoming the prime concern among researchers, policy planners, and decision makers. It requires knowledge that goes beyond the boundary of any single discipline and covers multiple objectives of researchers from different disciplines. There are numerous integrated interdisciplinary approaches for understanding the complexities of natural systems. Many model realizations which have been built using the integrated framework have successfully been implemented and evaluated in several countries in Africa, South America, Europe, and Asia as well as on the global scale (Kaechele and Dabbart 2002; Schaldach et al. 2006; Lapola et al. 2010; Schladach et al. 2011a; Nautiyal et al. 2010, 2011) and have focus on different aspects of land use systems (urban development, agriculture, animal husbandry, supply and demand for ecological resources, etc.). Thus, the main objective is to analyze sustainable socio-ecological development pathways particularly regarding the relationship between human land use activities and environmental processes.

The aim of research studies should be to build and to apply an integrated interdisciplinary model that supports the decision-making of stakeholders (farmers, policy makers, decision takers, researchers etc.) in the geo-climatic regions they are located. Consequently, the modeling of farmer decision-making processes stands in the centre of research efforts. These processes are not unidirectional but always influenced by several driving forces such as:

- Micro-macro environmental parameters (e.g., climate, soils).
- Socioeconomic resources—location and region specific.
- Various Policies.
- Development of regional and national economy.

These driving forces influence farmer to decide on the landscape and farmers have an available range of action alternatives where they have to make their choice of action (Fig. 2). For future sustainability of the farmer and land use, the decisions

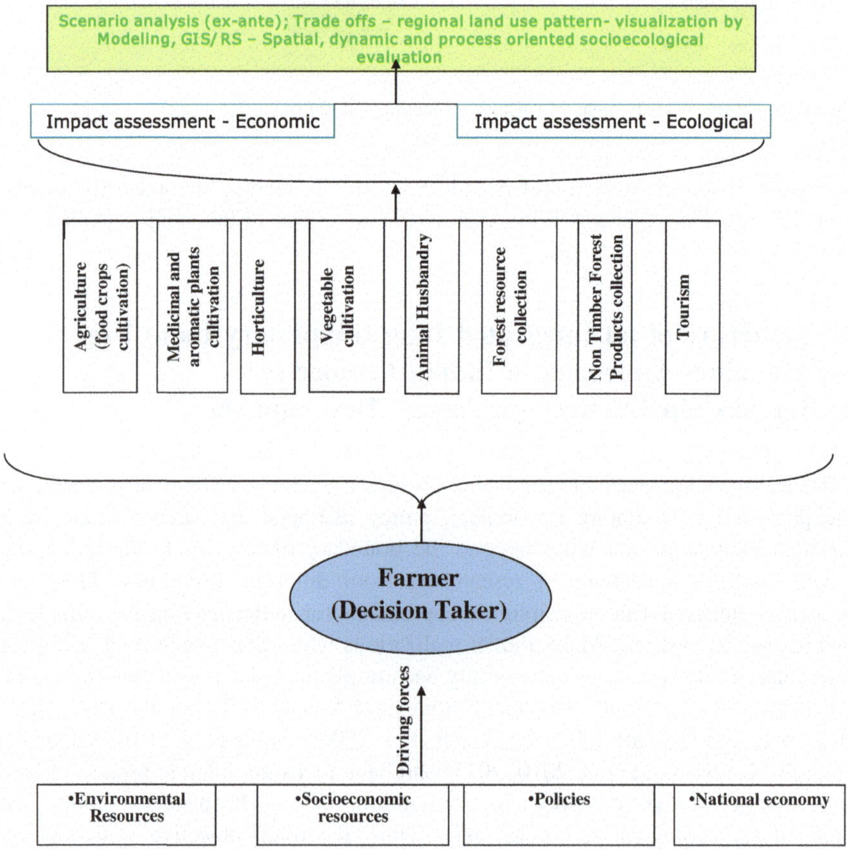

Fig. 2 Theoretical background of the rural landscape in India (Nautiyal 2011a)

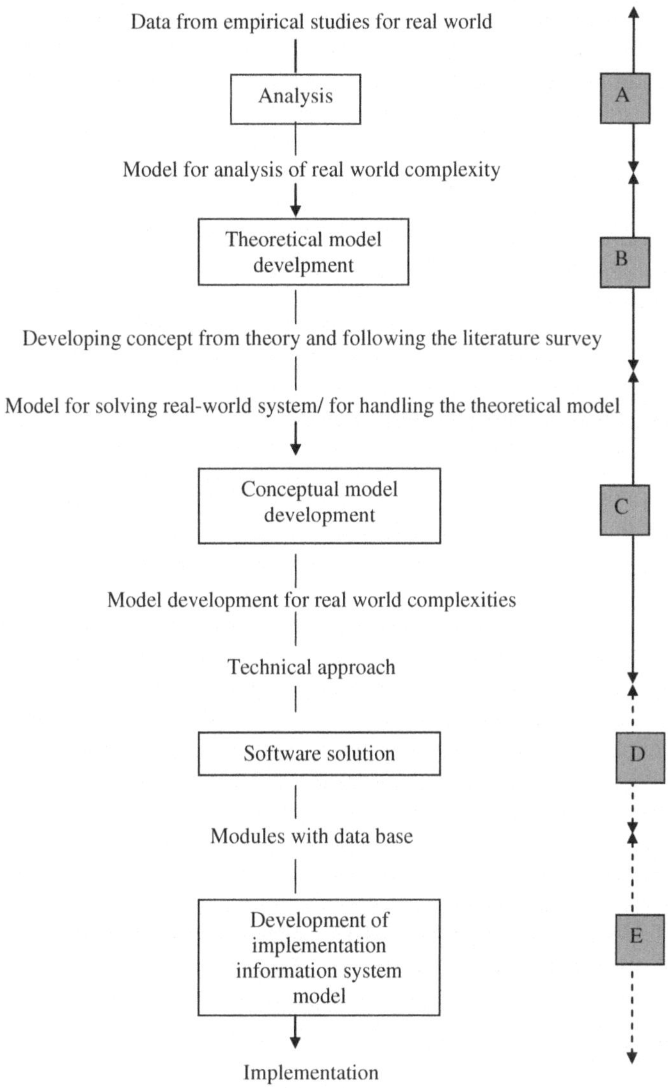

Fig. 3 Research design for real-world modeling (Nautiyal et al. 2010, 2011)

of farmer on every branch have to be assessed in economic and ecological perspectives. The land use in case of Indian rural ecosystems is not an independent sector and this is interwoven with forests. Thus, change in land use makes significant impact on other branches of the ecosystems. To understand the entire change process, it requires adequate information and scientific methodologies to evaluate scenario results of the system functioning.

Thus the integrated modeling approach facilitates the assessment of both the socioeconomic and ecological perspectives—such as—demand on labor, capital, farm land, and the benefit the activity provides for income generation of farmers; biodiversity, soil quality, erosion, water quality, nutrient cycling, and air pollution, ecological resource flow, and several other variables.

The steps to develop the model realization within the interdisciplinary research framework are depicted in Fig. 3. The scheme will also serve as a guideline for the research activities undertaken for developing robust approaches.

The requirement of the integrated interdisciplinary climate-land use model should be as follows.

8 Monitoring System Development

A database system for the storage and structuring of monitoring information is the backbone of the integrated modeling efforts. Monitoring information comprises the field-based data collection and queries, available data based on the empirical field work, data from secondary sources (Census; Statistical Handbook; Food and Agriculture Organization (FAO) and many other data sources), and remote sensing data. Remote sensing is one of the potential approaches to analyze the spatiotemporal dimension of the landscape development. This is a precondition for ex-post simulation analysis and model evaluation. In this direction the data has to be collected for past three decades which can be used and applied. The database can be realized with different freely available software products such as PostgreSQL or MySQL.

9 After the Development of the Monitoring System the Following Three Steps Are Important in the Perspective of Interdisciplinary Research on Land Use-Climate Modeling

1. Model development in a region that is well known to the researchers.
2. Model validation in representative area
3. Model application in a new area.

As discussed with the pervious section of this article, as a basic requirement to developing the model for real word, the representative sites (10 km × 10 km grid size) in some of the geo-climatic regions have to be chosen for initiation of work. As a basic requirement for model development, the regions should be well known to the research team. The expected benefit of this approach is that—basic understanding of landscape is a precondition in contributing to a sustainable land use land cover development and its adaptation to changing climate.

10 Model Validation

- Validation will mainly be done through model development for the well known area with data from past '$t - x$' ($x = 1$, up to 30 years into the past); model validation for well known area with present data on the basis of data from year '$t - x$'; model application—Model application of the model to year t and comparing with data from year t.
- After validation ($t - x$), the model needs to be applied for the future '$t + x$' ($x = 1$, towards future)

11 Overall Scenario Analysis

Through the process of model development the linkages of the modules of the integrated interdisciplinary modeling research efforts are to be associated with each other by means of dynamic LandSHIFT model (Schaldach et al. 2011a, b; Alcamo et al. 2010; Lapola et al. 2010). For multidisciplinary problem of the real world, we argue for bio-economic models which are helpful to solve the problems of economics and ecological complexities simultaneously. The bio-economic models evaluate possible changes in landscape (land use land cover) under different sets of technological and socioeconomic and environmental conditions. The complex rural ecosystem in Indian situation is depicted in Fig. 4. In the centre of this complexity

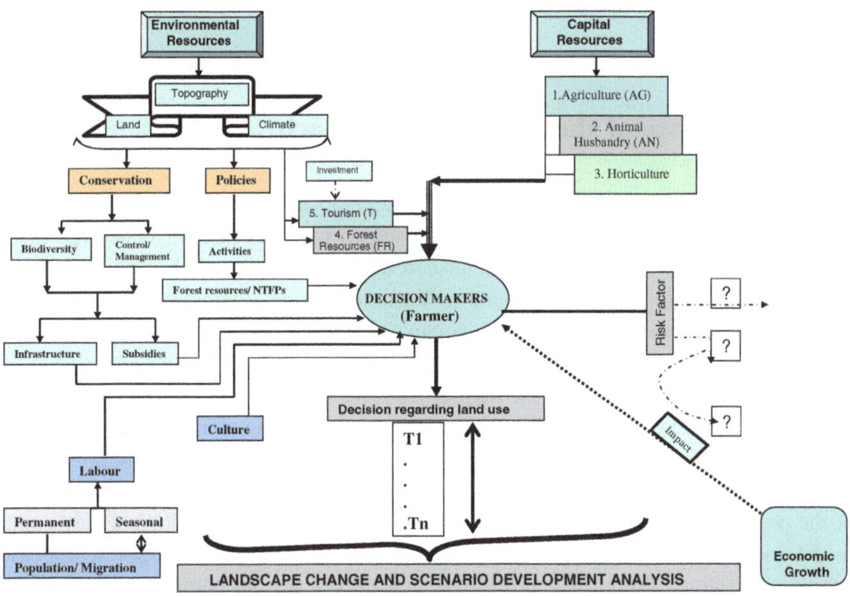

Fig. 4 Complex rural ecosystem in India influences the decision on land use (Nautiyal 2011b)

there are decision makers (farmer), who need to be supported in the decision-making process. In the entire process the approach will encompass relations between researchers and other actors in the policy process, allowing for exchanges, co-evolution, and joint construction of knowledge with the aim of enriching the decision-making process at different levels. Thus, scientific information (*scenario results, trade-offs, regional land use pattern, visualization by modeling, and socioecological evaluation*) from the research will support the requirements of policy demand and have to be easily accessible to the different stake holders. This approach could help evolve effective responses for land use management (i.e., impact, causes, effects, adaptation, conservation and management of biodiversity) at local, regional levels.

12 Validation and Implementation

After developing a LandSHIFT model for various agroecological regions, model validation in randomly selected areas need to be studied for implementation. Data collection have to be followed the structure of present research and meanwhile the parameters and codes need to adjusted based on region-specific conditions (if required). The model has to be implemented in few representative areas. This will offer the opportunity to gain additional knowledge about other landscapes and this is needed for future interdisciplinary research. Preparation data framework and models in the whole approach taken for various agroecological regions (theoretical, policy, technical and research experiments based) are important component of the research emphasizing real-world modeling.

13 Importance of the Modeling Approach

At the very important aspect of the study to develop the land use model for ex-ante analysis in the field of ecological modeling which is so far lacking in viewpoint of when there is need to invite the scientists/researchers from multidiscipline to work together. The land use management in current climate dilemma and sustainable landscape development is prime concern in any region. Therefore, current research will focus to analyze the complexity of the human and ecosystem interaction to develop the tool which would be helpful to solve the problem and can predict the scenario of the future landscape development under changing climatic conditions. The other main goal of the proposed research is to provide appropriate solution to the farmers, policy makers, decision takers while practically evaluating the consequences of the proposed solutions (existing and hidden) in economic and ecological viewpoint. In the endeavor of such integrated interdisciplinary modeling there is not actual case example to quote. Therefore, current research efforts are the commencement in the direction of development of integrated interdisciplinary modeling approach for the evaluation of landscape sustainability in India. In whole process of model development the joint endeavors need to be made by the scientists of various research groups. This will lay down the stepping stones for long-term interdisciplinary landscape research and certainly enhance the understanding of mutual interests desired for integrated interdisciplinary environmental related scientific research between the countries and of course in viewpoint global climate dilemma.

The research results need to be disseminated among the different stakeholders (i.e. people/gender; government bodies/decentralized systems; non governmental organizations/community based organizations/community based social organizations; experts etc.) while organizing workshop/s at various study regions chosen for the study related to climate change. Finally researchers and policy makers need to be informed for the detailed discussion on the research results. Concurrently final research results can be processed in the form of report, books and research articles for the quick diffusion of the research findings and reference for upcoming studies in the field of climate and land use change.

14 Expected Outcomes

- Database on land use land cover change and important driving factors.
- Documentation of TEK and peoples' experiences about the pattern of climate change and its impact on forest, agriculture, livestock and humans.
- To plan possible adaptation measures for vulnerability to climate change towards sustainable landscape development and socioeconomic upliftment of the rural people.
- Effect of climate on seasonal variability and reliability and climate extremes affecting agriculture production, forestry and water resources at micro level.
- Simulation models explaining the changes due to change in climate and information enabling sustainable development strategies for policy formulation and designing and developing the sound database for future research and policy intervention.
- Catalogue of adapted land use management systems and socio-economic development.
- Identification of areas/themes for further research.

15 Approaches/Methodologies for Primary and Secondary Data Collection

Methods in brief for data collection are summarized below. The details of field data collection, data collection from secondary sources (census hand books, statistical reports) and analysis, remote sensing data analysis and spatiotemporal analysis given in Rao and Pant (2001); Farrow and Winograd (2001); Semwal et al. (2004); Manson (2005, 2006); Schaldach et al. (2006); Koch and Schaldach 2006; van den Hove (2007); Hellden (2008); Nautiyal and Kaechele (2007a, b, c, 2008); Nautiyal et al. 2007); Schladach and Koch 2009; Lapola et al. (2010); Schladach et al. (2009, 2011a, b); Fuhrer and Leifeld (2010). To evaluate the complexity of rural ecosystem there is a need to develop bio-physical indicators that would provide quantified information and helps us to explain how things are changing over time. Well chosen indicators able to permit summarizing the complex information (methodology is given in Nautiyal et al. (2003, 2007, 2010, 2011); Nautiyal and Nidamanuri (2010). To understand the process of how the people have been changing their lives and activities due to a variety of factors (such as environmental, policy, socioeconomic), and therefore, how the natural ecosystems also change with human activities, a long-term study of the region is needed. Therefore, the framework for this study is prepared for three years (2010-11–2013-14).

- Village-wise data on population from different study regions would be obtained from the census/district/block records and also from the personal observation (as and when required). This data would enable computation of population growth rates.
- All the households have to be surveyed to determine average land holding size, area under different crops, crop compositions, cropping patterns, crop rotations, animal husbandry and forest resource collection in view of their understanding about the change in climatic conditions.
- In addition, investigations will also be made into the attitude of the farmer towards their thought about factors responsible for the conditions at the time—such as change in climate—consequently implementation of policies, socioeconomic change, population growth, national economy, limitations for land use, resource, infrastructure etc.
- To gather this information, sampled household must be visited at least 5–6 times during the study period (2010-11 to 2013-14). To measure crop diversity more accurately, the total land used for agriculture must be surveyed to assess the actual area under cultivation of different crops during winter (October/November–April/May) and rainy (May/June–September/October) seasons.
- The area that is under cultivation of each crop during the study period and in the recent past (at various points of time) due to change in climate and/or because of other driving forces, have to be calculated for all the crops by interviewing the villagers of the region. In this process the collected information need cross-checking by taking personal observations in field conditions.
- Data on livestock population need to be collected from the livestock census for the years of land use land cover mapping. In case of unavailability of data from the secondary sources then elders from the households in each village can be asked to recall the livestock population to estimate changes in grazing pressure, resource quantification.
- The different food production systems (i.e. agriculture, medicinal plants cultivation, kitchen gardens, animal husbandry, flow of resources and their

interlinkages and interdependencies with domestic and forest ecosystem) of village ecosystem have to be analyzed in order to evaluate the sustainability of each study region using ecological and economic indicators.

- To conduct the in-depth study related to land use land cover change (LUCC) in the study areas data have to be gathered at different spatial and temporal scales for the detail analysis. In order to evaluate sustainable resource management and economic development, the theoretical models can be designed based on the complex interaction between human and ecosystem in the selected agroecological zones of the country (i.e. the Himalaya; the Western Ghats and Arid region).
- This approach is important to maintaining and storing information from different aspects of the project in one place and in reducing the risk of segregation or fragmentation. This research approach can be advantageous in that it simplify the complex human ecosystem interactions while analyzing and evaluating the entire system.
- Remote sensing techniques, including visual observations and image interpretation, are useful in terms of providing valuable information, in a short time, on large areas. This helps in measurement of biophysical parameters in multi-temporal dimensions and a very important tool in case of missing of historical ground data. Therefore, in the empirical field investigations, the use of satellite data can be considered as importance component of a research framework.
- Spatial datasets such as topographical maps, digital elevation data, and satellite imagery pertaining to different time periods (1972 onwards at 5 years interval) need to be analyzed for preparing various land cover maps depicting the impact of various factors which have shaped the present landscape.
- Data have to be mapped followed by geo-referencing topographic map sheets in 1:25,000 scales which are most actual ones available. Precisely geo-referenced and radiometrically calibrated (Level-1G products) satellite images can be procured for the same season for the years after regular intervals.
- In order to remove the effects of atmospheric influences on the radiometric compatibility of the time-series satellite data, all the satellite data require to be atmospherically corrected, using the 6S radiative transfer code. Further, the spatial resolution of the satellite images need maintenance at the native resolution of 30 m for the Landsat-TM, and ETM+ by resampling the Landsat-MSS (75 m) and the SPOT-5 (10 m) images to 30 m spatial resolution.
- Various classification methods can be used in determining the extent of land cover. Ground-based reference data used for selecting training pixels for the satellite imagery can be identified by using a combination of automatic end-member (spectrally pure pixels) extraction method and ISODATA automatic clustering algorithm. The classified land cover classes need identification and validation based on the available ancillary ground truth data and in-depth field knowledge of the socio-ecological system.

References

Alcamo, J., Schaldach, R., Koch, J., Kölking, C., Lapola, D., Priess, J. (2010). LandSHIFT: Modeling land use change on the continental and global scale—Part II. Application to continental Africa: Testing, scenario analysis, sensitivity analysis. *Environmental Modelling and Software* (in review).

Arndt, C., Strzepeck, K., Tarp, F. Thurlow, J., Fant, C., Wright, L. (2010). Adapting to climate change: An integrated biophysical and economic assessment for Mozambique. *Sustainability Science*. doi:10.1007/s11625-010-0118-9 (online first).

Bauer, S., & Scholz, I. (2010). Adaptation to climate change in Southern Africa: New boundaries for sustainable development? *Climate and Development, 2*, 83–93.

Brown, O. (2008, January). *Migration and climate change* (pp. 1–64). Geneva: International Organization for Migration Geneva.

Brown, D. A. (2010, December 24). An ethical analysis of the Cancun climate negotiations outcome. *Climate Ethics*.

Challinor, A., Wheeler, T., Garforth, C., Craufurd, P., & Kassam, A. (2007). Assessing the vulnerability of food crop system in Africa to climate change. *Climate Change, 83*, 381–399.

Farrow, A., & Winograd, M. (2001). Land use modelling at the regional scale: An input to rural sustainability indicators for Central America. *Agriculture, Ecosystems and Environment, 85*, 249–268.

Findley, S. E. (1994). Does drought increase migration? A study of migration from rural Mali during the 1983–1985 droughts. *International Migration Review, 28*(3), 539–553.

Fischer G., Shah, M., Tubiello, F. N., van Velhuizen, H. (2005). *Socio-economic and climate change impacts for the environment*. Oxford: University of Oxford. Available at: http://african-environments.ouce.ox.ac.uk/events/2007/070312workshop.php

Fuessel, H. M. (2007). Vulnerability: A generally applicable conceptual framework for climate change research. *Global Environmental Change, 17*, 155–167.

Haberl, H., Winiwarter, V., Andersson, K., Ayres, R. U., Boone, C., Castillo, A., et al. (2006). From LTER to LTSER: Conceptualizing the socioeconomic dimension of long-term socioecological research. *Ecology and Society, 11*(2), 13. [online] http://www.ecologyandsociety.org/vol11/iss2/art13/

Halsnæs, K., & Verhagen, J. (2007). Development based climate change adaptation and mitigation—Conceptual issues and lessons learned in studies in developing countries. *Mitigation and Adaptation Strategies for Global Change, 12*, 665–684.

He, X., Zhou, J., Zhang, X., & Tang, K. (2006). Soil erosion response to climate change and human activity during the quaternary on the Loess Plateau, China. *Regional Environmental Change, 6*, 62–70.

Hellden, U. (2008). A coupled human-environment model for desertification simulation and impact studies. *Global and Planetary Change, 64*, 158–168.

Henry, S., Bruno, S., & Beauchemin, C. (2004). The impact of rainfall on the first outmigration: a multi-level event-history analysis in Burkina Faso. *Population and Environment, 25*, 423–460.

IPCC (Intergovernmental Panel on Climate Change). (2001). *Third assessment report. Cambridge report*. Cambridge: Cambridge University Press.

IPCC. (2007). *Working Group II contribution to the fourth assessment report, climate change: Climate change impacts, adaptation and vulnerability* (2007). Geneva: WHO and UNEP IPCC.

Kaechele, H., & Dabbert, S. (2002). An economic approach for a better understanding of conflicts between farmers and nature conservationists—an application of the decision support system MODAM to the Lower Odra Valley National Park. *Agricultural Systems, 74*, 241–255.

Kalkstein, L. S., & Smoyer, K. E. (1993). The impact of climate change on human health: Some international implications. *Experientia, 49*, 969–979.

Klabbers, J. H. G., Swart, R. J., Janssen, R., Vellinga, P., & Van Ulden, A. P. (1996). Climate science and climate policy: Improving the science/policy interface. *Mitigation and Adaptation Strategies for Global Change, 1*, 73–93.

Klein, R. J. T., Nicholls, R. J., & Mimura, N. (2004). Coastal adaptation to climate change: Can the IPCC technical guidelines be applied? *Mitigation and Adaptation Strategies for Global Change, 4*, 239–252.

Koch, J., Schaldach, R. (2006). Adapting a land use change model to the Jordan River region. In Y. Ünal, C. Kahya & D. Demirhan Bari (Eds.), *Proceedings of the International Conference on Climate Change and the Middle East: Past, Present and Future*, pp. 358–364.

Lapola, D., Schaldach, R., Alcamo, J., Bondeau, A., Koch, J., Kölking, C., et al. (2010). Indirect land-use changes can overcome carbon savings by biofuels in Brazil. In *Proceedings of the National Academy of Sciences of the United States of America (PNAS)*. [online] www.pnas.org/cgi/doi/10.1073/pnas.0907318107

Leifeld, J., & Fuhrer, J. (2010). Organic farming and soil carbon sequestration: What do we really know about the benefits. *Ambio, 39*, 585–599.

Li, Y., Chen, D., Barker-Reid, F., & Eckard, R. (2008). Simulation of N_2O emissions from rain-fed wheat and the impact of climate variation in southeastern Australia. *Plant and Soil, 309*, 239–251.

Manson, S. M. (2005). Agent based modeling and genetic programming for modeling land use change in the Southern Yucatan Peninsular Region of Mexico. *Agriculture, Ecosystems and Environment, 111*, 47–62.

Manson, S. M. (2006). Land use in Southern Yucatan Peninsular Region of Mexico: Scenarios of population and institutional change. *Computers, Environment and Urban Systems, 30*, 230–253.

McCarthy, J. J., Canziani, O. F., Leary, N. A., Dokken, D. J., & White, K. S. (2001). Climate change 2001: Impacts, adaptation, and vulnerability. *Contribution of working group II to the third assessment report of the intergovernmental panel on climate change*. Cambridge, UK: Cambridge University Press.

Meze-Hausken, E. (2000). Migration caused by climate change: how vulnerable are people in dryland areas? *Mitigation and Adaptation Strategies, 5*, 379–406.

Nath, P. K., Behera, B. (2011). A critical review of impact of and adaptation to climate change in developed and developing economies. *Environment Development and Sustainability, 13*, 141–162.

National Agricultural Research Project (NARP). Delineation of Agroclimatic Zones of India, http://imdagrimet.org/a170.htm, Viewed on 23 December 2009.

Nautiyal, S. (2010). *Climate change: Copenhagen to Karnataka, State Macro Scan* (p. 6), June 2010.

Nautiyal, S. (2011a). Can conservation and development interventions in the Indian Central Himalaya ensure environmental sustainability? A socioecological evaluation. *Sustainability Science, 6*(2), 151–167.

Nautiyal, Sunil. (2011b). Macro to micro viewpoint of climate change-linking Karnataka to global issue. *Natural Resources, 2*, 22–27.

Nautiyal, S., & Kaechele, H. (2007a). Adverse impact of pasture abandonment in Himalayas of India: Testing efficiency of a natural resource management plan (NRMP). *Environmental Impact Assessment Review, 27*, 109–125.

Nautiyal, S., & Kaechele, H. (2007b). Conservation of crop diversity for sustainable landscape development in the mountains of the Indian Himalayan region. *Management of Environmental Quality: An International Journal, 18*(5), 514–530.

Nautiyal, S., & Kaechele, H. (2007c). Conserving the Himalayan forests: Approaches and implications of different conservation regimes. *Biodiversity and Conservation, 16*, 3737–3754.

Nautiyal, S., & Kaechele, H. (2008). A modeling approach for complex interactions between humans and eco-systems for natural resource management in nature protection areas of Himalayas, India. *Management of Environmental Quality–An International Journal (Emerald). 19*(3), 335–352.

Nautiyal, S., & Kaechele, H. (2009). Natural resource management in a protected area of Indian Himalaya: A modeling approach for anthropogenic interactions on ecosystems. *Environmental Monitoring and Assessment, 153*, 253–271.

Nautiyal, S., Maikhuri, R. K., Rao, K. S., Semwal, R. L., & Saxena, K. G. (2003). Agroecosystem function around a Himalayan biosphere reserve. *Journal of Environmental Systems, 29*, 71–100.

Nautiyal, S., & Nayak, B. P. (Eds.). (2009). Climate change: Data requirement and availability (p. 188). ISEC and MoSPI publication.

Nautiyal, S., Nayak, B. P. (Eds.). (2010). Ecological economics: An approach towards socioeconomic and environmental sustainability (p. 383). ISEC& NIE Publication.

Nautiyal, S., & Nidamanuri, R. R. (2010). Conserving biodiversity in protected area of biodiversity hotspot in India: A case study. *International Journal of Ecology and Environmental Sciences., 36*, 195–200.

Nautiyal, S., Rao, K. S., Kaechele, H., & Zander, P. (2010). Conceptual model development for landscape management in the mountains of the Indian Himalayan region: An approach for sustainable socio-ecological development. *Landscape Online, Journal of International Association of Landscape Ecology, 18*, 1–19.

Nautiyal, S., Maikhuri, R. K., Rao, K. S., & Saxena, K. G. (2011). Livelihood and natural resources is Asian cold desert: A case study from selected villages in Central Himalaya, India. In: K. G. Saxena, L. Liang & X. Xue (Eds.), *Global change, biodiversity and livelihoods in cold desert region of Asia* (pp. 161–176). Dehradun: Bishen Singh Mahendra Pal Singh.

Nautiyal, S., Rao, K. S., Maikhuri, R. K., & Saxena, K. G. (2007). Energy and economic analysis of traditional versus introduced crops cultivation in the mountains of the Indian Himalayas: A case study. *Energy, 32*, 2321–2335.

Rao, K. S., & Pant, R. (2001). Land use dynamics and landscape change pattern in a typical micro watershed in the mid elevation zone of central Himalaya, India. *Agriculture, Ecosystems and Environment, 86*, 113–123.

Schaldach, R., Alcamo, J., Heistermann, M. (2006). The multiplescale land use change model LandShift: A scenario analysis of land use change and environmental consequences in Africa. In: A. Voinov, A. J. Jakeman, A. E. Rizzoli (Eds.), *Proceedings of the iEMSs third biennial meeting: "Summit on Environmental Modelling and Software"*. International Environmental Modelling and Software Society, Burlington, USA, July 2006. CD ROM. Internet.

Schaldach, R., Alcamo, R., Koch J., Kölking, C., Lapola, D., Priess, J. (2011a). LandSHIFT: Modeling land use change on the continental and global scale—Part I. Model development. *Environmental Modelling and Software, 26*(8), 1017–1027.

Schaldach, R., Flörke, M., & Lapola, D. (2009). The potential role of irrigated area for bio-fuel production in Europe: A model-based assessment of irrigation-water requirements. *Advances in Geosciences, 21*, 85–90.

Schaldach, R., Koch, J. (2009). Conceptual design and implementation of a model for the integrated simulation of large-scale land-use systems. In: I. N. Athanasiadis, P. A. Mitkas, A. E. Rizzoli, J. Marx-Gómez (Eds.), *Information technologies in environmental engineering* (pp. 425–438). Berlin, Heidelberg: Springer.

Schaldach, R., Priess, J. A., & Alcamo, J. (2011b). Simulating the impact of biofuel development on country-wide land-use change in India. *Biomass and Bioenergy, 35*, 2401–2410. doi:10.1016/j.biombioe.2010.08.048.

Semwal, R. L., Nautiyal, S., Sen, K. K., Rana, U., Maikhuri, R. K., Rao, K. S., & Saxena, K. G. (2004). Patterns and ecological implications of agricultural land-use changes: A case study from central Himalaya, India. *Agriculture, Ecosystems and Environment, 102*, 81–92.

Sharma, A. (2007). *Enabling agriculture* (pp. 17–24). eAgriculture India, 1–2 August.

Shisanya, C. A., & Khayesi, M. (2007). How is climate change perceived in relation to other socioeconomic and environmental threats in Nairobi, Kenya. *Climate Change, 85*, 271–284.

Stern, N. (2006). Stern review on the economics of climate change, London: Her Majesty's Treasury and the Cabinet Office. Available from http://www.hmtreasury.gov.uk/independent_reviews/stern_review_economics_climate_change/sternreview_index.cfm

Swaminathan, M. S. (1999). *A centre of hope: Harmony with nature and freedom from hunger*. Madras: East West Books.

van den Hove, S. (2007). Science policy interface for biodiversity in Europe.

Vavrus, S., Ruddiman, W. F., & Kutzbach, J. E. (2008). Climate model tests of the anthropogenic influence on greenhouse-induced climate change: the role of early human agriculture, industrialization, and vegetation feedbacks. *Quaternary Science Reviews, 27*, 1410–1425.

Warren, R., Arnell, N., Nicholls, R., Levy, P. Price, J. (2006). Understanding the regional impacts of climate change. *Research report prepared for the Stern Review, Tyndall Centre Working Paper 90, Norwich: Tyndall Centre*. Available from http://www.tyndall.ac.uk/publications/working_papers/twp90.pdf

WBGU. (2009). German Advisory Council on Global Change (Wissenschaftlicher Beirat der Bundesregierung Globale Umweltveränderungen—WBGU). Berlin. http://www.wbgu.de/wbgu_home_engl.html

World Wide Fund for Nature (WWF). http://www.wwfindia.org/?7560/Regional-Policy-Action-Platform-on-Climate-Change-RPAPCC-Policy-Paper

Emerging Model of Ecosystem Services and Effective Governance Through Scalable Technological Solutions Around Protected Areas: Sharing Experiences from India

Sunil K. Agarwal

Abstract Present paper aims to discuss innovative features and "systems approach" adopted in implementing technology-driven developmental programme on "People and Protected Areas (PPA): Conservation and Sustainable Livelihoods in Partnership with Local Communities" in India. This programme was jointly implemented at pilot scale by Science for Equity, Empowerment & Development (SEED) Division, Department of Science and Technology (DST), GOI and WWF-India during 2009 to 2012–13 in a network mode promoting innovative process mechanisms to enhance local livelihoods for communities living around 13 Protected Areas (PAs) in diverse ecological areas. In such initiatives, on an average 40–45 % reduction in fuel wood consumption was reported at various sites by introduction of renewable energy technologies like bio-globules, and improved cooking devices etc. Such need-based interventions have helped to build confidence among the members of the community as they were able to adopt simple technologies providing livelihood benefits, reduced carbon footprints as well as better quality of life. Findings of field-based action research programme suggests that **participatory technological appropriation and transfer (PTAT) involving community** is a necessary process for effective adaptation and governance to develop diverse and sustainable livelihood models for replication at the local level. Such process mechanism will have an impact not only on improving quality of life with livelihoods diversification, but also enable adoption of better conservation practices for sustainable use of forest resources in and around PAs. Paper suggests that the "people centred approach" for need-based technological interventions by adopting "system design" starting with the people and with lateral contribution from Science & Technology (S&T) Knowledge Hubs and close interface with field-based voluntary organizations/conservation groups can led to better impact outreach of developmental projects to evolve and deliver innovative models for conservation and sustainable livelihoods gains at local as well as global level. Adoption of such

S.K. Agarwal (✉)
Science for Equity, Empowerment and Development (SEED) Division, Department of Science and Technology, Government of India, New Delhi 110016, India
e-mail: sunilag@nic.in

S. Nautiyal et al. (eds.), *Climate Change Challenge (3C) and Social-Economic-Ecological Interface-Building*, Environmental Science,
DOI 10.1007/978-3-319-31014-5_35

process mechanism of governance at local level not only help to integrate conservation and livelihood around PAs, but also shows that local/tribal communities and PAs can exist in a progressive society, while, ensuring mutual benefits toward ecosystem services and contributing toward mitigating climate change.

Keywords Protected areas · Sustainable livelihoods · Participatory technology appropriation and transfer · Ecosystem services

1 Introduction

India has a rapidly developing economy and looking for ways to address the interface of poverty, development, and conservation. It is home to over a billion people accounting for about a third of the worlds poor. About 300 million of these poor people are estimated to be dependent on forest resources for their livelihoods. The resources on which the poor depend for their livelihoods are being increasingly eroded due to a complex set of factors including lack of tenurial security, rapidly changing aspirations, external threats and pressures, and breakdown of traditional management systems. This leads to environmental degradation and climate change impact which deepens vulnerability of the poor leading to overexploitation of resources and rural poverty.

In such a scenario, need is being felt to facilitate and accelerate proactively action research in tackling climate change and its impact, and evolving practical strategies for contributing to a low-carbon global economy and addressing associated social dimensions of climate change including its connections with growth and sustainability. Besides, emphasis is being given to support adaptation to climate change by rural and urban people, particularly the most vulnerable, through action research supported with good quality science-based knowledge including the institutional arrangements for managing the same locally (Huq et al. 2003; Klein et al. 2007; APHDR 2012). This requires both systems approach and effective governance of innovation systems and technology delivery through localized strategies to ensure that efforts made in evolving required technological solutions meet the needs of the vulnerable population living in complex, diverse and difficult settings like Protected Areas (PA).

2 Emerging Need: Protected Areas and Effective Governance

The Protected Area network in India covers around 4.9 % of India's total geographical area. The primary objective of a PA is biodiversity conservation. Currently, few direct benefits of PAs accrue to the local communities. However, the cost of conservation is mostly borne by the local communities with limited access to use of resources from within PAs. This often leads to conflict between local/tribal communities and the PA managers. In some areas, the consequences of conflict between PA managers and the local communities lead to degradation of habitat and little or no support for conservation at the local level thus impacting opportunity, vulnerability, and the voice of the poor. Therefore, need is being felt at local as well as at global level for more inclusive approaches/models for effective governance of appropriate scale by strengthening institutions for the development and management of PAs and for generating benefits for the rural poor/communities located on the fringe of natural resources areas in recognition of their stewardship role for conservation actions (Borrini-Feyerabend 1996; Sanjayan et al. 1997; Larson et al. 1998; Bruner et al. 2004; UNEP 2004; McShane 2010; Treves et al. 2010) and for ecosystem services.

In order to demonstrate that local communities and PAs can exist in a progressive society while ensuring mutual benefits, there is a need to learn from ongoing initiatives, promote new ideas and scale-up the impacts of existing programmes so as to influence both policy and practice across the country. Various schemes aimed at balancing the costs and the accrual of benefits of PAs to the local communities have been initiated by the government. Such schemes include development of sustainable livelihoods, sharing of tourism revenues, employment generation, as well as enhanced compensation. These initiatives are being implemented by the forest department as well as NGOs with varying degrees of success. Globally, it is also being felt to design and deliver workable models for greater inclusion for protected area management with consideration for local socioeconomic development in view of mixed experiences with other models like community-based natural resource management (CBNRM) and integrated conservation and development project (ICDP) etc. In this context, strong need is now emerging for effective and sustainable resource management based on traditional knowledge systems/practices and integrating Science and Technology involving local community as major stakeholders (Hyes 2006; Failing et al. 2007; Kothari et al. 2015) for better livelihood options. In this endeavor to benefit the communities living around PAs the Science for Equity, Empowerment and Development (SEED) Division of Department of Science and Technology (DST), Government of India (GOI) and WWF-India designed and implemented a network programme "People and Protected Areas (PPA): Conservation and Sustainable livelihoods in Partnership with Local Communities" at pilot scale for reconciling conservation and enhancing livelihood benefits. This multi-sites network programme of three years duration was implemented involving 12 conservation groups (NGOs) across 13 PAs spread over various eco-geographical regions of India (Table 1).

Table 1 PPA programme: programme locations and partner project implementing agencies

S. No	Protected areas	District/State	Partner project implementing agency	Major communities
1	Kibber Sanctuary	Spiti, Himachal Pradesh	MUSE, Spiti	Bodhs, Sherpas
2	Senchal Sanctuary	Darjeeling, West Bengal	Darjeeling Earth Group (DAEG), Darjeeling	Lepchas, Bhotiyas
3	Srivilliputhur Sanctuary	Virudhunagar, Tamil Nadu	Covenant Center for Development (CCD), Madurai	Paliyans
4	Bhimashankar Sanctuary	Pune, Maharashtra	Applied Environmental Research Foundation (AERF), Pune	Mahadev Kolis, Thakars, Kataris
5	Sitamata Sanctuary	Chittorgarh, Rajasthan	PRAYAS	Bhils, Meena
6	Sanjay Gandhi National Park	Thane, Maharashtra	Rural Communes (RC), Mumbai	Warlis, Katkaris
7	Suhelwa Sanctuary	Sravasti, Uttar Pradesh	Raghvendra Rural Development and Research Organisation (RRDO), Lucknow	Tharus
8	Baisipalli Sanctuary	Nayagarh, Odisha	Vasundhra, Bhubneshwar	Kondhs
9	Nalabana Sanctuary (Chilika)	Puri, Odisha	Center for Action Research and Documentation (CARD), Bhubneshwar	Kandaras
10	Dalma Sanctuary	Saraikela, Jharkhand	Shramjivi Unnayan (SU), Jamshedpur	Santhals, Bhumij, Sabars, Pahariyas
11	Purna Sanctuary	The Dangs, Gujarat	WWF-India, N. Delhi	Warlis
12	Srisailam Sanctuary (Tiger Reserve)	Praakasam, Andhra Pradesh	SAKTI, Hyderabad	Chenchus
13	Kanha National Park (Tiger Reserve)	Balaghat, Madhya Pradesh	Community Development Center (CDC), Balaghat	Baigas, Gond

3 Methodology/Approach

The PPA programme was focused on coordinating and supporting the efforts of local and grassroots NGOs promoting innovative technologies with “**systems approach**” to strengthen livelihoods opportunities for local communities living around PAs. Systems approach was to bring together 'technology providers' and 'technology user groups' (partner agencies) to work and evolve workable field level models for ecosystem services with customization and delivery of appropriate technology for intended benefits involving community at large. The aim was to demonstrate how effectively participatory technological appropriation and transfer (**PTAT**) in network mode could help to enhance livelihood opportunities with scalable solutions to address the problems of conservation as well as livelihood issues together at the local level. Field partners were identified on the basis of their long association with the region and local communities around PAs from diverse ecosystems and those working on sustainable livelihoods and conservation-related problems. WWF-India as a coordinating agency of the programme along with SEED, DST facilitated in designing, implementation, and on site performance review at field sites periodically to track the progress at each site. This process involved consultative process to provide necessary guidance (technical back up support with social engineering aspects) needed at 13 different project sites covering 50 villages around PAs to overcome local problems related to ecological, social as well as livelihood issues involving community as well. To build technical and innovative capacities at the users' level, WWF-India alogwith SEED, DST also organized capacity building trainings with field-based S and T institutions including core groups[1] of DST like BAIF Development Research Foundation, Pune and Himalayan Environmental Studies and Conservation Organization, Dehradun and many other such supporting agencies. Technological linkages support from nearby S and T knowledge hubs helped in capacity building of NGO partners and community with hand-holding in technology adoption and usage under field conditions.

Besides, methods used in respect of outcome deliverables of the programme included thematic and chronological analysis of projects progress reports, interviews with project partner agencies and community groups, collection of primary data on the specific activities and focus groups discussions with different stakeholders of the programme about the adaptive capacity for various interventions linked with technology use and appropriation. Both qualitative and quantitative data were drawn up for better understanding of how technology interface with local institutional arrangements can impact the prospects of conservation challenges *vis-à-vis* livelihood needs around PAs.

[1]**Core group**: Core groups are Science based NGOs/field institutions supported by DST, GOI to promote and nurture them as "S&T Incubators"/"Active Field Laboratories" in rural and other disadvantaged areas of the country to work and provide technological solutions and effective delivery of technologies for livelihood generation and societal benefits (**Available at**: https://www.dsttara.in).

4 Results: Technological Intervention and Adoption for Ecological and Livelihood Gains

The network programme showcased significant social, ecological, and livelihood accomplishments by engaging communities in need-based technology intervention areas (Table 2 and Box 2). For instance, DAEG operating near Senchal Sanctuary, Darjeeling was able to involve village women in producing bio-globules[2] leading toward reduced fuel wood consumption and drudgery involved, thus utilizing time saved in other income generating activities like floriculture and vegetable cultivation with relevant technological inputs (Fig. 1 in Box 2). Similarly, MUSE, another partner agency operating in Spiti, Himachal Pradesh near Kibber Sanctuary introduced solar water heating systems, which had resulted in lesser extraction of fuel wood for heating purposes (Fig. 2 in Box 2) and engaged community in making clay crafts with improved design as one of the livelihood diversification activity. Likewise, Vasundhra from Odisha working with *Kondhs tribe* around Baisipali Wildlife Sanctuary who are engaged in Mahua flower collection was able to reduce forest fire incidences in the Mahua season after the introduction of nylon nets for the collection of Mahua flowers and making value-added products. Some of the partner NGOs also promoted sustainable harvesting and value addition of Minor Forest produce (MFPs) in scientific way which led to an increase in overall annual household income through group enterprise. Other areas of technological interventions promoted by partners' agencies toward livelihood diversification were nursery raising, vegetable cultivation, protected cultivation, composting, and homestead farming with 20–25 % average increase in annual household income. Ecotourism and handicraft development activities were also taken up by many of the partners' agencies as it facilitated in augmenting income of the people. Use of natural fibres by Tharu tribes living around Suhelwa Sanctuary in Uttar Pradesh to make Woolen Durries (Mats) using improved looms has enabled them to get alternative source of livelihood locally.

On ecological front, about 40–45 % reduction in fuel wood consumption was reported at various sites by introduction of renewable energy technologies like bio-globules, dhaba digestor, water heating system and improved cooking devices etc. Such technological solutions and delivery toward ecosystem services with institutional arrangements at local level (Effective governance) has *helped to build confidence among the members of the community as they were able to adopt simple technologies providing livelihood benefits, reduced carbon footprints as well as better quality of life as shown in Box* 1 *and* 2.

[2]**Bio-globules**: Bio-globules technology to make briquettes is a eco-friendly and sustainable alternative to firewood to reduce villager's dependence on the forest. Bio-globules are made with local available biomass, charcoal dust and clay.

Table 2 Technology intervention areas: engaging communities for livelihood and ecological gains at different sites in participatory technology appropriation and transfer (PTAT) mode

Technology intervention areas	Number of partners	Activities: technology adoption for livelihood benefits and diversification
Handicrafts	5	Paper bags making, rope and mat from coir, bamboo articles, spinning wool, Dhurrie(Mat) making, *Lantana*[a] furniture
Renewable energy	4	Biogas, Dhaba Digester[b], biomass briquette, bio-globules (see foot note 2), improved stove, solar energy
Agriculture and protected cultivation	8	Nursery raising, plantation—fuel wood and fruit species, vegetable cultivation, vermi-composting, medicinal plants cultivation, millet cultivation, organic farming, home gardens, mushroom cultivation, bee-keeping and floriculture
Animal husbandry and fodder cultivation	5	Fodder cultivation, cattle rearing(small ruminants)
Drinking water	3	Safe drinking water facility
Minor forest produce (MFP)	3	Collection, processing, cultivation—honey collection, gum collection and processing etc.
Ecotourism	2	Ecotourism training, village Biologist groups
Millet, pulse, cereal and oilseed processing	7	Revival and value addition of Kalamatar and Barley, value addition of Nagli
Others	2	Fishing and fishing nets, machine repair and support service

[a]**Lantana**: *Lantana camara* is an invasive weed which encroaches rapidly forest and agricultural land. Opportunistic use of this weed as a resource with simple technologies for various economic applications (for making cheap furniture, utility articles, storage structure, honey bee boxes from *Lantana* sticks) can help not only to control its spread but also to provide livelihoods options locally

[b]**Dhaba Digestor**: It is a technology system for production of clean fuel based on anaerobic fermentation technology to treat organic waste for cooking and other applications. System is constructed by using two plastic water tanks which can telescope into each other with inlet and outlet to digester and gas cock to gas holder

Box 1. Impact of Model PPA Programme: Effective Partnership for Engaging Community in Technology use for Social, Ecological and Livelihood Benefits around Protected Areas

Linkages and Effective Governance

- **13 Protected Areas** from different ecosystems along with 13 project partners.
- A total of **50 villages** were covered under PPA programme with a population of around **25,000**.
- More than **3750 tribal households** have been involved directly in the programme. Overall, the programme has engaged with over 66 existing village level institutions and created around **60 new groups/institutions**.

- Self-management of credit, intra-group loaning, and self procurement of raw materials.

Major Technology-Based Activities: Adoption and Outreach

- MFP Value Addition, Herbal Products, Nursery Raising and Medicinal Plant Cultivation = 1107.
- Floriculture, Agro-forestry, Composting, Vegetable and Mushroom Cultivation, Agriculture and Forest Home Garden = 1677.
- Animal Husbandry and Fodder = 237.
- Millet, Pulse, Cereal and Oilseed Processing = 212.
- Safe Drinking Water = 271 (does not include tourists and other indirect users).
- Weaving and Craft = 192.
- Energy efficient devices- Biogas, Dhaba digester, Bio-globule, Charcoal making = 159.
- Trainings on Seaweed cultivation and improved fishing = 250.
- Scientific bee-keeping and Sustainable harvesting of quality honey.

Capacity Building for Alternative Livelihoods

CAPACITY BUILDING FOR ALTERNATIVE LIVELIHOODS		

Use of natural fibres by Tharu tribes living around Suhelwa Sanctuary in Uttar Pradesh to make Woolen Durries (Mats) as an alternative source of livelihood.	**Value addition with quality control to make Jam from Mahua *(Madhuca indica)* flowers to provide livelihood sustenance for the tribes in Odisha.**	**Women engaged in making Bio-globules from waste biomass: Eco-friendly technology with reduced dependency on forest for fuel wood around Senchal Sanctuary, Darjeeling, WB.**

Green Livelihood Models of DAEG (Fig.1) & MUSE (Fig.2)

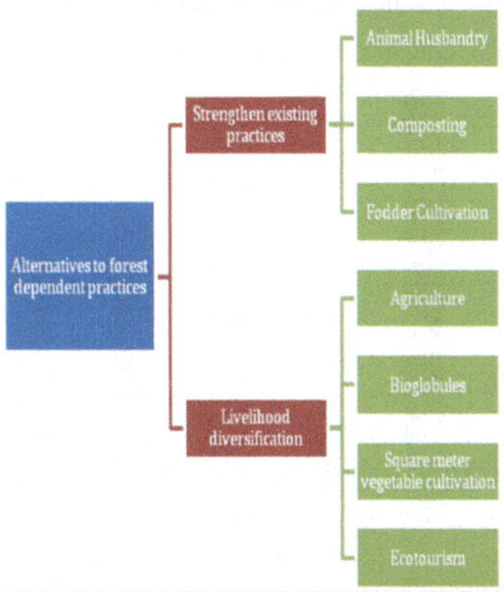

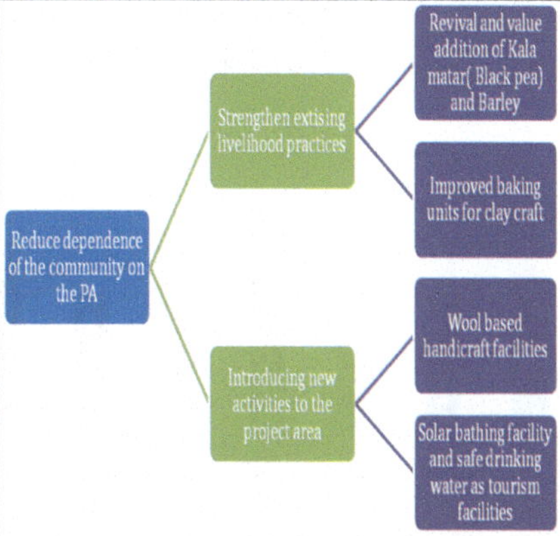

Ecological Impact	**Economic Impact**
Sustainable resource use and value addition of MFPs - better conservation practices. Significant reduction in fuel wood extraction and fodder collection from PAs in some project sites by adopting energy efficient devices and fodder development. Reduction in use of chemical fertilizers. Reduced carbon footprints with conservation gains.	Development of diverse models of sustainable livelihood for communities staying around PAs - enhanced household income. Development of value added products like Mahua jam, Tamarind cake, Ragi biscuits, mud pottery and coir mats etc. Establishment of linkage between local institutions and marketing partners.

Social Impact

Empowerment of local institutions for better facilities in their villages around PAs.

Improved Quality of life and reduced drudgery especially for women from fuel wood and fodder collection.

Empowerment of women groups to manage enterprises on their own and,

Better relations of local communities and institutions with the forest department, agriculture department, horticulture department, technical institutes, banks and marketing agencies.

Effective Management of PAs : Factors/Approaches for Local Solutions

Decentralized way of functioning transform institutional tools and enable people to be participants and beneficiaries of development governance: Socio-economic values.

Gender integrative participatory technology development/appropriation.

Eco-restoration and livelihood gain through judicious use and deliver of appropriate technology package customized with involvement of people at all levels.

Need is to adopt "systems management approach" for technology absorption with social and managerial inputs taking into account the perceptions of people - Access resource use factors.

Importance of S&T base (science based NGOs) in the actual area for effective technological intervention/ to develop location specific technology package with hand holding of community.

Strong institutional linkages between S&T based field groups/NGOs and R&D institutions to source and disseminate the proven technological package at grassroots which empowers and enable community to seek local solutions.

Equal sharing of benefits amongst different stakeholders.

Box.2. Model Network PPA Programme: Socio-Economic-Ecological Impact and Responsible Factors for Effective Governance

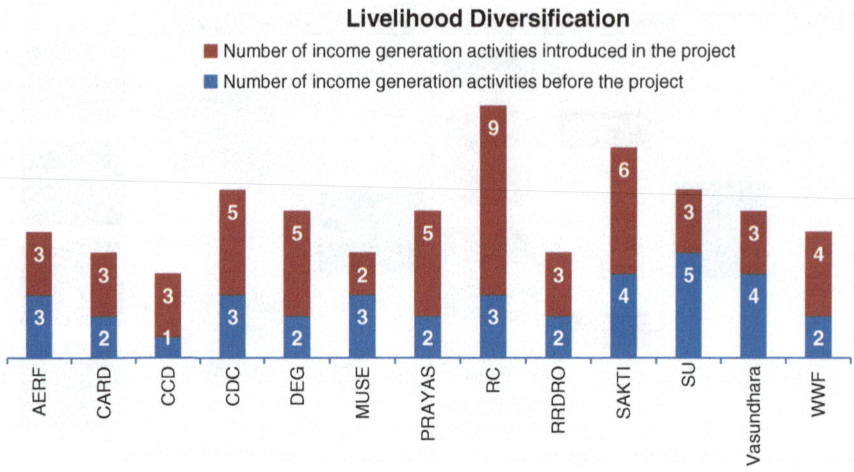

Fig. 3 PPA programme: livehihood diversification at different sites

Programme implementation at various sites also indicates that technologies should be such as would upgrade traditional skills and capabilities; be innovative and capable of easy assimilation; generate significant and assured added value to existing methods of operation; generate employment and use local resources; be capable of replication and adoption and should blend harmoniously with existing ecosystems leading to tangible improvements in the living conditions and self sustained development of the people as well as PAs (Gladwin et al. 2002; Agarwal and Joshi 2006).

Thus, analysis of data from project sites clearly suggests that technology-led livelihood diversification is an appropriate alternative strategy as adopted by households at all locations under the programme (Box 1). Results also show that all NGO partners on an average were able to introduce at least two new activities to the community (Fig. 3) for livelihood diversification. Findings of this model network programme evolved are in line with others who have advocated livelihood support strategies to conserve biological resources (McNeely 1988; GEF 2008; Ekpe 2012; Ekpe et al. 2015) and to understand the effective role of technology appropriateness, scaling up and adoption towards ecosystem services. Moreover, it indicates collective efforts are needed in such an endeavor with the involvement of different stakeholders from government, academic institutions, field level developmental agencies/civil societies, and community as well.

5 Discussion: Appropriate Technology and Scalable Solutions for Ecosystem Services

Programme planning, designing, and implementation at all programme sites evidently shows effectiveness of "people centered approach" involving community to decide need-based interventions for addressing conservation issues with livelihoods diversification for better quality of life. Building capacities and skill up-gradation with appropriate technological interventions and engaging rural youths has demonstrated confidence toward reduce dependency on forest resources and well being of people with scalable livelihood options in farm as well as non-farm sector, thus, indirectly contributing toward mitigation efforts for climate change impact. *Pilot scale intervention covering 50 villages around 13 PAs has clearly demonstrated the way technological innovations and scalable interventions with strong social engineering component as complete package/models could help to enhance livelihood opportunities for local communities as well as conserve the resources they depend upon* (*Box* 1). The programme was also able to establish institutional linkages with corporate, NGOs, academics, government, and some private companies. This process mechanism (Fig. 4) to evolve PPA-PTAT model toward ecosystem services in tune with livelihood gain is absolutely essential to bring better local participation and diverse actors together who have capacities for

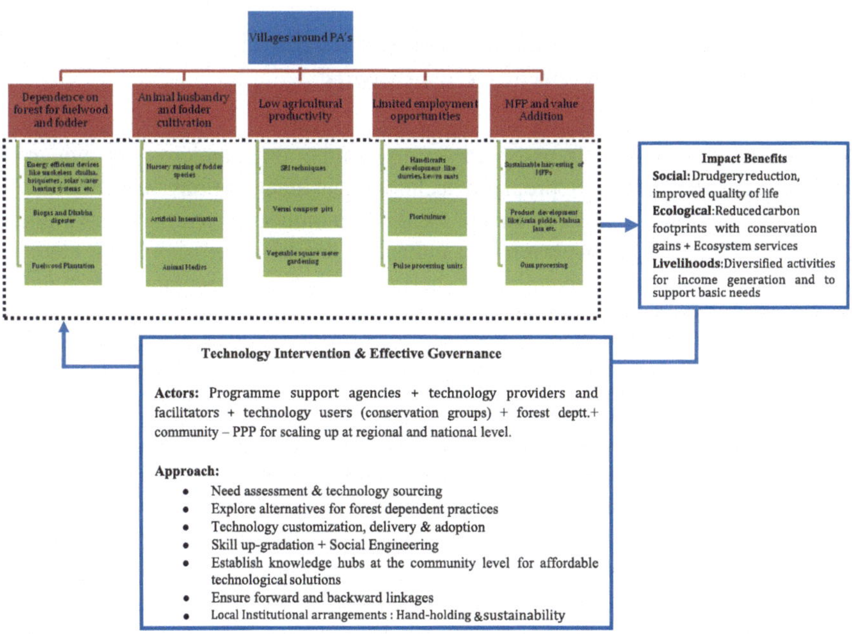

Fig. 4 Process mechanism of effective governance: PPA-PTAT model

innovation and to perform various roles/support knowledge inputs in sustaining developmental programmes (Kochendorfer-Lucius and van de Sand 2000; Padaki and Vaz 2003; Agarwal 2013) particularly for management of PAs.

Analysis of results and discussion with stakeholders clearly suggests innovations in technology modulation as per location specific needs, processes of delivery with local institutional arrangements, and social engineering component to ensure backward and forward linkages with hand-holding of community are essential components from sustainability point of view. Such management strategy in inclusive manner with system approach by engaging different actors can really address the problems of conservation vis-à-vis livelihoods issues in and around PAs at the local, national as well as global level. This is in agreement with suggestions made by Petursson et al. (2011); Feary et al. (2015) and Kothari et al. (2015) for current need of professionalism in protected area governance and management approaches and the fostering of skilled community conservation practitioners through ongoing learning and adaptation or transformation. In such an endeavor, field-based technological interventions cannot only lead to integrate conservation and livelihood around PAs, but also contribute significantly toward sustainable development while mitigating climate change. Strategically, it has to be ensured that such interventions are really location specific and affordable to the participating community to earn their livelihood while, conserving ecosystem, biodiversity, and environment.

6 Conclusions: Adaptations and Institutional Arrangements Involving Community

Analysis of the above discussed PPA programme interventions has evidently demonstrated the role of appropriate technological interventions with affordable solutions to the problems faced by the local communities for livelihood as well as conservation gains. Data with relevant indicators has indicated significant element of sustainability in implementation of such network model programme for social, ecological and livelihoods benefits wherein critical role of various actors and factors have been recognized (Box 1 and 2). Technology models evolved at micro level (like Figs. 1 and 2 in Box 2) have successfully demonstrated that local/tribal communities and PAs can exist in a progressive society, while, ensuring mutual benefits toward ecosystem services. As discussed above, this can be achieved through appropriate technology applications and investments planning in participatory mode. Such technology benefits will open new alternative livelihoods opportunities to local community with reduced dependency on forest resources through good management practices. In such efforts, both environmental and societal adaptations to climate change are programmatically to be embedded as evidently seen by introducing energy efficient devices for cooking and water heating, thus, contributing to reduced carbon foot prints. Analysis of work also suggests that capacity building to adapt toward likely climate change really requires local institutional arrangements on facilitating to identify suitable technological options and scale-up existing knowledge tools and approaches with appropriate one for adaptation (mitigating the effect of climate change) to tackle emerging issues primarily related to energy and natural resources as part of the development planning around PAs. Such field-based interventions also illustrates ways in which communities in climate-sensitive areas are vulnerable to changing conditions, and how climate change adaptation initiatives by enhancing livelihoods provide a practical means of improving people's immediate requirements, and provide them with increased capacities in terms of clean and affordable technological solutions. Further, such an approach of network model of technological intervention to address livelihoods *vis-a vis* conservation issues around PAs also indicates its contribution toward attainment of Millennium Development Goals (MDGs) particularly for improved income and livelihoods; gender equality and women's empowerment, environmental sustainability, and developing partnership and linkages.

To scale-up the impact, second phase of the programme is underway at 16 PAs —showing possibility of widespread application at macro level planning. Programme outcome and impact also suggest that improved synergies and better coordination amongst the wide array of stakeholders are needed for effective governance to meet the challenges of conserving PAs by connecting S&T with society and nature. Replication of such innovative **PPA-PTAT** models of different scale with collaborative strategies can make effective use of technology as a tool for natural resource policy planning and conservation in and around other PAs in a sustainable manner. This requires an enabling ecosystem that encourages innovation in process of governance of PAs—to manage and nurture natural resources and engaging communities in rural and social enterprise sector to meet their livelihood and energy needs for very basic end uses like lighting and cooking in eco-friendly manner.

Acknowledgments Many organizations and individuals have contributed toward the successful implementation of the Network Programme on PPA. The author would like to thanks all of them including DST, GOI, Delhi and WWF-India for valuable support and guidance to make programme more effective at field level. Special thanks are also due to partner project implementing agencies, who generously provided information and shared knowledge and data for the critical analysis of work presented in this discussion paper.

The views expressed in this information product are those of the author alone and do not necessarily reflect the views or policies of organization to which he belongs. Present paper is elaborated version of a voluntary paper accepted for XIV World Forestry Congress held at Durban, South Africa during 7–11 September, 2015.

References

Agarwal, S. K. (2013). Emerging technological intervention models with scalable solutions for adaptation to climate change and livelihood gains in Indian Himalayan Region: Case studies on action research at the grassroots level. In: S. Nautiyal et al. (Eds.), *Knowledge systems of societies for adaptation and mitigation of impacts of climate change* (pp. 575–600). Berlin: Springer. Environmental Science and Engineering (doi:10.1007/978-3-642-36143-2_33)

Agarwal, S. K., & Joshi, A. P. (2006). Technology on the move in the Himalayan region: From empowerment to sustainable livelihoods In: A. P. Joshi, S. K. Agarwal & R. Kumar (Eds.), *Mountain technology agenda: Status, gaps and possibilities* (pp. 140–160). Bishen Singh Mahendra Pal Singh, Dehradun.

Asia Pacific Human Development Report. (2012). *One planet to share: Sustaining human progress in a changing climate* (242 pp). UNDP Publication.

Borrini-Feyerabend, G. (1996). *Collaborative management of protected area: Tailoring the approach to the context*. Gland, Switzerland: IUCN.

Bruner, A., Gullison, R., & Balmford, A. (2004). Financial costs and shortfalls of managing and expanding protected area systems in developing countries. *Journal of Bioscience, 54*, 1119–1126.

Ekpe, E. K. (2012). A review of economic instruments employed for biodiversity conservation. Consilience. *The Journal of Sustainable Development, 9*, 16–32.

Ekpe, K. E., Hinke, C. R., Quigley, M. F., & Owusu, E. H. (2015). Natural resource and biodiversity conservation in Ghana: The use of livelihoods supports activities to achieve conservation objectives. *Journal of Biodiversity Science, Ecosystem Services & Management, 10*(4), 253–261.

Failing, L., Gregory, R., & Harstone, M. (2007). Integrating science and local knowledge in environmental risk management: A decision-focused approach. *Journal of Ecological Economics, 64*, 47–60.

Feary, S., Kothari, A., Lockwood, M., Pulsford, I., & Worboys, G. L. (2015). Conclusion. In G. L. Worboys, M. Lockwood, A. Kothari, S. Feary, & I. Pulsford (Eds.), *Protected area governance and management* (pp. 929–932). Canberra: ANU Press.

Gladwin, C., Peterson, J., & Mwale, A. (2002). The quality of science in participatory research: A case study from Eastern Zambia. *Journal of World Development, 30*(4), 523–543.

Global Environment Facility (GEF). (2008). *Joint evaluation of the GEF small grants programme* (p. 89). Washington, DC: GEF Evaluation Office.

Huq, S., Rahman, A., Konate, M., Sokona, Y., & Reid, H. (2003). *Mainstreaming adaptation to climate change in least developed countries*. London, UK: International Institute for Environment and Development Climate Change Programme.

Hyes, T. M. (2006). Parks, people, and forest protection: An institutional assessment of the effectiveness of protected areas. *Journal of World Development, 34*(12), 2064–2075.

Klein, R. J. T., Eriksen, S. E. H., Naes, L. O., Hammill, A., Tanner, T. M., Robledo, C., & O'Brien, K. L. (2007). Portfolio screening to support the mainstreaming of adaptation to climate change into development assistance. *Climate Change, 84*, 23–44.

Kochendorfer-Lucius, G., & van de Sand, K. (2000). New ways towards poverty alleviation: Institutional development to supersede conventional project approach. *Journal of Development Cooperation, 200*, 21–24.

Kothari, A., Cooney, R., Hunter, D., Mackinnon, K., Muller, E., Nelson, F., et al. (2015). Managing resource use and development. In G. L. Worboys, M. Lockwood, A. Kothari, S. Feary, & I. Pulsford (Eds.), *Protected area governance and management* (pp. 789–822). Canberra: ANU Press.

Larson, P., Freudenberger, M., & Wyckoff-Baird, B. (1998). *WWF integrated conservation and development projects: Ten lessons from the field 1985–1996*. Washington, DC: World Wildlife Fund.

McNeely, J. A. (1988). *Economics and biological diversity: Developing and using economic incentives to conserve biological resources*. Gland: IUCN.

McShane, T. O. (2010). Protected areas and poverty—The linkages and how to address them. In: D. Roe & J. Elliott (Eds.), *The earthscan reader in poverty and biodiversity conservation* (pp. 145–146). Earthscan, London.

Padaki, V., & Vaz, M. (2003). *Institutional development in social interventions*. New Delhi: SAGE Publications.

Petursson, J. E., Vedeld, P., & Kaboggoza, J. (2011). Transboundary biodiversity management: Institutions, local stakeholders, and protected areas: A case study from Mt. Elgon, Uganda and Kenya. *Journal of Society and Natural Resources, 24*, 1304–1321.

Sanjayan, M., Shen, S., & Jansen, M. (1997). *Experience with integrated conservation development projects in Asia*. World Bank Technical Paper 388. World Bank, Washington, DC.

Treves, L. N., Holland, M. B., & Brabdon, K. (2010). The role of protected areas in conserving biodiversity and sustaining local livelihoods. In: D. Roe & J. Elliott (Eds.), *The earthscan reader in poverty and biodiversity conservation* (pp. 157–164). Earthscan, London.

United Nations Environmental Programe. (2004). *Economic instruments in biodiversity related multi-lateral environmental agreements*. New York: United Nations Publications.

Socioeconomic and Ecological Modeling for Sustainable Landscape Management in Indian Himalayan Perspective

Sunil Nautiyal, Harald Kaechele, Peter Zander and K.S. Rao

Abstract This study presents the development and evaluation of a conceptual model for landscape management in the Himalayas of India following several years' research in the region. Data were gathered through empirical studies and remote sensing in space and time and subsequently analyzed, and some examples are presented. The theory is described and the resultant theoretical model through which we were able to analyze the consequences of changes in landscape and development of whole region via scenario development. The theoretical model was developed in view of the relationship between the people demand and use of the resources in a landscape of the Himalayas of India. The outcome of model was evaluated based on socioeconomic and ecological indicators. Concurrently, we have projected the need to handle the theoretical model (which is qualitative/descriptive) with a quantitative/mathematical modeling approach. For this we have stemmed own approach in viewpoint of our requirements while also undertaking a literature survey on the conceptual approach of interdisciplinary landscape research for additional methodological perspectives. This was intended to explore further options for adopting relevant methodologies to assist the development of the conceptual model for the mountain landscape. The overall process helped us to be transferred our theory to concepts and to understand the science of the whole landscape. Consequently, the notion for conceptual model system MODAM (Multiple Objectives Decision Support Tools for Landscape Management) was materialized and was developed while emphasizing a linear programming approach. Its purpose is to integrate interdisciplinary research planning and to carry out landscape management-related research in the mountains of

S. Nautiyal (✉)
Centre for Ecological Economics and Natural Resources, Institute for Social and Economic Change, Dr. V.K.R.V. Rao Road, Nagarabhavi, Bangalore 560072, India
e-mail: nautiyal_sunil@yahoo.com

H. Kaechele · P. Zander
Leibniz-Centre for Agricultural Landscape Research (ZALF), Institute of Socio-Economics, Eberswalder Str. 84, 15374 Müncheberg, Germany

K.S. Rao
Department of Botany, University of Delhi, New Delhi 110007, India

S. Nautiyal et al. (eds.), *Climate Change Challenge (3C) and Social-Economic-Ecological Interface-Building*, Environmental Science,
DOI 10.1007/978-3-319-31014-5_36

the Indian Himalayan region where landscape complexity is comparatively high. The concepts and insights described in this article are not unique among the disciplines; however, they should provide the basis for a discussion of decision-making issues among multidisciplinary experts working on issues of sustainable development within complex environments.

Keywords Empirical studies · Theoretical model · Conceptual model · Sustainable development · Trade-offs · Landscape · Himalayas

1 Introduction

Landscape degradation is the major problem recognized in many parts of the world. The conservation and development of the Himalayan landscape is one of the most significant areas among the biodiversity hotspots of the world (Singh 2002; Saxena et al. 2005; Nautiyal and Kaechele 2007a, b). In a global context this is important for existing biodiversity; however, at a national and regional level it is the important for the people actually living in the Himalayan region as well as those residing in the adjoining great Gangetic plains (Ives and Messerli 1989; Saxena et al. 2001, 2005; Sen et al. 2002; Semwal et al. 2004). Therefore, sustainable landscape management and development is the crucial issue for the long-term sustainability of the Himalayan landscape and existing natural resources (Nautiyal and Kaechele 2008; Nautiyal et al. 2007).

During the 1970s, the landscape management and conservation approach received considerable recognition under the Man and Biosphere (MAB) program of the United Nations Educational and Cultural Organization (UNESCO). The main aim of this agenda was to support an ecosystem conservation approach along with the development of the local economy and the people. There have been a number of studies conducted on issues related to landscape management, biodiversity conservation, and sustainable land use development in the mountains of the Indian Himalayan region (Singh and Kaur 1983; Naithani et al. 1992; Virgo and Subba 1994; Thapa and Weber 1995; Schweik et al. 1997; Jackson et al. 1998; Nautiyal 1998; Maikhuri et al. 2000; Silori and Badola 2000; Silori 2001; Singh 2002; Rao and Pant 2001; Sen et al. 2002; Gautam et al. 2002; Saxena et al. 2001, 2005; Kala 2004; Semwal et al. 2004; Rees et al. 2006; Hjortso et al. 2006) where >10 % of the landscape is segregated for conservation and equally expected for the development of the local people (Nautiyal and Kaechele 2007a, b, c). This has also been the case in many other parts of the world (Pimbert and Pretty 1997; Dale et al. 1998; Turner 2000; Jackson et al. 2000; Farrow and Winograd 2001; Kaechele and Dabbert 2002; Schuler and Kaechele 2003; Rees 2003; Zander 2003; Pischke and Cashmore 2006; Huggett 2005; Herrick et al. 2006; Pinto-Correia et al. 2006; Prato 2007). Landscape management is becoming the prime concern among researchers, policy planners, and decision takers. This approach goes beyond the boundary of the single discipline and covers multiple objectives of researchers from different disciplines. Therefore, it would be noteworthy to analyze the complexity of the human and ecosystem interaction and to consequently propose a tool that would be helpful to understand the science behind whole landscape management. Such a tool could also provide feasible solutions via a problem-solving approach and consequently assist the goal of sustainable landscape management. It is important to recognize the potential effectiveness of such research but, in the Himalayan context, limited efforts have so far been made on the integrated real-world modeling (Rees et al. 2006; Nautiyal and Kaechele 2008). While visualizing the complexity it is important to develop a theoretical model with guidelines for empirical studies and to recognize that the conceptual model needs to be developed in view of its suitability for the complex Himalayan environment. Therefore, this study endeavors to determine the most suitable methodology for handling the theoretical approach (Nautiyal and Kaechele 2008) from a qualitative/descriptive analysis to a quantitative, mathematical modeling approach.

2 Study Area and Climate

The study area is in the central Indian Himalayan region and located between 30° 17′N-30° 41′N latitude and 79° 40′E-80° 5′E longitude. The whole region is divided into three agro-ecological zones (Lower elevation: >1000 m asl, middle elevation: between 1000 and 1800 m asl, and higher elevation: >1800 m asl). The current study was completed in the higher elevational zone of the Central Himalayas and represents the whole higher elevational zone in which the area is located. The area has great ecological importance for its rich biodiversity and ecosystem services (Singh 2002; Singh et al. 2004; Saxena et al. 2005). The world famous national park "Valley of Flowers" and a Himalayan Biosphere Reserve "Nanda Devi" are established in the region for the ecosystem conservation and the development of the local economy and its people. The overview of the area is shown in Fig. 1. There are 34 villages currently in the boundary of the whole region and 10 of them were studied in detail. The total population of the villages was 2762 (947 male adults, 781 female adults, and 877 children below 15 years of age). The villages located in the higher elevational zone were similar (Nautiyal 1998; Nautiyal et al. 2007) and considered as a functional unit of development in the mountains of the Central Himalayan region (Nautiyal 1998). Therefore, considering the cluster of villages or a single village for long-term empirical study is equally important to design and develop new strategies for sustainable landscape development.

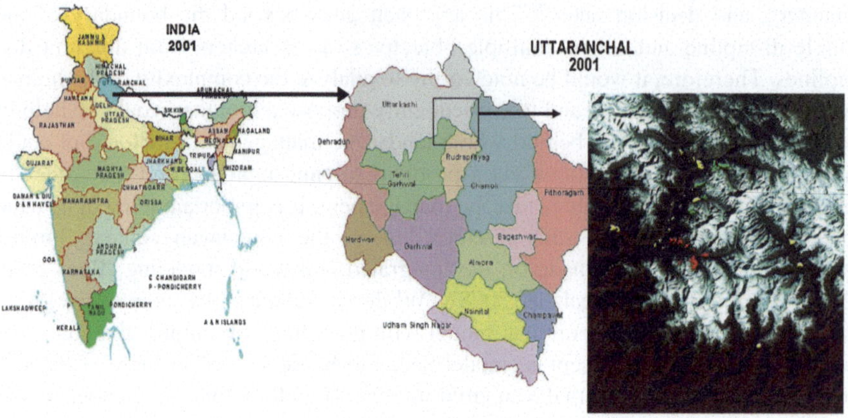

Fig. 1 Location of the study area in Indian Himalaya (Landsat-7 ETM+ image of the area covering)

3 Methodology

3.1 Design the Methodological Framework

Data from empirical studies and data from remote sensing were analyzed to explore the whole landscape in its multi-temporal and multidisciplinary dimensions. To determine the impact of scenario development in the region we have collected the data sets for the last three to four decades through the empirical studies and have also gathered some information from secondary sources such as the Census of India (2001). The theoretical model was developed while evaluating the whole system in space and time (Nautiyal and Kaechele 2009). To understand the process of how the people have been changing their activities due to a variety of factors (such as environmental, policy, socioeconomic), and therefore, how the natural ecosystems have also changed with human activities, a long-term study of a region is needed. The work conducted during the past decade in the Himalayan region of India facilitates the start of work in this direction, and thus the work plan for the data analysis was developed (Fig. 2). The whole framework is categorized in four parts (A, B, C, and D). Part ‘A’ comprises long-term empirical studies and theory to understand the science of whole landscape management and development of the theoretical model. Part ‘B’ is associated with the development of the conceptual model in the framework and has now been completed. Part ‘C’ is ongoing with the development of the software solution; however, part ‘D’ is the upcoming activities needed to implement for our results and outcome in the real-world scenario. The advantage of our research done thus far is to invite researchers from multiple disciplines for their suggestions in our ongoing activities so that it would be possible to make necessary amendment (if needed) in the procedure. This approach is

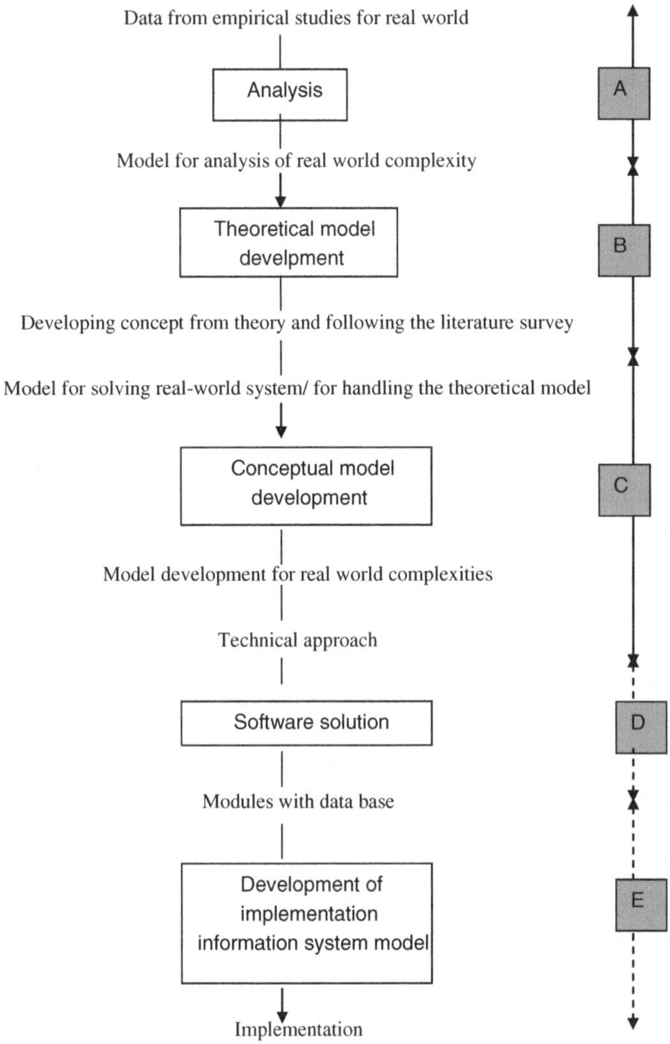

Fig. 2 Research design for conceptual model development needed for the real-world modeling (Nautiyal et al. 2010)

much more valuable for analyzing real-world complexities (Wand et al. 1995; Bettinger and Boston 2001; Farrow and Winograd 2001; Kaechele and Dabbert 2002, Gemino and Wand 2004). Omitting this approach in designing and developing the theory and concepts has been identified as a main reason for the lack of success of models (Herrick et al. 2006; Gonzalez-Perez and Henderson-Sellers 2007; Nautiyal and Kaechele 2009). Therefore, it is necessary that the theory and concepts are well defined at the very beginning of the project, and that they are designed and developed for tangible execution of the results in a real-world scenario (Nautiyal and Kaechele 2007a). During the development of theoretical model we have analyzed the whole landscape characteristics and interaction among the different branches of the landscape/ecosystems management.

3.2 Analysis of Existing Data Sets

This was an interdisciplinary study to bring together ecological sciences with social sciences. The data analysis was conducted in a detailed way for the whole region over three points of time; this enabled us to form the study from a microscopic viewpoint to understand the basic idea of how farmers make their own decisions on the landscape. The analysis approach was particularly exemplary because it has tried to analyze different food production systems of the landscape such as agriculture, medicinal plants cultivation, forest resource collection, and animal husbandry using the currencies in ecological, economical, and social terms of scenario development. Further, it also analyzes the interlinkages and interdependencies of all the branches of the landscape management. The methodology pertaining to the study of agriculture, animal husbandry, forest, and natural resource utilization pattern, and the whole in Sects. 4.2 and 4.3 village ecosystem function is described in detail in Nautiyal (1998), Nautiyal et al. (2003a, b, 2007), Rao et al. (2005), Nautiyal and Kaechele (2007a, b, c).

3.3 Analysis of Remote Sensing Imagery

To understand that how the sustainable ecosystems (including the interaction between anthropogenic and natural resource systems) are altered from past decades, we need to emphasize a holistic or integrating approach. The analysis of remote sensing imagery gives geographic and temporal information for land use and land cover change analysis and is helpful in understanding the degree and extent of human dimensions on environmental change and the outcome of human actions across the landscape. The imagery for the current study is used twice a year, particularly during a two-crop calendar (winter and summer) of the region in a decade interval starting in 1972–2005. The remote sensing data are in use to analyze the spatiotemporal dimension of the Himalayan landscape. Remote sensing provides opportunities to measure the biophysical parameters in multi-temporal dimensions (Holz 1985; Lo 1986; Jensen 1996; Campbell 1997; Wang and Moskovits 2001) and is especially noteworthy when empirical historical ground data are not available (Nautiyal and Kaechele 2007b). A detailed description on remote sensing data analysis for landscape is presented in Nautiyal and Kaechele (2006, 2007a, b, c).

3.4 Development of the Conceptual Model

A literature survey was conducted to follow the methodology for the development of the conceptual model for the Himalayan landscape development (Belovsky 1994; Wand et al. 1995; Balasubramamiam et al. 1996; Dale et al. 1998; Edwards et al. 1999; Tulloch 1999; Bettinger 1999; Martinez-Fernandez et al. 2000; Jackson et al. 2000; Bettinger and Boston 2001; Farrow and Winograd 2001; Baja et al. 2002; Kaechele and Dabbert 2002; Gomez-Sal et al. 2003; Zander 2003; Moody 2005; Petalas et al. 2005; Hillman et al. 2005; Odening et al. 2005; Rees et al. 2006; Leleu 2006; Gonzalez-Perez and Henderson-Sellers 2007, Zander et al. 2007, Rossing et al. 2007). A comparison was made in view of our requirement for handing the theoretical model and concurrently synthesized the appropriate methodology to develop the concepts concurrently for the outcome of the conceptual model.

4 Results and Discussion

4.1 Outcome from Empirical Studies

The empirical studies have considered every sector of the ecosystems of the Himalayan region and analysis was done using the identified indicators. Yet, in a rural ecosystem study, it is impossible to measure every variable of the system. Therefore, we analyzed the indicators which were succinctly defined as "quantified information which help to explain how things are changing over time (DoE 1996) and well-chosen indicators permit to summarize the complex information" (Bossel 1999; Hardi and Zdan 1997; Moldan and Billharz 1997; for further detail see Farrow and Winograd, under indicator section). The results of our empirical study are presented here with the few examples, such as the data on some of the branches of the village ecosystem function—agriculture productivity, resource collection, income of farm management, etc. The empirical studies were undertaken with a view to understand the human and ecosystem interaction under changing environmental, political, and economic conditions. Following are the results from our studies with a few basic examples which are described briefly in view of multi-functionality of the landscape research. For more detail, please see Nautiyal (1998), Nautiyal et al. (2001a, 2003a, b, 2007), Nautiyal and Kaechele (2007a, b, c).

4.2 Innovation and People Behavior

We have studied why and how do people change their way of earnings (including land use) under changing environment (natural, technological) and socioeconomic conditions. We have considered what would be the implications of such changes in

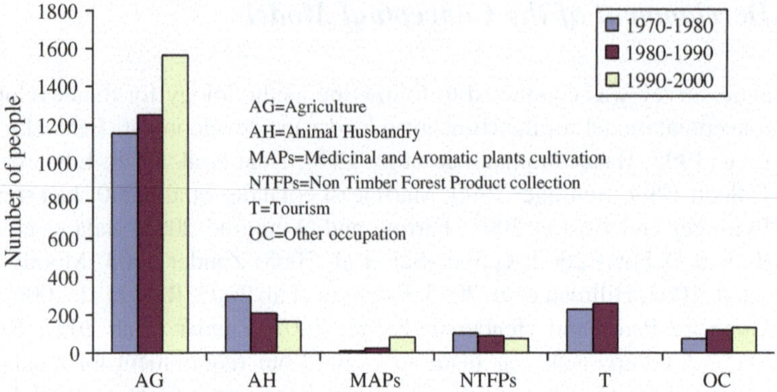

Fig. 3 The activities of the people in farm management at different points of time

long term? The demographic examination at the very beginning of the study was helpful to design the work plan for further activities. The working population for an agriculture sector is increasing, while people are also looking for other occupations besides agriculture. This is because every household has agricultural land and women, in particular, contribute to this sector throughout the year although the land use has drastically changed in the last three decades. We noticed that farmers are trying to explore different sources of income to sustain their life in these fragile environmental conditions. For the current analysis, the period of 1970–80 was the starting period to study agent behavior and ecosystem functioning of the region. Therefore, in the context of time (t1), it is important to analyze the system's functioning and consequently the behavior of the farmers of the area. Meanwhile, the activities are emphasized where a local economy is centered. The temporal data show the activities of people in the region (Fig. 3).

Due to strict conservation policies, such as land segregation for national parks and biosphere reserve, tourism is no longer present in the region. However, the involvement in animal husbandry is declining and interest in other occupations (such as daily wage labor and service) is increasing. Farmers are trying to explore the new possibilities to sustain life in changing circumstances, while also replenishing the loss from a variety of factors. The expansion of cash crops in large agricultural land areas, the domestication of wild medicinal and aromatic plants, and introduction of horticulture are innovations made by farmers and that have spread over time (Nautiyal et al. 2001a, b). Studies of household behavior and projections are gaining more importance. This holds not only for the projections, but also the decision-making behavior for the scenario development. Therefore, there is an increasing interest in micro-oriented interdisciplinary research, which proves that household processes have a large influence on economical and ecological processes (Klevmarken 1983; Nelissen 1991; Rao et al. 2005; Bisht et al. 2006).

4.3 Decision on Land Use to Secure the Livelihood

In the region, the rainfed agriculture on steep terraces is the predominant form of land use and a majority of the population is involved in agriculture. At present, the average per capita land holding ranges from 0.15 to 0.21 ha. The average of decadal growth rate for the study region for the last two to three decades is 15.74 % (Census of India 2001). Therefore, the per capita land holding decreased from 0.023 to 0.033 ha at decadal intervals. The land use change trend was analyzed from 1970 to the present. It is very difficult to analyze the land use change pattern before 1970 due to some flaw in the methodology, missing historical ground data at our level and, more particularly, only using the conventional methods. Therefore, we have taken the land use under each crop during 1970 to evaluate and to understand the land use change trends in the region. Based on the study, we found that the traditional land use system is changing very fast in this region. The magnitude of land use change in the region comes from two types: (a) where traditional land use is totally changed and farmers have introduced modern crops, such as tomato and bell pepper cultivation, to fetch more economic benefit; (b) where traditional land race-based cultivation is being replaced by the high yielding varieties of paddy and wheat and other cash crops (Nautiyal et al. 2007). From the period of 1970–80 and 1980–90, the land use under many traditional crops was reduced between 15 and 60 %. However, this trend continued, and in the year 2000, the decline in land use under traditional crops increased between 50 and 96 % (Nautiyal and Kaechele 2007c). Consequently, the productivity per ha per year from a hectare of agricultural land has increased from 286 kg in 1970–80 to 394 kg in 1980–90 to 579 kg in 1990–2000 (r^2 = 0.9763) (Fig. 4).

In Fig. 5 the decision-making behavior of farmers from an economic viewpoint is shown. Such analysis is important to highlight the dynamic characteristics of ecosystem and socioeconomic change, as well their interaction with each other (Turner 2000). The spatial distribution of farmer activities and related land uses in the region has been documented for decades. During 1970–80 the maximum income was coming from animal husbandry, followed by tourism-related activities and the fewest farmers were found involved in agriculture-related activities in commercial viewpoint [Policy-related issues (Nautiyal and Kaechele 2007a)]. However, gradually the more and more farmers turn toward agriculture for the

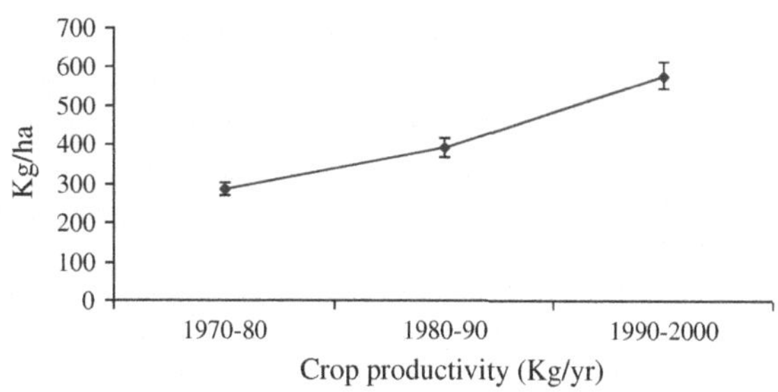

Fig. 4 The per capita per year productivity from the agriculture land use at multi-temporal dimensions in the study region

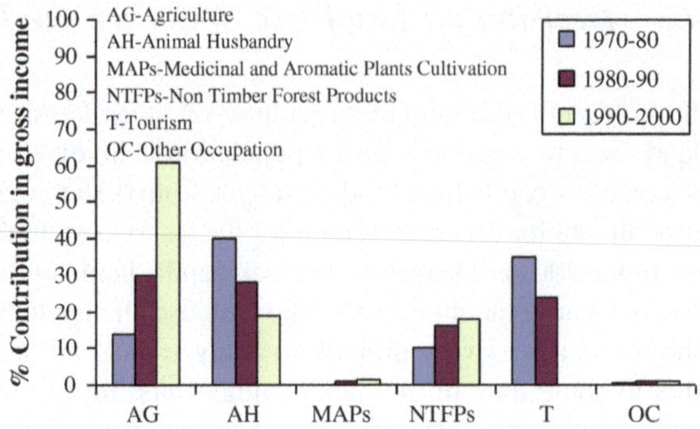

Fig. 5 The contribution of income from different branches of farm management in total gross income of the mountain farmer

commercial use and, from 2000 onwards, we have noticed that maximum income is coming from agriculture and consequently the land intensification is very high. The domestication of wild medicinal and aromatic plants is new paradigm shift in the Himalayan region. The people have adopted this activity as a problem-solving component in their economics (Nautiyal 1998; Maikhuri et al. 1998; Nautiyal et al. 1998, 2001a). During the period of 1970 this sector was only contributing 0.2 % toward the total income of the household. We noticed that currently this sector is contributing 2–3 % in the total gross income of the farmer coming from different branches of farm management. However, the agriculture sector contributes more than 60 % to the total gross margin, though this sector was once only contributing 14 % toward the total income of the farmer (Fig. 5) three decades ago. The environmental pressure builds up via these socioeconomic driving forces and augmented natural system variability, which stimulates changes in environmental systems (Turner 2000).

4.4 The Forest Ecosystem and Resource Collection

The change in production systems (Fig. 5) as well as increase in productivity (Fig. 4) brings about a change concerning the input especially stemming from the forest ecosystems. The production and consumption activities (input/output) resulting from different branches in landscape management interlink to each other, and hence they influence the whole functioning of the ecosystems. The traditional agriculture is dependent on the surrounding forest resources. The perseverance of farmer to secure the optimum output from agriculture is reflected in the per capita per year collection of leaf litter from the forest increasing to 222 kg/year (t3) from 100 kg/year (t1) (Nautiyal and Kaechele 2009). This is because mountain farming in not an independent production system. To maximize the production from each unit of land the dependency for input flow in terms of resource collection from the forests has increased tremendously (Sen et al. 2002; Semwal et al. 2004). There are several other reasons, such as the fodder (crop by product) yield decreasing up to 35–70 % (Nautiyal et al. 2007). The replacement of traditional varieties with the introduction of high yielding varieties is one of the factors found responsible for the high exploitation of the forest resources (Maikhuri et al. 1996; Sen et al. 2002; Bisht et al. 2006; Nautiyal and Kaechele 2009).

Other resource demands, such as non-timber forest product collection (NTFPs) for the forests and alpine pastures, have been quantified and make a significant contribution to the local economy. We have noticed that three decades ago (1970–1980) the NTFPs collection in the form of medicinal and aromatic plants and other wild edibles and for a variety of other uses was contributing 10 % toward the total gross income of the farmers, which was reported 16 % during the period of 1980–1990. Continual incremental change was further noticed from this sector/branch and during 1990–2000 this sector is again making significant contribution, with 18 % toward the total gross income of the local farmer.

4.5 *Landscape Change and Scenario Development*

The effects on the forest ecosystem structure in multi-temporal dimension are depicted in Fig. 6. We have noticed the vegetal cover changes in the forests and have reported that the open land in the forests has increased over time. The value for this class for the forest was noted to have increased from 5 to 14 % for the

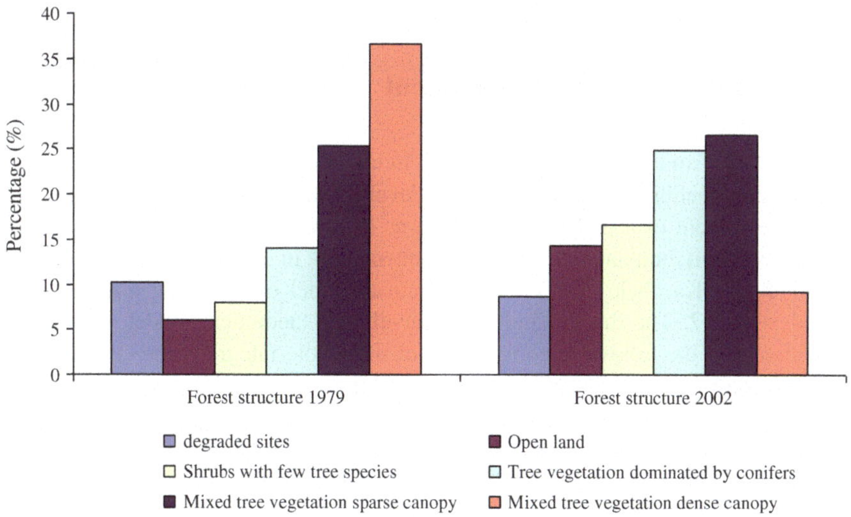

Fig. 6 Consequences of the scenario development, an example for forest cover change in the study region

period of 1979 and 2002, respectively. The analysis showed that in 2002 only 9.16 % of this vegetation class remained dense; however, it was once about 36.61 % in this forest in our research area. Consequently, the spatial extent of the vegetation is dominated by conifers in this area and is increasing. Studies have reported the degradation of broad-leaved forests due to variety of factors (Wakeel et al. 2005) and, unfortunately, the rate of forest loss and fragmentation in the Himalayas of India is very high (Pandit et al. 2007). Thus, the conservation of the Himalayan landscape is a major concern in view of the national as well as the global context (Saxena et al. 2005). The analysis shows that the vegetal cover loss is much higher from the forests and such analysis is crucial in landscape ecological research (Schroeder and Seppelt 2006). The amount, the rate, and the intensity of land use and land cover change are very high in developing countries (Rao and Pant 2001) and assessment of causes and consequences is the first step in developing a successful conservation and management scheme (Brandt and Townsend 2006) to be implemented or refinement in existing policies if needed. This is one scenario of the landscape; therefore, there is an urgent need to understand the future scenario development of the whole landscape in the mountains of the Indian Himalayan region. At the core of such analysis is the development of the theory which enables us to understand the link between the ecosystem processes, composition, and functions with output of the goods and services from the ecosystems and which can be assigned with economic importance as well with ecological values. This approach facilitated us to develop the theoretical model for the Himalayan region to understand the whole landscape scenario.

The few examples presented above based on the empirical studies in Himalayan region envisage the multifunctionality of the landscape in complex Himalayan environment which need to be handled in an integrated interdisciplinary way.

5 Theoretical Model Development

The long-term studies in the Himalayas of India succeeded us toward development of the theory behind integrated scenario development in the mountain landscape. The development of a theoretical model was commenced to understand the complexity between human and ecosystem interaction in the Himalayan landscape (Nautiyal and Kaechele 2009). The theoretical model of the study region is presented in Fig. 7. The theory, and consequently the theoretical model, was developed based on the following components of the whole integrated process: drivers, farmers' options, rules of farmers' decision making, simulate farmers' decision with respect to drivers, indicators to measure the farmers' decisions, and the overall trend of landscape development. The theoretical model (Fig. 7) developed for Himalayan landscape has been developed with six modules/components (T^1–T^6) described as follows:

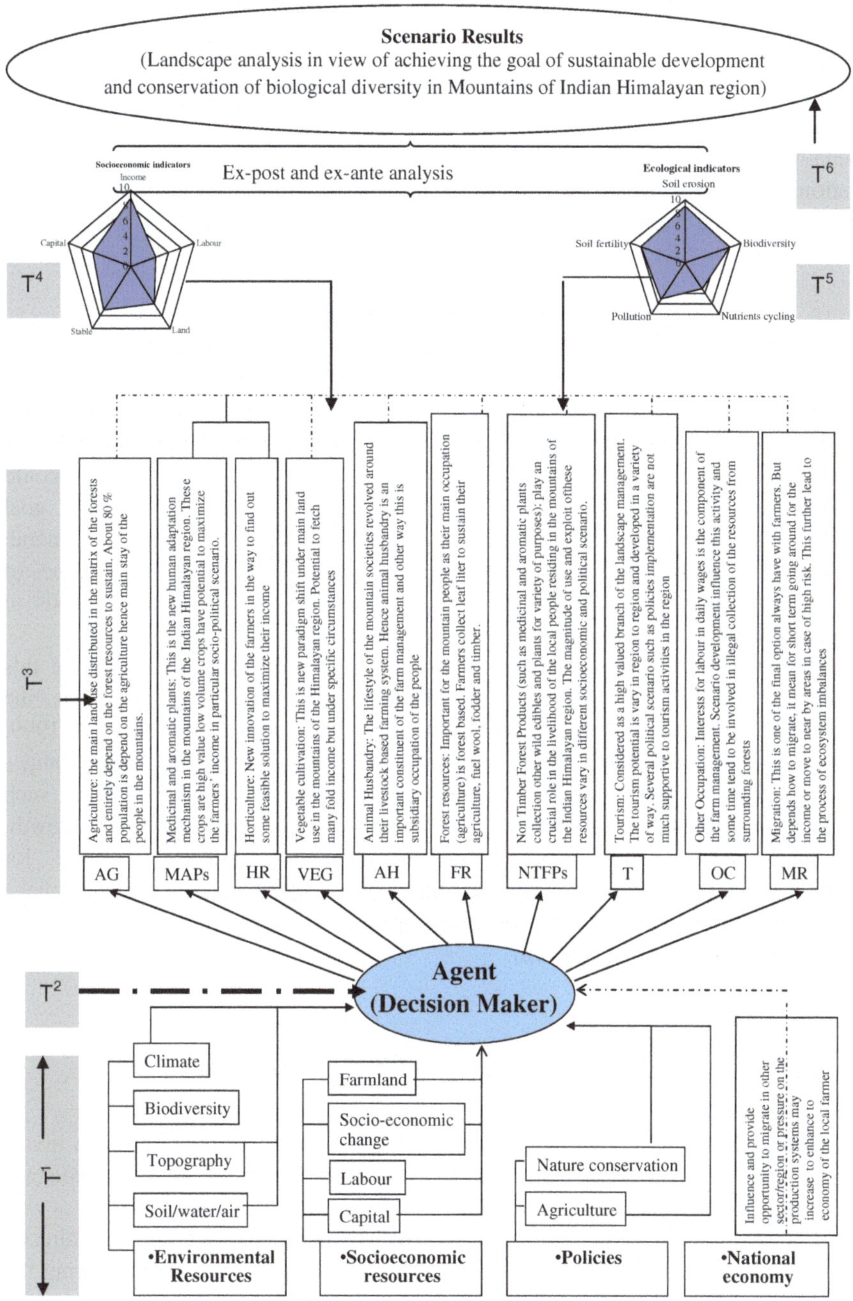

Fig. 7 Development of theoretical model for 1990 onwards is beginning for development of whole model system MODAM for Himalayan perspective (Nautiyal and Kaechele 2008, 2009). (T^1, T^2, … T^6: modules of theoretical model development)

5.1 Drivers (T^1)

The theoretical model was developed based on standard driving forces which influence the farmer's decision. The main driving forces which influence the farmer behaviors are environmental resources, socioeconomic resources, policies, and national economy; these are responsible for making farmers insecure. These are widely accepted from a global to a local scale (Lonergans et al. 2000; IPCC 2001).

5.2 Farmers' Decision (T^2)

Farmers' decisions adjudicate for economic success as well as the ecological performance of the chosen management systems. The theory says that for landscape management in the mountains of the Indian Himalayan region, farmers could choose one or several criteria from different goals, such as from the economic goal and/or ecological goal. The achievement of profitable economic goals, in general, has found to be the priority of farmers decisions and this is one of the outcomes of theoretical model development. In prevalence, the ecosystem analysis leads to evaluation of the efficiency of the system in term of economic goals.

The decision-making behavior of farmers is centered on maximization of this income from the existing resources available to his ease of access. Thus, the investigations were made to judge the attitude of the farmer (Schnell et al. 1999; Atteslander 2003; Pischke and Cashmore 2006) toward overall scenario changes and development and how farmers assess the different conditions at the time, such as the implementation of policies, socioeconomic change, population growth, national economy, limitations for land use, resource, and infrastructure, for example. Finally, we have noticed that among all the options, farmers always make best use of only three to four potential options in viewpoint to multiply their monetary output from the system.

5.3 Farmers' Options to Sustain Their Livelihood (T^3)

Farmers' options vary from region to region, and hence there is a need to investigate all the associated practices/backgrounds which directly or indirectly influence the farmers' behaviors. Options comprehend the current/traditional practices as well as new development tested by scientist. Accentuating this fact we have already adopted the religious, cultural, and technical background of the people during the empirical studies in our research area. Our empirical studies in the mountains of the Himalayas helped us to segregate the farm management where farmers always make experiments to maximize their income to secure returns, at least for next 2 to 3 years. The outcome of our empirical studies has categorized the landscape management (farm management) in the Himalayas of India into ten different branches. They are (1) Agriculture (AG), (2) Medicinal and aromatic plants cultivation (MAPs), (3) Horticulture (HR), (4) Vegetable cultivation (VEG), (5) Animal Husbandry (AH), (6) Forest resource (FR), (7) Non-timber forest products collection (NTFPs), (8) Tourism (T), (9) Other occupation (OC), and (10) Migration to outside (MR). Among all the stated options, farmers have to choose a few of them at the higher scale, while visualizing their potential as a way to secure their livelihood in times to come.

5.4 Indicators to Measure the Farmer's Decision (T^4, T^5)

The farmers' decisions make the overall impact on the landscape (Nautiyal and Kaechele 2009). Hence, all possible activities should be evaluated from a socioeconomic and ecological point of view to measure all possible indicators thereby offering the results to farmers so they may choose the most feasible options. This approach enables the development of the general trend in landscape development and examines the magnitude of use of the resources, such as land, and changes in use of resource dynamics over time (for example, collection of forest resources to sustain the traditional land use). The farm management in the region is assessed by farmers in view of economics. Therefore, all the decisions by the farmers are dominated by expected economic returns, tempered by attitude to the risk. At present, empirical evidence shows that the agriculture sector seems very remunerative to farmers and therefore, they tend to focus their attention on agriculture-related activities. To understand this phenomenon in our study region a glimpse of the trade-off analysis of landscape scenario is presented in economic (T^4) and ecological (T^5) perspectives as an outcome of farmers' decisions on the landscape and is briefly illustrated in Fig. 7. At present, relinquishment of the benefit/advantage from land use has considerable ecological consequences. To facilitate the whole evaluation process we could refer to them as sustainability indicators (Stoorvogel et al. 2004) that make assessment criteria of the economic evaluation, as well sustainability of the whole ecosystems. Thus, the hypotheses framed regarding the relationships between socioeconomic indicators (such as income, labor (female and male), land, and capital) and ecological indicators (such as soil erosion, biodiversity, nutrients cycling, pollution, and soil fertility). In general, we could conclude here that farmers' economic intuition has negative consequences, if envisaging the ecological angle of the landscape and, to some extent, the same results relate to the socioeconomic scenario. Therefore, the analysis of trade-offs of this nature means that multiple indicators need to be monitored simultaneously to assess the management strategies; this is the key requirement to understand the science behind the whole scenario results (Tittonell et al. 2007; Nautiyal and Kaechele 2009),

5.5 *Overall Trend of Landscape Development (T^6)*

Drivers influence farmers to select a few out of enormous options. Simultaneously, farmers decide based on the economic, as well as the ecologic performance of their management systems. A change in driving forces results in different (new) decisions for farmers, and therefore new economic and ecologic performances emerge. The theoretical model developed here enhanced our general understanding to be able to measure the changes in the whole scenario from the past to the present situation, using the present as the baseline for comparison. The model assessed the flows among the resources, influenced by farmers' actions, which are enforced by driving forces in the complex mountain environment. There are several ways to evaluate the overall trend of scenario development and that are possibly generated by the empirical evidences and data collection. We have recognized that efforts to generate the theoretical model are noteworthy as this allows us (a) articulation, which measures the complexity of the landscape and evaluates the resource use dynamics in space and time, and (b) accuracy, which combines measures of modules with interlinkages to each other. But the choice of evaluation of the theoretical model for future perspectives is constrained by existing data sets, and the methods in simulations are in question; therefore, we aim to use the theoretical model as a baseline for a broad perspective of this research.

6 State of Arts of Modeling Approach

After having developed the theoretical approach we tried to identify a modeling approach that enables us to handle the complex issue relating to human and ecosystem interactions in the Himalaya. We use the above derived requirements to assess or evaluate the present bio-economic landscape modeling approaches. Our review suggests that methods and theory testing approaches stemming from our own research were primarily geared toward construction of further conceptual models through the rationale of converging findings in line with empirical studies concurrently with the process of development of our theoretical model. There are several other methods focused on conceptual model development and the implementation of sustainable natural resource management (Wand et al. 1995; Edwards et al. 1999; Martinez-Fernandez et al. 2000; Castillo 2000; Baja et al. 2002; Petalas et al. 2005), but we have focused on some specific methodologies where we have found identical approaches to our concepts as is the case of the Himalayas of India. To accomplish this a literature survey was conducted where research was done on solving the problems of real-world complexities in the form of conceptual model development (Tulloch 1999; Bettinger 1999; Jackson et al. 2000; Bettinger and Boston 2001; Farrow and Winograd 2001; Baja et al. 2002; Kaechele and Dabbert 2002; Zander 2003; Gomez-Sal et al. 2003; Hillman et al. 2005; Moody 2005; Gonzalez-Perez and Henderson-Sellers 2007). A comparison of few models was undertaken to assist the evolution of methodological perspectives. We have articulated that every model has its own prospective and limitation (Table 1). For example, Tulloch's model (1999) has the ultimate approach to database development, record keeping for future perspectives but inadequate for ex ante analysis as well as having limited efficiency for handling real-world complexities. Bettinger

Table 1 Comparison between process models from the literature and proposed conceptual model system MODAM development in this research

Models	Method	Model description	Limitations*
Tulloch (1999)	Theory development (hypothetical)	Model developed for multiple uses of land use systems and emerged with the considerable comparison among the existing models for landscape management	1. Handling the complexities of real world such as interaction among the sectors, etc. 2. Policy scenario analysis 3 Integrated interdisciplinary research 4. Ex ante analysis 5. Potential for the changes as per change in real-world scenario 6. Risk/uncertainty
Van Riet and Cooks (1990)	Dynamic	Ecological planning of natural resource management and landscape development in view of the interaction between human and nature	1. Policies development analysis 2. Scope for integrated interdisciplinary approach 3. Conceptual development 4. Scenario development analysis 5. Risk/uncertainty
Bettinger and Boston (2001)	Dynamic/ conceptual	Focused on four components of the model such as people; databases; technology and organizational commitment to be executed	1. Real-world analysis 2. Ecological perspectives 3. Policy implement analysis 4. Integrated interdisciplinary approach 5. Potential for the changes as per change in real-world scenario 6. Risk/uncertainty

(continued)

Table 1 (continued)

Models	Method	Model description	Limitations*
Gomez-Sal et al. (2003)	Multivariate analysis	Assessment of landscape in view points of multi dimension such as social, cultural, economic, ecological, and production system analysis	1. Policy implementation analysis 2. Scenario development (ex ante simulation) 3. Potential for the changes as per change in real-world scenario 4. Integrated interdisciplinary research 5. Risk/uncertainty
Hillman et al. (2005)	Theoretical/ hypothetical	Multidisciplinary approaches for natural resource management	1. Policies scenario analysis 2. Concepts framework and tool to solve the complexity 3. Scenario development analysis 4. Potential for the changes as per change in real-world scenario 5. Risk/uncertainty
Janssen and van Ittersum (2007)	Mechanistic	Bio-economic farm models Easy understandable Simple and less demanding	1. Potential for the changes as per change in real-world scenario 2. Multidimensional 3. Integrated interdisciplinary research 4. Risk/uncertainty
Model system MODAM Kaechele and Dabbert (2002), Zander (2003)	Linear programming	Integrated interdisciplinary landscape research; Trade-offs/planning, ex ante analysis and scenario analysis and potential for the modification	1. Risk/uncertainty 2. Dynamic aspect 3. Additional external variables

**Note* Limitations are described from the viewpoint of our requirements, not in general. For example, many models are interdisciplinary but we use the term integrated interdisciplinary and this is one of our requirements to handle the complexity of our theoretical model (for detail please see Fig. 7). (Nautiyal et al. 2010)

(1999) and Bettinger and Boston (2001) described four systems—society, database development, technology—and the organizational commitment to implement these in a competent way, but this would be ideal for the simple situation to evaluate outcome of projects but lacks the ability to handle the complex natural environment. Likewise, for several other issues, every model has its own advantages as well its own limitations.

We have evaluated methods for model development considering our own requirements in terms of potential (model description) and limitations (where methods do not cover the specific obligation in view of our theoretical approach), as shown in Table 1.

7 Baseline for Conceptual Model Development

Based on the literature survey we noticed that at present there is no such model fits in a way that we can copy 1:1 in view of the complexity of our research area. Therefore, we have to develop a new model but wish to follow the approach developed for the model system MODAM. The methodological approaches developed by Kaechele and Dabbert (2002) and Zander (2003) for the development of the model system MODAM (Multiple Objectives Decision Support Tools for Agroecosystem Management) fit best in view of solving the mountain complexities in the Himalayas of India. However, this approach has a few limitations but these might be standardized in our new version of the Himalaya. The main issues which the model system (MODAM) emphasizes are field knowledge, empirical evidences, and real-world analysis, and underpinning of theory and concept in an integrated interdisciplinary approach; hence, it is suitable for ex post analysis and significant for ex ante planning problems. The outcomes need to be presented in such a way that addresses the needs of stakeholders into an indefinite future. Thus the strategy must satisfy social, economic, and ecological needs and the fact that these outcomes should continue into the future and imply the need to support sustainable landscape management (Kaechele and Dabbert 2002; Hillman et al. 2005). For handling the complexity of the mountain environment (Nautiyal et al. 2007), it is necessary to demonstrate the framework in view of its multi-objective system function. Long research experience with his endeavor suggests that a single objective goal would be inadequate to address the problem and prospects of the whole landscape study in the Himalayas. Concurrently, the bases for integrated interdisciplinary research were designed in MODAM framework. The major limitations of model system MODAM are (i) it cannot control the risk/uncertainty, (ii) it faces the problems in cases of additional external variables, and (iii) it does not have much efficiency in the dynamic aspects of the model. Besides these few limitations, the model system MODAM offers us immense opportunity to follow the approach developed, as this could be helpful to contribute feasible solutions for sustainable landscape management. The problem related to risk/uncertainty could be handled during the process of development as, in our conceptual model, flexibility is one of the components which offer us further development in the model system. For example, changes in land use activities and resource input are proportionate to production-related parameters of farm management. The model should always offer the chance for improvement in the different modules of the whole model, according to the changes in the real-world scenario (Kaechele and Dabbert 2002).

MODAM focuses on the multifunctionality of the land use/landscape of the region. This article presents only the background from which the concept for transmission of MODAM has been emerged and which, at present, is running in developed countries (such as European countries, North America) and is now going ahead to focusing on the multifunctionality of landscape in developing countries like India (Zander and Kächele 1999; Kaechele and Dabbert 2002; Zander 2003; Schuler and Kaechele 2003). But, in view of the complexity of the Indian Himalayan landscape *we have been developing the model and restructuring the tool for the Himalayan landscape* while following the methodological approaches and major characteristics of the model developed by the researchers. *Overall, we wish to generate sound methodological approaches for the Himalayan landscape which is diverse and has great importance in a global context.* This also describes the factors that determine the forces and relationship between different branches of the ecosystems and concurrently provides a graphical representation of those relationships. *Visualizing the great use of the remote sensing, we have applied and successfully testified forest and pasture ecosystem analysis in multi-temporal dimensions. We have found this approach supports the empirical evidence with more precision in the framework of our current modeling approach MODAM* (Nautiyal and Kaechele 2007a, b, c).

8 An Overview of Conceptual Model Development

The whole MODAM consists of several database and linear programming (LP) modules (Kaechele and Dabbert 2002; Zander 2003). As we are predominately interested in the MODAM approach, hence, we have concentrated on the conceptual core of MODAM—the LP modules. Database development to LP is interesting from a software point of view. Choosing LP seems to be adequate, and hence we refer to the comments of some authors concerning this approach. Before going in detail we would like to provide a basic overview of linear programming in landscape modeling.

For multidisciplinary problem of the real world, we argue for bio-economic models which are helpful to solve the problems of economics and ecological complexities simultaneously. In the real world, when models are used to evaluate possible changes in land use under different sets of technological and socioeconomic and ecological conditions, such models are known as mechanistic bio-economic farm models (Ruben et al. 1998, cited in Janssen and van Ittersum 2007). Often, when such mechanistic models are implemented, linear programming (LP) or some derivatives of LP are also used (Balasubramamiam et al. 1996;

Janssen and van Ittersum 2007). The LP model for farm analysis is explained in detail by Ten Berge et al. (2000) and the use of LP is very sophisticated in modeling without high additional costs in terms of programming development, and therefore could be used smoothly in producing the results from the models (Leleu 2006).

Linear programming models are big equation systems. In mathematics, LP problems are optimization problems in which the objective function and the constraints are linear—such as to integrate capacity of farms and possible activities of farm management. In this equation system the capacities are positioned on left-hand side of the equation system and the demand of the activities (expressed in so-called technical coefficients) concerning each capacity is placed on the right-hand side of the equation system. This is expressed as follows:

$$\begin{pmatrix} y_1 \\ y_2 \\ y_3 \end{pmatrix} \begin{matrix} \geq \\ = \\ \leq \end{matrix} \begin{pmatrix} a_1 b_1 \ldots z_1 \\ a_2 b_2 \ldots z_2 \\ a_3 b_3 \ldots z_3 \end{pmatrix} \times \begin{pmatrix} X_1 \\ X_2 \\ X_3 \end{pmatrix}$$

The y vector defines capacity, a, $b \ldots z$ technical coefficients in the matrix, and target function; the x vector describes the activities part in the matrix.

The description is made by Ten Berge et al. (2000) for each farm activity (called operation or production practices) with corresponding inputs and outputs. To obtain certain output levels different activities need to be defined to model for the different input levels, for example, the availability of labor and available quantity of the resources (vary from region to region and in case of mountain farming we can say the quantity of farmyard manure, collection of the resources from the forests, capital and stables, etc.). For the farm activities the inputs such as labor and manure are limited resources; the constraints for these resources need to be taken into consideration to understand that how the minimum or maximum amount of that particular input can be used for this production system. The activities and constraints are then optimized in view of specific objective function; for example, in production systems we can use the productivity or net profit from the field (Kaechele and Dabbert 2002; Zander 2003; Janssen and van Ittersum 2007). The standard mathematical notation for the LP model can be represented by max ≥ 0.

$$\begin{aligned} &* \text{e.g. to maximize the gross margin} \\ &\max Z = \sum_{i=1}^{m} C_i x_i^- \\ &\text{Subject to } ax \leq b \\ &x \geq 0 \end{aligned}$$

where Z is the objective function: a linear function of the n production activities (x) and their respective contributions (c–coefficients) to the objective.

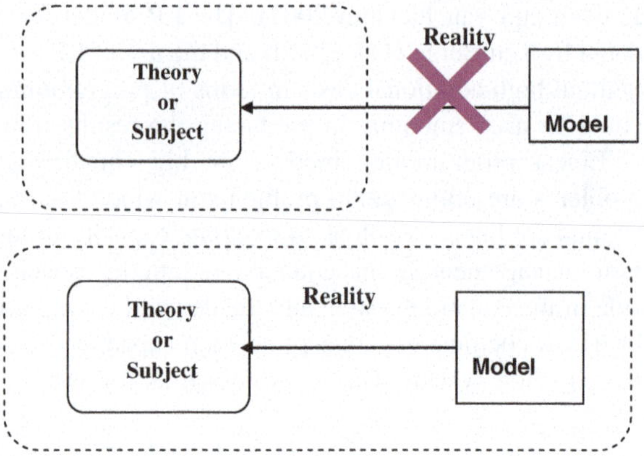

Fig. 8 Model is the part of its reality as its subject or theory (Gonzalez-Perez and Henderson-Sellers 2007)

$ax < b$ representing m linear constraints with right-hand side b. a is an $m \times n$ matrix with total coefficients (Hazell and Norton 1986) (Fig. 8).

From the perspective of superimposing the theoretical model image on the conceptual model image, the terminology in view of mathematical modeling is described in the following sections that how the theoretical components T^1–T^6 (Fig. 7) become part of conceptual model system MODAM C^1–C^6 (Fig. 9).

8.1 *Restrictions (T^1 C^1)*

These are constraints and represent the drivers; visualizing their presence in the landscape scenario, the farmer has to act accordingly and to opt for the feasible alternatives where these restrictions are making minimal hurdles. These are expressed as external variables such as environmental resources (land availability, climatic conditions), socioeconomic resources (labor and capital), policies (subsidies, nature conservation strategies), and national economy (competition for labor and capital). The quantification of variables delivers the left-hand side of the LP equation system. In an LP modeling system the ecological restrictions are derived from the environmental resources, while socioeconomic, political, and national economy factors become the economic restrictions.

8.2 *Goal Function (T^2 C^2)*

Goal function simulates the decision making of farmers in terms of maximizing their income. Therefore, the model chooses several activities out of manifold activities that best utilize given resources and which are restricted by external factors such as environmental or political factors. This procedure leads to maximization of income level.

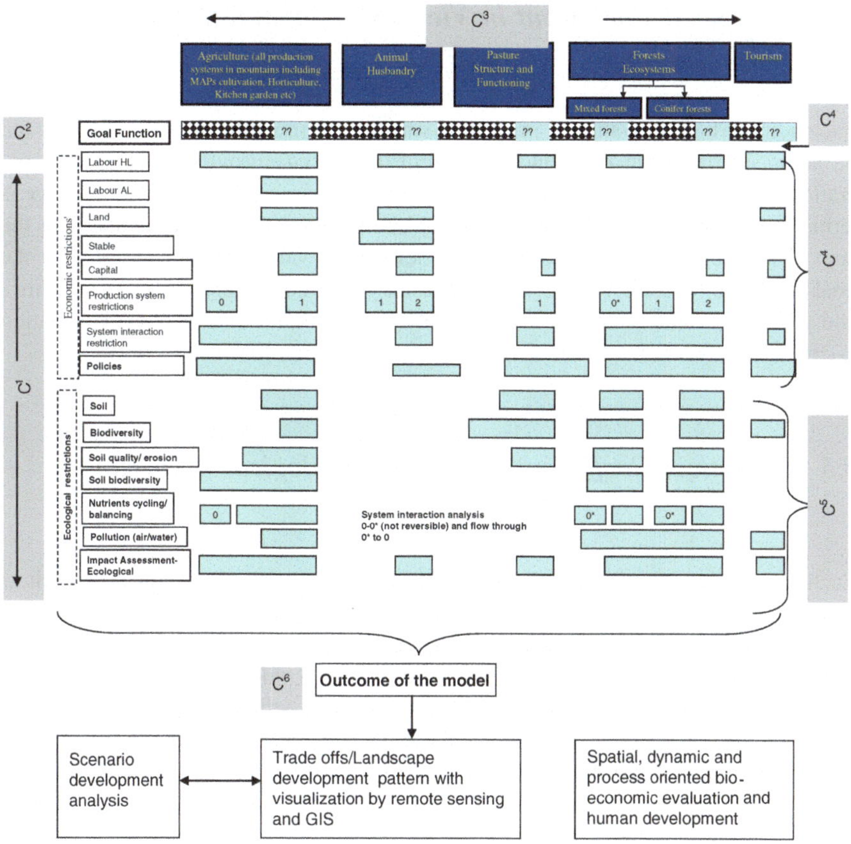

Fig. 9 The conceptual model system MODAM for mountains of the Indian Himalayan region (C^1, C^2, … C^7: modules of conceptual model development) (Nautiyal et al. 2010)

8.3 *Activities in Landscape Management (T^3 C^3)*

In landscape management there are a lot of options available regarding the farmers' ease of access in each branch. For example, in agriculture it is possible for farmers to cultivate several varieties of wheat, paddy, and finger millet under different management practices. In animal husbandry farmers have options to choose any kind of livestock husbandry; further, they have the option to select the different races for different management practices. Likewise, in all the other branches they have many options to perform. In the terminology of LP modeling, options are termed activities. Hence, our approach is to provide as many activities as possible to the farmers covering all the branches. They might choose a few of them in view to sustaining their livelihood (see C^2).

8.4 Technical Coefficient in View of Economic Restriction (T^4 C^4)

Each activity has be evaluated in view of socioeconomic indicators such as demand on labor, capital, farm land, and the benefit the activity provides for income generation of farmers. The outcome of this evaluation in our LP modeling language is termed as technical coefficient. Technical coefficients address the demand of the activities concerning the restrictions and deliver the values of each activity for the goal function. Therefore, in other words, technical coefficients link the activities and restrictions and deliver the values for the right-hand side of the LP equation system (for left-hand side see C^1).

8.5 Technical Coefficient in View of Ecological Restriction (T^5 C^5)

Each activity has also been well evaluated in view of ecological indicators such as biodiversity, soil quality, erosion, water quality, nutrient cycling, and air pollution. The outcome of this evaluation in our LP modeling language is termed as bio-technical coefficients. Bio-technical coefficients address the effects of the activities on the ecological parameters. Bio-technical coefficients in the LP equation systems link the activities to ecological parameters and deliver the values for the right-hand side (for left-hand side see C^1).

8.6 Scenario Results/Outcome of the Model (T^6 C^6)

Scenario results in terms of production, economic, and ecologic viewpoints which emerged from the farmers' decisions under the specific scenario that represents a specific set of drivers (for detail please see C^1). Results show which production system farmers would choose under specific circumstances, what is the income anticipated with this decision, and what is the ecological performance of the resulting landscape management. Results would show the ex post analysis as well as ex ante analysis; the latter has great importance in analysis of the impact of policy options. Economic results could be produced in the form of tables and figures, while spatial and temporal dimensions of the ecological results could be produced in GIS maps. The linkage between economic and ecological outcome could be demonstrated by means of trade-offs. Superimposing of theory and concepts is summarized in Table 2 concerning the theoretical and conceptual model dimensions.

Table 2 Synthetic vision of theoretical and conceptual models developed for the mountains of the Indian Himalayan region

Code	Evaluative dimension	Aim	Character	Theoretical model (theory)	Conceptual model (concept)
T^1 C^1	Driving forces	Aim to understanding the influence in the farmer/agent behavior/decision	Influence human activities	These are the driving forces influence farmer to behave	These are the economic and ecological restrictions and farmer acts accordingly
T^2 C^2	Decision maker	To understand his behavior and model his decision	Mind setup of the farmer to behave	He behaves according to his requirements (socioeconomic and ecological goals) and influences of the driving forces	Solution could be provided to maximize his income and control of farm resources and other constraints (regulate in view of the ecological perspective)
T^3 C^3	Options in farm management	Scope of activities	Choice to choose/Offer all options	These are the options available with farmer	These are the farmers' activities and he has to maximize the possibilities for the best output among them
T^4 C^4	Contribution to income and socioeconomic indicators	To assess the impact of farmers' options from an economic viewpoint	Evaluation of all socioeconomic parameters of each option	To evaluate the monetary benefit from each option and demand on farm resources	Provide technical coefficients of activities (right-hand side of equation system)
T^5 C^5	Assessment of ecological indicators	To assess farmers' options from an ecological perspective	Evaluation of all ecological parameters	Impact assessment of farmers' options on environment (indicator based)	Proved bio-technical coefficients of farmers' activities (right-hand side of equation system)
T^6 C^6	Integrated evaluation	To simulate production, economic, and ecological effects of farmers' decision under specific circumstances	Integrated analysis	Scenario results	Model outcome in form of tables, figures, GIS map, and trade-offs

Code 'T' for theoretical model and 'C' for conceptual model (Nautiyal et al. 2010)
T^1, T^2, T^3 … T^6: modules of theoretical model development as described in Fig. 7; C^1, C^2, C^3 … C^6: modules of conceptual model development as described in Fig. 9

9 Conclusion

From our empirical study we have found that the real world is a complex system concerning landscape management. The key for understanding the changes in land management is the knowledge of farmers' decisions. With these decisions farmers decide not only for their economic perspective but also determine the ecological performance of the resulting land management system. Farmers are influenced by several drivers and follow different goals that are economic, cultural, religious, or ecological or region-specific based. Farmers have many options for action of which some are already implementing and about many other options the farmers are currently unaware. Science has to deal with both options/activities—those that are already established in the landscape and those are in the process of the development by scientists to become visible for the real world. To understand the effects of alternatives options they must be evaluated from a socioeconomic as well as ecological viewpoint. Until now much scientific work is done on investigating parts of this complex system, but few models are available to integrate all of the components of this complex system. Our work is concentrating on understanding this complexity for sustainable Himalayan landscape management and delivering a modeling system that integrates all relevant components. After literature survey we decided for linear programming (LP) approach to integrate all the aspects of multifunctional landscape management. Therefore, our research approach has illustrated in detail the theory and development of the theoretical model and consequently the conversion of theoretical model into a conceptual model. The analysis and the procedure up to the development of the conceptual model was based on detailed field-scale data collected from the study region in the Himalayas of India over a period of several years. This noteworthy as excluding this approach in designing and developing the theory and concepts has been identified a main reason to be unsuccessful of the models. Yet, to be useful to understand the real-world complexities, we followed the integrated interdisciplinary approach. The conceptual model system MODAM which we developed for the Himalayas is more a philosophy rather than a pure software solution. In complex landscape research, as in the landscape of the Himalayan region, it might be difficult to make an integrated approach from the segregated datasets. Therefore, the outlook of MODAM has enough potential to make multidisciplinary research into integrated interdisciplinary nature. Figure 9 shows the structure of model system MODAM for the Himalayan landscape which is intended to handle the theoretical model (Fig. 7) and which we developed for the landscape fits extremely well in view of the landscape complexities (please see Table 2). Our research experience in this endeavor indicates that the modeling is not synonymous with the formulation of a software solution. This is because that the modeling starts much earlier when thinking process about the model starts and comprehends a complex process of preceding research coordination. Hence, from a scientific point of view, the challenge is to link the natural as well as social and economic disciplines together to make the research integrated and interdisciplinary. Developing a software solution is the end of a long modeling process. Landscape research is, of course, a point of multiple objectives planning but ultimately the task should be defined in an integrated interdisciplinary approach. Thus, another important topic for the next step in our research is the development of a software solution to parameterize and simulate the nature of the conceptual model needed for scenario results (theoretical model) in terms of production, economic, and ecologic viewpoints which emerge from the farmers' decisions under specific scenarios that represent specific set of drivers in the real-world scenario.

Acknowledgements SN extends sincere thanks to Alexander von Humboldt Foundation for award of the fellowship. Views expressed here are of the authors only and not of the organizations with which they are affiliated.

References

Atteslander, P. (2003). *Methoden der empirischen Sozialforschung*. Berlin: Walter de Gruyter, 407 pp.

Baja, S., Chapman, D. M., & Dragovich, D. (2002). A conceptual model for defining and assessing land management units using a fuzzy modeling approach in GIS environment. *Environmental Management, 29*, 647–661.

Balasubramamiam, C., Somasundaram, M. V., & Pundarikanthan, N. V. (1996). LP analysis in a tank irrigation system for near real representation and optimal allocation of area. *Irrigation and Drainage System, 10*, 143–158.

Belovsky, G. E. (1994). How good must models and data be in ecology? *Oecologia, 100*, 475–480.

Bettinger, P. (1999). Distributing geographic information systems capabilities to field offices: Benefits and challenges. *Journal of Forestry, 97*, 22–26.

Bettinger, P., & Boston, K. (2001). A conceptual model for describing decision making situations in integrated natural resource planning and modeling projects. *Environmental Management, 28*, 1–7.

Bisht, I. S., Rao, K. S., Bhandari, D. C., Nautiyal, S., Maikhuri, R. K., & Dhillon, B. S. (2006). A suitable site for in situ (on-farm) management of plant diversity in traditional agroecosystems of western Himalaya in Uttaranchal state: A case study. *Genetic Resources and Crop Evolution, 53*, 1333–1350.

Bossel, H. (1999). *Indicators for sustainable development: Theory, method, applications*. Winnipeg, Canada: International Institute for Sustainable Development.

Brandt, J. S., & Townsend, P. A. (2006). Land use-land cover conversion, regeneration and degradation in the high elevation Bolivian Andes. *Landscape Ecology, 21*, 607–623.

Campbell, J. B. (1997). *Introduction to remote sensing*. New York: Guilford Press.

Castillo, A. (2000). Ecological information system: Analysing the communication and utilization of scientific information in Mexico. *Environmental Management, 25*, 383–392.

Census of India. (2001). *Office of the Registrar General, Government of India Press, New Delhi*. India, 2A, Mansingh Road, New Delhi 110011, India.

Dale, V. H., King, A. W., Mann, L. K., Washington-Allen, R. A., & McCord, R. A. (1998). Assessing land use impacts on natural resources. *Environmental Management, 22*, 203–211.

DoE (UK Department of the Environment). (1996). *Indicators of sustainable development for the United Kingdom*. London: HMSO.

Edwards, P. J., Kollmann, J., Gurnell, A. M., Petts, G. E., Tockner, K., & Ward, J. V. (1999). A conceptual model of vegetation dynamics on gravel bars of a large Alpine river. *Wetlands Ecology and Management, 7*, 141–153.

Farrow, A., & Winograd, M. (2001). Land use modelling at the regional scale: An input to rural sustainability indicators for Central America. *Agriculture, Ecosystem & Environment, 85*, 249–268.

Gautam, A. P., Webb, E. L., & Eiumnoh, A. (2002). GIS assessment of land use/land cover changes associated with community forestry implementation in the Middle Hills of Nepal. *Mountain Research and Development, 2002*(22), 63–69.

Gemino, A., & Wand, Y. (2004). A framework for empirical evaluation of conceptual modelling techniques. *Requirements Engineering, 9*, 248–260.

Gomez-Sal, A., Belmontes, J. A., & Nicolau, J. M. (2003). Assessing landscape values: A proposal for a multidisciplinary conceptual model. *Ecological Modelling, 168*, 319–341.

Gonzalez-Perez, C., & Henderson-Sellers, B. (2007). Modelling software development methodologies: A conceptual foundation. *Journal of Systems and Software*. doi:10.1016/j.jss.2007.02.048.

Hardi, P., & Zdan, T. (Eds.). (1997). *Assessing sustainable development: Principles in practice*. Winnipeg, Canada: International Institute for Sustainable Development.

Hazell, P. B. R., & Norton, R. D. (1986). *Mathematical programming for economic analysis in agriculture*. London: Macmillan Publishing Company, 389 pp.

Herrick, J. E., Bestelmeyer, B. T., Archer, S., Tugel, A. J., & Brown, J. R. (2006). An integrated framework for science based arid land management. *Journal of Arid Environments, 65*, 319–335.

Hillman, T., Crase, L., Furze, B., Ananda, J., & Maybery, D. (2005). Multidisciplinary approaches to natural resource management. *Hydrobiologia, 552*, 99–108.

Hjortso, C. N., Steffen, S., & Finn, H. (2006). Applying multi-criteria decision-making to protected areas and buffer zone management: A case study in the Royal Chitwan National Park, Nepal. *Journal of Forest Economics, 12*, 91–108.

Holz, R. K. (1985). *The surveillant science: Remote sensing of environment* (2nd ed.). New York: Wiley.

Huggett, A. J. (2005). The concept and utility of ecological thresholds in biodiversity conservation. *Biological Conservation, 124*, 301–310.

IPCC. (2001). *Third assessment report of the intergovermental panel on climate change*. Cambridge: CUP.

Ives, J. D., & Messerli, B. (1989). *The Himalayan dilemma: Reconciling development and conservation*. London: Routledge.

Jackson, L. J., Trebitz, A. S., & Cottingham, K. L. (2000). An introduction to the practice of ecological modeling. *BioScience, 50*, 694–706.

Jackson, W. J., Tamarkar, R. M., Hunt, S., & Shepherd, K. R. (1998). Land use changes in two middle hill districts of Nepal. *Mountain Research and Development, 18*, 193–212.

Janssen, S., & van Ittersum, M. K. (2007). *Assessing farmer behaviour as affected by policy and technological innovations: Bio-economic farm models*. Report No. 24, SEAMLESS integrated project, EU 6th Framework Programme, contract no. 010036-2. www.SEAMLESS-IP.org 86 pp, ISBN 90-8585-112-2 and 978-90-8585-112-7.

Jensen, J. R. (1996). *Introductory digital image processing: A remote sensing perspective*. Upper Saddle River, New Jersey: Prentice-Hall.

Kaechele, H., & Dabbert, S. (2002). An economic approach for a better understanding of conflicts between farmers and nature conservationists—an application of the decision support system MODAM to the Lower Odra Valley National Park. *Agricultural Systems, 74*, 241–255.

Kala, C. P. (2004). Pastoralism, plant conservation, and conflicts on proliferation of Himalayan knotweed in high altitude protected areas of the Western Himalaya, India. *Biodiversity & Conservation, 13*, 985–995.

Klevmarken, A. (1983). There is a need for applied micro-econometrics: Survey research about the household. In G. Eliasson & H. Fries (Eds.), *Microeconomics: IUI Yearbook 1982–1983*. Stockholm: Industrial Institute for Economic and Social Research.

Leleu, H. (2006). A linear programming framework for free disposal hull technologies and cost functions: Primal and dual models. *European Journal of Operational Research, 168*, 340–344.

Lo, C. P. (1986). *Applied remote sensing*. New York: Longman.

Lonergans, S., Gustavson, K., & Carter, B. (2000). *The index of human insecurity* (Vol. 6). India: AVISO.

Maikhuri, R. K., Nautiyal, S., Rao, K. S., Chandrasekhar, K., Gavali, R., & Saxena, K. G. (2000). Analysis and resolution of protected area—people conflicts in Nanda Devi Biosphere Reserve, India. *Environmental Conservation, 27*, 43–53.

Maikhuri, R. K., Nautiyal, S., Rao, K. S., & Saxena, K. G. (1998). Medicinal plant cultivation and biosphere reserve management: A case study from Nanda Devi Biosphere Reserve. *Current Science, 74*, 157–163.

Maikhuri, R. K., Rao, K. S., & Saxena, K. G. (1996). Traditional crop diversity for sustainable development of central Himalayan agroecosystems. *The International Journal of Sustainable Development and World Ecology, 3*, 8–31.

Martinez-Fernandez, J., Esteve-Selma, M. A., & Calvo-Sendin, J. F. (2000). Environmental and socioeconomic interactions in the evolution of traditional irrigated lands: A dynamic system model. *Human Ecology, 28*, 279–299.

Moldan, B., & Billharz, S. (Eds.). (1997). *Sustainability indicators: Report of the project on indicators of sustainable development*. Sussex, UK: Wiley.

Moody, D. L. (2005). Theoretical and practical issues in evaluating the quality of conceptual models: Current state and future directions. *Data & Knowledge Engineering, 55*, 243–276.

Naithani, H. B., Negi, J. D. S., Thapliyal, R. C., & Pokhriyal, T. C. (1992). Valley of flowers: Needs for conservation or preservation. *Indian Forester, 118*, 371–378.

Nautiyal, S. (1998). *Ecosystem function of buffer zone villages of Nanda Devi Biosphere Reserve*. Ph.D. thesis, H.N.B. Garhwal University Srinagar Garhwal, 227 p.

Nautiyal, S., & Kaechele, H. (2006). Traditional crop diversity needs institutional and policy support for their conservation and sustainable land use development in Himalayas of India 2006. Available at http://web.fu-berlin.de/ffu/akumwelt/bc2006/. Accessed January 10, 2007.

Nautiyal, S., & Kaechele, H. (2007a). Adverse impact of pasture abandonment in Himalayas of India: Testing efficiency of a natural resource management plan (NRMP). *Environmental Impact Assessment Review, 27*, 109–125.

Nautiyal, S., & Kaechele, H. (2007b). Conserving the Himalayan forests: Approaches and implications of different conservation regimes. *Biodiversity and Conservation, 16*, 3737–3754.

Nautiyal, S., & Kaechele, H. (2007c). Management of crop diversity in traditional agroecosystems and conservation of associated indigenous knowledge for sustainable landscape development in Mountains of Indian Himalaya. *An International Journal of Management of Environmental Quality, 18*(5), 514–530.

Nautiyal, S., & Kaechele, H. (2008). A modeling approach for complex interactions between humans and eco-systems for natural resource management in nature protection areas of Himalayas, India. *Management of Environmental Quality-An International Journal (Emerald), 19*(3), 335–352.

Nautiyal, S., & Kaechele, H. (2009). Natural resource management in a protected area of Indian Himalaya: A modeling approach for anthropogenic interactions on ecosystems. *Environmental Monitoring and Assessment, 153*, 253–271.

Nautiyal, S., Kaechele, H., Rao, K. S., Maikhuri, R. K., & Saxena, K. G. (2007). Energy use and economic analysis of traditional versus introduced crops cultivation in mountains of Indian Himalaya: A case study. *Energy, 32*(12), 2321–2335.

Nautiyal, S., Maikhuri, R. K., Rao, K. S., & Saxena, K. G. (2001a). Medicinal plant resources in Nanda Devi Biosphere Reserve in the Central Himalaya. *Journal of Herbs, Spices and Medicinal Plant, 8*, 47–64.

Nautiyal, S., Maikhuri, R. K., Rao, K. S., Semwal, R. L., & Saxena, K. G. (2003a). Agroecosystem function around a Himalayan Biosphere Reserve. *Journal of Environmental Systems, 29*, 71–100.

Nautiyal, S., Maikhuri, R. K., Semwal, R. L., & Rao, K. S. (1998). Conservation through cultivation: A case study of medicinal plants in buffer zone villages of NDBR. In *Research for*

Mountain Development. Some Initiatives and Accomplishments. Gyanodaya Prakashan, Nainital. Himaparyavaran Publication No. 12, 342–357.

Nautiyal, S., Rao, K. S., Kaechele, H., & Zander, P. (2010). Conceptual model development for landscape management in the mountains of the Indian Himalayan region: An approach for sustainable socio-ecological development. *Landscape Online, Journal of International Association of Landscape Ecology, 18*, 1–19.

Nautiyal, S., Rao, K. S., Maikhuri, R. K., & Saxena, K. G. (2003b). Transhumant pastoralism in the Nanda Devi Biosphere Reserve, India: A case study in the Buffer Zone. *Mountain Research Development 23*, 255–262.

Nautiyal, S., Rao, K. S., Maikhuri, R. K., Semwal, R. L., & Saxena, K. G. (2001b). Traditional knowledge related to medicinal and aromatic plants in tribal societies in a part of Himalaya. *Journal of Medical Arom Plan Science, 22*(23), 528–541.

Nelissen, J. H. M. (1991). Household and education projections by means of a microsimulation model. *Economic Modelling, 8*, 480–511.

Odening, M., Musshoff, O., & Balmann, A. (2005). Investment decision in hog finishing: An application of the real options approach. *Agricultural Economics, 32*, 47–60.

Pandit, M. K., Sodhi, N. S., Koh, L. P., Bhaskar, A., & Brook, B. W. (2007). Unreported yet massive deforestation driving loss of endemic biodiversityin Indian Himalaya. Biodiversity and Conservation, 16, 153–163.

Petalas, C., Pliakas, F., Diamantis, I., & Kallioras, A. (2005). Development of an integrated conceptual model for the rational management of the transboundary Nestos River, Greece. *Environmental Geology, 48*, 941–954.

Pimbert, M. P., & Pretty, J. N. (1997). Parks, peoples and professionals: Putting 'participation' into protected area management. In K. B. Chimire & M. P. Pimbert (Eds.), *Social change and conservation* (pp. 297–330). London: Earthscan.

Pinto-Correia, T., Gustavsson, R., & Pirnat, J. (2006). Bridging the gap between centrally defined policies and local decisions—towards more sensitive and creative rural landscape management. *Landscape Ecology, 21*, 333–346.

Pischke, F., & Cashmore, M. (2006). Decision-oriented environmental assessment: An empirical study of its theory and methods. *Environmental Impact Assessment Review, 26*, 643–662.

Prato, T. (2007). Assessing ecosystem sustainability and management using fuzzy logic. *Ecological Economics, 61*, 171–177.

Rao, K. S., Nautiyal, S., Maikhuri, R. K., & Saxena, K. G. (2005). Material and energy flows in village ecosystems with contrasting lifestyles in Nanda Devi Biosphere Reserve, Central Himalaya, India. *Journal of Mountain Science, 2*, 271–293.

Rao, K. S., & Pant, R. (2001). Land use dynamics and landscape change pattern in a typical micro watershed in the mid elevation zone of central Himalaya, India. *Agriculture, Ecosystems and Environment, 86*, 113–123.

Rees, H. G., Holmes, M. G. R., Fry, M. J., Young, A. R., Pitson, D. G., & Kansakar, S. R. (2006). An integrated water resource management tool for the Himalayan region. *Environmental Modelling and Software, 21*, 1001–1012.

Rees, W. E. (2003). Economic development and environmental protection an ecological economic perspective. *Environmental Monitoring and Assessment, 86*, 29–45.

Rossing, W. A. H., Zander, P., Josien, E., Groot, J. C. J., Meyer, B. C., & Knierim, A. (2007). Integrative modelling approaches for analysis of impact of multifunctional agriculture: A review for France, Germany and The Netherlands. *Agriculture, Ecosystem and Environmental, 120*, 41–57.

Ruben, R., Moll, H., & Kuyvenhoven, A. (1998). Integrating agricultural research and policy analysis: Analytical framework and policy applications for bio-economic modelling. *Agricultural System, 58*, 331–349.

Saxena, K. G., Maikhuri, R. K., & Rao, K. S. (2005). Changes in agricultural biodiversity: Implications for sustainable livelihood in the Himalaya. *Journal of Mountain Science, 2*, 23–31.

Saxena, K. G., Rao, K. S., Sen, K. K., Maikhuri, R. K., & Semwal, R. L. (2001). Integrated natural resource management: Approaches and lessons from the Himalaya. *Conservational Ecology, 5* (2), 14. (Online) URL: http://www.consecol.org/vol5/iss2/art14/

Schnell, R., Hill, P. B., & Esser, E. (1999). *Methoden der empirischen Sozialforschung*. Germany: De Gruyter Oldenbourg.

Schroeder, B., & Seppelt, R. (2006). Analysis of pattern-process interactions based on landscape models-overview, general concepts, and methodological issues. *Ecological Modelling*. doi:10.1016/j.ecolmodel.2006.05.036.

Schuler, J., & Kaechele, H. (2003). Modelling on-farm costs of soil conservation policies with MODAM. *Environmental Science and Policy, 1*, 51–55.

Schweik, C. M., Adhikari, K., & Pandit, K. N. (1997). Land-cover change and forest institutions: A comparison of two sub-basins in the southern Siwalik Hills of Nepal. *Mountain Research and Development, 17*, 99–116.

Semwal, R. L., Nautiyal, S., Sen, K. K., Rana, U., Maikhuri, R. K., Rao, K. S., et al. (2004). Patterns and ecological implications of agricultural land-use changes: A case study from central Himalaya, India. *Agricultural Ecosystem and Environment, 102*, 81–92.

Sen, K. K., Semwal, R. L., Rana, U., Nautiyal, S., Maikhuri, R. K., Rao, K. S., et al. (2002). Patterns and implications for land use/cover change: A case study in Pranmati Watershed (Garhwal Himalaya, India). *Mountain Research and Development, 22*, 56–62.

Silori, C. S. (2001). Biosphere reserve management in theory and practice: Case of Nanda Devi Biosphere Reserve, Western Himalaya, India. *Journal of International Wildlife Law and Policy, 4*, 205–219.

Silori, C. S., & Badola, R. (2000). Medicinal plant cultivation and sustainable development: A case study in the Buffer Zone of the Nanda Devi Biosphere Reserve, Western Himalaya, India. *Mountain Research and Development, 20*, 272–279.

Singh, S. P. (2002). Balancing the approaches of environmental conservation by considering ecosystem services as well as biodiversity. *Current Science, 82*, 1331–1335.

Singh, S. P., Tewari, A., & Jina, B. S. (2004). Carbon sequestration in forests of Central Himalayas: Considerations for soil carbon storage and carbon trade. *International Journal of Ecology and Environmental Science, 31*(45), 48.

Singh, T. V., & Kaur, J. (1983). *Studies in eco-development: Himalayan Mountains and men*. Lucknow, India: Print House.

Stoorvogel, J. J., Antle, J. M., Crissman, C. C., & Bowen, W. (2004). The tradeoff analysis model: Integrated bio-physical and economic modeling of agricultural production systems. *Agricultural System, 80*, 43–66.

Ten Berge, H. F. M., van Ittersum, M. K., Rossing, W. A. H., van de Ven, G. W. J., Schans, J., & van de Sanden, P. A. C. M. (2000). Farming options for The Netherlands explored by multi-objective modelling. *European Journal of Agronomy, 13*, 263–277.

Thapa, G. B., & Weber, K. E. (1995). Status and management of watersheds in the Upper Pokhara Valley, Nepal. *Environmental Management, 19*, 497–513.

Tittonell, P., van Wijk, M. T., Rufino, M. C., Vrugt, J. A., & Giller, K. E. (2007). Analysing trade-offs in resource and labour allocation by smallholder farmers using inverse modelling techniques: A case-study from Kakamega district, Western Kenya. *Agricultural System*. doi:10.1016/j.agsy.2007.04.002).

Tulloch, D. L. (1999). Theoretical model of multipurpose land information systems development. *Transactions in GIS, 3*, 259–283.

Turner, R. K. (2000). Integrating natural and socioeconomic science in coastal management. *Journal of Marine Systems, 25*, 447–460.

Van Riet, W. F., & Cooks, J. (1990). An ecological planning model. *Environmental Management, 14*, 339–348.

Virgo, K. J., & Subba, K. J. (1994). Land-use change between 1978 and 1990 in Dhankuta district, Koshi Hills, Eastern Nepal. *Mountain Research and Development, 14*, 159–170.

Wakeel, A., Rao, K. S., Maikhuri, R. K., & Saxena, K. G. (2005). Forest management and land use/cover changes in a typical micro watershed in the mid elevation zone of central Himalaya, India. *Forest Ecology and Management, 213*, 229–242.

Wand, Y., Monarchi, D. E., Parsons, J., & Woo, C. C. (1995). Theoretical foundations for conceptual modelling in information systems development. *Decision Support System, 15*, 285–304.

Wang, Y., & Moskovits, D. K. (2001). Tracking fragmentation of natural communities and changes in land cover: Applications of landsat data for conservation in an urban landscape (Chicago Wilderness). *Conservation Biology, 15*, 835–843.

Zander, P. (2003). *Agricultural land use and conservation options: A modelling approach*. Ph.D. thesis, Department of Land use Systems and Landscapes Ecology, Center for Agricultural Landscape and Land Use Research 2003. Available at: http://www.gcw.nl/dissertations/3372/dis3372.pdf Accessed February 4, 2004.

Zander, P., & Kächele, H. (1999). Modelling multiple objectives of land use for sustainable development. *Agricultural System, 59*, 311–325.

Zander, P., Knierim, A., Groot, J. C. J., & Rossing, W. A. H. (2007). Multifunctionality of agriculture: Tools and methods for impact assessment and valuation. *Agriculture, Ecosystem and Environment, 120*, 1–4.

Part V
Epilogue

Climate Change Challenge (3C) and Social-Economic-Ecological Interface-Building—Exploring Potential Adaptation Strategies for Bio-resource Conservation and Livelihood Development: Epilogue

Sunil Nautiyal, Ruediger Schaldah, K.V. Raju, Harald Kaechele, Bill Pritchard and K.S. Rao

Climate change is arguably the single most dominant environmental threat facing humanity. Its manifestations, particularly through rising temperatures, changing rainfall, sea-level rise and increasing droughts and floods have the potential to adversely impact natural ecosystems (such as forests, grasslands, rivers and oceans) and socioeconomic systems (such as food production, fisheries and coastal settlements). This is adding additional stresses to the ecosystem services, which form a substantial source of income to the rural inhabitants. It is most proximate and inextricably linked to well-being, development and economic growth which are part of the eight Millennium Development Goals (MDGs), which ran from 2000 to 2015. Addressing climate change requires policy formulation, research, technology transfer and diffusion, financing and enhancing adaptive capacity of the poor at national, regional as well as

S. Nautiyal (✉)
Centre for Ecological Economics and Natural Resources (CEENR), Institute for Social and Economic Change (ISEC), Dr. VKRV Rao Road, Nagarabhavi, Bangalore 560072, India
e-mail: nautiyal_sunil@yahoo.com

R. Schaldah
Centre for Environmental System Research (CESR), University of Kassel, Kassel, Germany

K.V. Raju
ICRISAT, Patancheru, India

H. Kaechele
Leibniz-Centre for Agricultural Landscape Research (ZALF), Institute of Socio-Economics, Müncheberg, Germany

B. Pritchard
School of Geosciences, University of Sydney, Sydney, Australia

B. Pritchard
Institute for Geography, University of Cologne, Cologne, Germany

K.S. Rao
Department of Botany, University of Delhi, Delhi, India

S. Nautiyal et al. (eds.), *Climate Change Challenge (3C) and Social-Economic-Ecological Interface-Building*, Environmental Science,
DOI 10.1007/978-3-319-31014-5_37

local levels. As identified by the UNFCCC, the people whose lives are most threatened by climate change are vulnerable groups particularly in developing countries, whose livelihood is traditional crop husbandry or livestock rearing. Therefore, mitigating climate change is an ethical concern. The chapters by Meenakshi Rajeev et al., Arun B. Chandran and Anushri, Letha Devi et al., Sreenivasaiah, K. (Chaps. 14, 23, 26, 32) addressed this issue through agricultural development. According to the report of the Intergovernmental Panel on Climate Change (IPCC 2014), India's agricultural sector would be one of the worst hit of any country in the world. With an erratic and extreme monsoon, the report states that by 2030 India would face an agricultural loss of over US $7 billion, affecting income of 10 % of the people (Neeshad online). But if climate resilience measures in the form adaptive strategies are implemented, 80 % of the losses could be averted, the report adds. In India, several Missions and strategies (Green India Mission, REDD+ etc.) have been implemented at various levels under International initiatives. It is crucial to look at the need of those initiatives and mould responses accordingly to satisfy the requirement for climate change adaptation and sustainable development at local level. UNFCCC has also given strategies on the development of national adaptation and programmes and the support by regional centres; which endows a platform for a bottom-up approach to confer and adapt with climate change impacts at regional level. Pandit chapter (Chap. 10) explores the question how humanity might find the solutions to these global problems. Strengthening livelihoods, developing sustainable land use policy, etc., have been increasingly seen as a critical strategy for supporting adaptation. The chapters by Bill Pritchard, Muhammad Haroon et al., Swamy and Nagaraju, Raju et al., Barun Deb Pal (Chaps. 2, 4, 5, 11, 13) looked at various facets of livelihood sustainability which will help the adaptation of the stakeholders to the possible impacts of climate change. In the recent COP 21 held in Paris, India committed towards creating an additional carbon sink of 2.5–3.0 billion tonnes of CO_2 equivalent through additional forest and tree cover (an increase of about 680–817 million tonnes of carbon stock) by 2030, for which 5 million hectares will be brought under forest cover. This will enhance carbon sequestration by about 100 million tonnes CO_2 equivalent annually (MoEF&CC). However, the Green India Mission is expected to deliver 50–60 % of this target, and therefore, there is a need for developing the plans and strategies to achieve the remaining goals along with introducing other instruments in for example creating climate resilient systems, adopting good practices, developing the strategies for emission reductions and financial provision etc. Therefore, there is a need to focus equally on adaptation as well as on mitigation to cope up with the impact of climate change across ecological regions of country.

1 Holistic Outlook: Integrated Approach

Anthropogenic changes in land use and land cover are global phenomena which are having intensifying consequences on food production, forest and water resources, in addition to climate change. Understanding the future vulnerability, exposure and responses required for interlinked human and natural systems is critical and complex. But the challenge is to integrate huge number of parameters interacting in and among social, economic, and cultural sub-systems, which are not included holistically in most of analyses. Hence, appropriate methods and tools along with field-based case studies on human and biophysical environment with the intervention of climate change would provide better understanding of how paths are altered and how goals relating to sustainability under a changing climate can be achieved. As evidenced by the chapters by Anu Susan Sam et al. (Chap. 25), Himani Prakash (Chap. 21), Parmod Kumar (Chap. 30), P.S. Swain et al. (Chap. 6), Koppad and Tikhile (Chap. 27), Lakshmi and Indumathi (Chap. 33), Nautiyal and Schaldach (Chap. 34) and Gawan and Sen (Chap. 18) transdisciplinary and interdisciplinary approaches are in the forefront of this need.

The development of better land use policy (such as land use development at regional, state or national level) depends on the perceived risks of uncertainties due to variety of factors—such as climate change, socioeconomic, ecological cultural characteristics of the regions involved, and questions relating to technical feasibility and policy measures (Klabbers et al. 1996). For effective implementation, the scientific and technological research should support the policy-making processes. When science and policy differ then outcomes in the form of policy communication is often problematic. The chapter by Sunil K Agarwal (Chap. 35) addresses this issue of science policy and solutions. The research on land use and climate modelling (ex-post; ex-ante) will aid in making effective science-policy recommendations for climate change, and land use development (i.e. impact, causes, effects, adaptation and mitigation) at various spatial scales, all of which will further will help to support better policy formulations and galvanize institutional innovations. What is required is that scientific information should meet the requirements of policy demand and should be easily accessible to policy makers and decision takers. The integrated modelling approach strongly supports this viewpoint (chapter by Nautiyal and Schaldach (Chap. 34) and Schaldach et al. 2010; Nautiyal et al. 2010, 2013). On the other hand, policy makers and decision takers should formulate requisite information such a way that is easily understandable for researchers to provide available scientific information in their deliberations (van den Hove 2007). Landscape modelling helps to construct frameworks and to organize ideas and data to understand the complex human-ecological system and specially the spatial dynamics and processes over different temporal scales. Scenarios used in modelling involve a hypothetical sequence of future events that consider the fundamental uncertainties of the future. The research endeavours should focus on developing recommendations for micro levels depicting various geo-climatic regions in country as evidenced by the chapters by Saikia et al. (Chap. 8) Ankita Mitra et al. (Chap. 20) Kataktalware et al. (Chap. 31) Chand and Garita (Chap. 15) Kumar et al. (Chap. 24). There is a need not only for developing strategies at regional or national level (which is largely a top-down approach) but more importantly, to develop a bottom-up approach to address climate change, sustainable land use and linked socio-ecological development. Integrated approach should be encouraged to aim at adopting a bottom-up approach for sustainable land use development, climate mitigation and adaptation strategies. The need for pilot studies of mitigation and adaptation projects in various agro-climatic regions is addressed in the chapter by Nautiyal et al. (Chap. 36).

2 Need of a Proactive Response to Climate Change and Associated Bioresources Conservation

The response to climate change does not relate to the confines of the environment alone, but has multiple constituencies. Therefore, climate change assessment has to be done in viewpoint of impact, vulnerability and adaptation in different agro-ecological regions. As the scholarly fraternity are making efforts at global, nation, state and even at local levels to address climate change in the post 2015 agenda through mitigation and adaptation strategies, the information/knowledge base about ecosystem/land use modelling requires enhancement, especially in India. The chapters by Saikia et al. (Chap. 8), Khuda Bakhsh et al. (Chap. 12), P.J. Dilip Kumar (Chap. 3), Mansi and Jamwal (Chap. 28), Suresh Nadagoudar (Chap. 9), Kala S Sridhar (Chap. 16) address the bio-resource conservation and sustainable livelihood development under a changing climate and further providing connections with land use land cover policy making.

3 Climate Change and Food Security—The Global and Indian Contexts

Bill Pritchard and Muhammad Haroon et al. (Chaps. 2 and 4) addressed the food security of India and Pakistan under climate change scenarios. There is incontrovertible evidence that global temperatures have increased during the past century, and that the role of humans (anthropogenic forcing) is centrally implicated in this. The study of climate change is extensive in nature, looking at each component individually and then as a complete unit. It involves specialists from across the natural and social sciences. In recent years, considerable attention has been given to the implications of climate change for global food security. The Fifth Assessment Report of the International Panel on Climate Change (IPCC), released in 2014, concluded that global food security would be dramatically affected by climate change. In accordance with FAO definitions, the IPCC considers food security to include four dimensions—food availability, food accessibility, food utilization and food systems stability. Also, it needs to be taken into account that the food system by itself is a major contributor of greenhouse gases, and hence, reforming the global food system needs to be a major plank of climate change mitigation. The impact of climate change on food security will be felt at global, national and local scales. Vulnerable people and communities are at the frontline of these threats, and India faces major impacts including potential changes to the timing and strength of monsoons, retreat of Himalayan glaciers and sea-level rise. India is also vulnerable to the way climate change may affect neighbouring countries. Bangladesh is widely recognised as one the world' s most vulnerable nations to climate change, and climate migrants/refugees may seek to relocate to India, which may aggravate geopolitical tensions. The aim is to enhance our information base about the threats, and adaptive potential, facing India. These adaptive actions can include enhanced food system supply and management via improved agricultural practices (both horticulture and livestock) and changed livelihood arrangements.

4 Climate Change and Vulnerability

The vulnerability of human populations and natural systems to climate change differs substantially across regions and across populations within regions (He et al. 2006; Torresan et al. 2008). Climate stress in particular can compromise the ability of the different branches of the ecosystem to sustain productivity that influences livelihood of the local people (Archer et al. 2007; Nautiyal and Kaechele 2008). Such a situation is particularly concerning in the light of projected increasing climate stress under future climate change due to, for example, increased frequency of extreme precipitation events (IPCC 2001, 2007; Nautiyal et al. 2013). The natural and social systems of different regions have varied characteristics, resources and institutions, and are subject to varied pressures that give rise to differences in sensitivity and adaptive capacity (Torresan et al. 2008). From these differences emerge different key concerns for each of the conditions in regional level in different parts of the world. Within regions, however, impacts, adaptive capacity and vulnerability will vary and that depicts the particular local environment (Olsson et al. 2004). Therefore, the impact level varies at macro and micro level. In emerging economies like India, where economic and institutional circumstances are less favourable, socioeconomic systems are much vulnerable to climate change and such vulnerability is highest where adaptability is much less than the sensitivity (Wainger et al. 2004). From an Indian perspective, the Himalayan region, the east coast and west coast are highly sensitive to climate change, though in different ways. However, adaptive capacities are comparatively weak. Threatened Himalayan and coastal environmental systems are exposed to range of hazards connected to climate change (i.e., receding of glaciers, sea-level rise, increased level of inundation and storm flooding, decrease in rainy days and high rain fall intensity) that may further lead to landscape degradation and consequently appears in a suite of socioecological impacts (such as loss of habitation, soil loss, loss of tourism, decline in productivity of the system and migration) (Saxena et al. 2001, 2005; Ives and Messerli 1989; Olsson et al. 2004; Torresan et al. 2008; Wills et al. 2008). Olmos (2001) stated that in developing countries adapting to climate change is an urgent issue as the poor are expected to disproportionately suffer the impacts of climate change so developing countries should be focus on vulnerability and adaptation programmes. The chapters by M. Srinivasa Rao et al. (Chap. 19), Barun Deb Pal (Chap. 13), Arun Kumar et al. (Chap. 22), Mansi et al. (Chap. 29) Kala S. Sridhar (Chap. 16), Arun B. Chandran and Anushree (Chap. 23) C.M. Lakshmana (Chap. 7) address the issues of climate change and livelihood supporting activities. Agriculture which currently accounts for 24 % of world output uses only 40 % of land area (FAO) and is highly dependent on the climate.

Human dependence on agricultural livelihoods, particularly by the poor, is high and so land use/land cover change needs much attention to monitor the impact of climate change in developing countries. It is therefore, required that responses to climate change can either seek to be reduced the level or rate of change (*mitigation*) or need to manage its consequences (*adaptation*) (Halsnaes and Verhagen 2007) at various level. There are a lot of reports devoted to evaluate vulnerability of climate change on a national and global level (Easterling 1996; Smit and Smithers 1994; Ikerd 1997; FAO 2003; Stern 2006; Barnosky 2008; Fischer et al. 2005; IPCC 2007; Rosenzweig et al. 2002; Warren et al. 2006; Reilly et al. 2001; Reinsborough 2003; Willis et al. 2004; Wilson and Tyrchniewicz 1995; Milestad and Darnhofer 2003; McCarthy et al. 2001) but unfortunately more comprehensive and site specific vulnerability assessment that would have been suitable to plan possible adaptation measures at local/regional scales are yet to be carried out.

Finally, Vogel et al. (2007) suggests that vulnerability assessment is required across spatial and varying temporal scales, e.g. assessing present vulnerability at varying scales (e.g. local to household level as well as national assessments) using the Human Development Index or other indicators and mapping of vulnerability by the study of Vulnerability and global environmental change in the year 2001. Technical adaptations to climate change in many developing countries where the issue of climate change is overshadowed by a number of immediate development priorities such as poverty eradication; food and water security; health and natural resource management; local air—water pollution (Chapter by Surender Kumar and Parmod Kumar (Chap. 17); Klein et al. 2004; Brown 2008; Downing et al. 1997; Guilmoto 1998; Ezra and Kiros 2001; Verchot et al. 2007). This has become a key issue in climate change negotiations, where much of the attention focused on sources of funding, as well as the emerging issues of equity and compensation (Yin et al. 2000) are going to bear the brunt of climate change and suffer most from its negative impacts (Verchot et al. 2007). Looking at vulnerability as an end point has played a useful role in measuring the extent of the climate change problem, and weighing the costs of impacts and adaptations against the costs of greenhouse gas mitigation (Xinhua et al. 1999; Vavrus et al. 2008). Future climate change scenarios and estimates of impacts can provide a useful contextual frame for studies that take climate change as a starting point (Nordlund 2008; Li et al. 2008). However, to understand vulnerability as well as adaptation, "greater insights can be gained from closely looking around and closely looking back at microscopic level that how farmers changes their way of living in changing climatic condition rather than from looking forward" (Adger 2003; Fuessel 2007; Thomas 2008; Abbas 2015). Thus, transdisciplinary and interdisciplinary approaches require for developing the climate resilient society and ecology.

References

Abbas, A. (2015). *Socioeconomic impacts of climatic extremes and their management: A case study of floods in Pakistan with special reference to aspects of mitigation, coping strategies and adaptation of farming sector* (Ph.D. thesis) Humboldt University, Berlin, Germany.

Adger, W. N. (2003). Social capital collective action and adaptation to climate change. *Economic Geography, 79*, 387–404.

Archer, E., Mukhala, E., Walker, S., Dilley, M., & Masamvu, K. (2007). Sustaining agricultural production and food security in Southern Africa: An improved role for climate prediction? *Climate Change, 83*, 287–300.

Barnosky, A. D. (2008). Climate change, refugia, and biodiversity: Where do we go from here? An editorial comment. *Climate Change, 86*, 29–32.

Brown, O. (2008, January). *Migration and climate change* (pp. 1–64). International Organization for Migration Geneva.

Downing, T. E., Ringius, L., Hulme, M., & Waughray, D. (1997). Adapting to climate change in Africa: Prospects and guidelines. *Mitigation and Adaptation Strategies for Global Change, 2*, 19–44.

Easterling, W. I. I. I. (1996). Adapting North American agriculture to climate change in review. *Agricultural and Forest Meteorology, 80*, 1–54.

Ezra, M., & Kiros, G. (2001). Rural out-migration in the drought prone areas of Ethiopia: A multilevel analysis. *International Migration Review, 35*(3), 749–771.

FAO. (2003). *World agriculture towards 2015/2030: An FAO perspective*. Rome: Food and Agriculture Organisation of the United Nations.

Fischer, G., Shah, M., Tubiello, F. N., & van Velhuizen, H. (2005). *Socio-economic and climate change impacts for the environment*. University of Oxford. Available at http://african-environments.ouce.ox.ac.uk/events/2007/070312workshop.php

Fuessel, H. M. (2007). Vulnerability: A generally applicable conceptual framework for climate change research. *Global Environmental Change, 17*, 155–167.

Guilmoto, C. (1998). Institutions and migrations: Short term versus long term moves in rural West Africa. *Population Studies, 52*(1), 85–103.

Halsnæs, K., & Verhagen, J. (2007). Development based climate change adaptation and mitigation—conceptual issues and lessons learned in studies in developing countries. *Mitigation and Adaptation Strategies for Global Change, 12*, 665–684.

He, X., Zhou, J., Zhang, X., & Tang, K. (2006). Soil erosion response to climate change and human activity during the quaternary on the Loess Plateau, China. *Regional Environmental Change, 6*, 62–70.

Ikerd, J. (1997, January). Understanding and managing the multi-dimensions of sustainable agriculture, Southern Regional sustainable agriculture professional development program. Gainesville, FL: SARE Regional Training Consortium.

IPCC (Intergovernmental Panel on Climate Change). (2001). *Third assessment report. Cambridge report*. Cambridge: Cambridge University Press.

IPCC. (2007). *Working group II contribution to the fourth assessment report, climate change: Climate change impacts, adaptation and vulnerability*. Geneva: WHO and UNEP IPCC

IPCC (Intergovernmental Panel on Climate Change). (2014). *Fifth assessment report. Cambridge report*. Cambridge: Cambridge University Press.

Ives, J. D., & Messerli, B. (1989). *The Himalayan dilemma: Reconciling development and conservation*. London: Routledge.

Klabbers, J. H. G., Swart, R. J., Janssen, R., Vellinga, P., & van Ulden, A. P. (1996). Climate science and climate policy: Improving the science/policy interface. *Mitigation and Adaptation Strategies for Global Change, 1*, 73–93.

Klein, R. J. T., Nicholls, R. J., & Mimura, N. (2004). Coastal adaptation to climate change: Can the IPCC technical guidelines be applied? *Mitigation and Adaptation Strategies for Global Change, 4*, 239–252.

Li, Y., Chen, D., Barker-Reid, F., & Eckard, R. (2008). Simulation of N_2O emissions from rain-fed wheat and the impact of climate variation in southeastern Australia. *Plant and Soil, 309*, 239–251.

McCarthy, J. J., Canziani, O. F., Leary, N. A., Dokken, D. J., & White, K. S. (2001). *Climate change 2001: Impacts, adaptation, and vulnerability. Contribution of working group II to the third assessment report of the Intergovernmental Panel on Climate Change*. Cambridge, UK: Cambridge University Press.

Milestad, R., & Darnhofer, I. (2003). Building farm resilience: The prospect and challenges of organic farming. *Journal of Sustainable Agriculture, 22*, 81–97.

Nautiyal, S., & Kaechele, H. (2008). Natural resource management in a protected area of Indian Himalaya: A modeling approach for anthropogenic interactions on ecosystems. *Environmental Monitoring and Assessment, 153*(1–4), 253–271.

Nautiyal, S., Rao. K.S., Kaechele, H., & Zander, P. (2010). Conceptual model development for landscape management in the mountains of the Indian Himalayan region: an approach for sustainable socio-ecological development. Landscape Online. *Journal of International Association of Landscape Ecology, 18*, 1–19.

Nautiyal, S., Rao, K. S., Kaechele, H., Raju, K. V., & Schaldach, R. (2013). *Knowledge systems of societies for adaptation and mitigation of impacts of climate change, environmental science and engineering*. Berlin: Springer. 720 pp.

Neeshad (online). https://www.careaboutclimate.org/climate_stories_05.html

Nordlund, G. (2008). Futures research and the IPCC assessment study on the effects of climate change. *Futures, 40*(10), 873–876.

Olmos, S. (2001). Vulnerability and adaptation to climate change: Concepts, issues, assessment methods. In *Climate change knowledge network* (21 pp).

Olsson, P., Folke, C., & Berkes, F. (2004). Adaptive comanagement for building resilience in socio-ecological systems. *Environmental Management, 34*, 75–90.

Reilly, J., Graham, J., & Hrubovcak, J. (2001). *Agriculture: The potential consequences of climate variability and change for the United States, US national assessment of the potential consequences of climate variability and change, US Global Change Research Program*. New York: Cambridge University Press. http://www.usgcrp.gov/usgcrp/Library/nationalassessment/Agriculture.pdf

Reinsborough, M. (2003). A Ricardian model of climate change in Canada. *Canadian Journal of Economics, 36*, 21–40.

Rosenzweig, C. E., Tubiello, F., Goldberg, R., & Bloomfield, J. (2002). Increased crop damage in the US from excess precipitation under climate change. *Global Environmental Change, 12*, 197–202.

Saxena, K. G., Maikhuri, R. K., & Rao, K. S. (2005). Changes in agricultural biodiversity: Implications for sustainable livelihood in the Himalaya. *Journal of Mountain Science, 2*, 23–31.

Saxena, K. G., Rao, K. S., Sen, K. K., Maikhuri, R. K., & Semwal, R. L. (2001). Integrated natural resource management: Approaches and lessons from the Himalaya. *Conservation Ecology, 5*, 14 (Online). URL: http://www.consecol.org/vol5/iss2/art14/

Schaldach, R., Priess, J. A., & Alcamo, J. (2010). Simulating the impact of biofuel development on countrywide land-use change in India. *Biomass and Bioenergy, 35*, 1–10. doi:10.1016/j.biombioe.2010.08.048

Smit, B., & Smithers, J. (1994). Sustainable agriculture: Interpretations, analyses and prospects. *Canadian Journal of Regional Science, 16*(3), 499–524.

Stern, N. (2006). *Stern review on the economics of climate change, London: Her Majesty's Treasury and the Cabinet Office*. Available from http://www.hmtreasury.gov.uk/independent_reviews/stern_review_economics_climate_change/sternreview_index.cfm

Thomas, R. J. (2008). Opportunities to reduce the vulnerability of dryland farmers in Central and West Asia and North Africa to climate change. *Agriculture, Ecosystems & Environment, 126*, 36–45.

Torresan, S., Critto, A., Valle, M. D., Harvey, N., & Marcomini, A. (2008). Assessing coastal vulnerability to climate change: Comparing segmentation at global and regional scales. *Sustainability Science, 3*(45), 65.

van den Hove, S. (2007). Science policy interface for biodiversity in Europe.

Vavrus, S., Ruddiman, W. F., & Kutzbach, J. E. (2008). Climate model tests of the anthropogenic influence on greenhouse-induced climate change: The role of early human agriculture, industrialization, and vegetation feedbacks. *Quaternary Science Reviews, 27*, 1410–1425.

Verchot, L. V., Noordwijk, M. V., Kandji, S., Tomich, T., Ong, C., Albrecht, A., et al. (2007). Climate change: Linking adaptation and mitigation through agroforestry. *Mitigation and Adaptation Strategies for Global Change, 12*, 901–918.

Vogel, Coleen, Moser, S. C., Kasperson, R. E., & Geoffrey, D. (2007). Linking vulnerability, adaptation and resilience science to practice: Players, pathways and partnerships. *Global Environmental Change, 17*, 349–364.

Wainger, L. A., King, D. M., Cantrell, J. A., & Bird, S. L. (2004). Development of indicators to assess economic vulnerability to changes in ecosystem services: Case study of Counties in Maryland, USA. *Environmental Management, 34*, 730–747.

Warren, R., Arnell, N., Nicholls, R., Levy, P., & Price, J. (2006). *Understanding the regional impacts of climate change'. Research report prepared for the Stern Review, Tyndall Centre Working Paper 90*. Norwich: Tyndall Centre. Available from http://www.tyndall.ac.uk/publications/working_papers/twp90.pdf

Willis, J. K., Chambers, D. P., & Nerem, R. S. (2008). Assessing the globally averaged sea level budget on seasonal to interannual timescales. *Journal Geophysical Research, 113*, C06015. doi:10.1029/2007JC004517

Willis, S. G., Hole, D. G., Collingham, Y. C., Hilton, G., Rahbek, C., & Huntley, B. (2004). Assessing the impacts of future climate change on protected area networks: A method to simulate individual species' responses. *Environmental Management, 43*(5), 836–845.

Wilson, A., & Tyrchniewicz, A. (1995). *Agriculture and sustainable development: Policy analysis on the Great Plains* (p. 108). Winnipeg: International Institute for Sustainable Development, Winnipeg.

Xinhua, X., Dahui, W., Hong, J., & Huixiang, S. (1999). Study on greenhouse gas emissions in Jiangsu Province. *Water, Air, and Soil Pollution, 109*, 293–301.

Yin, Y., Cohen, S., & Huang, G. H. (2000). Global climate change and regional sustainable development: The case of Mackenzie basin in Canada. *Integrated Assessment, 1*, 21–36.